The Carcinogenic Effects *of* Polycyclic Aromatic Hydrocarbons

The

Carcinogenic Effects *of* Polycyclic Aromatic Hydrocarbons

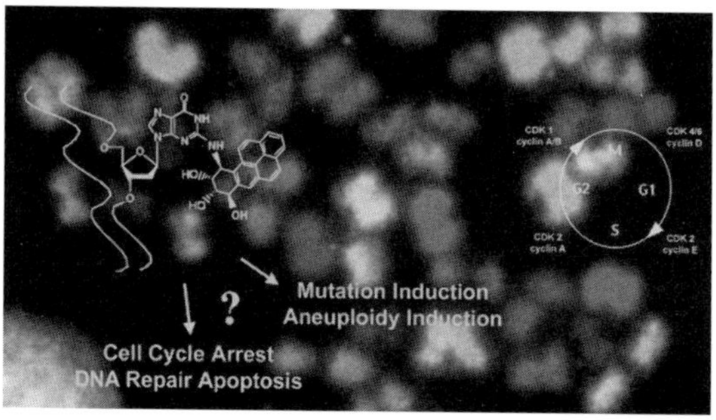

Editor

Andreas Luch

Massachusetts Institute of Technology, USA

ICP

Imperial College Press

Published by

Imperial College Press
57 Shelton Street
Covent Garden
London WC2H 9HE

Distributed by

World Scientific Publishing Co. Pte. Ltd.
5 Toh Tuck Link, Singapore 596224
USA office: 27 Warren Street, Suite 401-402, Hackensack, NJ 07601
UK office: 57 Shelton Street, Covent Garden, London WC2H 9HE

Library of Congress Cataloging-in-Publication
The carcinogenic effects of polycyclic aromatic hydrocarbons / editor, Andreas Luch.
p. cm.
Includes bibliographical references and index.
ISBN 1-86094-417-5 (alk. paper)
1. Polycyclic aromatic hydrocarbons--Carcinogenicity. 2. Polycyclic aromatic hydrocarbons--Toxicology. 3. Chemical carcinogenesis. 4. Genetic toxicology. J. Luch, Andreas.

RC268.7.P64C374 2004
616.99'4071--dc22 2004056977

British Library Cataloguing-in-Publication Data
A catalogue record for this book is available from the British Library.

Typeset by Stallion Press
Email: sales@stallionpress.com

Printed in Singapore by B & JO Enterprise

For my two beloved children,

Alina & Rubin

who were born amidst the work on this book

Preface

When I agreed in August 2002 to work on a book entitled *The Carcinogenic Effects of Polycyclic Aromatic Hydrocarbons* I finally overcame a period of almost a year of thinking and hesitation. Besides arranging my move from Munich to Boston, I was not sure about whether it would be worthwhile to invest a great amount of time and effort into a project that most researchers these days would rather consider somewhat 'old-fashioned' and boring. Their credo, "Cancer is a disease of the genes", has developed over the past decades as a result of the rise of molecular genetics and the discovery of genetic traits underlying tumorigenesis. But is it really? The relative contribution of heritable genetic constitution vs. environmental factors in cancer causation has been a matter of debate ever since the discovery of 'oncogenes', and even long before. Evidence from epidemiological observations, working place or migration studies however, points to environmental factors as the major players. In the age of cancer genetics it therefore seems reasonable to recall the importance of *chemical carcinogenesis* and to outline our present knowledge on the molecular mode of action of a very important and ubiquitously present group of tumorigenic compounds, the polycyclic aromatic hydrocarbons.

The scientific work on carcinogenic polycyclic aromatic hydrocarbons as a part of *Environmental and Molecular Toxicology* requires knowledge and input not only from experts in the field of 'molecular' biology or 'molecular' pathology, but primarily the analytical skills and achievements of chemists and biochemists well-educated and trained to think in dimensions of molecules and their three-dimensional occurrence and interactions (stereochemistry). Work on carcinogenic chemicals indeed is *molecular* toxicology in its strictest sense, not to be mixed up with *macromolecular* biology and genetics which an increasing number of toxicology departments are mainly focusing on these days. In the country where I come from, *Molecular Toxicology* will only survive when it saves its roots and keeps its interdisciplinary character as a highly interactive field of research stimulated and further strengthened by *all* natural sciences and human medicine. This interplay of many disciplines is what makes the work on biological effects of chemicals interesting and exciting but also rather difficult and complex. But to return to the main point: assessing the biological effects of *chemicals* is the central issue, the mainstay in toxicology! — and that's true *per definitionem*. Research that avoids the application of chemicals and the investigation of their conversion, reactivity and fate within the living organism has nothing to do with toxicology.

There were essentially these two issues that inspired me to get into this project: to recall the importance of chemically-induced carcinogenesis in human cancer etiology, and to demonstrate that the work on biologically active compounds may still be interesting and, to use a more modern word, 'en vogue'. It opens the doors to a highly interdisciplinary field of research and provides fascinating insights into biological (and pathological) processes at a truly molecular level. Carcinogenic polycyclic aromatic hydrocarbons are among the most interesting and important compounds in environmental toxicology. The main goal of this book is to outline and to communicate the progress that has been achieved during the past decade in this field. I deeply hope that it will find its place and acceptance in the scientific community.

This book is dedicated to Anthony Dipple. I decided to ask some of his long-term friends and colleagues to write a paragraph to honor him as an outstanding researcher and an amiable human being. To my knowledge, this book is the first monograph on carcinogenic polycyclic aromatic hydrocarbons that has been published since his premature death in 1999. Although I saw Tony only occasionally during scientific meetings, he left a huge impression on me. I was equally fascinated by his scientific achievements and personality. Since he committed his lifetime work to the investigation of carcinogenic hydrocarbons, this book may be an appropriate place to remember him and to recall the importance of his contributions to our research field.

Finally I want to thank all my co-authors for their great work and the time they invested to nicely cover a specific topic in the field of carcinogenic polycyclic aromatic hydrocarbons. On my personal side, most credit goes to my lovely wife, Ina, who never complained about the additional burden and strain during a time when everything changed dramatically due to the arrival of two new souls to our small family. To help to make this book project possible despite the culminating workload far away from our home town and families, that surely is her invaluable merit.

Andreas Luch
Cambridge, MA, Summer 2004

In Memoriam:

Anthony Dipple

(1940 – 1999)

Anthony (Tony) Dipple dedicated most of his scientific life to a better understanding of chemical carcinogenesis and specifically the mechanisms underlying the actions of polycyclic aromatic hydrocarbons. Over the span of his career, he displayed wide interests ranging from mechanistic organic chemistry to molecular biology. Tony was born on February 9, 1940 in Mansfield, England. After finishing Queen Elizabeth Grammar School for Boys in Mansfield, where he was a chess captain, a member of the football (soccer) team and a prefect in his senior year, Tony was awarded a university scholarship. In 1961 he received his B.Sc. in Chemistry at the University of Birmingham. In 1964, Tony submitted his Ph.D. thesis work, "Studies on Chemical Degradation of Ribonucleic Acids", performed under the supervision of Dr. A. S. Jones, and a week after receiving his Ph.D. in Biological Chemistry from the University of Birmingham, Tony and his wife, Hilary, were on their way to the United States on the Queen Mary for Tony's postdoctoral studies.

From 1964 – 1966, as a Damon Runyon Cancer Research Fellow, Tony worked on synthesis of fluoropyrimidine nucleosides as potential anti-cancer drugs in the McArdle Laboratory, University of Wisconsin, Madison. During this time, his daughter Joanne was born in 1965. The small family returned to England in 1966, when Tony started as a lecturer at the Institute of Cancer Research, Chester Beatty Research Institute in London, where he stayed until 1975. Here, Tony began research on polycyclic aromatic hydrocarbon carcinogen interactions with DNA, a field of interest he continued to pursue until his death. He synthesized reactive derivatives of polycyclic aromatic hydrocarbons such as 7-bromomethylbenz[*a*]anthracene, and found that they reacted specifically with the amino groups of the purine DNA bases. These DNA-reactive model compounds were made available by Tony to many colleagues for studies of mutagenesis, toxicity and cell transformation. During the years in London, Tony's family grew, with the births of Geoffrey and Christopher in 1967 and 1969.

In 1975, Tony moved back to the United States to join the newly created NCI-Frederick Cancer Research and Development Center in Frederick, Maryland. As head of the Molecular Aspects of Chemical Carcinogenesis Section and later also as director of the Chemistry of Carcinogenesis Laboratory, Tony continued his studies of mechanisms involved in the initiation of cancer by chemicals. Over the years, Tony mentored numerous postdoctoral fellows and assistants in his laboratory. The success of the laboratory benefited from contributions from his colleagues C. Anita Bigger, Robert C. Moschel and Karen H. Vousden in Frederick as well as from numerous collaborators including William M. Baird, Ronald G. Harvey, Shantu Amin, Donald M. Jerina and their associates. The laboratory was highly productive, and highlights of these investigations are described below.

A major area of study involved the chemistry of alkylation of DNA by benzylating agents and styrene oxides. Simple alkylating agents were found to modify primarily the N-7 ring nitrogens and to a lesser extent the exocyclic oxygen atoms, whereas the benzylic alkylating agents derived from polycyclic aromatic hydrocarbons modified the exocyclic amino groups, often quite extensively. These investigations led to the conclusion that reactions on the exocyclic amino groups of the purine bases require a considerable degree of S_N1 character, whereas reactions on the ring nitrogens (specifically N-7) are favored for agents which react via a pathway with greater S_N2 character. Studies of a deoxyadenosine adduct generated by reaction with the directly alkylating 7-bromomethyl-12-methylbenz[*a*]anthracene led to the first crystal structure of a polycyclic aromatic hydrocarbon-nucleoside adduct.

7,12-Dimethylbenz[*a*]anthracene is one of the most potent tumor initiators among the polycyclic aromatic hydrocarbons. The finding in Tony's laboratory that there is poor repair of DNA damage caused by this carcinogen may be associated with its high potency. Since the highly hindered fjord-region 3,4-diol 1,2-epoxides of 7,12-dimethylbenz[*a*]anthracene and benzo[c]phenanthrene both react extensively with deoxyadenosine, it was proposed that the poor repair of the deoxyadenosine adducts was responsible for their high tumorigenicity. Less hindered bay-region diol epoxides derived from carcinogens such as benzo[*a*]pyrene, chrysene and benz[*a*]anthracene modify deoxyguanosine to a greater extent than deoxyadenosine and are less tumorigenic.

To investigate the mutagenic properties of the reactive diol epoxide metabolites of the polycyclic aromatic hydrocarbons, molecular biology approaches were used. A shuttle vector system (pS189) was used and showed that the mutational spectra for diol epoxides from benzo[c]phenanthrene, 5-methylchrysene and 7-methylbenz[*a*]anthracene were not identical, although some similarities among the types of mutations were found. To examine further the correlation between adduct structure and mutations, site-specific mutagenesis was investigated with diol epoxide adducts derived from benzo[c]phenanthrene and benzo[*a*]pyrene in two different DNA sequence contexts using a bacteriophage (M13) as a vector in *E. coli*. Both sequence context and adduct structure had complex and interdependent effects on mutational frequencies and distributions. This complexity may have resulted from the involvement of multiple DNA polymerases in *E. coli* with different specificities for nucleotide incorporation opposite individual adducts.

Tony's most recent area of research interest involved the effects of polycyclic aromatic hydrocarbon diol epoxides on the cell cycle. These potent carcinogens were found to damage DNA in human cells without inducing the expected cell cycle arrest in the G1 phase. Normally DNA damage triggers G1 cell cycle arrest to allow for DNA repair before the cell enters S (synthesis) phase during which DNA replication occurs. This lack of G1 arrest is likely to enhance error-prone insertions opposite the unrepaired adducts in the S phase and lead to enhanced mutations compared to other types of DNA damage.

As Executive Editor and co-founder in 1980 with Dr. Colin Garner of the journal *Carcinogenesis*, Tony was well-known within the scientific community. Dedicated efforts by the two editors were rewarded as *Carcinogenesis* became a highly respected, premier journal. He was a member of the editorial boards of *Chemical Research in Toxicology*, *Woman and Cancer* and of the Editorial Academy of the *International Journal of Oncology*. In addition, Tony served on NIH site visit panels and review boards, such as the NIH Chemical Pathology Study Section, the NIEHS Environmental Health Sciences Review Committee, the American Institute for Cancer Research Review Panel and the American Cancer Society Advisory Committee for Biochemistry and Carcinogenesis. In 1987, Tony was awarded a Doctor of Science in Biological Chemistry by the University of Birmingham for his research accomplishments and scientific excellence in his field.

Tony is remembered by his colleagues as "a pleasure to work with on professional matters". He had an "ability to make a point without diminishing his opponent which resulted from clear thinking and remarkable personal graciousness". They "had great respect for Tony because of his outstanding and seminal contributions to our field, and for his qualities as a person", "a towering intellect". "As editor of *Carcinogenesis* Tony gained his greatest professional recognition… his forbearance was enormous…" "His thinking was focused on the problem in question. He was kind and compassionate and had a very pleasant demeanor. Even when he had to reject a paper submitted to *Carcinogenesis* he did so humanely. He was noted for writing rather long sentences some of which filled an entire paragraph that still retained absolute clarity." "I saw Tony as a positive, forward looking, matter-of-fact person." "Tony was a kind, gracious person, always there for those who needed advice, always patient, as if he had all the time in the world to listen." "All those who worked with him will miss his patience as a teacher, and the goodwill and support he gave to all."

As the quotations above demonstrate, Tony was a warm and generous person who enjoyed life. Although his health did not allow challenging physical activities after his two serious health incidents (a heart attack in 1984 and a kidney failure and transplant in 1990), he found ways to exercise and at the same time enjoy the outdoors. He could spend hours biking on the Chesapeake and Ohio Canal tow path and traversed the entire 185 miles from Washington, DC to Cumberland, MD in bits and pieces. Most recently he began playing golf as a particularly enjoyable pastime. Another favorite recreation was camping on the beach at Assateague State Park, where he found peace and rest from the stress at work, and where the main worry would be what to prepare for dinner. He found cooking at the campsite over an open fire as much fun as preparing a feast for friends at home. There, he would delight in playing the organ and coax them into singing together after dinner. Tony is missed by his many friends and colleagues for his analytical mind and ability to dissect a problem, but most of all for the cheerful times they shared.

Ingrid Pontén,[1] Jane M. Sayer[2] and Donald M. Jerina[2]

[1] Safety Assessment, AstraZeneca R&D Södertälje, S-15185 Södertälje, Sweden
Email: ingrid.ponten@astrazeneca.com

[2] Laboratory of Bioorganic Chemistry, NIDDK, NIH, DHHS, Bethesda MD 20892, USA
Emails: jmsayer@helix.nih.gov & dmjerina@nih.gov

Acknowledgement:
Information for this article was obtained from Dipple A, *Curriculum vitae*; Dipple A, "Mechanisms of action of chemical carcinogens", D.Sc. thesis, 1987; Dipple A (1999) *Int. J. Oncol.* 14: 1019 – 1020; and Bigger C A H (1999) *Environ. Mol. Mutagen.* 34: 227 – 232.

List of Contributors

Shantu Amin
American Health Foundation Cancer Center, Institute for Cancer Prevention, Valhalla, NY, USA
E-mail: samin@ifcp.us

William M. Baird
Oregon State University, Department of Environmental and Molecular Toxicology, Corvallis, OR, USA
E-mail: william.baird@orst.edu

Ahmad Besaratinia
Division of Biology, Beckman Research Institute of the City of Hope National Medical Center, Duarte, CA, USA
E-mail: ania@coh.org

Karam El-Bayoumy
American Health Foundation Cancer Center, Institute for Cancer Prevention, Valhalla, NY, USA
E-mail: kelbayou@ifcp.us

Nicholas E. Geacintov
Chemistry Department, New York University, New York, NY, USA
E-mail: nicholas.geacintov@nyu.edu

Hansruedi Glatt
German Institute of Human Nutrition, Department of Toxicology, Potsdam-Rehbrücke, Germany
E-mail: glatt@mail.dife.de

Ari Hirvonen
Laboratory of Biomonitoring, Department of Industrial Hygiene and Toxicology, Finnish Institute of Occupational Health, Helsinki, Finland
E-mail: ari.hirvonen@ttl.fi

Andreas Luch
Massachusetts Institute of Technology, Center for Cancer Research, Cambridge, MA, USA
E-mail: luch@mit.edu

Hanspeter Naegeli
Institute of Pharmacology and Toxicology, University of Zürich-Tierspital, Zürich, Switzerland
E-mail: naegelih@vetpharm.unizh.ch

Gerd P. Pfeifer
Division of Biology, Beckman Research Institute of the City of Hope National Medical Center, Duarte, CA, USA
E-mail: gpfeifer@coh.org

David H. Phillips
Institute of Cancer Research, Sutton, Surrey SM2 5NG, UK
E-mail: david.phillips@icr.ac.uk

Albrecht Seidel
Biochemical Institute for Environmental Carcinogens, Prof. Dr. Gernot Grimmer-Foundation, Grosshansdorf, Germany
E-mail: albrecht.seidel@biu-grimmer.de

Pablo Steinberg
Institute of Nutritional Science, University of Potsdam, Bergholz-Rehbrücke, Germany
E-mail: steinber@rz.uni-potsdam.de

Contents

1 Polycyclic Aromatic Hydrocarbon-Induced Carcinogenesis – An Introduction

Andreas Luch

Massachusetts Institute of Technology,
Center for Cancer Research, Cambridge, MA, USA
E-mail: luch@mit.edu

The association of human cancer with the exposure to polycyclic aromatic hydrocarbons (PAHs) dates back to Percivall Pott's observation of chimney sweeps' cancer in 1775.[1] Pott (1714–1788), who was surgeon to St. Bartholomew's Hospital in London, described the occurrence of scrotal cancer in chimney sweeps, and traced it to the contamination of the skin by soot. The interest of this observation lay in the first proof of the environmental origin of one particular form of cancer. About 100 years later, Volkmann and Bell confirmed the early observation made by Pott by describing several cases of scrotal skin tumors among workers in the German and Scottish paraffin industry, respectively.[2,3] In 1907, the following definition was included into the Workmen's Compensation Act of Great Britain: "*scrotal epithelioma occurring in chimney sweeps and epitheliomatous cancer or ulceration of the skin occurring in the handling or use of pitch, tar or tarry compounds*".[4] With this addendum it has been officially acknowledged for the first time that cancer of any cutaneous site could be caused by pitch, tar or tarry compounds. A few years later, bitumen, mineral oil and paraffin were also included into the Compensation Act. All of these regulations resulted from the knowledge that was gained in the second half of the nineteenth and the beginning of the twentieth century, when the skin carcinogenicity of soot, tar and related PAH-containing mixtures was confirmed under various work place conditions.[4] Until then, physicians collected the unforeseen outcome of an undesigned and undesirable grand-scale 'natural experiment' based on the rise of industrialization. The imperative next step was that of systematic inquiry and reproduction of the diseases at will. After many failures to reproduce the human outcome in laboratory animals, success was finally achieved by Yamagiwa and Ichikawa in 1915. Their report on the production of malignant epithelial tumors by repetitive application of coal tar to the ear skin of rabbits[5,6] marked the transition into the modern era of PAH-related experimental cancer research. Shortly thereafter, painting of the back of mice was introduced as a method of biological testing of carcinogenic tars,[7] and subsequently ethereal extracts of soot were confirmed to be carcinogenic in this mouse model.[8]

After the first successful production of cancer under experimental conditions, the scientific interest naturally shifted to the identification of the nature of the chemical(s) responsible. It was in this field more than any other that the greatest strides have been made in the 1920s and 30s, in large measures as an outcome of the studies of Sir Ernest Kennaway and his co-workers at the Royal Cancer Hospital in London. When Kennaway started with his attempt to identify the cancer-producing compound(s) in coal tar, it was known from previous work by Bloch and Dreifuss that the active fraction was of high boiling-point, neutral, nitrogen- and sulphur-free, and capable of forming a picrate complex.[9] In 1925, Kennaway reported on the generation of carcinogenic tars by heating of petroleum, isoprene and acetylene up to 700–900°C in a hydrogen-containing atmosphere.[10] As already known at this time from initial work of Berthelot in 1866, 'pyrolytic' conditions such as those applied would cause molecular condensation and rearrangement reactions that ultimately lead to the generation of PAHs. The carcinogenic tars produced under these conditions as well as carcinogenic products derived from the incubation of tetralin (1,2,3,4-tetrahydronaphthalene) with aluminum chloride at moderate temperatures (30–40°C; 'Schroeter reaction')[11] were found to exhibit strong and characteristic fluorescence spectra with several distinct bands in the blue and violet region (at 4000, 4180 and 4400 Å) — a quality that finally turned out to be "*… the single thread that led all through the labyrinth*" (Kennaway) towards identifying the carcinogenic species. It was the work of Mayneord and Hieger from the Cancer Hospital that confirmed the identity of the characteristic fluorescence bands ('cancer bands') found in the spectra of carcinogenic tars and in carcinogenic substances produced via the 'Schroeter reaction'. Among all available PAHs synthesized and characterized at this time, 1,2-benzanthracene (benz[*a*]anthracene, B[*a*]A, Figure 1.1) was found to give a fluorescence spectrum similar (but not identical) to that of the carcinogenic mixtures.[12] About the same time, synthetic

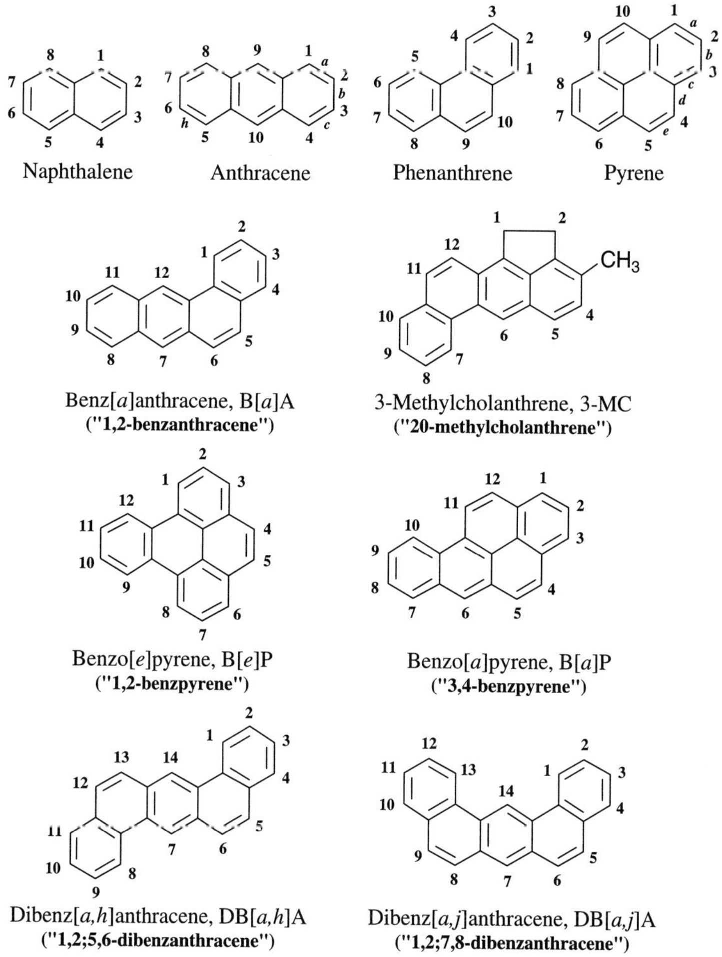

Figure 1.1: Polycyclic aromatic hydrocarbons. (The older nomenclature system used prior to 1966 is written in brackets: see text for explanation of inconsistencies between older and IUPAC-based nomenclature.)

preparations of higher molecular PAHs such as 1,2;5,6-dibenzanthracene (dibenz[*a*,*h*]anthracene, DB[*a*,*h*]A) and 1,2;7,8-dibenzanthracene (dibenz-[*a*, *j*]anthracene, DB[*a*, *j*]A) (Figure 1.1) have been described;[13] and subsequently these newly synthesized and pure compounds were tested for carcinogenic activity by Kennaway and Hieger.[14,15] Since DB[*a*,*h*]A, DB-[*a*, *j*]A and 3-methyl-DB[*a*,*h*]A scored positive in the mouse skin bioassay, these pentacyclic hydrocarbons were actually the first pure compounds proved to act independently as real and strong carcinogens. In addition to their biological activity, the carcinogenic dibenzanthracenes displayed fluorescence spectra with similar (but again not identical) features as those of carcinogenic tars or carcinogenic 'Schroeter products'. Beginning in 1930, and with the help of the British Gas, Light and Coke Company, Hieger isolated about 7 g of a yellow powder out of two tons (!) of coal tar pitch by means of repetitive steps of fractional distillation, extraction and crystallization. The product showed both strong carcinogenic activity and high fluorescence in the spectral positions designated as the 'cancer bands'.[16,17] Further fractionation of the carcinogenic powder by Hewett and Cook afforded two pure crystalline products with melting points of 176 and 187°C. Both compounds were shown to be isomeric with the pentacyclic perylene ($C_{20}H_{12}$), but only the lower melting major component displayed the characteristic 'cancer bands' in its fluorescence spectrum. Subsequent synthetic preparation of '3,4-benzpyrene' (benzo[*a*]pyrene, B[*a*]P) and '1,2-benzpyrene' (benzo[*e*]pyrene, B[*e*]P) (Figure 1.1) unequivocally revealed that the major component of the crystals prepared from the pitch distillate was identical to B[*a*]P, which was also proven to be highly carcinogenic[18] (see also Cook *et al.*[19]). The synthetic B[*e*]P was identical with the minor and non-carcinogenic component of the crystallate melting at 187°C. In 1939, Kennaway, Cook, Hewett, Hieger and Mayneord were awarded with the first Anna Fuller Memorial Prize "... *in recognition of their notable accomplishments in the fields of Cancer Research, and specifically for the isolation and synthesis of cancer-producing hydrocarbons from coal tar, the identification by fluorescence spectroscopy, and for the study of the biological effects of these substances*".[20]

At this point it should be noted that the nomenclature used in the classical literature mentioned above is based on older, sometimes confusing and contradictory ring numbering systems, all of which have been replaced by the IUPAC rules established in 1966.[21] For that reason, the old and now obsolete names for B[*e*]P ("1,2-benzpyrene") and B[*a*]P ("3,4-benzpyrene") generally used prior to 1966 are actually misleading given the modern numbering system and applied at their common basic structure, the tetracyclic pyrene moiety (Figure 1.1). The currently accepted IUPAC-based nomenclature system of PAHs is explained in greater detail by Grimmer[22] and Harvey.[23,24]

Soon after isolation of B[*a*]P from coal tar and the proof that pure PAHs are able to induce skin tumors in animals, urinary metabolites of small PAHs such as anthracene have been detected and reported along with the suggestion that "*... the (toxic hydrocarbons) might either be converted (intravitally) into more active pathogenic substances or be detoxicated by conversion into some harmless compound(s)....*"[25] Today it is well established that PAHs would not be carcinogenic if they were not metabolized. First evidence came from the observation that radioactively labeled PAHs, which became available in the late 1940s,[26–28] bound to both the protein and DNA fraction in epidermal cells after administration of the compound onto the back of mice.[29–31] Furthermore, DNA was proposed to be the essential 'cellular receptor for carcinogenesis' as the carcinogenic potencies of a series of PAHs and the extents to which these are bound to DNA *in vivo* were roughly correlating.[31] Later, it was first anticipated that the appearance of PAH-induced tumors would depend upon a series of additional 'cellular events',[32] before it was actually shown that DNA binding of PAHs required the presence of activating enzymes residing in the endoplasmatic reticulum ('microsomal fraction') of cells.[33] Although the PAHs were among the first carcinogens to be studied with regard to their biotransformation, the discovery of their active DNA-binding intermediates has taken quite a long time. Some 20–25 years after the first reports on urinary metabolites of anthracene[25] and DB[*a*,*h*]A,[34] epoxides (arene oxides) were shown to be the intermediates,[35,36] and another 10 years passed until the first arene oxide intermediate was isolated.[37] In 1973, Borgen *et al.* reported that a metabolite of B[*a*]P, the 7,8-dihydrodiol, binds to a 10-fold greater extent to DNA *in vitro* than its parent compound. In either case, activating microsomal preparations were required.[38] This led Sims *et al.* to propose that a secondary metabolite, the 7,8-dihydrodiol 9,10-epoxide ('diol-epoxide') derivative of B[*a*]P, is actually the chemical species covalently interacting with DNA.[39] Subsequent work confirmed the central role of diol-epoxide metabolites in mediating the DNA binding of B[*a*]P and other carcinogenic PAHs (see Chapter 2). Since they were also confirmed to be highly mutagenic and carcinogenic, diol-epoxides are regarded as 'ultimate carcinogenic' metabolites initiating the process of PAH tumorigenesis (Figure 1.2).

Studies on cancer incidences or cancer induction in relation to age, dose and time of carcinogenic onset in both humans and animals principially indicated that the underlying processes involve multiple stages.[40,41] With regard to PAH-induced tumorigenesis it was already noticed in the early 1940s that application of low doses of these compounds onto the back of mice usually failed to produce any tumors. However, once combined with

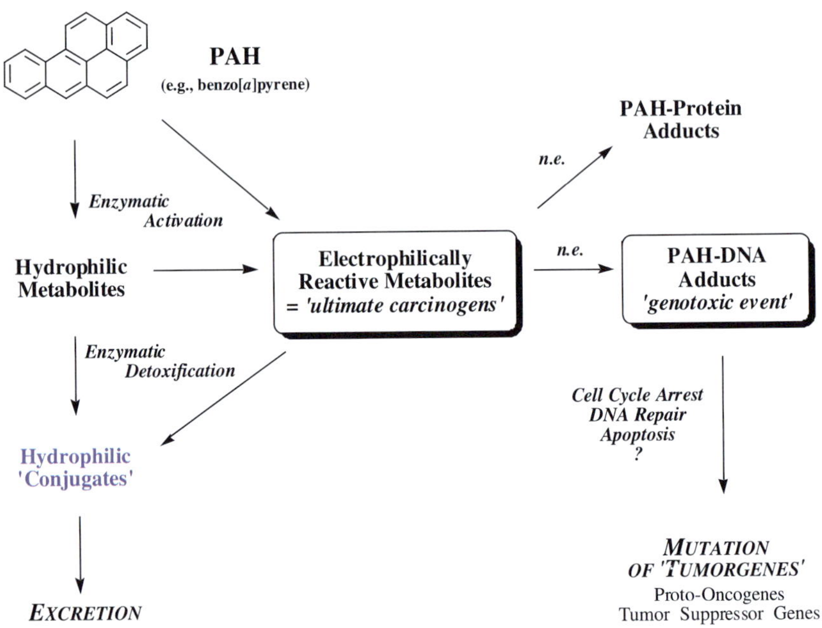

Figure 1.2: Polycyclic aromatic hydrocarbons: early events in chemical carcinogenesis.

croton oil, a preparation from the seeds of *Croton tiglium* (Euphorbiaceae) which by itself is non-carcinogenic, tumor production could be efficiently induced.[42] Based on this observation, Berenblum and Shubik developed the two-stage concept of mouse skin tumorigenesis in 1947.[43,44] Some years earlier, Friedewald and Rous had already proposed to divide the process of chemical carcinogenesis into the stages of 'initiation' and 'promotion'.[45,46] Since then, and mainly based on experimental skin tumorigenesis, both stages have been operationally defined in the context of each other. It has been suggested that the 'initiating stage' consists of a specific and irreversible conversion of normal cells into latent tumor cells that remain dormant until further stimulation by 'promoting agents' would occur. The onset of the 'promotional stage' would then lead to an outgrowth of initiated cells resulting in the proliferation of clones of altered cells and enhanced tumor formation. In animal tumor models this stage always requires a prolonged period of action over weeks or months and may be reversible after termination of the treatment. In addition, promotion before or without initiation has no tumorigenic effect, but promotor application can be delayed for weeks after the PAH treatment without loss of tumor production.[47] In the 1960s, work by Hecker[48,49] and Van Duuren *et al.*[50–52] succeeded in the characterization of the structure of the principal promoting agents in croton oil, which are tricyclic diterpene alcohols esterified with different fatty acids at positions 12 and 13 (phorbol esters). The most active ester was found to be 12-*O*-tetradecanoylphorbol 13-acetate (TPA). Today it is well established that skin tumor promoting agents such as TPA and others (e.g. teleocidin, okadaic acid) act mainly through receptor-mediated 'epigenetic' (non-genotoxic) mechanisms,[53–55] although induction of chromosomal aberrations such as DNA strand breaks may also be seen under certain circumstances *in vitro*.[56,57] On the other hand, initiating compounds, such as PAHs, are genotoxic agents that covalently interact with DNA (see above). If not removed properly by repair enzymes, the formation of PAH-DNA adducts may result in nucleobase-mispairing and the formation of mutations (see Chapters 5 and 6). Induction of mutations in cancer-related genes, such as proto-oncogenes or tumor suppressor genes, is therefore generally assumed as being the crucial event during the process of tumor initiation (Figure 1.2). While all carcinogenic PAHs are initiators, many of them have both initiating and promoting activity and are therefore considered as 'complete carcinogens'. Repeated treatments of animals with high doses of potent PAHs such as B[*a*]P or DB[*a*,*h*]A over extended periods of time always produce tumors without any further need for the application of typical promotors such as TPA.

Although first isolated from coal tar, which is a residual product derived from carbonization of bituminous coal at 1000–1200°C, thermal decomposition of virtually any organic material may lead to the generation of PAHs. The formation is based upon pyrolysis (incomplete combustion), intermolecular condensation and cyclization reactions, and was found to be most productive in the temperature range of 660–740°C.[58,59] According upon the results from a series of pyrolysis experiments, Badger suggested the stepwise synthesis of PAHs such as B[*a*]P from C_2 species involving free radical pathways.[60,61] This mechanism could help to explain the observation that similar PAH profiles always emerge — irrespective of the kind of starting material.[22,62] Even under different conditions of decomposition, such as nitrogen at temperatures between 700 and 1000°C, or at a temperature of 700°C in the presence of air, similar PAH profiles were found to be formed from different types of organic material. However, the quantities of individual PAHs may differ considerably depending on the starting material. For instance, pyrolysis of 1 gram glucose, dry tobacco or paraffin wax at 700°C under nitrogen gas resulted in the formation of 886 ng, 752 ng and 66.6 μg B[*a*]P, respectively.[22] On the other hand, the level of temperature present or applied during combustion of the organic matter turned out to be the main determining factor for the specific PAH profile pattern produced. Temperatures of 1000°C and higher, as present during carbonization (coking) of coal, would mainly produce unsubstituted PAHs and heteroatom-containing analogs ('heteroarenes'). Alkyl-substituted PAHs and heteroarenes, on the other hand, are predominating in crude mineral oils that are formed during geological periods from the decay of plants and their terpenoid and steroid constituents at relatively low temperatures (150–200°C; 'low temperature pyrolysis').[62] Under these conditions, the less temperature-stable alkyl side chains of the molecules, already present in their terpene and steroid precursors, usually persist and are not cleaved off. In some way, the early generation and discovery of the strong carcinogen 3-methylcholanthrene ("20-methylcholanthrene", Figure 1.1) from pyrolysis of the bile acid 12-ketocholanic acid and of its derivative, dehydronorcholene, at 150–320°C in the presence of selenium[63,64] can be regarded as an experiment resembling the natural geosynthesis of alkylated PAHs. According to the aforementioned outcome, intermediate temperatures would produce complex mixtures of unsubstituted and alkyl-substituted PAHs varying only gradually depending on the kind of origin.[22,62]

PAHs present in the environment usually originate from anthropogenic sources. Only a small percentage of the overall release into the atmosphere is caused by natural events such as forest fires and volcanic activities. Coal tar products, derived from the coking of bituminous coal, are among the most important PAH-containing sources in the occupational environment. Evaporation during heating of raw and oily PAH-containing matter, or formation by pyrolysis and incomplete combustion are the processes during which PAHs are being emitted at the work place.[58] Some of the most important industries with considerable PAH exposures are the coke production and aluminum smelting, iron and steel sintering and foundry operations, petrochemical processing/petroleum catalytic cracking, coal gasification, shale oil and asphalt production (see Chapters 3 and 4). In addition to those industrial sources, PAHs are released into the environment during combustion for residential heating, power and heat generation, incineration, open fires, and combustion of gasoline or diesel fuel in motor vehicles.[59,65–67] Due to these different kinds of emission sources, PAHs are ubiquitously present in the environment. Modern analytical methods enable their detection in the atmosphere, in sediments, soil, water and — as an inevitable consequence from that — also in living organisms ('biosphere'), including human food sources.[59,62,66,68,69] Airborne PAHs are largely present as aerosols due to their low vapor pressure and high melting points. They either exist as more or less pure particles, or are adsorbed onto particulate matter such as soot and dust. Since most of these PAH-containing particles are smaller than 5 μm in diameter, this material is able to penetrate into the lower respiratory tract including gas-exchanging alveoli.[70] Hence, most of the PAHs present in the atmosphere may pose a potential health risk to humans.

Modern analytical methods allow the detection of more than hundred unsubstituted and alkylated PAHs in airborne particulates,[71] from which more than two dozens of PAHs with varying carcinogenic potencies are commonly analyzed. The best known and characterized compound is B[*a*]P. However, others such as pyrene, B[*a*]A, chrysene, 5-methylchrysene, benzo[*b*]fluoranthene, benzo[*k*]fluoranthene, DB[*a*,*h*]A, cyclopenta[*c*,*d*]pyrene, indeno[1,2,3-*cd*]pyrene or anthanthrene are also frequently detected in ambient air samples in considerable amounts, and therefore most of them have been prioritized by the US-Environmental Protection Acency.[59,72] Background concentrations of airborne PAHs were estimated to be in the range of 1–50 pg/m^3 and are known to extend to hundred- or even thousand-fold higher levels in urban or industrial areas depending on the season (summer/winter).[73–75] Although these levels still seem fairly low, studies from the 1970s and 80s estimated that the worldwide overall emission of B[*a*]P into the atmosphere may reach a value of 5000 metric tons per year.[76] For the U.S. alone, the total amount of B[*a*]P released was estimated between 300 and 1300 metric tons, and the overall mass of all PAHs together at around 11,000 metric tons annually.[67] PAHs are also widely distributed in human foodstuff as a result of contamination from air, soil and water (e.g. fruits, vegetables, cereals), or as a by-product of certain kinds of processing (e.g. frying, grilling, smoking/charbroiling of fish or meat).[62,66,68,77–79] Depending on the kind of food and the way of preparation, levels are highly variable but can reach microgram ranges per kilogram dry weight and may therefore actually be considered as the main sources of human exposure in the general population. The average total daily intake of PAHs by a member of the general human population has been estimated to be 0.207 μg from air, 0.027 μg from water, and 0.16–1.6 μg from food.[80] Tobacco smoke, on the other hand, which has been linked to about 90% of all lung cancer cases, other smoking-related cancer types, and approximately 1.2 million worldwide deaths annually,[81] may be one of the most important sources of PAHs in indoor air. About 150 different unsubstituted and methylated PAHs, many of them known as strong carcinogens, could be identified in tobacco smoke condensate, and about 10–50 ng of B[*a*]P was estimated in early studies to be inhaled into the lungs from each cigarette.[62,82,83]

Although levels of B[*a*]P were determined in a more recent study to be typically less than 10 ng in modern cigarettes,[84] the presence of a variety of potent PAHs indubitably contributes to the overall tumorigenicity of cigarette smoke in man. A compilation of all investigations on carcinogenic PAHs and other constituents present in cigarette smoke has been published recently.[85]

PAHs belong to the more general class of polycyclic aromatic compounds (PACs) that contain two or more aromatic rings fused together in a linear ('*cata*-condensed') or angular ('*peri*-condensed') configuration. Anthracene and phenanthrene may serve as simple examples for both different types of PAHs (Figure 1.1). In *sensu stricto*, PAHs consist of carbon and hydrogen only. Hence, those compounds containing heteroatoms (e.g. N, S, O) in their molecular structure would not belong to the group of PAHs in a more narrow sense of the chemical nomenclature. These heteroarenes are therefore usually not considered in the present book, unless they would be the product of the biotransformation of a parent PAH (e.g., hydroxy and epoxy group-containing reactive intermediates such as diol-epoxides; see above). One exception from this rule are the nitro-substituted derivatives of PAHs (NO_2-PAHs) that are formed in the atmosphere from PAH precursors.[86,87] Due to their widespread distribution in the environment, their occurrence together with unsubstituted PAHs and their biological relevance as possible human carcinogens, NO_2-PAHs will be discussed together with unsubstituted PAHs regarding their possible contribution in the etiology of human breast cancer (Chapter 9). In the following the reader will find articles reviewing the most recent knowledge on metabolism, genotoxicity and repair of PAH-induced DNA damage. The topics of biomonitoring the exposure to PAHs, epigenetic effects induced by PAHs and factors modulating the individual tumor susceptibility in the human population will also be covered by the present book.

Acknowledgment

I am very grateful to Dr. W. M. Baird for his critical reading and the comments he made on the manuscript. The work of the author was supported by the German Research Foundation (DFG: LU 841/2-1).

References

1. Pott P (1775) *Chirurgical Observations Relative to the Cataract, The Polypus of the Nose, The Cancer of the Scrotum, The Different Kinds of Ruptures, and The Mortification of the Toes and Feet.* Hawes, Clarke, and Collins, London.
2. Volkmann R (1874) Ueber Theer- und Russkrebs. *Berl. Klin. Wochenschr.* 11: 218.
3. Bell J (1876) Paraffin epithelioma of the scrotum. *Edinb. Med. J.* 22: 135–137.
4. Henry SA (1947) Occupational cutaneous cancer attributable to certain chemicals in industry. *Br. Med. Bull.* 4: 389–401.
5. Yamagiwa K and Ichikawa K (1915) Experimentelle Studie über die Pathogenese der Epithelialgeschwülste. *Mitt. Med. Fak. Kaiserl. Univ. Tokio* 15: 295–344.
6. Yamagiwa K and Ichikawa K (1918) Experimental study of the pathogenesis of carcinoma. *J. Cancer Res.* 3: 1–29.
7. Tsutsui H (1918) Über das künstlich erzeugte Cancroid bei der Maus. *Gann* 12: 17–21.
8. Passey RD (1922) Experimental soot cancer. *Br. Med. J.* 2: 1112–1113.
9. Bloch B and Dreifuss W (1921) Ueber die experimentelle Erzeugung von Carcinomen mit Lymphdrüsen- und Lungenmetastasen durch Teerbestandteile. *Schweiz. Med. Wochenschr.* 51: 1033–1037.
10. Kennaway E (1925) Experiments on cancer-producing substances. *Br. Med. J.* 2: 1–4.
11. Schroeter G (1924) Über den Chemismus der Auf- und Abbaureaktionen mittels Aluminiumchlorids. *Ber. Dtsch. Chem. Ges.* 57: 1990–2003.
12. Hieger I (1930) The spectra of cancer-producing tars and oils and of related substances. *Biochem. J.* 24: 505–511.
13. Clar E (1929) Zur Kenntnis mehrkerniger aromatischer Kohlenwasserstoffe und ihrer Abkömmlinge, I. Mitt.: Dibenzanthracene und ihre Chinone. *Ber. Dtsch. Chem. Ges.* 62: 350–359.
14. Kennaway E and Hieger I (1930) Carcinogenic substances and their fluorescence spectra. *Br. Med. J.* 1: 1044–1046.
15. Kennaway E (1930) Further experiments on cancer-producing substances. *Biochem. J.* 24: 497–504.
16. Hieger I (1933) Isolation of a cancer-producing hydrocarbon from coal tar. I. Concentration of the active substance. *J. Chem. Soc.* 395–396.
17. Hieger I (1937) The fluorescence spectrum of 3,4-benzpyrene. *Am. J. Cancer* 29: 705–714.

18. Cook JW, Hewett CL and Hieger I (1933) The isolation of a cancer-producing hydrocarbon from coal tar. Parts I, II, and III. *J. Chem. Soc.* 395–405.
19. Cook JW, Haslewood GAD, Hewett CL, Hieger I, Kennaway EL and Mayncord WV (1937) Chemical compounds as carcinogenic agents. *Am. J. Cancer* 29: 219–259.
20. Kennaway E (1955) The identification of a carcinogenic compound in coal tar. *Br. Med. J.* 2: 749–752.
21. IUPAC (1966) *International Union of Pure and Applied Chemistry: Nomenclature of Organic Chemistry.* Butterworths, London.
22. Grimmer G (1983) Chemistry. In: *Environmental Carcinogens: Polycyclic Aromatic Hydrocarbons* [Grimmer G (ed.)] pp 27–60, CRC Press, Boca Raton, FL.
23. Harvey RG (1991) Nomenclature. In: *Polycyclic Aromatic Hydrocarbons: Chemistry and Carcinogenicity* [Harvey RG (ed.)] pp 1–10, Cambridge University Press, Cambridge, UK.
24. Harvey RG (1997) Nomenclature. In: *Polycyclic Aromatic Hydrocarbons* [Harvey RG (ed.)] pp 12–17, Wiley-VCH, New York.
25. Boyland E and Levi AA (1935) Metabolism of polycyclic compounds. I. Production of dihydroxydihydroanthracene from anthracene. *Biochem. J.* 29: 2679–2683.
26. Heidelberger C and Jones HB (1948) The distribution of radioactivity in the mouse following administration of dibenzanthracene labeled in the 9 and 10 positions with carbon 14. *Cancer* 1: 252–260.
27. Heidelberger C, Kirk MR and Perkins MS (1948) The metabolic degradation in the mouse of dibenzanthracene labeled in the 9 and 10 positions with carbon 14. *Cancer* 1: 261–275.
28. Heidelberger C (1953) Application of radioisotopes to studies of carcinogenesis and tumour metabolism. *Adv. Cancer Res.* 1: 273–338.
29. Miller EC (1951) Studies on the formation of protein-bound derivatives of 3,4-benzpyrene in the epidermal fraction of mouse skin. *Cancer Res.* 11: 100–108.
30. Heidelberger C and Davenport GR (1961) Local functional components of carcinogenesis. *Acta Unio Int. Contra Cancrum* 17: 55–63.
31. Brookes P and Lawley PD (1964) Evidence for the binding of polynuclear aromatic hydrocarbons to the nucleic acid of mouse skin: relation between carcinogenic power of hydrocarbons and their binding to deoxyribonucleic acid. *Nature* 202: 781–784.
32. Goshman LM and Heidelberger C (1967) Binding of tritium-labeled polycyclic hydrocarbons to DNA of mouse skin. *Cancer Res.* 27: 1678–1688.

33. Grover PL and Sims P (1968) Enzyme-catalyzed reactions of polycyclic hydrocarbons with deoxyribonucleic acid and protein *in vitro*. *Biochem. J.* 110: 159–160.
34. Boyland E, Levi AA, Mawson EH and Roe E (1941) Metabolism of polycyclic compounds. IV. Production of a dihydroxy-1,2,5,6-dibenzanthracene from 1,2,5,6-dibenzanthracene. *Biochem. J.* 35: 184–191.
35. Booth J, Boyland E, Sato T and Sims P (1960) Metabolism of polycyclic compounds. 17. The reaction of 1:2-dihydronaphthalene and 1:2-epoxy-1:2:3:4-tetrahydronaphthalene with glutathione catalysed by tissue preparations. *Biochem. J.* 77: 182–186.
36. Boyland E and Sims P (1960) Metabolism of polycyclic compounds. 16. The metabolism of 1:2-dihydronaphthalene and 1:2-epoxy-1:2:3:4-tetrahydronaphthalene. *Biochem. J.* 77: 175–181.
37. Jerina DM, Daly JW, Witkop B, Zaltzman-Nirenberg P and Udenfriend S (1970) 1,2-Naphthalene oxide as an intermediate in the microsomal hydroxylation of naphthalene. *Biochemistry (Washington)* 9: 147–156.
38. Borgen A, Darvey H, Castagnoli N, Crocker TT, Rasmussen RE and Wang IY (1973) Metabolic conversion of benzo[*a*]pyrene by Syrian hamster liver microsomes and binding of metabolites to deoxyribonucleic acid. *J. Med. Chem.* 16: 502–506.
39. Sims P, Grover PL, Swaisland A, Pal K and Hewer A (1974) Metabolic activation of benzo[*a*]pyrene proceeds by a diol-epoxide. *Nature* 252: 326–328.
40. Berenblum I (1954) A speculative review: the probable nature of promoting action and its significance in the understanding of the mechanism of carcinogenesis. *Cancer Res.* 14: 471–477.
41. Farber E and Cameron R (1980) The sequential analysis of cancer development. *Adv. Cancer Res.* 31: 125–226.
42. Mottram JC (1944) Developing factor in experimental blastogenesis. *J. Pathol. Bacteriol.* 56: 181–187.
43. Berenblum I and Shubik P (1947) The role of croton oil applications, associated with a single painting of a carcinogen, in tumour induction of the mouse's skin. *Br. J. Cancer* 1: 379–382.
44. Berenblum I and Shubik P (1947) A new, quantitative, approach to the study of the stages of chemical carcinogenesis in the mouse's skin. *Br. J. Cancer* 1: 383–391.
45. Friedewald WF and Rous P (1944) The initiating and promoting elements in tumor production. An analysis of the effects of tar, benzpyrene, and methylcholanthrene on rabbit skin. *J. Exp. Med.* 80: 101–126.

46. Friedewald WF and Rous P (1944) The determining influence of tar, benzpyrene, and methylcholanthrene on the character of the benign tumors induced therewith in rabbit skin. *J. Exp. Med.* 80: 127–144.
47. Berenblum I and Shubik P (1949) The persistence of latent tumour cells induced in the mouse's skin by a single application of 9:10-dimethyl-1:2-benzanthracene. *Br. J. Cancer* 3: 384–386.
48. Hecker E (1962) Über das toxische, entzündliche und cocarcinogene Prinzip des Crotonöls. *Angew. Chem.* 74: 722.
49. Hecker E (1968) Cocarcinogenic principles from the seed oil of *Croton tiglium* and from other Euphorbiaceae. *Cancer Res.* 28: 2338–2349.
50. Van Duuren BL, Orris L and Arroyo E (1963) Tumor-enhancing activity of the active principles of *Croton tiglium* L. *Nature* 200: 1115–1116.
51. Van Duuren BL and Orris L (1965) The tumor-enhancing principles of *Croton tiglium* L. *Cancer Res.* 25: 1871–1875.
52. Van Duuren BL (1969) Tumor-promoting agents in two-stage carcinogenesis. *Prog. Exp. Tumor Res.* 11: 31–68.
53. Slaga TJ (1984) *Mechanisms of Tumor Promotion.* CRC Press, Boca Raton, FL.
54. Pitot HC (1993) The molecular biology of carcinogenesis. *Cancer* 72: 962–970.
55. Schulte-Hermann R, Marian B and Bursch W (1999) Tumor promotion. In: *Toxicology* [Marquardt H, Schäfer SG, McClellan RO, and Welsch F (eds.)] pp 179–215, Academic Press, San Diego.
56. Birnboim HC (1982) DNA strand breakage in human leukocytes exposed to a tumor promotor, phorbol myristate acetate. *Science* 215: 1247–1249.
57. Dzarlieva-Petrusevska RT and Fusenig NE (1985) Tumor promoter 12-*O*-tetradecanoylphorbol-13-acetate (TPA)-induced chromosome aberrations in mouse keratinocyte cell lines: a possible genetic mechanism of tumor promotion. *Carcinogenesis* 6: 1447–1456.
58. Bjørseth A and Becher G (1986) *PAH in Work Atmospheres: Occurrence and Determination.* CRC Press, Boca Raton, FL.
59. Bayram A and Müezzinoglu A (1996) Environmental aspects of polycyclic aromatic hydrocarbons (PAHs) originating mainly from coal-fired combustion systems and their monitoring requirements. In: *Environmental Xenobiotics* [Richardson M (ed.)] pp 333–354, Taylor & Francis, London, UK.
60. Badger GM and Novotny J (1963) Mode of formation of 3,4-benzopyrene at high temperatures. *Nature* 198: 1086.

85. Rodgman A (2003) The composition of cigarette smoke: problems with lists of tumorigens. *Beitr. Tabakforsch. Int.* 20: 402–437.
86. Pitts JNJ, van Cauwenberghe KA, Frosjean D, Schmidt JP, Fitz DR, Belser WL, Knudson GB and Nynds PM (1978) Atmospheric reaction of polycyclic aromatic hydrocarbons: facile formation of mutagenic nitro derivatives. *Science* 202: 515–519.
87. Beland FA, Heflich RH, Howard PC and Fu PP (1985) The *in vitro* metabolic activation of nitro polycyclic aromatic hydrocarbons. In: *Polycyclic Hydrocarbons and Carcinogenesis* [Harvey RG (ed.)] ACS Monograph 283, pp 371–376, American Chemical Society, Washington D.C.

2 Metabolic Activation and Detoxification of Polycyclic Aromatic Hydrocarbons

Andreas Luch[1] and William M. Baird[2]

[1] *Massachusetts Institute of Technology, Center for Cancer Research, Cambridge, MA, USA*
E-mail: luch@mit.edu
[2] *Oregon State University, Department of Environmental and Molecular Toxicology, Corvallis, OR, USA*
E-mail: william.baird@orst.edu

2.1 Introduction

Polycyclic aromatic hydrocarbons (PAHs) exert their carcinogenic activity through binding to genomic DNA. However, compounds from this class of chemicals are composed of complex polycondensed benzo ring systems (*polyarenes*, Figure 2.1) that confer highly inert and lipophilic (lipid-soluble) physico-chemical properties. PAHs are therefore incapable of acting as *per se* electrophiles that would directly target cellular nucleophiles such as amino-functional bases within DNA. In order to exert their genotoxic effects, PAHs require enzymatic conversion into electrophilically

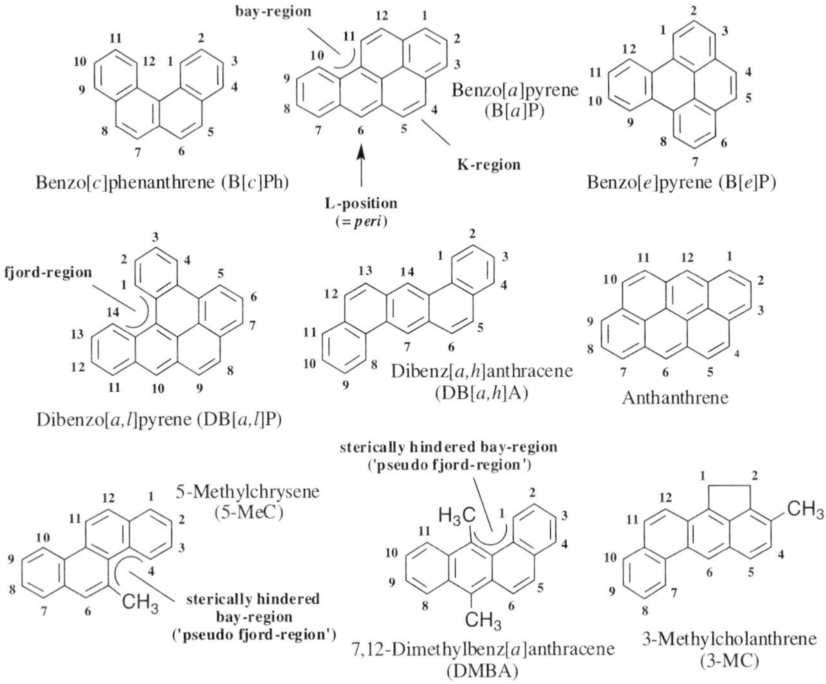

Figure 2.1: Polycyclic aromatic hydrocarbons and toxicologically relevant molecule regions (cf. Figure 9.1 in Chapter 9).

reactive derivatives (Figure 2.2).[1–5] On the other hand, biotransformation of foreign compounds (xenobiotics) that invade living organisms by passing lipid barriers is aimed to result in detoxification and subsequent elimination via excretion pathways such as bile → feces, rather than in generation of more toxic descendants. How may this apparent contradiction be explained?

Conversion of chemically inert xenobiotics such as PAHs into hydrophilic (water-soluble) and excretable derivatives is a multi-step task, intravitally promoted by different enzymes. In principle, this process proceeds through 'activated' (reactive) intermediates that are capable of undergoing further 'conjugation' reactions with usually highly charged and polar functional groups or molecules. In this context, formation of activated intermediates through oxidation (epoxidation) reactions, mainly catalyzed by oxidoreductases, is considered as being 'phase-I' of biotransformation.

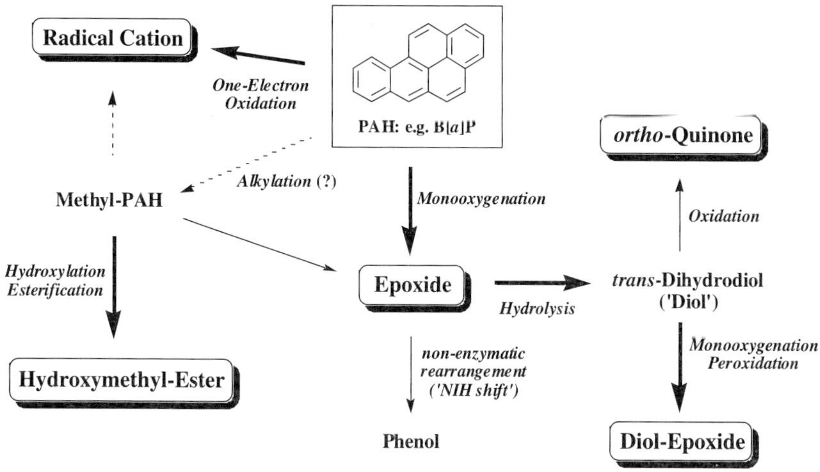

Figure 2.2: Metabolic generation of electrophilic metabolites of PAHs.

Further modification steps catalyzed by hydrolases, dehydrogenases, peroxidases and reductases may also contribute to this phase of biotransformation. The resulting 'functional-group'-containing derivatives then may or may not enter 'phase-II' of biotransformation in which transferases catalyze subsequent conjugation to polar molecules such as glutathione (GSH) and glucuronic acid, or to small residues such as sulfate or acetic acid. Unfortunately, generation of highly reactive intermediates may also lead to covalent binding to cellular constituents such as proteins or DNA in an 'alternative' pathway.[6–9] Since this approach poses an inherent risk to the genomic integrity of living cells, it can be considered as the Achilles' heel in detoxification of certain xenobiotics such as PAHs.

2.2 Structure-Activity Relationships

Although the number of theoretically possible PAHs is quite large, only a small fraction of these have been as yet chemically synthesized in the laboratory. From the examples studied for their biological effects *in vivo* investigators realized that the most potent carcinogenic members contain four to six benzo rings and often possess certain structural features such as sterically crowded bay- or fjord-regions (Figure 2.1[10,11]). Although these regions were soon attributed to high carcinogenic potency (see below), some examples exist where compounds either lack considerable activity despite the presence of such intramolecular regions (e.g., phenanthrene or benzo[*e*]pyrene, B[*e*]P[12–14]), or have been found to be carcinogenic at least under certain circumstances yet not possessing any sterically crowded molecule region at all (e.g., anthanthrene,[15,16] Figure 2.1).

As already found in the late 1930s and comprehensively reviewed by Badger in 1948,[17] methyl substitution of PAHs can further modify the biological activity. Depending on the site of methyl substitution in the molecule, the carcinogenic potency can be reduced, not changed, or dramatically enhanced.[10,18–24] For example, the activity strongly increases in case of methylation of position 5 in chrysene or positions 7 and 12 in benz[*a*]anthracene (B[*a*]A) to yield 5-methylchrysene (5-MeC) and 7,12-dimethylbenz[*a*]anthracene (DMBA), respectively (Figure 2.1). Both kinds of derivatization create a methyl-substituted and, thus, a sterically hindered bay-region ('pseudo fjord-region') that was found to be generally associated with increased biological activity of the particular PAH. Accordingly, 5-MeC and DMBA are much more potent carcinogens compared to the unsubstituted parent compounds, which themselves are only weak tumor inducers. In the case of DMBA, the carcinogenic activity is extremely high and even exceeds that of benzo[*a*]pyrene (B[*a*]P), the most well characterized carcinogenic PAH to date.[25–28]

2.3 Enzymatic Activation

2.3.1 Monooxygenation and Dihydrodiol Epoxide Pathway

Soon after pure dibenz[*a,h*]anthracene (DB[*a,h*]A) was synthesized and B[*a*]P was isolated from coal tar (Figure 2.1, cf. Chapter 1), extensive studies on metabolic activation of PAHs were carried out to answer the question of how PAHs might be enzymatically converted *in vivo*. The strong fluorescence of these compounds facilitated early achievements in detection and characterization of several hydroxy, dihydrodiol and quinone metabolites, as well as various conjugates in the urine and bile of treated animals[29] (cf. Chapter 3). However, none of these metabolites were found to be carcinogenic. In 1950, Boyland proposed that PAHs might be activated through formation of reactive epoxides (arene oxides) as the carcinogenic intermediates directly interacting with cellular constituents.[30] Theoretical studies[31] then suggested that the carcinogenic activity of PAHs might be correlated to the presence of an electronically dense (partly olefinic) so-called K-region (*K* for *Krebs*, the German word for cancer; Figure 2.1) that would be the site of conversion into reactive epoxides. The resulting 'K-region theory' proposed by Pulman and Pulman in 1955[32] was an attempt to correlate the carcinogenicity of PAHs with the extent of their olefinic character (electronic density) in this particular molecule region (e.g., position 4–5 in B[*a*]P, or 5–6 and 12–13 in DB[*a,h*]A; Figure 2.1). Soon after the first successful synthesis of various K-region epoxides[33] it became clear, however, that these derivatives showed little, if any, carcinogenic activity and had much weaker effects in animals than the parent hydrocarbons.[10,18,34–37] Later work finally revealed that DNA-reactive PAH metabolites formed *in vitro* or *in vivo* did not originate from oxidation at their K-regions.[10,37,38]

In the mid 1950s, first evidence emerged that PAHs may stimulate their own metabolism. Expression of a liver oxidase, that was found to be derived from the endoplasmatic reticulum and that was originally designated as the aryl hydrocarbon hydroxylase ('AHH') or cytochrome P_1-450 ('P-488'), could be induced *in vivo* by administration of certain PAHs such as 3-methylcholanthrene (3-MC) or B[*a*]P (Figure 2.1).[39–44] In addition, binding of PAHs to DNA *in vitro* was shown to be dependent on the presence of 'microsomal' enzymes originating from the endoplasmatic reticulum (cf. below).[45,46] Synthetic preparation of various regioisomeric dihydrodiols of B[*a*]P and investigation of their biotransformation catalyzed by microsomal preparations subsequently revealed that the 7,8-dihydrodiol binds to a much greater extent to DNA *in vitro* compared to B[*a*]P and other B[*a*]P dihydrodiols (at 4,5- and 9,10-positions).[47] This finding led to the suggestion that B[*a*]P-7,8-dihydrodiol was further activated at the adjacent (vicinal) double bond to form the 7,8-dihydrodiol 9,10-epoxide of B[*a*]P (B[*a*]PDE = 7,8-dihydroxy-9,10-epoxy-7,8,9,10-tetrahydro-B[*a*]P), and that this highly electrophilic diol-epoxide would be responsible for the binding of B[*a*]P to DNA *in vitro*[37,48,49] and in mouse skin.[50] In the following, extensive studies on the biological activity of PAH metabolites both in cell culture[51–59] and *in vivo*[59–63] provided further support for the assumption that a multistep enzymatic activation pathway with a sequence of non-K-region epoxidation, hydrolysis of the primary epoxide to a dihydrodiol, and further epoxidation at the adjacent double bond produces vicinal diol-epoxides as the carcinogenic metabolites of PAHs (Figure 2.3). All of these investigations supported the notion that the critical diol-epoxides in PAH-induced carcinogenesis would be the ones adjacent to the 'bay-region' of these compounds. (The bay-region is formed by a non-linear/angular condensation of benzo rings; see Figure 2.1.) Synthesis of the authentic 7,8-diol-9,10-epoxides of B[*a*]P[53,54,64–66] finally provided the proof that binding of B[*a*]P to DNA in cells in culture and *in vivo* occurs mainly through these chemical species. Based on quantum mechanical molecular orbital calculations, the 'bay-region theory' proposed by Jerina and Lehr in 1977[67–70] suggested that an epoxide ring adjacent to the bay-region of a hydrocarbon (e.g., benzylic position 10 in B[*a*]PDE, Figure 2.3) would have the highest electrophilic reactivity and, therefore, would be the preferred position of covalent interaction with nucleophiles such as purine bases in genomic DNA. This assumption is based on the fact that intermediate formation of a carbonium ion by an opening epoxide ring occurs more easily at benzylic positions of saturated rings adjacent to a bay-region than at non-bay-region positions of PAHs (Figure 2.4). The notion of a carbonium ion-like transition state during interaction of diol-epoxides with cellular nucleophiles resulted in semi-quantitative calculations between the inherent ability of PAHs to electronically stabilize a putative benzylic carbonium ion at its bay-region (amount of delocalization energy released) and its genotoxic (DNA-binding) potency.

Initial oxidation (epoxidation) of PAHs is catalyzed by cytochrome P450-dependent monooxygenases (CYP, EC 1.14.14.1).[71–77] These enzymes, also called mixed-function oxidases, can be ubiquitously found in multicellular organisms and are expressed (or induced) in most mammalian tissues.[78,79] They are mainly localized in the endoplasmatic reticulum, i.e., subcellular membranes retrievable from tissue homogenates by ultracentrifugation as the so-called 'microsomal fraction'. CYP enzymes

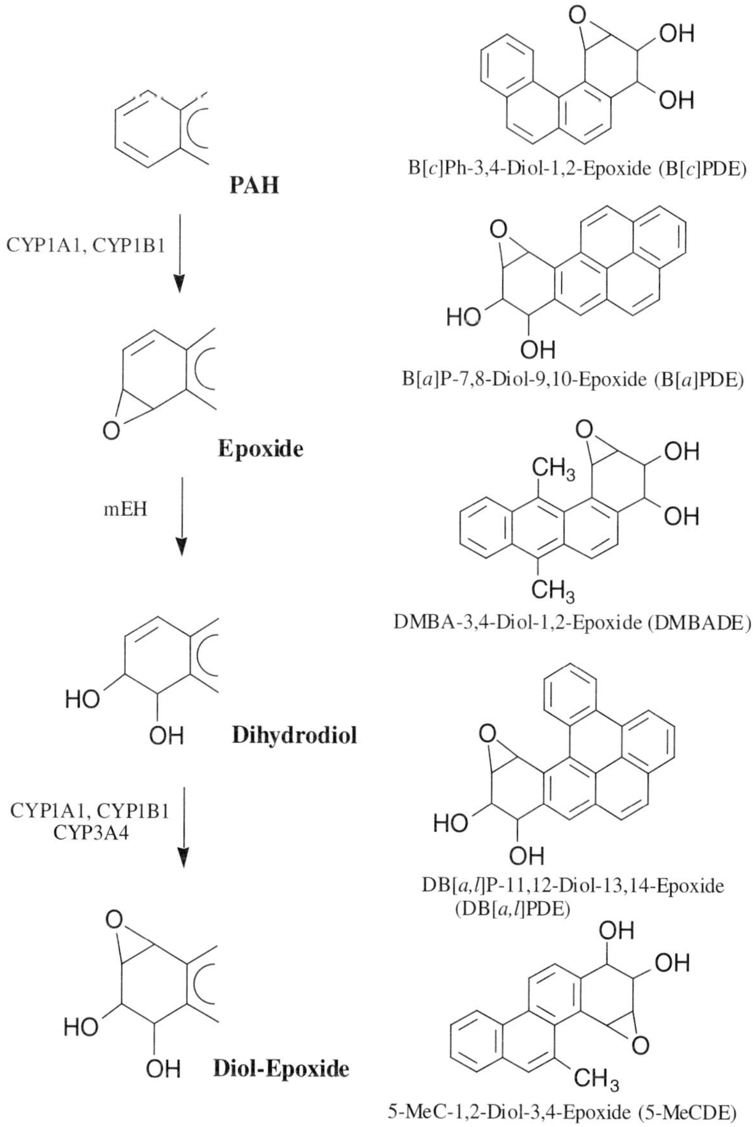

Figure 2.3: Monooxygenation pathway towards bay- or fjord-region diol-epoxides of PAHs.

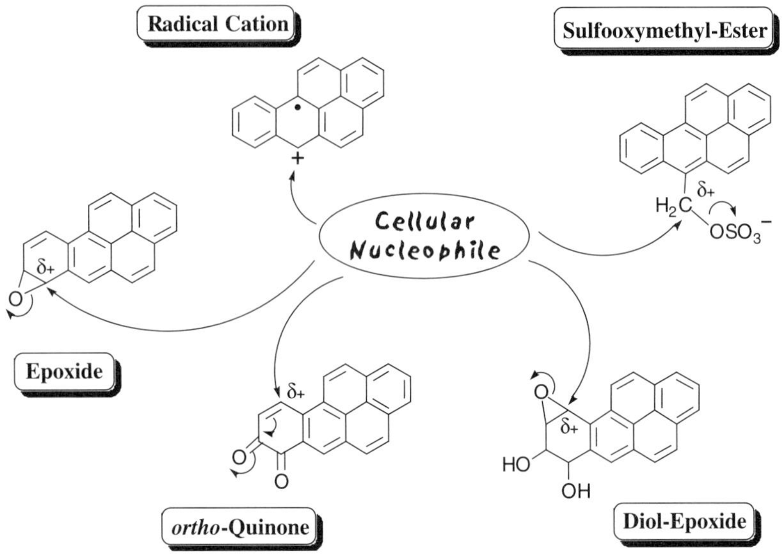

Figure 2.4: Potential reactions of cellular nucleophiles with electrophilically reactive metabolites of benzo[*a*]pyrene.

are hemoproteins that associate with another membrane-bound enzyme, the NADPH-dependent flavoprotein cytochrome P450 reductase (EC 1.6.2.4). The functional multimeric enzyme complex constituted in the lipid bilayer incorporates one atom of molecular oxygen into the polycyclic substrate, whereas the second oxygen is reduced to give one molecule of water (*mixed-function oxidase*). CYP enzymes are highly expressed in liver but can be also detected in various other organs such as lung, kidney, gastrointestinal tract etc. To date, 57 different *CYP* genes have been discovered in humans,[80,81] and more than 400 proteins are known to belong to this enzyme superfamily. Based on their amino acid sequences, they are grouped into single families (indicated by an arabic number; more that 40% sequence homology) or subfamilies (indicated by a capital letter; more that 55% sequence homology). CYP enzymes are considered as the mainstay in biotransformation of a large number of foreign compounds such as drugs and environmental pollutants, thereby possessing a broad and overlapping substrate specificity. On the other hand, about half of all human CYP enzymes are also involved in endogenous metabolism such as steroidogenesis (e.g., hydroxylation of estrogens).

Monooxygenation of xenobiotics in mammalian species including man are mainly catalyzed by CYP enzymes from families 1 to 3. Among them, the individual forms CYP1A1 and CYP1B1 (predominantly expressed in extrahepatic tissues) and — to a much lower extent — the forms CYP1A2 and CYP3A4 (predominantly expressed in the liver) are the most important enzymes carrying out PAH oxidation reactions (Figure 2.3).[73,82–88] Whereas CYP1A1 is virtually absent or only detectable at very low levels in mammalian tissues under normal conditions, CYP1B1 is expressed at significant levels in almost all organs except liver and lung.[89–92] On the other hand, CYP3A4 is the most abundant form in mammalian liver tissue (up to 60% of the total), followed by CYP1A2. As mentioned above, PAHs are able to induce their own biotransformation. A broad range of carcinogenic PAHs such as B[*a*]P, DMBA, dibenzo[*a,l*]pyrene (DB[*a,l*]P), 3-MC and others are able to significantly increase the expression levels of CYP1A1 (identical to the form originally termed as 'AHH' or 'P-488') and CYP1B1 in liver, lung and most extrahepatic tissues through binding to a cytosolic receptor protein, the arylhydrocarbon receptor (AhR).[92–94] The AhR, which is encoded by the single autosomal gene locus termed *Ah* locus (for arylhydrocarbon responsiveness), mediates expression of CYP1A1 and CYP1B1 (as well as CYP1A2) and other xenobiotic-metabolizing enzymes after forming a complex with one of its ligands. In order to sufficiently bind to the receptor, these compounds have to meet certain structural requirements such as a large molecular area/depth ratio and a particular molecular dimension. These requirements are fulfilled by a wide range of xenobiotics such as many of the carcinogenic PAHs, polyhalogenated biphenyls or dioxins such as the 2,3,7,8-tetrachlorodibenzo-*p*-dioxin (TCDD).[94,95] Upon formation, the AhR-ligand complex translocates from the cytosol into the nuclear compartment where it then forms a complex with the AhR nuclear translocator (ARNT) protein. Within the nucleus, the resulting ligand-activated AhR-ARNT heterodimer then becomes competent to bind to specific *Ah*- or *xenobiotic-responsive element* (*AhRE* or *XRE*) sequences and to drive transcription from adjacent target promoters.[96] *AhRE* or *XRE* sequences are enhancer elements upstream of genomic target genes encoding a diverse set of enzymes such as the aforementioned CYP1 proteins and other xenobiotic-metabolizing enzymes such as glutathione *S*-transferases (GST), UDP-glucuronosyltransferases (UGT), or the NAD(P)H-dependent quinone oxidoreductase 1 (NQO1 = DT-diaphorase). A further detailed and a more elaborated description of the AhR-driven adaption pathway of induction of xenobiotic-metabolizing enzymes is provided by a number of excellent reviews on this topic.[95,97–102]

Formation of diol-epoxides at bay- or fjord-regions is considered as one of the major, if not, the predominant pathway of metabolic activation of most mutagenic and carcinogenic PAHs (cf. above; Figure 2.3). Fjord-regions arise from additional *peri*-condensation (angular arrangement) of a single benzo ring at the bay-region of a precursor PAH. For example, an additional aromatic ring at position 3–4 in phenanthrene and 11–12 in B[*a*]P creates the fjord-region PAHs benzo[*c*]phenanthrene (B[*c*]Ph) and DB[*a,l*]P, respectively (Figure 2.1). Due to repulsive electronic interactions ('steric hindrance') between hydrogen atoms at the edges of these regions (i.e., positions 1 and 12 in B[*c*]Ph or 1 and 14 in DB[*a,l*]P), fjord-region PAHs are no longer perfectly planar; instead, they display an out-of plane distortion.[103–105] As will be outlined later, extended steric hindrance in bay-regions, no matter if produced by methyl substitution (as in 5-MeC or DMBA) or by an additional aromatic benzo ring (as in B[*c*]Ph or DB[*a,l*]P), is often associated with increased biological activity.

The activation pathway leading to bay- or fjord-region diol-epoxides requires two additional enzymatic steps after initial monooxygenation (epoxidation) at a non-K-region double bond, e.g., position 7,8 in B[*a*]P or position 11,12 in DB[*a,l*]P (Figures 2.2 and 2.3). Firstly, the arene oxides initially produced undergo enzymatic hydrolysis to *trans*-dihydrodiols. In mammalian species, there are at least five different epoxide hydrolases detectable.[106] However, only two forms, the microsomal and cytosolic epoxide hydrolases, are involved in xenobiotic metabolism. Both enzymes lack significant sequence similarities and possess quite different substrate specificities. Whereas the cytosolic form catalyzes hydrolysis of some *trans*-substituted epoxides such as *trans*-stilbene oxide or various aliphatic epoxides, the microsomal expoxide hydrolase (mEH; EC 3.3.2.3) is involved in metabolism of arene oxides originating from PAHs.[107–110]

This microsomal form is highly expressed in mammalian liver, but can also be detected in most other organs. In contrast to the CYP enzymes mentioned above and conjugation enzymes such as UGT and GST, however, only little effects on the expression levels of mEH were observed in human hepatocyte cultures or in rodent tissues after treatment with chemical inducers such as 3-MC.[107,111–113]

Since the very early experiments with phenanthrene and anthracene[29,114] it has been known that dihydrodiols detected as metabolites *in vivo* would have *trans* configuration (that is, both hydroxy groups are *trans*-oriented to each other). Subsequent studies revealed that this is universally true for dihydrodiol metabolites generated from all other PAHs tested thereafter.[5,115,116] In more recent years, it became clear that the stereoselective mechanism involved in generation of *trans*-dihydrodiols is a two-step process executed by a catalytic amino acid triad within the catalytic center of the mEH.[117,118] In the first step, a nucleophilic aspartate residue attacks the epoxide ring to form an intermediate substrate-enzyme ester adduct at the allylic position of the PAH. The ester bond is then hydrolyzed by a single molecule of water pre-activated by histidine-driven deprotonation. This enzymatic reaction selectively generates vicinal *trans*-dihydrodiols that may then undergo further metabolic derivatization.

CYP enzymes are again responsible for subsequent epoxidation of *trans*-dihydrodiols to vicinal bay- or fjord-region diol-epoxides (Figure 2.3).[74,119] Besides CYP1A1 and 1B1, which are the predominant forms involved in both initial oxidation of PAHs to arene oxides and further oxidation of *trans*-dihydrodiols to diol-epoxides, evidence from cell cultures indicate that also CYP3A4 and — to a very small extent — CYP1A2 and other isoforms are capable in catalyzing monooxygenation at olefinic double bonds adjacent to dihydrodiol moieties.[82,84,120] In additon to CYP-mediated catalysis, vicinal diol-epoxides might be also produced from oxidation of *trans*-dihydrodiols by peroxyl radicals (Figure 2.2). In the early 1980s, Marnett and others demonstrated that hematin-catalyzed generation of peroxyl radicals from fatty acid hydroperoxides may eventually lead to epoxidation ('co-oxidation') of *trans*-dihydrodiols at their vicinal double bond to yield diol-epoxides *in vitro* and possibly also *in vivo*.[121–128] Enzymes such as prostaglandin(-endoperoxide) H synthases (EC 1.14.99.1) and lipoxygenase

(EC 1.13.11.12) were shown to be involved not only in this pathway, but also in activation of a wide range of different chemical carcinogens to DNA-binding metabolites.[129–131] Prostaglandin H synthases possess both a cyclooxygenase (COX) and a hydroperoxidase activity that generate hydroperoxy and hydroxy endoperoxides as early products in metabolism of poly-unsaturated fatty acids such as arachidonic acid or eicosapentanoic and linolenic acid. Some years later, evidence was presented that other human peroxidases such as lactoperoxidase from mammary epithelial cells, a previous unknown placental peroxidase, or myeloperoxidase (MPO; EC 1.11.1.7) may also be capable of promoting the generation of DNA-reactive diol-epoxides from *trans*-dihydrodiol precursors under certain circumstances both *in vitro* and *in vivo*.[132–138] From all of these enzymes, the isoform 2 of prostaglandin H synthase (PGHS-2 $\approx$ COX-2) was found to be highly inducible by mitogens, cytokines, inflammatory agents such as tumor promotors (e.g., TCDD, phorbol esters, deoxycholate) and PAHs (e.g., B[*a*]P),[130,139,140] and the MPO was shown to be constitutively expressed in lysosomes of polymorphonuclear leukocytes (neutrophilic granulocytes).[130] These expression patterns may therefore point to a promoting effect of tissue inflammation during initial stages of chemical tumorigenesis.[134–136] As will be outlined in more detail below, other peroxidase-dependent activation pathways such as one-electron oxidation of PAHs to radical cation intermediates,[130,141,142] or formation of reactive dialdehyde derivatives through ring-opening of *trans*-dihydrodiols[143,144] may also contribute to cancer-promoting (or initiating) effects of tissue inflammation. Although some possible synergistic effects presently cannot be ruled out, the exact contribution of peroxidase-dependent activation pathways in PAH-induced tumorigenesis remains elusive. In particular, strong and concurrent evidence for a considerable contribution to PAH-DNA adduct damage in animal models *in vivo* is still lacking.[145–147] Recent studies applying inhibitors of the COX activity of both isoforms of PGHS (e.g., resveratrol for COX-1, and SC-58125 or celecoxib for COX-2), however, demonstrated a protective role of these agents in PAH-induced skin cancer.[148–150] It has also been shown that overexpression of COX-2 sensitizes mice for DMBA-induced skin carcinogenesis,[151] and deficiency of either COX-1 or COX-2 led to a reduction of DMBA-induced skin tumors by 75%.[152] Interestingly, these effects were accompanied by an approximately 2-fold higher DMBA-DNA adduct level in COX-deficient animals compared to wild-type mice.[152] These observations are somehow in contrast to early studies with aspirin and indomethacin that revealed no protective effects on B[*a*]P-induced DNA damage and lung tumor formation in mice and guinea pigs.[145,153] On the other hand, anti-tumorigenic activities of inhibitors of COX-2 (PGHS-2), also seen in the human population with respect to colorectal cancer and other kinds of malignancies,[140,154] are most likely due to interference with COX-2-induced angiogenesis or COX-2-inhibited apoptosis and its resulting nutrient-driven acceleration of tumor growth, rather than to inhibitory effects on the bioactivation of carcinogenic agents.[154–157] In the case of MPO, the strongest support for an important role in activation of B[*a*]P to ultimate carcinogenic intermediates (i.e., bay-region diol-epoxides) *in vivo* presently derives from a study with atopic dermatitis patients topically treated with PAH-containing ointments.[138] Patients carrying a mutant MPO genotype ($-463G \rightarrow A$ within the promotor region) with decreased enzyme activity were found to have significantly reduced levels of B[*a*]PDE-DNA adducts in their skin tissue (5- to 6-fold compared to control). In addition, recent epidemiological studies provide some evidence for a protective role of the same genotype against lung cancer in humans[158–160] (cf. Chapter 10). Together, these data may point to an important role of MPO during initiation of PAH-induced tumorigenesis, particularly under conditions when chronic tissue inflammation promotes recruitment and local infiltration of cells that constitutively express high levels of this enzyme.

CYP1A1 and CYP1B1 are generally considered to be the predominant enzymes responsible for initial epoxidation of PAHs and further oxidation of non-K-region *trans*-dihydrodiols to ultimate carcinogenic bay- or fjord-region diol-epoxides (see above). In agreement with its pivotal role in PAH-induced expression of these enzymes, genetically engineered mice lacking the AhR protein (AhR-null genotype: $AhR^{-/-}$) were found to be resistant to B[*a*]P-induced skin tumorigenesis under conditions where almost all wild-type mice were affected.[161] Most likely, resistance to B[*a*]P-induced tumors is mainly due to the inability of AhR-null mice to express significant levels of CYP1A1 and CYP1B1 upon exposure to B[*a*]P.[94,161]

Conversely, mice with an AhR wild-type background express high levels of both CYP enzymes not only in the target tissue exposed to B[*a*]P in this study (skin), but also in almost all other organs investigated (i.e., liver, lung, thymus, heart, kidney, intestine, testis, uterus, ovary, and brain); and upon exposure to a wide range of different carcinogenic PAHs such as B[*a*]P, DMBA, DB[*a,l*]P, 3-MC and others.[89,94,161] In accordance with these most recent studies, and also with more early findings on the relationship between AhR-mediated enzyme induction and PAH-induced lethality,[162] mice lacking functional CYP1A1 were more resistant to B[*a*]P-induced hepatotoxicity compared to CYP1A1 heterozygotes ($CYP1A1^{+/-}$).[163] On the other hand, CYP1B1-deficient mice did not show any signs of DMBA-induced bone marrow toxicity as did there wild-type counterparts,[164] an effect possibly due to the requirement of this enzyme in pre-B cell apoptosis.[165] Mice lacking functional CYP1B1 were also found to be at significantly lower risk to develop lymphomas, skin hyperplasias, and stromal cell-derived ovary tumors after gastric gavage of DMBA or DB[*a,l*]P.[166–168] These data, along with a comparable abrogation of DMBA-mediated tumor formation in internal organs and in skin of mice after targeted disruption of the *mEH* gene,[169] provide convincing evidence for the predominant role of CYP1A1, CYP1B1, and mEH in activation of carcinogenic PAHs *in vivo*. It also indirectly points to the importance of the monooxygenation activation pathway in PAH-induced carcinogenesis and the central role of bay- or fjord-region diol-epoxides generated herein. On the other hand, detection of considerable levels of hepatic DNA adducts in AhR-null or CYP1A1-null mice after intraperitoneal administration of B[*a*]P[163,170] may be judged in view of the presence of an AhR-independent 'constitutive aryl hydrocarbon hydroxylase activity' in mouse liver corresponding to a hepatic CYP2C enzyme,[163,171] or may be due to low levels of constitutively expressed CYP1B1 and CYP1A2.[163,170] Alternatively, systemically applied PAHs may also be initially activated in extrahepatic tissues that express high levels of CYP1B1. Subsequent distribution of proximate carcinogenic intermediates (e.g., *trans*-dihydrodiols) throughout the body and further activation by constitutive liver enzymes such as CYP3A4 may then contribute to the hepatic DNA damage observed in AhR-deficient mice.[163] Occasionally, the limitations in the predictability of the '*in vivo* situation' become obvious for complex 'toxicological endpoints' such as organ-specific toxicity and tumorigenicity that not always perfectly correlate with the overall levels of DNA damage observed in these organs after challenging the entire organism with a systemic burden of carcinogenic PAHs.[163,168]

2.3.2 Stereochemistry of Activation

Enzymatic reactions usually proceed with high stereoselectivity. Soon after the first identification of vicinal diol-epoxides as the crucial metabolites responsible for DNA binding of B[*a*]P *in vitro* and *in vivo*,[48,50] it was realized that not only the reactions of enzymes involved in this pathway were highly stereoselective,[172–174] but also the biological activity of individual stereoisomers differed markedly from each other.[175–177]

In principle, CYP-mediated integration of an oxygen atom can occur from both (stereoheterotopic) sides of the aromatic ring system. However, initial epoxidation of B[*a*]P at its 7,8-position catalyzed by liver (microsomal) preparations from rats, for example, leads to formation of the optically active (+)-7*R*,8*S*-oxide in high enantiomeric excess (Figure 2.5). Subsequent mEH-mediated hydrolysis, initiated by enzymatic attack at the allylic position (carbon atom 8 in B[*a*]P), selectively affords the dihydrodiol with *trans*-oriented hydroxy groups, i.e., the (−)-*trans*-7*R*,8*R*-dihydrodiol as the solvolysis product of the predominating (+)-7*R*,8*S*-oxide (Figure 2.5). In contrast, the (−)-7*S*,8*R*-oxide and the corresponding (+)-7*S*,8*S*-dihydrodiol are only generated in small amounts. The preponderance of (+)-7*R*,8*S*-oxide and (−)-7*R*,8*R*-dihydrodiol in activation of B[*a*]P was found to vary slightly depending on the kind of chemical inducer used to increase the levels of CYP enzymes in the liver of rats. Microsomal preparations obtained with strong inducers of CYP1A1 (and 1B1), such as 3-MC or polyhalogenated biphenyls, resulted in the highest enantiomeric excess of about 92–98%.[173,178] Similar extends of selectivity have also been observed in case of the regioisomeric 4,5- and 9,10-dihydrodiols, the two additional *trans*-dihydrodiols formed during biotransformation of B[*a*]P (see below). Hence, initial metabolism of B[*a*]P by rat liver microsomes enantioselectively produces the *trans*-4,5-,

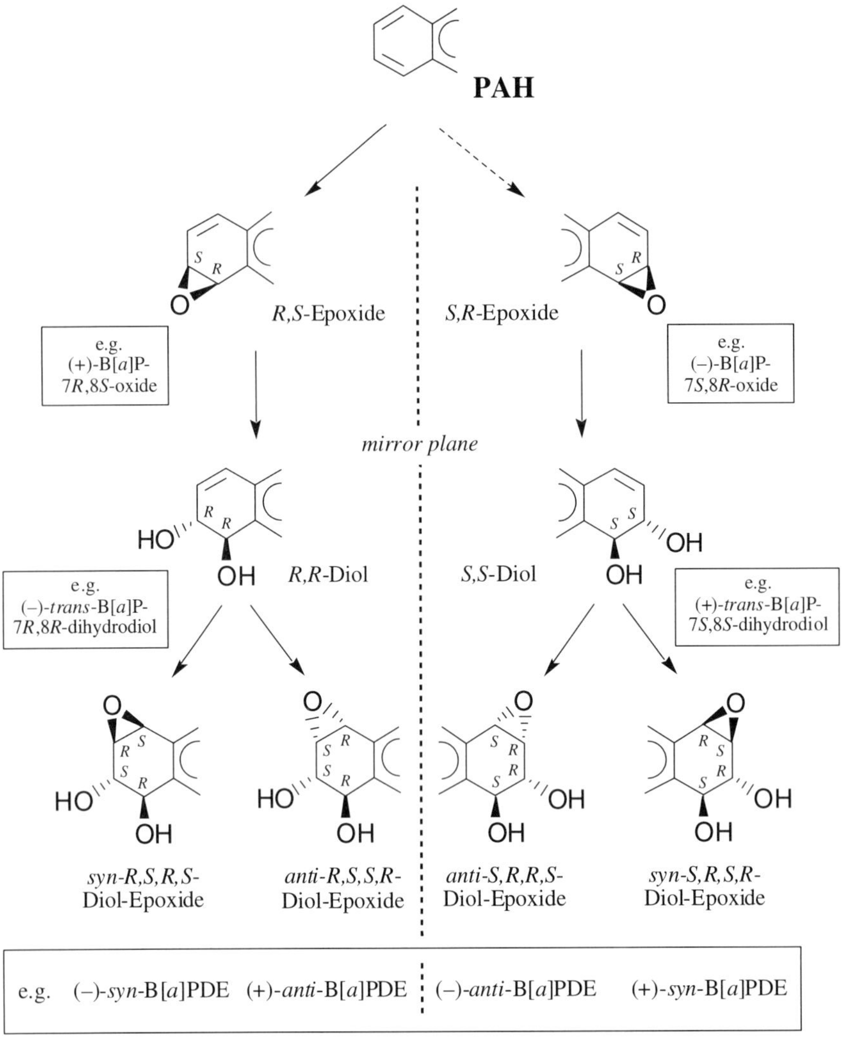

Figure 2.5: Stereoselective activation to bay- or fjord-region diol-epoxides of PAHs.

trans-7,8- and *trans*-9,10-dihydrodiols with *R,R*-configuration in high excesses.[10,178,179]

Theoretically, secondary epoxidation of the non-K-region 7,8-dihydrodiol of B[*a*]P at its vicinal double bond again can occur by introduction of the oxygen atom from either side of the molecule. Each dihydrodiol enantiomer could therefore give rise to a set of two diastereomeric diol-epoxides that would have the epoxide ring either at the same (*syn*) or the opposite (*anti*) side of the aromatic plane as the hydroxy group in the benzylic position (carbon atom 7 in B[*a*]P) (Figure 2.5). In contrast to these theoretical assumptions however, the (−)-7,8-dihydrodiol of B[*a*]P is preferentially converted to the (+)-*anti*-7*R*,8*S*-diol-9*S*,10*R*-epoxide [(+)-*anti*-B[*a*]PDE]. On the other hand, the small amount of (+)-7,8-dihydrodiol metabolically formed during primary metabolism is further oxidized mainly to the diastereomeric (+)-*syn*-7*S*,8*R*-diol-9*S*,10*R*-epoxide [(+)-*syn*-B[*a*]PDE] (Figure 2.5).[10,173,179] Considering the high enantioselectivity during formation of the *trans*-dihydrodiols and the diastereoselectivity during formation of the diol-epoxides, the main product in bay-region metabolism of B[*a*]P catalyzed by rat liver microsomes appears to be (+)-*anti*-B[*a*]PDE. In contrast, the three other diastereomeric diol-epoxides are only formed in minor amounts.

Follow up studies with various other PAHs such as B[*a*]A, B[*c*]Ph, and DMBA revealed that primary metabolism at the non-K-region double bond adjacent to the bay- or fjord-region (that is, the 3,4-position in all of these compounds; cf. Figure 2.1) in general affords the *trans*-dihydrodiol with *R,R*-configuration in high excess.[179,180] Although only a small percentage of the overall microsomal metabolism takes place at its 3,4-position (about 9%), B[*c*]Ph, the prototypic PAH possessing a fjord-region (Figure 2.1) is preferentially converted by 3-MC-induced liver microsomes to the (−)-3*R*,4*R*-*trans*-dihydrodiol (enantiomeric excess 88%).[181] Subsequent epoxidation at the adjacent 1,2-position again selectively generates the diol-epoxide with '*R,S,S,R*'-configuration, the (−)-*anti*-4*R*,3*S*-diol-2*S*,1*R*-epoxide [(−)-*anti*-B[*c*]PhDE]. Conversely, oxidation of the minor amounts of (+)-3*S*,4*S*-dihydrodiol produces the diastereomeric fjord-region diol-epoxide with '*S,R,S,R*'-configuration, the (+)-*syn*-4*S*,3*R*-diol-2*S*,1*R*-epoxide [(+)-*syn*-B[*c*]PhDE] (Figure 2.6).[182]

Based on the analysis of covalent and stable adducts formed with nuclear DNA, a similar stereoselective course of bay- and fjord-region activation of B[*a*]P and B[*c*]Ph, respectively, was detected in mammalian cells in culture and in mouse skin.[147,183–188] Exactly these stereoisomeric diol-epoxides, the (+)-*anti*-B[*a*]PDE and the (−)-*anti*-B[*c*]PhDE, have also been identified as

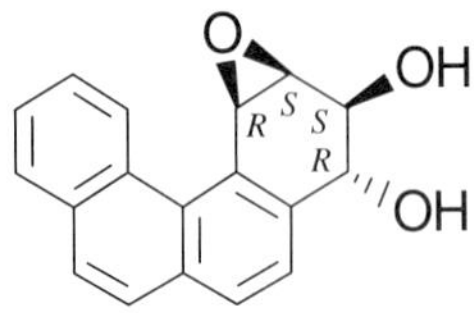

(–)-*anti*-B[*c*]PhDE

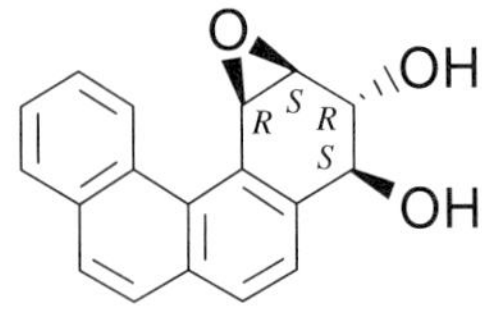

(+)-*syn*-B[*c*]PhDE

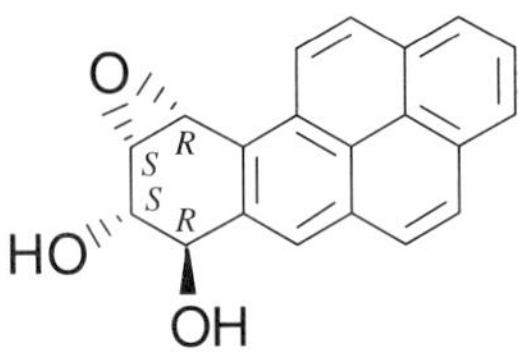

(+)-*anti*-B[*a*]PDE

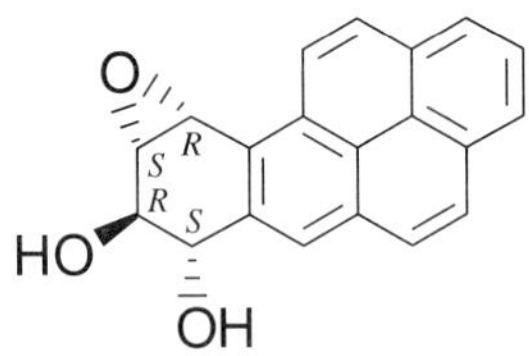

(+)-*syn*-B[*a*]PDE

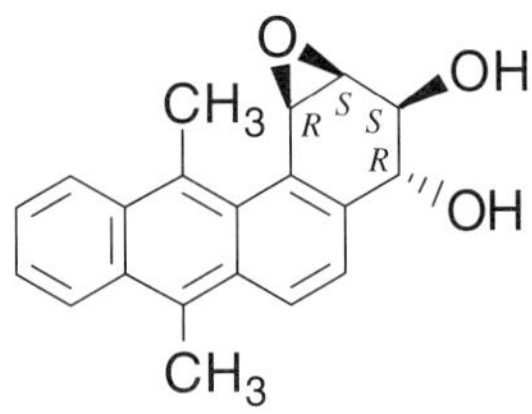

anti-DMBADE*

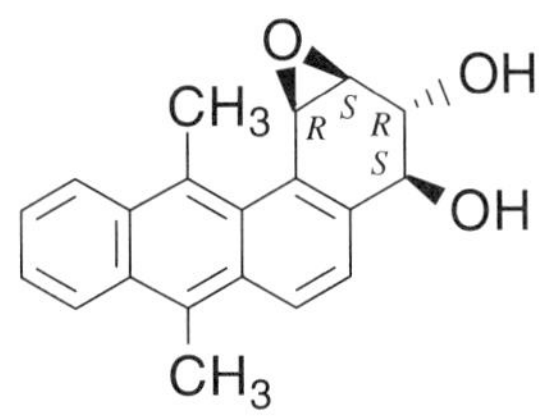

syn-DMBADE*

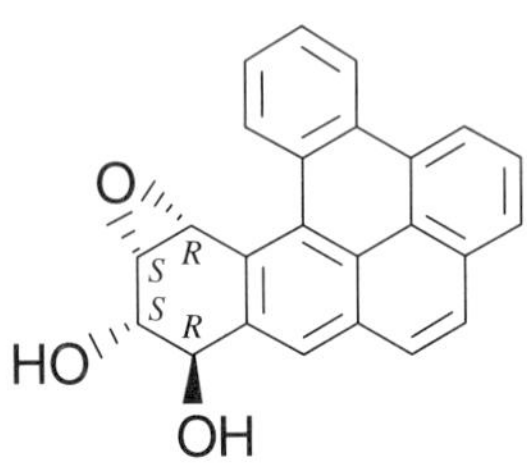

(–)-*anti*-DB[*a,l*]PDE

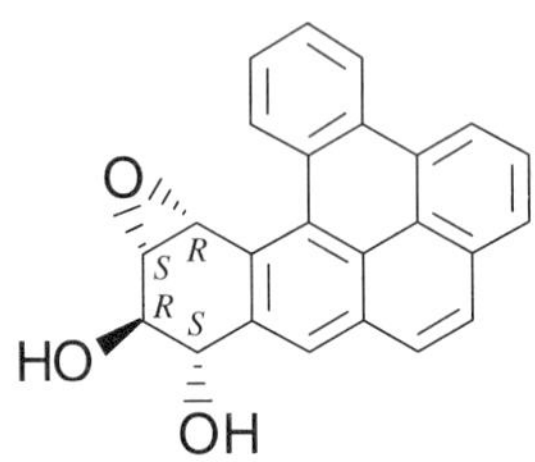

(+)-*syn*-DB[*a,l*]PDE

Figure 2.6: Bay- and fjord-region diol-epoxide diastereomers.

*Due to their instability in solution, the optical rotation of DMBADE diastereomers could not be determined.

the isomers exhibiting the strongest mutagenicity in mammalian cells and the highest carcinogenic potency in tumor mouse models compared to the other three diastereomers.[3,44,175,189–191] The limited overall tumorigenicity observed in case of the parent hydrocarbon B[*c*]Ph in mice[191,192] may therefore be mainly the result of the limited primary metabolism of this compound towards its 3,4-dihydrodiol. Consistently, only small amounts of DNA adducts are formed from B[*c*]Ph in mouse skin, whereas treatment with the *trans*-3,4-dihydrodiol affords a much higher level of DNA damage.[186]

Bay- or fjord-region diol-epoxides of various other carcinogenic PAHs were also identified as the metabolites that mediate cellular binding to DNA and, thus, the genotoxic effects of their parent hydrocarbons. Particularly strong evidence was collected for the sterically hindered bay-region PAHs 5-MeC[193–195] and DMBA,[196–198] as well as for the fjord-region PAH DB[*a,l*]P[199–201] (cf. Figure 2.6). In the case of B[*a*]P and 5-MeC, DNA binding is preferentially mediated through their bay-region *anti*-(*R,S,S,R*) diol-epoxides that bind to the exocyclic amino group of 2′-deoxyguanosine residues (N^2-dG) (Figure 2.7).[183,195,202,203] Only some minor products were found to arise from diastereomeric *syn*-diol-epoxides or from reactions at different positions in nucleobases. In contrast, both the *anti*-(*R,S,S,R*) and the *syn*-(*S,R,S,R*) 3,4-diol-1,2-epoxides are more equally contributing to the DNA binding of DMBA and B[*c*]Ph (Figure 2.6).[185,186,204,205] Whereas DMBA and B[*c*]Ph diol-epoxides bind to both N^6-dA (exocylic amino group of 2′-deoxyadenosine) or N^2-dG in considerable amounts (with N^6-dA prevailing), the hexacyclic DB[*a,l*]P predominantly targets N^6-dA and, to a much lesser extent, N^2-dG residues in DNA as well.[200,201] DB[*a,l*]P is considered the most potent carcinogenic PAH tested to date in mouse skin and rat mammary gland.[27,28,206] The predominant DNA adduct detected in mouse skin after topical administration of DB[*a,l*]P appears to arise from covalent interaction of the (−)-*anti*-11*R*,12*S*-diol-13*S*,14*R*-epoxide [(−)-*anti*-DB[*a,l*]PDE] (Figure 2.6) with N^6-dA.[147,199] Based on the exceptionally high reactivity of the strong carcinogen DMBA at N^6-dA residues in DNA it had been already anticipated in 1983 by Dipple and colleagues that PAH-N^6-dA adducts might have the greater potency for tumor induction as

in those particular cases (e.g., strong carcinogens such as DB[*a,h*]A and 5-MeC with high ionization potentials of >7.35 eV).[208]

It is now known for quite a long time that primary metabolism of PAHs not only results in formation of *trans*-dihydrodiols via arene oxides, but also in generation of high amounts of phenol and quinone derivatives depending on the activating conditions and the metabolism system used.[44,222–224] In the case of B[*a*]P, for example, the 3-, 7- and 9-hydroxy derivatives as well as the *trans*-4,5-, -7,8- and -9,10-dihydrodiols and several quinones (such as 1,6-, 3,6- and 6,12-quinones) may be detectable.[222–227] Originally, metabolic conversion of B[*a*]P catalyzed by CYP-containing enzyme preparations was quantified via the so-called 'AHH' activity (see above), which was based on fluorometric determination of the 3-hydroxy derivative of B[*a*]P, one of its major phenols.[43] Whereas *trans*-dihydrodiols are generated through monooxygenation and subsequent hydrolysis of the intermediate arene oxide in a two-step route as explained above, phenols and quinones may arise from more than only one mechanism (Figure 2.8). In addition to the three aforementioned phenols, 6-hydroxy-B[*a*]P, which itself is unstable, is known to be a major but only transient metabolite that undergoes further autoxidation to yield the three stable polynuclear 1,6-, 3,6- and 6,12-quinones of B[*a*]P (Figure 2.8).[208,228–234] Herein, the unstable 6-phenol most likely is formed through intermediate generation of a radical cation. In contrast, the 3-, 7- and 9-hydroxy derivatives of B[*a*]P were thought to be formed from spontaneous (non-enzymatic) rearrangement ('NIH shift') of the respective arene oxide precursors, i.e., the 2,3-, 7,8- and 9,10-oxides of B[*a*]P (Figure 2.8).[44,49,224,235] Despite early evidence for a 2,3-oxide as an intermediate precursor undergoing non-enzymatic isomerization to produce the 3-phenol,[236] more recent work again points to a one-electron oxidation pathway that may be involved in the generation of the 3-hydroxy (as well as the 1-hydroxy) derivative of B[*a*]P.[237] Intermediate occurrence of radical cations that react with water to form phenols and quinones of PAHs under physiological conditions provides strongest support for the notion that these highly reactive species may also covalently interact with other cellular nucleophiles such as DNA (Figures 2.4 and 2.7).

Cavalieri and co-workers provided evidence that hydrocarbon radical cations mainly bind to N7-, N3- or C8-positions in purine bases of

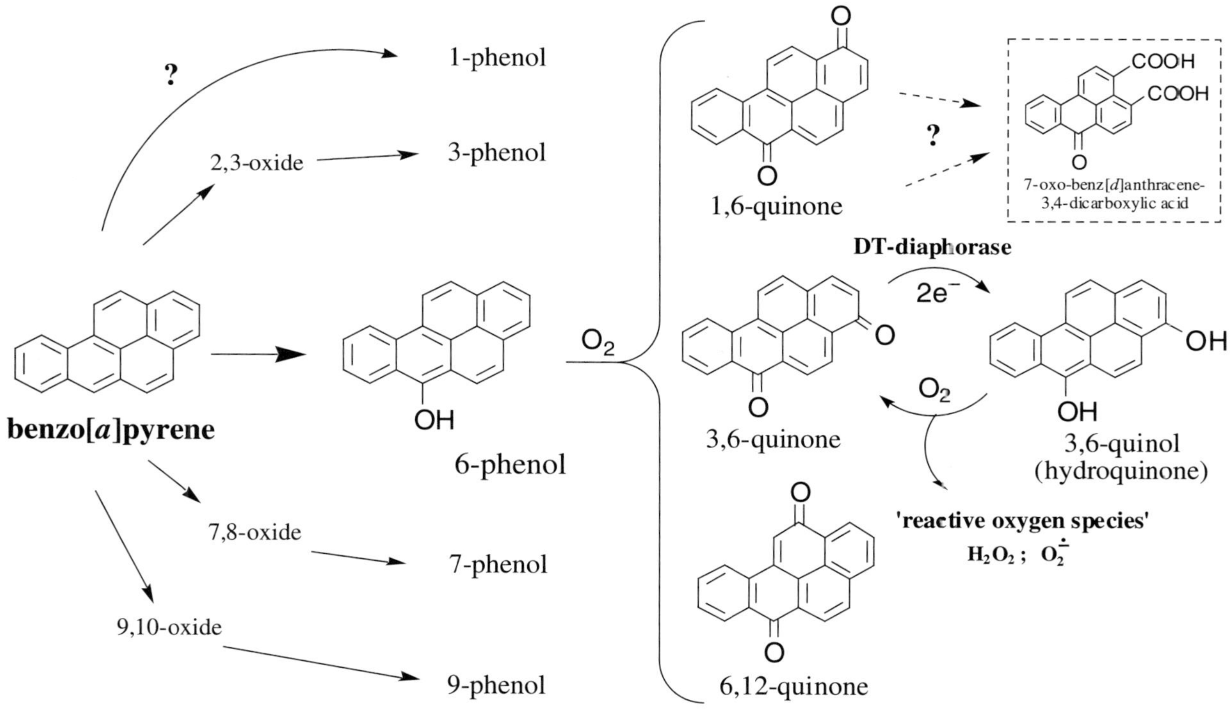

Figure 2.8: Phenols, quinols and polynuclear quinones of benzo[*a*]pyrene.

DNA.[215–218] Due to a labile N-glycosidic bond, the resulting DNA adducts subsequently undergo spontaneous depurination to yield apurinic sites within the sugar-phosphodiester backbone (Figure 2.7).[209] In accordance with their suggestion of a 'central role' of radical cations, the authors reported that the majority of all DNA lesions formed by strongly carcinogenic PAHs such as B[*a*]P, DMBA or DB[*a,l*]P *in vitro* (microsomal or peroxidase-mediated activation) and in mouse skin are unstable and released from the DNA by depurination.[214–219,238–240] Based on their results, it was proposed that apurinic sites generated by depurinating PAH-DNA adducts from radical cations, rather than stable PAH-DNA adducts from diol-epoxides, may be responsible for the induction of mutations in critical target genes leading to cancer initiation.[209,241–243] Depurinating adducts were reported to form at least a 10-fold higher level of apurinic sites compared to the level of these lesions formed in cells by spontaneous depurination (about 2,000 apurinic sites over the course of 4 hours).[241] Therefore, it was hypothesized that apurinic sites generated by depurinating PAH-DNA adducts in excess of the repair capacity of the cell,[241,242] or the repair activity of the cell itself[240,244] may lead to transforming mutations and may be responsible for tumor initiation *in vivo*.

2.3.4 Formation of Quinones

Biotransformation of PAHs may also lead to formation of polynuclear quinones, such as the 1,6-, 3,6- and 6,12-quinones in the case of B[*a*]P (Figure 2.8). These quinones are ubiquitous metabolites of B[*a*]P that possibly arise from autoxidation of phenolic derivatives (i.e., 6-hydroxy-B[*a*]P; see above).[208,228–234] As originally suggested by Ts'o and co-workers in 1975,[229–231,245] autoxidation of phenols may proceed through hydroxy and semiquinone radicals and concomitant formation of 'reactive oxygen species'. As with *ortho*-quinone and catechol derivatives of PAHs (see below), redox-cycling between polynuclear quinones and their corresponding hydroquinones/quinols (e.g., 3,6-quinone and 3,6-dihydroxy-B[*a*]P; Figure 2.8) most likely is a major source of 'reactive oxygen species' such as superoxide anion radicals ($O_2^-\bullet$) and hydrogen peroxide (H_2O_2). Subsequent interaction between $O_2^-\bullet$ and H_2O_2, or decomposition of H_2O_2 in the presence of transition metal ions (e.g., $Fe^{2+/3+}$ or $Cu^{+/2+}$), referred to as 'Fenton reaction',[246–248] may give rise to highly reactive hydroxy radicals (HO •). These species are known to cause oxidative DNA damage that results in the formation of 8-hydroxy-2′-deoxyguanosine (8-OH-dG) or thymine glykol (Figure 2.9). Accordingly, exposure to PAHs *in vitro* and *in vivo* has been found to be accompanied by an increase in the levels of these kind of DNA lesions.[246,249–252]

On the other hand, the 3,6-quinone of B[*a*]P itself was also found to give rise to stable and mutagenic DNA adducts in cells transfected with cytochrome P450 reductase.[253] The presence of this enzyme promotes one-electron reduction of quinones to DNA-damaging semiquinone anion radicals (see below). Coexpression of NQO1 (DT-diaphorase; EC 1.6.99.2), however, significantly reduced the quinone-mediated DNA damage and mutagenicity.[253,254] DT-diaphorase catalyzes the two-electron reduction of quinones to corresponding hydroquinones/quinols (Figure 2.8) which may then be further modified (detoxified) by phase-II enzyme-mediated conjugation reactions. The importance of NQO1 has also been recently demonstrated *in vivo* with mice deficient for the corresponding gene. In particular, female knock-out mice were at significantly higher risk for developing B[*a*]P- as well as DMBA-induced skin tumors.[255,256] However, the mechanistic basis for this observation remains unclear since the 3,6-quinone of B[*a*]P did not produce any tumors in these mice[257] — a finding in line with early reports on the lack of carcinogenic potency in case of this metabolite.[258] On the other hand, polynuclear quinones such as the 1,6- or 3,6-quinones of B[*a*]P may also be degraded to ring-opened products such as 7-oxo-benz[*d*]anthracene-3,4-dicarboxylic acid (Figure 2.8). This compound was most recently discovered in high quantities in the urine of rats upon treatment with B[*a*]P (up to 30% of the total metabolites in urine!).[259] Although its anhydride has been proven to be mutagenic *in vitro*,[260] the enzymes involved in this pathway, its mechanistic basis and its biological importance are presently not understood.

The monooxygenation pathway of metabolic activation of PAHs proceeds through CYP- and mEH-mediated formation of non-K-region *trans*-dihydrodiols as intermediates on the way towards vicinal diol-epoxides (Figures 2.3 and 2.5). In competition with the secondary epoxidation

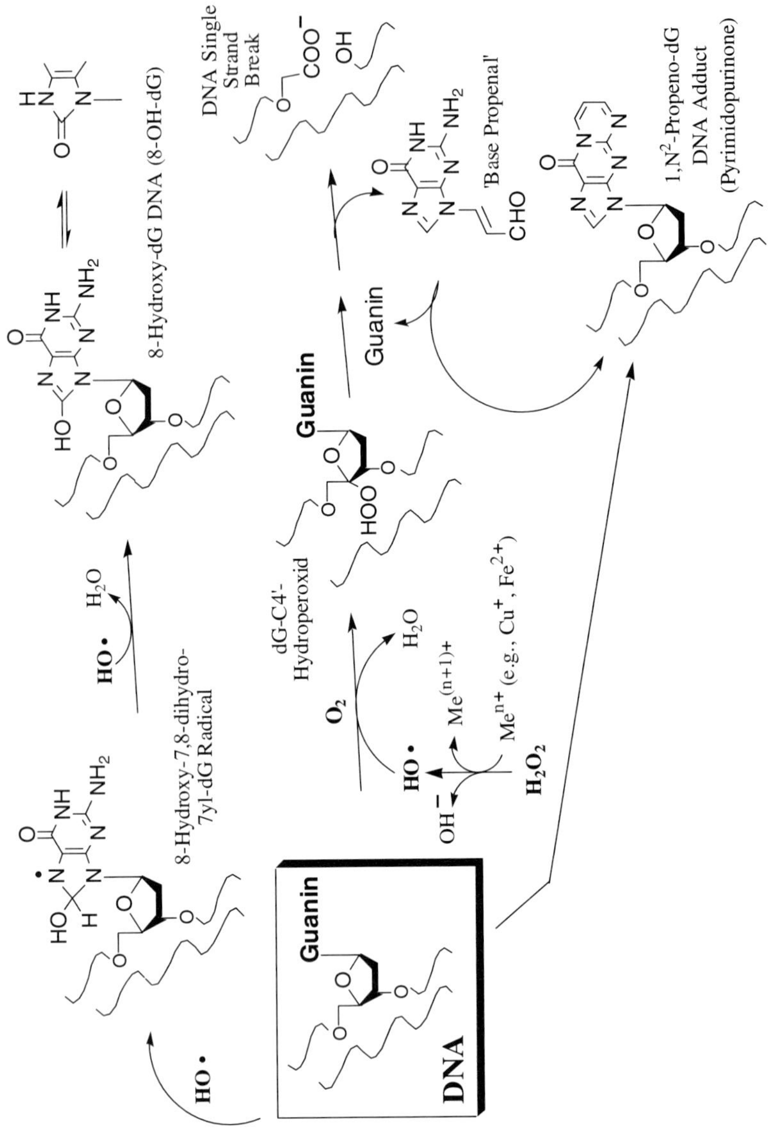

Figure 2.9: DNA damage induced by 'reactive oxygen species'.

reaction adjacent to the diol moiety, *trans*-dihydrodiols may also be enzymatically converted into *ortho*-quinones (Figures 2.2 and 2.10). As published in the mid 1980s, rat liver dihydrodiol dehydrogenase (EC 1.1.1.213) was found to convert non-K-region *trans*-dihydrodiols of PAHs into their corresponding *ortho*-quinones.[261,262] Accordingly, the presence of this

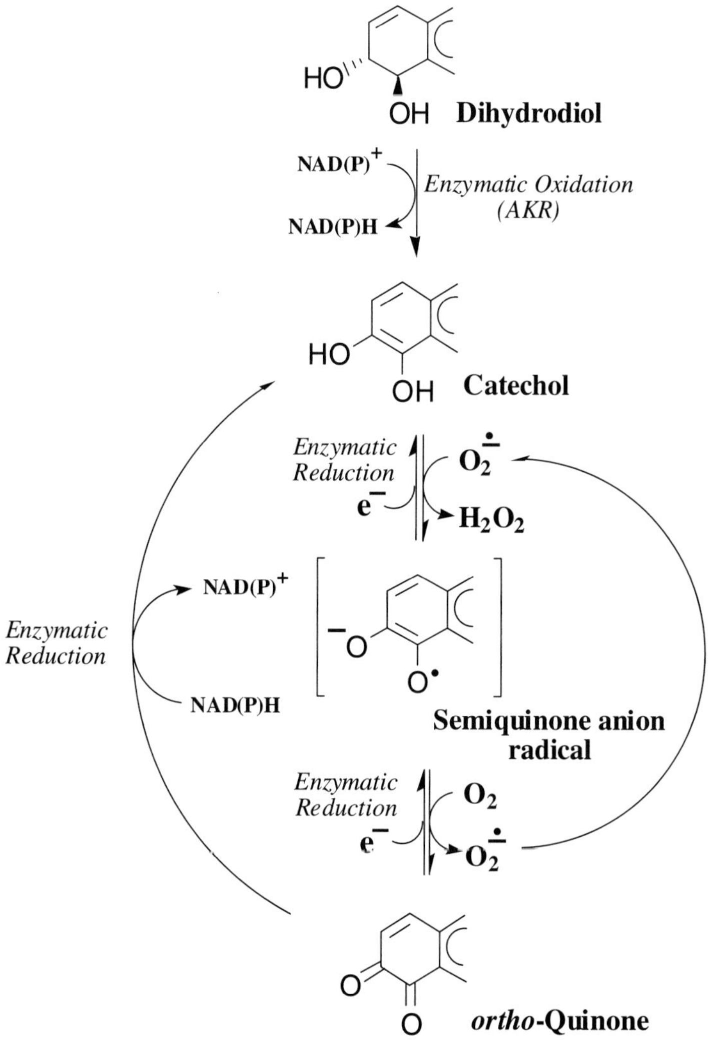

Figure 2.10: PAH catechol-quinone redox-cycling.

enzyme in microsomal incubation mixtures of B[*a*]P suppressed the formation of bay-region diol-epoxides and their inherent mutagenicity.[263] Dihydrodiol dehydrogenase was first purified in 1980 from rat liver using benzene dihydrodiol as the substrate,[264] and was later found to co-purify with 3α-hydroxysteroid dehydrogenase.[265] Today, rat liver dihydrodiol dehydrogenase (termed AKR1C9) is known as a member of the growing superfamily of NAD(P)H-dependent aldo-keto reductases (AKR). AKR are cytosolic and monomeric oxidoreductases that encompass more than 110 proteins distributed in 14 families.[266,267] As with rat liver dihydrodiol dehydrogenase, the human orthologues AKR1C1-1C4 (EC 1.3.1.20; >80% nucleotide identity within the subfamily) were found to be active against a wide range of non-K-region *trans*-dihydrodiols of carcinogenic PAHs.[268–270] From these four isoforms, human AKR1C1 and AKR1C2 displayed the highest catalytic activity towards B[*a*]P-7,8-dihydrodiol. However, only AKR1C1 is both constitutively expressed as well as highly induced in cell lines exposed to the parent hydrocarbon (B[*a*]P) or its 7,8-quinone.[271] More recent work now provides evidence that AKR1A1, the constitutively and widely expressed human aldehyde reductase (EC 1.1.1.2), may also be involved in NAD(P)H-dependent generation of PAH *ortho*-quinones from non-K-region *trans*-dihydrodiol precursors.[272,273] In contrast, none of these AKR isoforms eliciting dihydrodiol dehydrogenase activity was found to accept K-region *trans*-dihydrodiols such as B[*a*]P-4,5-dihydrodiol as convertible substrates.[269,272]

NAD(P)H-dependent enzymatic oxidation of non-K-region *trans*-dihydrodiols of PAHs initially produces ketol derivatives which spontaneously rearrange to form catechols (Figure 2.10). Similar to polynuclear hydroquinones (Figure 2.8), the catechols are unstable and undergo autoxidation in the presence of oxygen.[262,266] Autoxidation of catechols again proceeds through one-electron steps and intermediate formation of PAH semiquinone anion radicals and 'reactive oxygen species' such as the superoxide anion radical and hydrogen peroxide (cf. above; Figure 2.10).[274] The PAH *ortho*-quinones formed via this pathway are not only highly reactive electrophiles ('Michael addition' reaction acceptors) that potentially contribute to DNA damage (see below). Similar to polynuclear quinones, *ortho*-quinone derivatives are redox-active species that may undergo both one- or two-electron reduction steps to re-form the semiquinone anion radical and the catechol (Figure 2.10). Since NAD(P)H-dependent NQO1 (DT-diaphorase) was found not to be active towards *ortho*-quinones of the majority of PAHs (cf. polynuclear quinones above),[266,275] these substrates rather may prefer to enter one-electron redox cycles catalyzed by other NAD(P)H-dependent enzymes such as microsomal cytochrome P450 reductase, microsomal cytochrome-b_5 reductase (EC 1.6.2.2), or mitochondrial ubiquinone oxidoreductase (EC 1.6.5.3).[266,269,274,275] One-electron reduction and autoxidation steps establish futile redox cycles in which generation of 'reactive oxygen species' may be amplified multiple times. As a consequence, the quantities of quinones and catechols actually produced within a cell are only of minor importance with regard to the levels of oxygen species generated under these conditions. Theoretically, only one molecule of catechol could constantly produce these species as long as cellular reducing equivalents are available and not depleted.

Once produced within a cell, 'reactive oxygen species' such as $O_2^{-}\bullet$, H_2O_2, and HO• may cause oxidative DNA damage such as 8-OH-dG (Figure 2.9). Due to base mispairing with dA during DNA replication, these lesions are mutagenic (dG$\rightarrow$dT transversions) *in vitro* and possibly also *in vivo*.[276,277] On the other side, hydroxy radicals may also induce DNA single strand scissions — most likely through *Criegee rearrangement* of the sugar-phosphodiester backbone upon radical attack at the 2′-deoxyribose moiety (Figure 2.9).[266,278,279] However, PAH *ortho*-quinone-induced DNA cleavage in a cell-free system was found to require 'redox-cycling conditions', i.e., an excess of reducing equivalents (NADPH) and unphysiologically high concentrations of transition metal ions (10 μM Cu^{2+}).[279] Even higher metal concentrations (100 μM Cu^{2+}) were necessary to score in a cell-free *p53* mutation assay,[280] though the major bulk of B[*a*]P-7,8-dione-induced (point) mutations seen in this assay were attributed to an intermediate formation of 8-OH-dG lesions. It therefore remains unclear whether Cu^{+}-mediated production of hydroxy radicals or intermediate generation of copper hydroperoxide radicals [Cu(OOH•)] ultimately leads to

DNA oxidation.[280,281] In both studies, however, the causative role of 'reactive oxygen species' was established as radical scavengers (e.g., mannitol, formic acid) or inactivating enzymes (superoxide dismutase, catalase) were able to significantly attenuate DNA strand scission and *p53* mutagenesis. Interestingly, *ortho*-quinone-induced DNA cleavage in rat hepatocytes, or generation of $O_2^{-}\bullet$ and H_2O_2 during oxidation of non-K-region *trans*-dihydrodiols by rat liver dihydrodiol dehydrogenase did not require any fortification with metal ions.[274,282] As a by-product of radical-induced *Criegee rearrangements* within DNA, 'base propenals' potentially could be released at damaged sites (Figure 2.9). These nucleobase derivatives can be considered as structural analogues of malondialdehyde that by their own have the potency to induce stable propeno DNA adducts (e.g., pyrimidopurinone) via oxypropenylation of purine bases (Figure 2.9).[266,279] Pyrimidopurinone (pyrimido[1,2α]purin-10(3H)-one) is the major reaction product between malondialdehyde (the hydrolysis product of base propenals) and 2′-deoxyguanosine. On the other hand, the highly reactive mutagen malondialdehyde may also be released during oxidative degradation of membrane lipids — another consequence of the occurrence of 'reactive oxygen species'.[283–285]

In addition to being redox-active, non-K-region *ortho*-quinones of PAHs are also intrinsically reactive due to their α,β-unsaturated carbonyl function. Since they were found to undergo '1,4-Michael addition' reactions with thiol-functional nucleophiles such as the amino acid cysteine or the tripeptide GSH,[270,286] similar reactions with amino-groups of nucleobases are conceivable (Figures 2.4 and 2.7). Penning and colleagues provided some preliminary evidence that *ortho*-quinones may be able to induce stable as well as depurinating DNA adducts under certain circumstances *in vitro*.[266,287–289] Similar to the situation with vicinal diol-epoxides and radical cations of PAHs, stable or depurinating DNA adducts with *ortho*-quinones would arise from electrophilic attack either at an exocyclic amino-group or at the N7 imino function in purine bases. Both kinds of nucleophiles would attack the β-position of the α,β-unsaturated carbonyl moiety (Figure 2.4), thereby producing an intermediate catechol-purine adduct ('Michael addition') which subsequently undergoes autoxidation to the corresponding *ortho*-quinone-purine adduct (Figure 2.7). In contrast to the predicted DNA reactivity, however, non-K-region *ortho*-quinones of carcinogenic PAHs displayed only little genotoxic effects in a bacterial mutagenicity test system in the absence of 'redox-cycling conditions' (see above).[269,282,290] The mutagenic activity of B[*a*]P-7,8-quinone, for example, was found to be several orders of magnitude lower compared to the bay-region diol-epoxide of B[*a*]P (about 10-5,500-fold, depending on the bacterial tester strain used)[282] (cf. Chapter 8).

2.3.5 'Bioalkylation' and Benzylic Ester Pathway

Methyl substitution of PAHs was found to be frequently accompanied by an increase in carcinogenic activity (cf. above). Early work from Hoffmann and colleagues on methylchrysenes suggested that alkylation of the bay-region and a free *peri*-position (adjacent to the angular benzo ring) would be the structural requirements favoring an increase in biological activity, such as seen in the case of 5-MeC (Figure 2.1).[23,291] On the other hand, the strong carcinogenicity of compounds such as 6-methyl-B[*a*]P and 7-methylbenz[*a*]anthracene (7-methyl-B[*a*]A) demonstrates that substitution in the *peri*-position or in non-bay-region positions does not necessarily reduce biological activity.[24] Methylation of the 7 position in B[*a*]A even creates the most active species among all mono-methylated derivatives of this PAH.

The biological effects of methylation of PAHs might be explained as a result of conformational changes induced in their DNA-binding metabolites (i.e., the bay-region diol-epoxides) or in their DNA adducts.[23,292,293] However, additional metabolic routes are also conceivable, such as an enzymatic attack at the alkyl side chain leading to formation of hydroxymethyl derivatives and, subsequently, to electrophilically reactive benzylic esters ('benzylic ester pathway'; Figures 2.2 and 2.4).[19,294–297] Since these benzylic ester derivatives would react in an S_N1-related mechanism (owing to appropriate 'leaving groups' such as sulfate or acetate), in 1968 Dipple *et al.* attempted to correlate relative stabilities of the putatively formed arylmethyl carbonium ion intermediates with carcinogenic potency of the corresponding parent PAHs.[19] The compilation of some series of methylated B[*a*]A, B[*a*]P, B[*c*]Ph, and DB[*a,h*]A led to the conclusion that *"although some exceptions are found, the overall positive correlation* (...) *is good"*. These early suggestions of an alternative toxification pathway via esterification of hydroxymethyl metabolites of carcinogenic PAHs prompted investigations on the DNA binding activity of 7-methyl-B[*a*]A, DMBA or 6-methyl-B[*a*]P and their hydroxymethyl, bromomethyl and acetoxymethyl derivatives.[10,298–302] All of these compounds were found to react extensively with DNA *in vitro* — either directly (bromomethyl and acetoxymethyl derivatives), or upon activation by microsomal enzymes (hydroxymethyl derivatives; Figure 2.7). In addition, non-enzymatic DNA binding of 7-hydroxymethyl-B[*a*]A or 6-hydroxymethyl-B[*a*]P was detected when incubation mixtures were supplemented with ATP, pointing to the spontaneous formation of phosphate esters as DNA-reactive intermediates.[300,301,303] In the case of 7-methyl-B[*a*]A and DMBA, however, it was demonstrated that DNA binding of 7-bromomethyl and 7-hydroxymethyl-12-methyl derivatives of B[*a*]A did not produce the same DNA adducts as those found after treatment of cells with the parent hydrocarbon, thus ruling out an important role of the benzylic ester pathway in activation of these compounds.[298,304] Despite these findings, subsequent work provided evidence that cytosolic sulfotransferase activity in the liver of rodents may be capable of catalyzing DNA binding of 7-hydroxymethyl- and 7-hydroxymethyl-12-methyl-B[*a*]A, 5-hydroxy-MeC as well as 6-hydroxymethyl-B[*a*]P through formation of sulfate ester intermediates.[296,305–311] Although later work from Surh *et al.* ruled out that the 7-sulfooxymethyl-12-methyl derivative would play an important role in DMBA-induced carcinogenesis,[312] the work related to 6-methyl-B[*a*]P demonstrated that hydroxymethyl sulfate esters of certain PAHs may be indeed biologically active not only *in vitro*, but also *in vivo*.[310,311,313]

6-Methyl-, 6-hydroxymethyl- and 6-sulfooxymethyl-B[*a*]P were found to be strong carcinogens in rodent bioassays.[295,311,314–316] Based on the detection of 6-hydroxymethyl-B[*a*]P as a metabolite of B[*a*]P and 6-methyl-B[*a*]P in some of their early work,[295] Flesher and co-workers proposed that 'bioalkylation' of B[*a*]P in its L-region (carbon atom 6 = *peri*-position; Figure 2.1) would be the first activation step required in carcinogenesis. Subsequent hydroxylation and conjugation, mediated by CYP enzymes and sulfo- or acetyltransferases, respectively, would then produce electrophilic species (sulfate or acetoxy esters) responsible for the genotoxic activity of the parent PAH.[317] Later work suggested S-adenosyl-L-methionine as being the carbon donor in the first step of this pathway, that is, the methyltransferase-dependent conversion of B[*a*]P to its 6-methyl derivative[318,319] (Figure 2.2). Studying various other carcinogenic and noncarcinogenic PAHs, Flesher and colleagues proposed the general concept that a necessary first step in carcinogenesis by most methyl-substituted PAHs is the conversion to a hydroxymethyl metabolite, whereas most hydrocarbons lacking methyl substituents undergo initial bioalkylation reaction in the *meso*-anthracenic or L-region position.[320–324] According to this unified view, all carcinogenic PAHs must either, *a priori*, bear a *meso*-anthracenic methyl substituent, or, if not, undergo a bioalkylation substitution reaction to fulfill this requirement. To support this concept, some preliminary evidence for bioalkylation reactions *in vivo* had been provided based on thin-layer co-chromatography experiments of synthetic standards with DNA adducts isolated from subcutaneous tissue of rats upon injection of the compound under consideration.[321–323,325] The small contribution of metabolically formed 6-methyl-B[*a*]P to the overall DNA binding of the parent hydrocarbon[325] and the limited carcinogenic potency of PAHs that contain suitable (*meso*-anthracenic or L-region) bioalkylation sites (e.g., anthracene, B[*a*]A), however, rather suggest that genuine biomethylation of PAHs may be only of minor importance compared with other metabolic routes (Figure 2.2).

DNA binding, mutagenicity and carcinogenicity studies rather supported the conclusion that carcinogenic methylated PAHs such as 7-methyl-B[*a*]A, DMBA and 5-MeC are also mainly activated through formation of diol-epoxides at the terminal benzo ring[5,9] (Figure 2.6). In contrast, the benzylic ester pathway most likely only contributes little to the overall genotoxicity of compounds that possess a sterically hindered bay-region. In the case of carcinogenic PAHs that lack a terminal benzo ring and a bay-region however, hydroxylation of alkyl side chains and subsequent enzymatic generation of benzylic sulfate esters might play a more crucial role in bioactivation. Accordingly, 1-methylpyrene, for example, a known environmental PAH possessing carcinogenic activity in newborn mice,[326] may readily undergo enzymatic conversion into its 1-sulfooxymethyl derivative[327] (cf. Chapter 8). Initial benzylic hydroxylation of this two-step activation pathway was found to be efficiently catalyzed *in vitro* by the same CYP enzymes that are involved in generation of bay-region diol-epoxides; that is CYP1A1, 1B1, and 3A4 (see above).[328] However, heterologously expressed CYP enzymes showed remarkable species differences in their regioselectivity. In the case of human origin, all three enzymes produced high quantities of the benzylic alcohol. CYP1A1 from rats, however, preferentially promoted ring oxidation rather than benzylic hydroxylation.[328] Unfortunately, CYP1B1 and 3A4 from rodent origin were not tested in this assay. Subsequent sulfation (*O*-sulfonation) is then carried out by enzymatic transfer of the terminal sulfo moiety from 5′-phosphoadenosine-3′-phosphosulfate (PAPS) onto the benzylic hydroxy group of the hydroxymethyl derivative.[329] PAPS serves as the co-substrate of cytosolic sulfotransferases that belong to the superfamily termed SULT (EC 2.8.2.1). In this superfamily, 11 human enzymes are known.[330] From all human sulfotransferases tested *in vitro* with respect to activation of 1-hydroxymethylpyrene, SULT1A1, 1E1 and 2A1 were the most active forms.[330] In addition, sulfotransferases from rat liver cytosol readily catalyze sulfation of 1-hydroxymethylpyrene.[331–333] More recent work with genetically constructed cell lines confirmed the capability of various rat isoforms in this regard.[329,334]

Further evidence for a role of benzylic sulfation in activation of 1-hydroxymethylpyrene comes from DNA adduct analyses *in vitro* and *in vivo*. The DNA adduct profile induced by 1-hydroxymethylpyrene in rat liver was found to be similar to that obtained with 1-sulfooxymethylpyrene in the same organ or with isolated DNA in solution.[331,335] However, administration of the sulfate ester resulted in a considerably higher DNA damage (about 15 times higher as with the precursor), which rather points to a minor role of benzylic activation *in vivo*. In fact, only about 0.05% of the total dose of radioactive 1-hydroxymethylpyrene, administered to rats by intraperitoneal injection, could be assigned to methylpyrenyl mercapturic acid.[335] This cysteine derivative is a known detoxification product of the sulfate ester and, thus, a follow-up product of the benzylic ester activation pathway. In contrast, 80% of the substrate was further oxidized at its side chain to yield 1-pyrenyl carboxylic acid and its conjugates. Despite its genotoxic effects *in vitro* and in rat liver, 1-hydroxymethylpyrene was not capable of inducing sarcomas in subcutaneous tissue of rats[336] and was also virtually inactive in mouse skin.[331] The fact that the same dose of 1-sulfooxymethylpyrene was highly sarcomagenic in rats and quite active in mouse skin again points to a quantitatively limited *O*-sulfonation reaction and underscores the need for further studies on the contribution of the 'benzylic ester pathway' in PAH-induced carcinogenesis *in vivo*.

2.4 Detoxification

Biotransformation of lipophilic xenobiotics is aimed to result in detoxification and elimination of water-soluble derivatives via excretion pathways (urine, bile → feces). Based on their chemical reactivity, PAH metabolites generated during enzyme-catalyzed metabolism may be characterized as nucleophilic (such as *trans*-dihydrodiols, phenols and quinols) or electrophilic (such as arene oxides, diol-epoxides and quinones).

In principle, nucleophilic metabolites of PAHs can undergo enzymatic detoxification by conjugation to glucuronic acid or sulfate[337–339] (Figure 2.11). Originally observed and described as early as 1903 in the case of phenanthrene and 1935 in the case of anthracene,[114] glucuronidation of PAH metabolites is catalyzed by UDP-glucuronosyltransferases (UGT; EC 2.4.1.17). UGT enzymes belong to a superfamily of microsomal proteins highly expressed in liver, kidney and intestine[340,341] that utilize uridine-5′-diphospho(UDP)-α-D-glucuronic acid as the co-substrate. Based on their amino acid sequence, currently two large families (UGT1 and UGT2) are distinguished.[342] They either contain nine and six (human), or each seven (rat) enzyme members. Using the 'Ames assay' as an indicator system, *in vitro* studies provided evidence for a protective role of UGT enzymes against B[*a*]P-mediated genotoxicity.[343,344] In UGT1-deficient cell lines or animals, more recent work was able to establish a correlation between UGT deficiency, attenuated glucuronidation of B[*a*]P metabolites, and increasing levels of B[*a*]P-induced DNA or chromosomal damage *in vitro* and *in vivo*.[345–347] Phenols and quinols (hydroquinones) of B[*a*]P were found to be excellent substrates for the 3-MC-inducible rat or human liver UGT known as 'phenol UGT' (and now named as

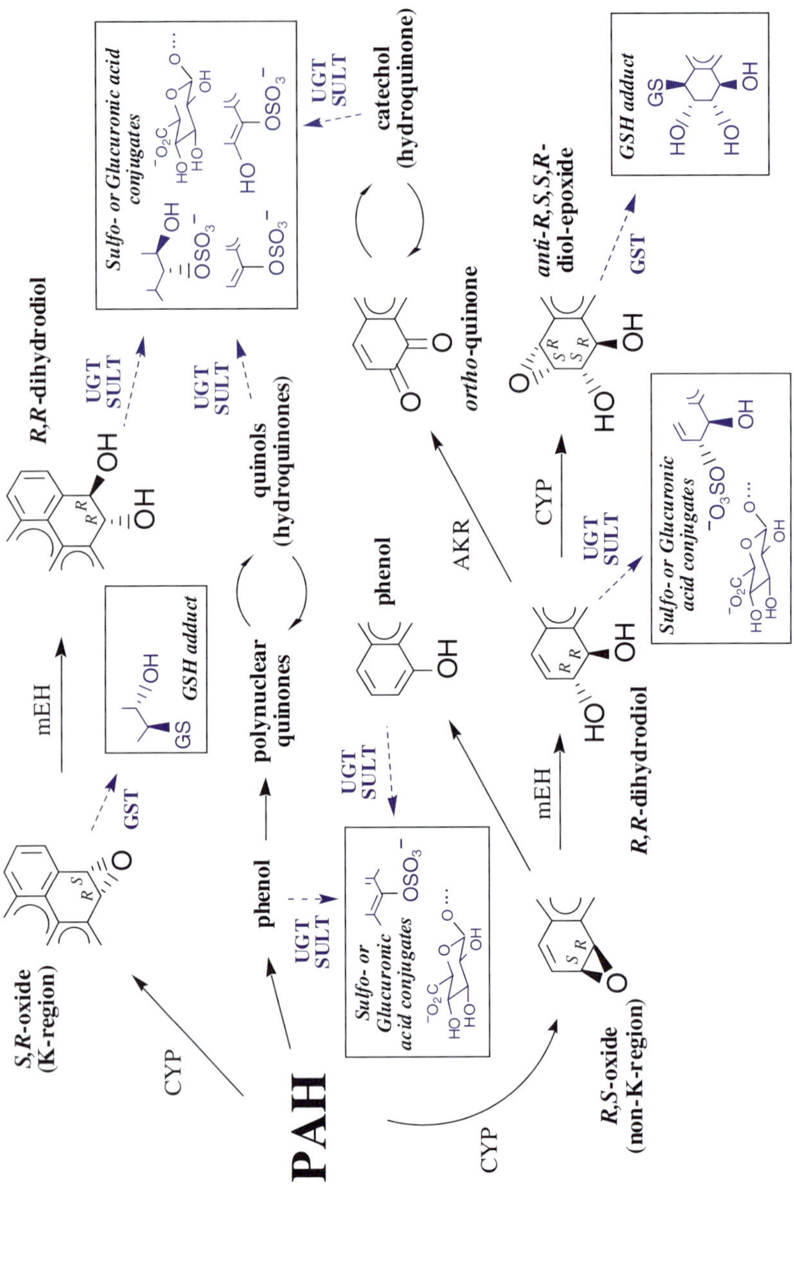

Figure 2.11: Detoxification of nucleophilic and electrophilic metabolites of PAHs.

'UGT1A1').[348–350] For example, the mutagenicity of B[*a*]P-3,6-dione can be reduced by glucuronidation of its redox-cycling partner, the 3,6-quinol (3,6-diphenol; Figure 2.8).[351,352] Conjugation of the quinol then prevents further redox-cycling with concomitant generation of semiquinone anion radicals and 'reactive oxygen species'. Today, it is well established that several different rat and human UGTs exert activity towards phenolic derivatives of B[*a*]P.[350,352,353] In addition, the bay-region diol-epoxide precursors, i.e., both enantiomeric *trans*-7,8-dihydrodiols, as well as other *trans*-dihydrodiols are also targeted by rat and human UGT enzymes (Figure 2.11).[354–356]

As an alternative detoxification reaction, arylsulfate ester formation of hydroxy derivatives of PAHs (i.e., phenols, dihydrodiols) was first described in the 1950s for naphthalene,[357] and later with phenanthrene.[358] Since SULT-catalyzed conjugation reactions usually occur with limited turnover rates, glucuronidation of nucleophilic metabolites prevails at high substrate concentrations.[115,337,359] On the other hand, sulfate ester formation dominates at lower concentrations, owing to the high substrate affinity of SULT enzymes. In the case of certain PAH metabolites such as hydroxymethyl derivatives (e.g., 6-hydroxymethyl-B[*a*]P), however, SULT-catalyzed conjugation reactions may result in activation rather than detoxification (see above). Activating sulfate ester conjugation reactions were also observed with PAHs containing benzylic secondary alcoholic functional groups such as 1-(α-hydroxyethyl)pyrene,[330] or with metabolites from nonalternant PAHs such as 4-hydroxy-3,4-dihydrocyclopenta[*c,d*]pyrene[360] or 4*H*-cyclopenta[*def*]chrysene-4-ol[330] (cf. Figure 9.1 in Chapter 9). Despite theoretical considerations and in contrast to the aforementioned compounds, no sulfate ester-mediated activation occurs at benzylic secondary alcoholic groups present in B[*a*]P *trans*-4,5- and 7,8-dihydrodiols, or in 7,8,9,10-tetrahydrotetraol derivatives (i.e., the hydrolysis products of bay-region B[*a*]P diol-epoxides).[334,361] Interesting mechanistic insights came from studies on sulfate esterification and activation of the synthetic derivatives 7- or 10-hydroxy-7,8,9,10-tetrahydro-B[*a*]P.[334,361,362] Whereas the 7-hydroxy derivative was activated by SULT enzymes to mutagenic intermediates in a concentration-dependent manner, the isomer containing the hydroxy group in the bay-region (10-position) failed to elicit genotoxicity upon SULT-dependent activation. So, obviously SULT enzymes do not conjugate B[*a*]P *trans*-7,8-dihydrodiol at its benzylic hydroxy group, but they do conjugate the 7-hydroxy group in the ring-saturated tetrahydro derivative. Although early studies demonstrated a SULT-mediated conjugation of B[*a*]P *trans*-dihydrodiols to water-soluble products,[363–365] no structural analysis of the conjugates formed was presented at this time. In addition, phenols of B[*a*]P were found to be much better substrates for SULT enzymes compared to *trans*-dihydrodiols. Accordingly, *trans*-dihydrodiol metabolites of B[*a*]P generated *in vivo* are preferentially conjugated to glucuronic acid unless they are sequestered as GSH adducts at the arene oxide precursor level (see below; Figure 2.11).

Spontaneous rearrangement and efficient enzymatic (mEH-mediated) hydrolysis of PAH arene oxides give rise to phenolic products and *trans*-dihydrodiols, respectively. Both kinds of nucleophilic follow-up products may then be efficiently terminated in their toxicity by glucuronic acid conjugation and/or sulfation that facilitates subsequent excretion via urine and feces.[115,337,366,367] Conversely, detoxification of electrophilic metabolites of PAHs such as arene oxides and diol-epoxides may be accomplished through enzymatic conjugation with GSH, a tripeptide (γ-L-Glu-L-Cys-Gly) that contains a reactive sulfhydryl group[115,337,368,369] (Figure 2.11). The resulting GSH adducts are either excreted immediately via bile → feces, or transported to the kidney where enzymatic degradation and acetylation ultimately leads to mercapturic acids (*N*-acetyl cysteinyl derivatives; cf. Chapter 3). Mercapturic acid derivatives of naphthalene or anthracene were the first detected in the urine of animals in the 1930s.[370,371] It was not before 1961 that GSH conjugation was discovered to be catalyzed by an enzyme present in rat liver preparations originally called glutathione kinase.[372] The glutathione-transfering enzymes (i.e., glutathione *S*-transferases, GST; EC 2.5.1.18) have since been found to consist of a superfamily of mostly cytosolic homo- or heterodimeric proteins that are highly expressed in liver, but also at detectable levels in most other mammalian tissues. Human GST enzymes, for example, are subdivided into the eight cytosolic families ('classes') designated α (GST-A), κ (GST-K), μ (GST-M), π (GST-P), σ (GST-S), θ (GST-T), ζ (GST-Z), and ω (GST-O).[373] In addition, a separate class of membrane-bound GST enzymes has been established.

GST enzymes usually display broad and overlapping substrate specifies. Various forms from different classes may be competent in catalyzing the GSH transfer onto arene oxides[374–377] or vicinal diol-epoxides[378–380] of PAHs. However, certain arene oxides or diol-epoxides may differ appreciably in their ability to serve as substrates. For instance, the K-region arene oxide of B[*a*]P (4,5-oxide) was found to be a good substrate for GST from rat and human liver,[376,381,382] but the non-K-region arene oxides (7,8- and 9,10-oxide) were not.[337,383] On the other hand, human GSTA1-1, GSTM1-1 and GSTP1-1 are highly active towards bay- and fjord-region diol-epoxides of various carcinogenic PAHs.[384,385] Cell culture studies provided evidence that enzyme-mediated GSH conjugation of diol-epoxides terminates their inherent toxicity, thereby protecting the genome from possible DNA damage and its detrimental consequences.[386–388] As seen from increased susceptibility to DMBA-induced skin tumors in mice deficient for expression of GSTP1-1 and GSTP1-2, the protective role of GST enzymes against PAH-mediated genotoxicity can also be verified in animals *in vivo*.[389]

2.5 Summary and Perspectives

B[*a*]P, the most extensively studied carcinogenic PAH to date, is converted by phase-I enzymes (CYP, mEH, AKR, peroxidases) to a large number of different metabolites including arene oxides, phenols, *trans*-dihydrodiols, quinones and diol-epoxides. All of these metabolites potentially can be further modified by phase-II enzyme-mediated conjugation reactions leading to water-soluble and, thus, excretable derivatives.[227,337] Due to their planarity and molecular dimension, B[*a*]P and other carcinogenic PAHs are able to bind to the AhR protein and initiate their own biotransformation through induction of both 'activating' enzymes (i.e., CYP1A1 and 1B1) as well as 'conjugating/deactivating' enzymes (i.e., various isoforms of UGT and GST). It might therefore be anticipated that the complex enzymic cellular defense system consisting of UGT, SULT and GST enzymes, along with mEH, is well suited to properly detoxify all different kinds of potentially harmful derivatives of carcinogenic PAHs that may be generated during initial phase–I metabolism (Figure 2.11). So why are these compounds tumorigenic in animals and — most likely — also in man?

PAHs are tumorigenic because they induce DNA damage and mutations in growth-controlling genes such as tumor suppressor or oncogenes (see Chapters 5–7). The principle DNA-damaging metabolites that occur during enzymatic metabolism of PAHs are summarized in Figure 2.4. The DNA damage they potentially induce consists mainly of DNA adducts at exocyclic amino groups of purine bases (Figure 2.7). In addition, depurinating adducts can be generated through interaction of metabolites (e.g., radical cations, benzylic sulfate esters and *ortho*-quinones) with N7 positions in purine bases leading to the formation of apurinic sites within the sugar-phosphodiester DNA backbone. Further, but more indirectly, redox-active metabolites may cause the generation of 'reactive oxygen species' that also damage DNA by inducing lesions such as 8-OH-dG, propeno adducts or DNA single strand breaks (Figure 2.9). Altogether, a great variety of different DNA lesions results from activating metabolism of carcinogenic PAHs. This knowledge was obtained mostly by *in vitro* studies with cells in culture, or by incubation studies with subcellular and purified enzyme preparations in the presence of DNA.

Extensive experimental evidence points to bay- or fjord-region diol-epoxides of carcinogenic PAHs with *R,S,S,R* configuration [e.g., (+)-*anti*-B[*a*]PDE and (−)-*anti*-B[*c*]PhDE] as the principal DNA-binding metabolites produced through CYP-mediated metabolism both in cells in culture and *in vivo* (Figures 2.3, 2.5 and 2.6). These metabolites mainly produce stable DNA adducts at exocyclic amino groups of purine bases that give rise to point mutations at dG and/or dA during error-prone DNA replication (cf. Chapter 5). Nucleotide transversions such as dG → dT and dA → dT detected in oncogenes (e.g., H-*ras*[390,391]) or in tumor suppressor genes (e.g., *p53*[392,393]) have been etiologically linked to PAH-induced carcinogenesis at target organ sites. For example, nucleotide transversions within codons 12 (dG → dT) or 61 (dA → dT) of cellular H-*ras* were found in mouse skin upon exposure to carcinogenic PAHs such as B[*a*]P,[394] DMBA[395] or DB[*a,l*]P.[240] Treatment with B[*a*]P predominantly resulted in the formation of dG → dT transversions, whereas DMBA and DB[*a,l*]P mainly caused dA → dT mutations. Thus, these mutational events correlate nicely with the preferential binding of metabolically formed diol-epoxides of B[*a*]P and DMBA or DB[*a,l*]P to dG [(+)-*anti*-B[*a*]PDE-dG adduct] and dA [*anti*-DMBADE-dA or (−)-*anti*-DB[*a,l*]PDE-dA adducts], respectively.[147] On the other hand, apurinic sites originating from depurinating PAH-dG or -dA adducts (Figure 2.7) are also known to give rise to dG → dT and dA → dT transversions — due to the preferential incorporation of dA opposite to apurinic sites ('A-rule'[241,396]). The kind of mutation formed is therefore not qualified to distinguish the nature of PAH-induced DNA damage responsible for its induction.

At present, the biological consequences of the formation of depurinating PAH-DNA adducts are unknown. Detection of small amounts of PAH-N7-purine adducts in the urine of animals[397,398] or humans[399,400] provided evidence for PAH-mediated generation of apurinic sites *in vivo*. However, apurinic sites are also spontaneously formed in high numbers (about 10^4–10^6 sites/cell/day)[401,402] and are rapidly repaired in mammalian cells due to a fast and efficient base excision repair pathway.[403,404] This repair pathway has been proven to be capable of removing apurinic sites induced by PAHs.[403,405] Although these DNA lesions may therefore be strongly mutagenic in the absence of an efficient repair system, and stable PAH-DNA adducts tend to be mutagenic only in a particular percentage depending upon the specific adduct structure and its sequence context (see Chapters 5 and 6), the absence of an increase in apurinic sites beyond the background level in DNA of skin tissue treated with cancer-initiating doses of B[*a*]P, DMBA and DB[*a,l*]P[147] suggests that the biological consequences of depurinating adducts are unlikely to be a main factor in tumor initiation by carcinogenic PAHs. In contrast, the levels of stable PAH diol-epoxide-DNA adducts usually increase with increasing carcinogenic potency (e.g., B[*a*]P < DMBA < DB[*a,l*]P in mouse skin[147]). Inhibition of bay-region metabolism by methylation of appropriate molecule positions was an early chemical approach to demonstrate the importance of the bay-region activation pathway in PAH tumorigenesis (e.g., any methylation of B[*a*]P at carbon positions 7–10 abolished activity[24]). The failure to properly induce tumors in mice lacking diol-epoxide generating enzymes (CYP1B1, mEH),[166–169] or being deficient in the induction of these enzymes (AhR-null mice),[161] is a more recent and elegant genetic approach that further supported the central role of bay- and fjord-region diol-epoxides in PAH-mediated carcinogenesis.

Acknowledgment

The work of the authors was supported by the German Research Foundation (DFG: LU 841/2–1), and by the National Cancer Institute (NCI) grant CA28825.

References

1. Miller EC (1978) Some current perspectives on chemical carcinogenesis in humans and experimental animals: Presidental address. *Cancer Res.* 38: 1479–1496.
2. Pelkonen O and Nebert DW (1982) Metabolism of polycyclic aromatic hydrocarbons: etiologic role in carcinogenesis. *Pharmacol. Rev.* 34: 189–222.
3. Thakker DR, Yagi H, Levin W, Wood AW, Conney AH and Jerina DM (1985) Polycyclic aromatic hydrocarbons: metabolic activation to ultimate carcinogens. In: *Bioactivation of Foreign Compounds* [Anders MW (ed.)] pp 177–242, Academic Press, London.
4. Guengerich FP (1988) Roles of cytochrome P-450 enzymes in chemical carcinogenesis and cancer chemotherapy. *Cancer Res.* 48: 2946–2954.
5. Harvey RG (1991) Metabolic activation, DNA binding, and mechanisms of carcinogenesis. In: *Polycyclic Aromatic Hydrocarbons: Chemistry and Carcinogenicity* [Harvey RG (ed.)] pp 50–95, Cambridge University Press, Cambridge, UK.
6. Harvey RG and Geacintov NE (1988) Intercalation and binding of carcinogenic hydrocarbon metabolites to nucleic acids. *Acc. Chem. Res.* 21: 66–73.
7. Skipper PL, Peng X, Soohoo CK and Tannenbaum SR (1994) Protein adducts as biomarkers of human carcinogen exposure. *Drug Metab. Rev.* 26: 111–124.
8. Garner CR (1998) The role of DNA adducts in chemical carcinogenesis. *Mutat. Res.* 402: 67–75.
9. Dipple A (1994) Reactions of polycyclic aromatic hydrocarbons with DNA. In: *DNA Adducts: Identification and Biological Significance* [Hemminki K, Dipple A, Shuker DEG, Kadlubar FF, Segerbäck D and Bartsch H (eds.)] IARC Scientific Publication No. 125, pp 107–129, International Agency for Research on Cancer, Lyon, France.

10. Dipple A, Moschel RC and Bigger CAH (1984) Polynuclear aromatic carcinogens. In: *Chemical Carcinogens* [Searle CE (ed.)] ACS Monograph 182, pp 41–163, American Chemical Society, Washington, D.C.
11. Dipple A (1985) Polycyclic aromatic hydrocarbon carcinogenesis. In: *Polycyclic Hydrocarbons and Carcinogenesis* [Harvey RG (ed.)] ACS Symposium Series 283, pp 1–17, American Chemical Society, Washington, D.C.
12. Chang RL, Levin W, Wood AW, Lehr RE, Kumar S, Yagi H, Jerina DM and Conney AH (1981) Tumorigenicity of the diastereomeric bay-region benzo[*e*]pyrene 9,10-diol-11,12-epoxides in newborn mice. *Cancer Res.* 41: 915–918.
13. Deutsch-Wenzel RP, Brune H, Grimmer G, Dettbarn G and Misfeld J (1983) Experimental studies in rat lungs of the carcinogenicity and dose-response relationship of eight frequently occurring PAHs. *J. Natl. Cancer Inst.* 71: 539–544.
14. Osborne MR and Crosby NT (1987) Benzo[*e*]pyrene. In: *Benzopyrenes* [Osborne MR and Crosby NT (eds.)] Cambridge Monographs on Cancer Research, pp 229–250, Cambridge University Press, Cambridge, UK.
15. Cavalieri E, Mailander P and Pelfrene A (1977) Carcinogenic activity of anthanthrene on mouse skin. *Z. Krebsforsch.* 89: 113–118.
16. Scribner JD (1973) Tumor initiation by apparently noncarcinogenic polycyclic aromatic hydrocarbons. *J. Natl. Cancer Inst.* 50: 1717–1719.
17. Badger GM (1948) The carcinogenic hydrocarbons: chemical constitution and carcinogenic activity. *Br. J. Cancer* 2: 309–350.
18. Boyland E and Sims P (1967) The carcinogenic activity in mice of compounds related to benz[*a*]anthracene. *Int. J. Cancer* 2: 500–504.
19. Dipple A, Lawley PD and Brookes P (1968) Theory of tumour initiation by chemical carcinogens: dependence of activity on structure of ultimate carcinogen. *Eur. J. Cancer* 4: 493–506.
20. Hoffmann D, Bondinell WE and Wynder EL (1974) Carcinogenicity of methylchrysenes. *Science* 183: 215–217.
21. Hecht SS, Loy M, Mazzarese R and Hoffmann D (1978) On the carcinogenicity of 5-methylchrysene: structure-activity studies and metabolism. In: *Polycyclic Hydrocarbons and Cancer* [Gelboin HV and Ts'o POP (eds.)] Volume I: Environment, Chemistry, and Metabolism, pp 119–130, Academic Press, New York.
22. Levin W, Wood AW, Chang RL, Newman MS, Thakker DR, Conney AH and Jerina DM (1983) The effect of steric strain in the bay-region of polycyclic aromatic hydrocarbons: tumorigenicity of alkyl-substituted benz[*a*]anthracenes. *Cancer Lett.* 20: 139–146.

23. Hecht SS, Amin S, Melikian AA, LaVoie EJ and Hoffmann D (1985) Effects of methyl and fluorine substitution on the metabolic activation and tumorigenicity of polycyclic aromatic hydrocarbons. In: *Polycyclic Hydrocarbons and Carcinogenesis* [Harvey RG (ed.)] ACS Symposium Series 283, pp 85–105, American Chemical Society, Washington, D.C.
24. Harvey RG (1991) Structure-activity relationships. In: *Polycyclic Aromatic Hydrocarbons: Chemistry and Carcinogenicity* [Harvey RG (ed.)] pp 26–49, Cambridge University Press, Cambridge, UK.
25. Wislocki PG, Juliana MM, MacDonald JS, Chou MW, Yang SK and Lu AY (1981) Tumorigenicity of 7,12-dimethylbenz[*a*]anthrancene, its hydroxymethylated derivatives and selected dihydrodiols in the newborn mouse. *Carcinogenesis* 2: 511–514.
26. Harvey RG, Cortez T, Sugiyama T, Ito Y, Sawyer TW and DiGiovanni J (1988) Biologically active dihydrodiol metabolites of polycyclic aromatic hydrocarbons structurally related to the potent carcinogenic hydrocarbon 7,12-dimethylbenz[*a*]anthracene. *J. Med. Chem.* 31: 154–159.
27. Cavalieri EL, Higginbotham S, RamaKrishna NVS, Devanesan PD, Todorovic R, Rogan EG and Salmasi S (1991) Comparative dose-response tumorigenicity studies of dibenzo[*a*,*l*]pyrene versus 7,12-dimethylbenz[*a*]-anthracene, benzo[*a*]pyrene and two dibenzo[*a*,*l*]pyrene dihydrodiols in mouse skin and rat mammary gland. *Carcinogenesis* 12: 1939–1944.
28. Higginbotham S, RamaKrishna NVS, Johansson SL, Rogan EG and Cavalieri EL (1993) Tumor-initiating activity and carcinogenicity of dibenzo-[*a*,*l*]pyrene versus 7,12-dimethylbenz[*a*]anthrancene and benzo[*a*]pyrene at low doses in mouse skin. *Carcinogenesis* 14: 875–878.
29. Boyland E and Weigert F (1947) Metabolism of carcinogenic compounds. *Br. Med. Bull.* 4: 354–359.
30. Boyland E (1950) The biological significance of metabolism of polycyclic compounds. *Biochem. Soc. Symp.* 5: 40–54.
31. Coulson CA (1953) Electronic configuration and carcinogenesis. *Adv. Cancer Res.* 1: 1–56.
32. Pullman A and Pullman B (1955) Electronic structure and carcinogenic activity of aromatic molecules. New developments. *Adv. Cancer Res.* 3: 117–169.
33. Newman MS and Blum S (1964) A new cyclization reaction leading to epoxides of aromatic hydrocarbons. *J. Am. Chem. Soc.* 86: 5598–5600.
34. Miller EC and Miller JA (1967) Low carcinogenicity of the K-region epoxides of 7-methylbenz[*a*]anthracene and benz[*a*]anthracene in the mouse and rat. *Proc. Soc. Exp. Biol. Med.* 124: 915–919.

35. Sims P (1967) The carcinogenic activities in mice of compounds related to 3-methylcholanthrene. *Int. J. Cancer* 2: 505–508.
36. Van Duuren BL, Langseth L, Goldschmidt BM and Orris L (1967) Carcinogenicity of epoxides, lactones, and peroxy compounds. VI. Structure and carcinogenic activity. *J. Natl. Cancer Inst.* 39: 1217–1228.
37. Sims P and Grover PL (1974) Epoxides in polycyclic aromatic hydrocarbon metabolism and carcinogenesis. *Adv. Cancer Res.* 20: 165–174.
38. Baird WM, Harvey RG and Brookes P (1975) Comparison of the cellular DNA-bound products of benzo[*a*]pyrene with the products formed by reaction of benzo[*a*]pyrene-4,5-oxide with DNA. *Cancer Res.* 35: 54–57.
39. Conney AH, Miller EC and Miller JA (1956) The metabolism of methylated aminoazo dyes. V. Evidence for induction of enzyme synthesis in the rat by 3-methylcholanthrene. *Cancer Res.* 16: 450–459.
40. Conney AH, Miller EC and Miller JA (1957) Substrate-induced synthesis and other properties of benzpyrene hydroxylase in rat liver. *J. Biol. Chem.* 228: 753–766.
41. Sladek NE and Mannering GJ (1966) Evidence for a new P-450 hemoprotein in hepatic microsomes from methylcholanthrene treated rats. *Biochem. Biophys. Res. Commun.* 24: 668–674.
42. Gelboin HV (1967) Carcinogens, enzyme induction, and gene action. *Adv. Cancer Res.* 10: 1–81.
43. Nebert DW and Gelboin HV (1968) Substrate-inducible microsomal aryl hydroxylase in mammalian cell culture. I. Assay and properties of induced enzyme. *J. Biol. Chem.* 243: 6242–6249.
44. Conney AH (1982) Induction of microsomal enzymes by foreign chemicals and carcinogenesis by polycyclic aromatic hydrocarbons: G.H.A. Clowes Memorial Lecture. *Cancer Res.* 42: 4875–4917.
45. Grover PL and Sims P (1968) Enzyme-catalyzed reactions of polycyclic hydrocarbons with deoxyribonucleic acid and protein *in vitro*. *Biochem. J.* 110: 159–160.
46. Gelboin HV (1969) A microsome-dependent binding of benzo[*a*]pyrene to DNA. *Cancer Res.* 29: 1272–1276.
47. Borgen A, Darvey H, Castagnoli N, Crocker TT, Rasmussen RE and Wang IY (1973) Metabolic conversion of benzo[*a*]pyrene by Syrian hamster liver microsomes and binding of metabolites to deoxyribonucleic acid. *J. Med. Chem.* 16: 502–506.
48. Sims P, Grover PL, Swaisland A, Pal K and Hewer A (1974) Metabolic activation of benzo[*a*]pyrene proceeds by a diol-epoxide. *Nature* 252: 326–328.

49. Jerina DM and Daly JW (1974) Arene oxides: a new aspect of drug metabolism. Metabolic formation of arene oxides explains many toxic and carcinogenic properties of aromatic hydrocarbons. *Science* 185: 573–582.
50. Daudel P, Duquesne M, Vigny P, Grover PL and Sims P (1975) Fluorescence spectral evidence that benzo[*a*]pyrene-DNA products in mouse skin arise from diol-epoxides. *FEBS Lett.* 57: 250–253.
51. Marquardt H, Grover PL and Sims P (1976) *In vitro* malignant transformation of mouse fibroblasts by non-K-region dihydrodiols derived from 7-methylbenz[*a*]anthracene, 7,12-dimethylbenz[*a*]anthracene, and benzo[*a*]pyrene. *Cancer Res.* 36: 2059–2064.
52. Wood AW, Levin W, Lu AY, Ryan D, West SB, Lehr RE, Schaefer-Ridder M, Jerina DM and Conney AH (1976) Mutagenicity of metabolically activated benz[*a*]anthracene 3,4-dihydrodiol: evidence for bay region activation of carcinogenic polycyclic aromatic hydrocarbons. *Biochem. Biophys. Res. Commun.* 72: 680–686.
53. King HWS, Osborne MR, Beland FA, Harvey RG and Brookes P (1976) (±)-7α,8β-dihydroxy-9β,10β-epoxy-7,8,9,10-tetrahydrobenzo[*a*]pyrene is an intermediate in the metabolism and binding to DNA of benzo[*a*]pyrene. *Proc. Natl. Acad. Sci. USA* 73: 2679–2681.
54. Osborne MR, Beland FA, Harvey RG and Brookes P (1976) The reaction of (±)-7α,8β-dihydroxy-9β,10β-epoxy-7,8,9,10-tetrahydrobenzo[*a*]pyrene with DNA. *Int. J. Cancer* 18: 362–368.
55. Weinstein I, Jeffrey AM, Jennette KW, Blobstein SH, Harvey RG, Harris C, Autrup H, Kasai H and Nakanishi K (1976) Benzo[*a*]pyrene diol epoxides as intermediates in nucleic acid binding *in vitro* and *in vivo*. *Science* 193: 592–595.
56. Newbold RF and Brookes P (1976) Exceptional mutagenicity of benzo-[*a*]pyrene diol epoxide in cultured mammalian cells. *Nature* 261: 52–54.
57. Huberman E, Sachs L, Yang SK and Gelboin HV (1976) Identification of mutagenic metabolites of benzo[*a*]pyrene in mammalian cells. *Proc. Natl. Acad. Sci. USA* 73: 607–611.
58. Jeffrey AM, Weinstein IB, Jennette KW, Grzeskowiak K, Nakanishi K, Harvey RG, Autrup H and Harris C (1977) Structures of benzo[*a*]pyrene-nucleic acid adducts formed in human and bovine bronchial explants. *Nature* 269: 348–350.
59. Levin W, Wood AW, Wislocki PG, Chang RL, Kapitulnik J, Mah HD, Yagi H, Jerina DM and Conney AH (1978) Mutagenicity and carcinogenicity

of benzo[*a*]pyrene and benzo[*a*]pyrene derivatives. In: *Polycyclic Hydrocarbons and Cancer* [Gelboin HV and Ts'o POP (eds.)] Volume I: Environment, Chemistry, and Metabolism, pp 189–202, Academic Press, New York.

60. Chouroulinkov I, Gentil A, Grover PL and Sims P (1976) Tumour initiating activities on mouse skin of dihydrodiols derived from benzo[*a*]pyrene. *Br. J. Cancer* 34: 523–532.
61. Levin W, Wood AW, Yagi H, Jerina DM and Conney AH (1976) (±)-*Trans*-7,8-dihydroxy-7,8-dihydrobenzo[*a*]pyrene: a potent skin carcinogen when applied topically to mice. *Proc. Natl. Acad. Sci. USA* 73: 3867–3871.
62. Kapitulnik J, Levin W, Conney AH, Yagi H and Jerina DM (1977) Benzo[*a*]pyrene 7,8-dihydrodiol is more carcinogenic than benzo[*a*]pyrene in newborn mice. *Nature* 266: 378–380.
63. Slaga TJ, Gleason GL, Mills G, Ewald L, Fu PP, Lee HM and Harvey RG (1980) Comparison of the skin tumor-initiating activities of dihydrodiols and diol-epoxides of various polycyclic aromatic hydrocarbons. *Cancer Res.* 40: 1981–1984.
64. Yagi H, Hernandez O and Jerina DM (1975) Synthesis of (±)-7α,8β-dihydroxy-9β,10β-epoxy-7,8,9,10-tetrahydrobenzo[*a*]pyrene, a potential metabolite of the carcinogen benzo[*a*]pyrene with stereochemistry related to the antileukemic tripolides. *J. Am. Chem. Soc.* 97: 6881–6883.
65. Yagi H, Thakker DR, Hernandez O, Koreeda M and Jerina DM (1977) Synthesis and reactions of the highly mutagenic 7,8-diol 9,10-epoxides of the carcinogen benzo[*a*]pyrene. *J. Am. Chem. Soc.* 99: 1604–1611.
66. Harvey RG and Fu PP (1978) Synthesis and reactions of diol epoxides and related metabolites of carcinogenic hydrocarbons. In: *Polycyclic Hydrocarbons and Cancer* [Gelboin HV and Ts'o POP (eds.)] Volume I: Environment, Chemistry, and Metabolism, pp 133–165, Academic Press, New York.
67. Jerina DM and Lehr RE (1977) The bay-region theory: a quantum mechanical approach to aromatic hydrocarbon-induced carcinogenicity. In: *Microsomes and Drug Oxidations* [Ullrich V, Roots I, Hildebrandt A, Estabrook RW and Conney AH (eds.)] pp 709–720, Pergamon Press, Oxford, UK.
68. Jerina DM, Lehr R, Schaefer-Ridder M, Yagi H, Karle JM, Thakker DR, Wood AW, Lu AYH, Ryan D, West S, Levin W and Conney AH (1977) Bay-region epoxides of dihydrodiols: a concept explaining the mutagenic and carcinogenic activity of benzo[*a*]pyrene and benz[*a*]anthracene. In: *Origins of Human Cancer* [Hiatt HH, Watson JD and Winstein JA (eds.)] Book B, pp 639–658, Cold Spring Habour, New York.

69. Jerina DM, Yagi H, Lehr RE, Thakker DR, Schaefer-Ridder M, Karle JM, Levin W, Wood AW, Chang RL and Conney AH (1978) The bay-region theory of carcinogenesis by polycyclic aromatic hydrocarbons. In: *Polycyclic Hydrocarbons and Cancer* [Gelboin HV and Ts'o POP (eds.)] Volume I: Environment, Chemistry, and Metabolism, pp 173–188, Academic Press, New York.
70. Lehr RE, Kumar S, Levin W, Wood AW, Chang RL, Conney AH, Yagi H, Sayer JM and Jerina DM (1985) The bay region theory of polycyclic aromatic hydrocarbon carcinogenesis. In: *Polycyclic Hydrocarbons and Carcinogenesis* [Harvey RG (ed.)] ACS Symposium Series 283, pp 63–84, American Chemical Society, Washington, D.C.
71. Hall M and Grover PL (1990) Polycyclic aromatic hydrocarbons: metabolism, activation and tumour initiation. In: *Chemical Carcinogenesis and Mutagenesis* [Cooper CS and Grover PL (eds.)] Volume I, pp 327–372, Springer-Verlag, Berlin.
72. Ioannides C and Parke DV (1990) The cytochrome P450 I gene family of microsomal hemoproteins and their role in the metabolic activation of chemicals. *Drug Metab. Rev.* 22: 1–85.
73. Guengerich FP (1991) Reactions and significance of cytochrome P-450 enzymes. *J. Biol. Chem.* 266: 10019–10022.
74. Guengerich FP and Shimada T (1991) Oxidation of toxic and carcinogenic chemicals by human cytochrome P-450 enzymes. *Chem. Res. Toxicol.* 4: 391–407.
75. Gonzalez FJ (1993) Cytochromes P450 enzymes. In: *Handbook of Experimental Pharmacology* [Schenkman JB and Greim H (eds.)] Volume 105: Cytochrome P450, pp 239–257, Springer-Verlag, Berlin.
76. Gonzalez FJ and Gelboin HV (1994) Role of human cytochromes P450 in the metabolic activation of chemical carcinogens and toxins. *Drug Metab. Rev.* 26: 165–183.
77. Guengerich FJ (1995) Human cytochrome P450 enzymes. In: *Cytochrome P450 — Structure, Mechanism, and Biochemistry* [Ortiz de Montellano PR (ed.)] pp 473–535, Plenum Press, New York.
78. Nebert DW and Gonzalez FJ (1987) P450 enzymes: structure, evolution, and regulation. *Annu. Rev. Biochem.* 56: 945–993.
79. Gonzalez FJ and Nebert DW (1990) Evolution of the P-450 gene superfamily. *Pharmacol. Ther.* 45: 1–38.
80. Nelson DR, Koymans L, Kamataki T, Stegeman JJ, Feyereisen R, Waxman DJ, Waterman MR, Gotoh O, Coon MJ, Estabrook RW, Gunsalus IC

and Nebert DW (1996) P450 superfamily: update on new sequences, gene mapping, accession numbers, and nomenclature. *Pharmacogenetics* 6: 1–42.
81. Nelson DR (2003) Cytochrome P450 gene superfamily, http://drnelson.utmem.edu/cytochromeP450.html.
82. Shou M, Korzekwa KR, Crespi CL, Gonzalez FJ and Gelboin HV (1994) The role of 12 cDNA-expressed human, rodent, rabbit cytochromes P450 in the metabolism of benzo[*a*]pyrene and benzo[*a*]pyrene *trans*-7,8-dihydrodiol. *Mol. Carcinog.* 10: 159–168.
83. Kawajiri K and Hayashi SI (1996) The CYP1 family. In: *Cytochromes P450 — Metabolic and Toxicological Aspects* [Ioannides C (ed.)] pp 77 97, CRC Press, Boca Raton, FL.
84. Shou M, Krausz KW, Gonzalez FJ and Gelboin HV (1996) Metabolic activation of the potent carcinogen dibenzo[*a,h*]anthracene by cDNA-expressed human cytochromes P450. *Arch. Biochem. Biophys.* 328: 201–207.
85. Maurel P (1996) The CYP3 family. In: *Cytochromes P450 — Metabolic and Toxicological Aspects* [Ioannides C (ed.)] pp 241–270, CRC Press, Boca Raton, FL.
86. Baird WM and Ralston SL (1997) Carcinogenic polycyclic aromatic hydrocarbons. In: *Comprehensive Toxicology* [Sipes IG, McQueen CA and Gandolfi AJ (eds.)] Volume 12: Chemical Carcinogens and Anticarcinogens, pp 171–200, Pergamon Press, Oxford, UK.
87. Shimada T (2000) Human cytochrome P450 1B1 and chemical carcinogenesis. *Rev. Toxicol.* 3: 139–166.
88. Williams JA and Phillips DH (2000) Mammary expression of xenobiotic metabolizing enzymes and their potential role in breast cancer. *Cancer Res.* 60: 4667–4677.
89. Shimada T, Hayes CL, Yamazaki H, Amin S, Hecht SS and Guengerich FP (1996) Activation of chemically diverse procarcinogens by human cytochrome P-450 1B1. *Cancer Res.* 56: 2979–2984.
90. Larsen MC, Angus WGR, Brake PB, Eltom SE, Sukow KA and Jefcoate CR (1998) Characterization of CYP1B1 and CYP1A1 expression in human mammary epithelial cells: role of the aryl hydrocarbon receptor in polycyclic aromatic hydrocarbon metabolism. *Cancer Res.* 58: 2366–2374.
91. Walker NJ, Portier CJ, Lax SF, Crofts FG, Li Y, Lucier GW and Sutter TR (1999) Characterization of the dose-response of CYP1B1, CYP1A1, and CYP1A2 in the liver of female Sprague-Dawley rats following chronic exposure to 2,3,7,8-tetrachlorodibenzo-*p*-dioxin. *Toxicol. Appl. Pharmacol.* 154: 279–286.

92. Shimada T, Sugie A, Shindo M, Nakajima T, Azuma E, Hashimoto M and Inoue K (2003) Tissue-specific induction of cytochromes P450 1A1 and 1B1 by polycyclic aromatic hydrocarbons and polychlorinated biphenyls in engineered C57BL/6J mice of arylhydrocarbon receptor gene. *Toxicol. Appl. Pharmacol.* 187: 1–10.
93. Ionannides C and Parke DV (1993) Induction of cytochrome P4501 as an indicator of potential chemical carcinogenesis. *Drug Metab. Rev.* 25: 485–501.
94. Shimada T, Inoue K, Suzuki Y, Kawai T, Azuma E, Nakajima T, Shindo M, Kurose K, Sugie A, Yamagishi Y, Fujii-Kuriyama Y and Hashimoto M (2002) Arylhydrocarbon receptor-dependent induction of liver and lung cytochromes P450 1A1, 1A2, and 1B1 by polycyclic aromatic hydrocarbons and polychlorinated biphenyls in genetically engineered C57BL/6J mice. *Carcinogenesis* 23: 1199–1207.
95. Whitlock JPJ (1999) Induction of cytochrome P-4501A1. *Annu. Rev. Pharmacol. Toxicol.* 39: 103–125.
96. Reyes H, Reisz-Porszasz S and Hankinson O (1992) Identification of the Ah receptor nuclear translocator protein (Arnt) as a component of the DNA binding form of the Ah receptor. *Science* 256: 1193–1195.
97. Nebert DW, Eisen HJ, Negishi M, Lang MA and Hjelmeland LM (1981) Genetic mechanisms controlling the induction of polysubstrate monooxygenase (P-450) activities. *Annu. Rev. Pharmacol. Toxicol.* 21: 431–462.
98. Nebert DW (1989) The Ah locus: genetic differences in toxicity, cancer, mutation, and birth defects. *Crit. Rev. Toxicol.* 2: 153–174.
99. Okey AB (1990) Enzyme induction in the cytochrome P450 system. *Pharmacol. Ther.* 45: 241–298.
100. Nebert DW, Puga A and Vasiliou V (1993) Role of the Ah receptor and the dioxin-inducible [*Ah*] gene battery in toxicity, cancer, and signal transduction. *Ann. N.Y. Acad. Sci.* 685: 624–640.
101. Hankinson O (1995) The aryl hydrocarbon receptor complex. *Annu. Rev. Pharmacol. Toxicol.* 35: 307–340.
102. Gu YZ, Hogenesch JB and Bradfield CA (2000) The PAS superfamily: sensors of environmental and developmental signals. *Annu. Rev. Pharmacol. Toxicol.* 40: 519–561.
103. Hirshfeld FL (1963) The structure of overcrowded aromatic compounds. Part VII. Out-of-plane deformation in benzo[*c*]phenanthrene and 1,12-dimethylbenzo[*c*]phenanthrene. *J. Chem. Soc.* 2126–2135.

104. Rabinowitz JR, Little SB and Lewis-Bevan L (1996) The effect of crowding in the bay/fjord region on the structure and reactivities of polycyclic aromatic hydrocarbons and their metabolites: quantum mechanical studies. *Polycycl. Aromat. Compds.* 11: 237–244.
105. Katz AK, Carrell HL and Glusker JP (1998) Dibenzo[*a,l*]pyrene (dibenzo-[*def,p*]chrysene): fjord-region distortions. *Carcinogenesis* 19: 1641–1648.
106. Fretland AJ and Omiecinski CJ (2000) Epoxide hydrolases: biochemistry and molecular biology. *Chem.-Biol. Interact.* 129: 41–59.
107. Oesch F (1973) Mammalian epoxide hydrases: inducible enzymes catalyzing the inactivation of carcinogenic and cytotoxic metabolites derived from aromatic and olefinic compounds. *Xenobiotica* 3: 305–340.
108. Bentley P, Schmassmann H, Sims P and Oesch F (1976) Epoxides derived from various polycyclic hydrocarbons as substrates of homogeneous and microsome-bound epoxide hydratase. A general assay and kinetic properties. *Eur. J. Biochem.* 69: 97–103.
109. Guenther TM and Oesch F (1981) Microsomal epoxide hydrolase and its role in polycyclic aromatic hydrocarbon biotransformation. In: *Polycyclic Hydrocarbons and Cancer* [Gelboin HV and Ts'o POP (eds.)] Volume III, pp 183–212, Academic Press, New York.
110. Guengerich FP (1982) Epoxide hydrolase: properties and metabolic roles. *Rev. Biochem. Toxicol.* 4: 5–30.
111. Raphael D, Glatt HR, Protic-Sabljic M and Oesch F (1982) Effects of various enzyme inducers on monooxygenase, glutathione *S*-transferase and epoxide hydrolase activities in cultured hepatoma cells. *Chem.-Biol. Interact.* 42: 27–43.
112. Astrom A, Maner S and DePierre JW (1987) Induction of liver microsomal epoxide hydrolase, UDP-glucuronyl transferase and cytosolic glutathione transferase in different rodent species by 2-acetylaminofluorene or 3-methylcholanthrene. *Xenobiotica* 17: 155–163.
113. Hassett C, Laurenzana EM, Sidhu JS and Omiecinski CJ (1998) Effects of chemical inducers on human microsomal epoxide hydrolase in primary hepatocyte cultures. *Biochem. Pharmacol.* 55: 1059–1069.
114. Boyland E and Levi AA (1935) Metabolism of polycyclic compounds. I. Production of dihydroxydihydroanthracene from anthracene. *Biochem. J.* 29: 2679–2683.
115. Armstrong RN (1987) Enzyme-catalyzed detoxication reactions: mechanisms and stereochemistry. *Crit. Rev. Biochem. Mol. Biol.* 22: 39–88.

116. Shou M, Gonzalez FJ and Gelboin HV (1996) Stereoselective epoxidation and hydration at the K-region of polycyclic aromatic hydrocarbons by cDNA-expressed cytochromes P450 1A1, 1A2, and epoxide hydrolase. *Biochemistry* 35: 15807–15813.
117. Lacourciere GM and Armstrong RN (1993) The catalytic mechanism of microsomal epoxide hydrolase involves an ester intermediate. *J. Am. Chem. Soc.* 115: 10466–10467.
118. Arand M, Müller F, Mecky A, Hinz W, Urban P, Pompon D, Kellner R and Oesch F (1999) Catalytic triad of microsomal epoxide hydrolase: replacement of Glu[404] with Asp leads to a strongly increased turnover rate. *Biochem. J.* 337: 37–43.
119. Phillips DH and Grover PL (1994) Polycyclic hydrocarbon activation: bay regions and beyond. *Drug Metab. Rev.* 26: 443–467.
120. Shimada T, Martin MV, Pruess-Schwartz D, Marnett L and Guengerich FP (1989) Roles of individual human cytochrome P-450 enzymes in the bioactivation of benzo[*a*]pyrene, 7,8-dihydroxy-7,8-dihydrobenzo[*a*]pyrene, and other dihydrodiol derivatives of polycyclic aromatic hydrocarbons. *Cancer Res.* 49: 6304–6312.
121. Marnett LJ and Bienkowski MJ (1980) Hydroperoxide-dependent oxygenation of *trans*-7,8-dihydroxy-7,8-dihydrobenzo[*a*]pyrene by ram seminal vesicle microsomes, source of oxygen. *Biochem. Biophys. Res. Commun.* 96: 639–647.
122. Sivarajah K, Lasker JM and Eling TA (1981) Prostaglandin synthetase-dependent cooxidation of (±)-benzo[*a*]pyrene-7,8-dihydrodiol by human lung and other mammalian tissues. *Cancer Res.* 41: 1834–1839.
123. Guthrie J, Robertson IGC, Zeiger E, Boyd JA and Eling TE (1982) Selective activation of some dihydrodiols of several polycyclic aromatic hydrocarbons to mutagenic products by prostaglandin synthetase. *Cancer Res.* 42: 1620–1623.
124. Reed GA and Marnett LJ (1982) Metabolism and activation of 7,8-dihydroxybenzo[*a*]pyrene during prostaglandin biosynthesis. Intermediacy of a bay-region epoxide. *J. Biol. Chem.* 257: 11368–11376.
125. Dix TA and Marnett LJ (1983) Metabolism of polycyclic aromatic hydrocarbon derivatives to ultimate carcinogens during lipid peroxidation. *Science* 221: 77–79.
126. Marnett LJ (1985) Hydroperoxide-dependent oxygenation of polycyclic aromatic hydrocarbons and their metabolites. In: *Polycyclic Hydrocarbons and Carcinogenesis* [Harvey RG (ed.)] ACS Symposium Series 283, pp 307–326, American Chemical Society, Washington, D.C.

127. Pruess-Schwartz D, Nimesheim A, and Marnett LJ (1989) Peroxyl radical- and cytochrome P450-dependent metabolic activation of (+)-7,8-dihydroxy-7,8-dihydrobenzo[*a*]pyrene in mouse skin *in vitro* and *in vivo*. *Cancer Res.* 49: 1732–1737.
128. Marnett LJ (1990) Prostaglandin synthase-mediated metabolism of carcinogens and a potential role for peroxyl radicals as reactive intermediates. *Environ. Health Perspect.* 88: 5–12.
129. Eling TE, Thompson DC, Foureman GL, Curtis JF and Hughes MF (1990) Prostaglandin H synthase and xenobiotic oxidation. *Annu. Rev. Pharmacol. Toxicol.* 30: 1–45.
130. O'Brien PJ (2000) Peroxidases. *Chem.-Biol. Interact.* 129: 113–139.
131. Wiese FW, Thompson PA and Kadlubar FF (2001) Carcinogen substrate specificity of human COX-1 and COX-2. *Carcinogenesis* 22: 5–10.
132. Mallet WG, Mosebrook DR and Trush MA (1991) Activation of (±)-*trans*-7,8-dihydroxy-7,8-dihydrobenzo[*a*]pyrene to diolepoxides by human polymorphonuclear leukocytes or myeloperoxidase. *Carcinogenesis* 12: 521–524.
133. Trush A, Esterline RL, Mallet WG, Mosebrook DR and Twerdok LE (1991) Further evidence for the role of myeloperoxidase in the activation of benzo[*a*]pyrene-7,8-dihydrodiol by polymorphonuclear leukocytes. *Adv. Exp. Med. Biol.* 283: 399–401.
134. Petruska JM, Mosebrook DR, Jakab GJ and Trush MA (1992) Myeloperoxidase-enhanced formation of (±)-*trans*-7,8-dihydroxy-7,8-dihydrobenzo[*a*]-pyrene-DNA adducts in lung tissue *in vitro*: a role of pulmonary inflammation in the bioactivation of a procarcinogen. *Carcinogenesis* 13: 1075–1081.
135. Borm PJA, Knaapen AM, Schins RPF, Godschalk RWL and van Schooten FJ (1997) Neutrophils amplify the formation of DNA adducts by benzo[*a*]pyrene in lung target cells. *Environ. Health Perspect.* 105, Suppl 5: 1089–1093.
136. Josephy PD and Coomber BL (1998) Current concepts: neutrophils and the activation of carcinogens in the breast and other organs. *Can. J. Physiol. Pharmacol.* 76: 693–700.
137. Madhavan ND and Naidu KA (2000) Purification and partial characterization of peroxidase from human term placenta of non-smokers: metabolism of benzo[*a*]pyrene-7,8-dihydrodiol. *Placenta* 21: 501–509.
138. Rojas M, Godschalk R, Alexandrov K, Cascorbi I, Kriek E, Ostertag J, van Schooten FJ and Bartsch H (2001) Myeloperoxidase −463A variant reduces benzo[*a*]pyrene diol epoxide DNA adducts in skin of coal tar treated patients. *Carcinogenesis* 22: 1015–1018.

139. Kelley DJ, Mestre JR, Subbarmaiah K, Sacks PG, Schantz SP, Tanabe T, Inoue H, Ramonettie JT and Dannenberg AJ (1997) Benzo[*a*]pyrene up-regulates cyclooxygenase-2 gene expression in oral epithelial cells. *Carcinogenesis* 18: 795–799.
140. Marks F and Fürstenberger G (2000) Cancer chemoprevention through interruption of multistage carcinogenesis: the lessons learnt by comparing mouse skin carcinogenesis and human large bowel cancer. *Eur. J. Cancer* 36: 314–329.
141. Devanesan P, Rogan E and Cavalieri E (1987) The relationship between ionisation potential and prostaglandin H synthase-catalyzed binding of aromatic hydrocarbons to DNA. *Chem.-Biol. Interact.* 61: 89–95.
142. Cavalieri EL and Rogan EG (1992) The approach to understanding aromatic hydrocarbon carcinogenesis: the central role of radical cations in metabolic activation. *Pharmacol. Ther.* 55: 183–199.
143. Stansbury KH and Trush MA (1999) Evidence for dialdehyde/imine intermediate formation during the metabolism of benzo[*a*]pyrene-7,8-dihydrodiol by horseradish peroxidase. *Polycycl. Aromat. Compds.* 16: 225–234.
144. Stansbury KH, Noll DM, Groopman JD and Trush MA (2000) Enzyme-mediated dialdehyde formation: an alternative pathway for benzo[*a*]pyrene 7,8-dihydrodiol bioactivation. *Chem. Res. Toxicol.* 13: 1174–1180.
145. Garattini E, Coccia P, Romano M, Jiritano L, Noseda A and Salmona M (1984) Prostaglandin-endoperoxide sythetase and the activation of benzo-[*a*]pyrene to reactive metabolites *in vivo* in guinea pigs. *Cancer Res.* 44: 5150–5155.
146. Reddy AP, Pruess-Schwartz D, Ji C, Gorycki P and Marnett LJ (1992) ^{32}P-postlabelling analysis of DNA adduction in mouse skin following topical administration of (+)-7,8-dihydroxy-7,8-dihydrobenzo[*a*]pyrene. *Chem. Res. Toxcicol.* 5: 26–33.
147. Melendez-Colon VJ, Luch A, Seidel A and Baird WM (1999) Cancer initiation by polycyclic aromatic hydrocarbons results from formation of stable DNA adducts rather than apurinic sites. *Carcinogenesis* 20: 1885–1891.
148. Jang M, Cai L, Udeani GO, Slowing KV, Thomas CF, Beecher CW, Fong HH, Farnsworth NR, Kinghorn AD, Mehta RG, Moon RC and Pezzuto JM (1997) Cancer chemopreventive activity of resveratrol, a natural product derived from grapes. *Science* 275: 218–220.
149. Müller-Decker K, Kopp-Schneider A, Marks F, Seibert K and Fürstenberger G (1998) Localization of prostaglandin H synthase isoenzymes in murine epidermal tumors: suppression of skin tumor promotion by inhibition of prostaglandin H synthase-2. *Mol. Carcinog.* 23: 36–44.

150. Abou-Issa HM, Alshafie GA, Seibert K, Koki AT, Masferrer JL and Harris RE (2001) Dose-response effects of the COX-2 inhibitor, celecoxib, on the chemoprevention of mammary carcinogenesis. *Anticancer Res.* 21: 3425–3432.
151. Müller-Decker K, Neufang G, Berger I, Neumann M, Marks F and Fürstenberger G (2002) Transgenic cyclooxygenase-2 overexpression sensitizes mouse skin for carcinogenesis. *Proc. Natl. Acad. Sci. USA* 99: 12483–12488.
152. Tiano HF, Loftin CD, Akunda J, Lee CA, Spalding J, Sessoms A, Dunson DB, Rogan EG, Morham SG, Smart RC and Langenbach R (2002) Deficiency of either cyclooxygenase (COX)-1 or COX-2 alters epidermal differentiation and reduces mouse skin tumorigenesis. *Cancer Res.* 62: 3395–3401.
153. Adriaenssens PI, Sivarajah K, Boorman GA, Eling TE and Anderson MW (1983) Effect of aspirin and indomethacin on the formation of benzo-[*a*]pyrene-induced pulmonary adenomas and DNA adducts in A/Hej mice. *Cancer Res.* 43: 4762–4767.
154. Gupta RA and DuBois RN (2001) Colorectal cancer prevention and treatment by inhibition of cyclooxygenase-2. *Nat. Rev. Cancer* 1: 11–21.
155. Williams CS, Mann M and DuBois RN (1999) The role of cyclooxygenase in inflammation, cancer and development. *Oncogene* 18: 7908–7916.
156. Masferrer JL, Leahy KM, Koki AT, Zweifel BS, Settle SL, Woerner BM, Edwards DA, Flickinger AG, Moore RJ and Seibert K (2000) Antiangiogenic and antitumor activities of cyclooxygenase-2 inhibitors. *Cancer Res.* 60: 1306–1311.
157. Leahy KM, Ornberg RL, Wang Y, Zweifel BS, Koki AT and Masferrer JL (2002) Cyclooxygenase-2 inhibition by celecoxib reduces proliferation and induces apoptosis in angiogenic endothelial cells *in vivo*. *Cancer Res.* 62: 625–631.
158. Le Marchand L, Seifried A, Lum A and Wilkens LR (2000) Association of myeloperoxidase $^{-463}G \rightarrow A$ polymorphism with lung cancer risk. *Cancer Epidemiol. Biomarkers Prev.* 9: 181–184.
159. Schabath MB, Spitz MR, Zhang X, Delclos GL and Wu X (2000) Genetic variants of myeloperoxidase and lung cancer risk. *Carcinogenesis* 21: 1163–1166.
160. Kiyohara C, Otsu A, Shirakawa T, Fukuda S and Hopkin JM (2002) Genetic polymorphisms and lung cancer susceptibility: a review. *Lung Cancer* 37: 241–256.
161. Shimizu Y, Nakatsuru Y, Ichinose M, Takahashi Y, Kume H, Mimura J, Fujii-Kuriyama Y and Ishikawa T (2000) Benzo[*a*]pyrene carcinogenicity

is lost in mice lacking the aryl hydrocarbon receptor. *Proc. Natl. Acad. Sci. USA* 97: 779–782.

162. Robinson JR, Felton JS, Levitt RC, Thorgeirsson SS and Nebert DW (1975) Relationship between "aromatic hydrocarbon responsiveness" and survival times in mice treated with various drugs and environmental compounds. *Mol. Pharmacol.* 11: 850–865.
163. Uno S, Dalton TP, Shertzer HG, Genter MB, Warshawsky D, Talaska G and Nebert DW (2001) Benzo[*a*]pyrene-induced toxicity: paradoxical protection in Cyp1a1(−/−) knockout mice having increased hepatic B*a*P-DNA adduct levels. *Biochem. Biophys. Res. Commun.* 289: 1049–1056.
164. Heidel SM, MacWilliams PS, Baird WM, Dashwood WM, Buters JTM, Gonzalez FJ, Larsen MC, Czuprynski CJ and Jefcoate CR (2000) Cytochrome P4501B1 mediates induction of bone marrow cytotoxicity and preleukemia cells in mice treated with 7,12-dimethylbenz[*a*]anthracene. *Cancer Res.* 60: 3454–3460.
165. Heidel SM, Holston K, Buters JTM, Gonzalez FJ, Jefcoate CR and Czuprynski CJ (1999) Bone marrow stromal cell cytochrome P4501B1 is required for pre-B cell apoptosis induced by 7,12-dimethylbenz[*a*]anthracene. *Mol. Pharmacol.* 56: 1317–1323.
166. Buters JTM, Sakai S, Richter T, Pineau T, Alexander DL, Savas U, Doehmer J, Ward JM, Jefcoate CR and Gonzalez FJ (1999) Cytochrome P450 CYP1B1 determines susceptibility to 7,12-dimethylbenz[*a*]anthracene-induced lymphomas. *Proc. Natl. Acad. Sci. USA* 96: 1977–1982.
167. Buters JTM, Mahadevan B, Quintanilla-Martinez L, Gonzalez FJ, Greim H, Baird WM and Luch A (2002) Cytochrome P450 1B1 determines susceptibility to dibenzo[*a,l*]pyrene-induced tumor formation. *Chem. Res. Toxicol.* 15: 1127–1135.
168. Buters J, Quintanilla-Martinez L, Schober W, Soballa VJ, Hintermair J, Wolff T, Gonzalez FJ and Greim H (2003) CYP1B1 determines susceptibility to low doses of 7,12-dimethylbenz[*a*]anthracene-induced ovarian cancers in mice: correlation of CYP1B1-mediated DNA adducts with carcinogenicity. *Carcinogenesis* 24: 327–334.
169. Miyata M, Kudo G, Lee YH, Yang TJ, Gelboin HV, Fernandez-Salguero P, Kimura S and Gonzalez FJ (1999) Targeted disruption of the microsomal epoxide hydrolase gene: microsomal epoxide hydrolase is required for the carcinogenic activity of 7,12-dimethylbenz[*a*]anthracene. *J. Biol. Chem.* 274: 23963–23968.

170. Kondraganti SR, Fernandez-Salguero P, Gonzalez FJ, Ramos KS, Jiang W and Moorthy B (2003) Polycyclic aromatic hydrocarbon-inducible DNA adducts: evidence by ^{32}P-postlabeling and use of knockout mice for Ah receptor-independent mechanisms of metabolic activation *in vivo*. *Int. J. Cancer* 103: 5–11.
171. Meehan RR, Speed RM, Gosden JR, Rout D, Hutton JJ, Taylor BA, Hilkens J, Kroezen V, Hilgers J, Adesnik M, Friedberg T, Hastie ND and Wolf CR (1988) Chromosomal organization of the cytochrome P450-2C gene family in the mouse: a locus associated with constitutive aryl hydrocarbon hydroxylase. *Proc. Natl. Acad. Sci. USA* 85: 2662–2666.
172. Yang SK, McCourt DW, Roller PP and Gelboin HV (1976) Enzymatic conversion of benzo[*a*]pyrene leading predominantly to the diol-epoxide *r*-7,*t*-8-dihydroxy-*t*-9,10-oxy-7,8,9,10-tetrahydrobenzo[*a*]pyrene through a single enantiomer of *r*-7,*t*-8-dihydroxy-7,8-dihydrobenzo[*a*]pyrene. *Proc. Natl. Acad. Sci. USA* 73: 2594–2598.
173. Thakker DR, Yagi H, Akagi H, Koreeda M, Lu AYH, Levin W, Wood AW, Conney AH and Jerina DM (1977) Metabolism of benzo[*a*]pyrene: VI. Stereoselective metabolism of benzo[*a*]pyrene and benzo[*a*]pyrene 7,8-dihydrodiol to diol epoxides. *Chem.-Biol. Interact.* 16: 281–300.
174. Yang SK, McCourt DW, Leutz JC and Gelboin HV (1977) Benzo[*a*]pyrene diol epoxides: mechanisms of enzymatic formation and optically active intermediates. *Science* 196: 1199–1201.
175. Wood AW, Chang RL, Levin W, Yagi H, Thakker DR, Jerina DM and Conney AH (1977) Differences in mutagenicity of the optical enantiomers of the diastereomeric benzo[*a*]pyrene 7,8-diol-9,10-epoxides. *Biochem. Biophys. Res. Commun.* 77: 1389–1396.
176. Buening MK, Wislocki PG, Levin W, Yagi H, Thakker DR, Akagi H, Koreeda M, Jerina DM and Conney AH (1978) Tumorigenicity of the optical enantiomers of the diastereomeric benzo[*a*]pyrene 7,8-diol-9,10-epoxides in newborn mice: exceptional activity of (+)-7β,8α-dihydroxy-9α,10α-epoxy-7,8,9,10-tetrahydrobenzo[*a*]pyrene. *Proc. Natl. Acad. Sci. USA* 75: 5358–5361.
177. Slaga TJ, Bracken WJ, Gleason G, Levin W, Yagi H, Jerina DM and Conney AH (1979) Marked differences in the skin tumor-initiating activities of the optical enantiomers of the diastereomeric benzo[*a*]pyrene 7,8-diol-9,10-epoxides. *Cancer Res.* 39: 67–71.
178. Yang SK, Mushtaq M and Chiu PL (1985) Stereoselective metabolism and activation of polycyclic aromatic hydrocarbons. In: *Polycyclic Hydrocarbons*

and Carcinogenesis [Harvey RG (ed.)] ACS Symposium Series 283, pp 19–34, American Chemical Society, Washington, D.C.

179. Yang SK (1988) Stereoselectivity of cytochrome P-450 isozymes and epoxide hydrolase in the metabolism of polycyclic aromatic hydrocarbons. *Biochem. Pharmacol.* 37: 61–70.
180. Thakker DR, Levin W, Yagi H, Wood AW, Conney AH and Jerina DM (1988) Stereoselective biotransformation of polycyclic aromatic hydrocarbons to ultimate carcinogens. In: *Stereochemical Aspects of Pharmacologically Active Compounds* [Wainer AW and Dryer D (eds.)] pp 271–296, Marcel Dekker, New York.
181. Ittah Y, Thakker DR, Levin W, Croisy-Delcey M, Ryan DE, Thomas PE, Conney AH and Jerina DM (1983) Metabolism of benzo[*c*]phenanthrene by rat liver microsomes and by a purified monooxygenase system reconstituted with different isozymes of cytochrome P-450. *Chem.-Biol. Interact.* 45: 15–28.
182. Thakker DR, Levin W, Yagi H, Yeh HJC, Ryan DE, Thomas PE, Conney AH and Jerina DM (1986) Stereoselective metabolism of the (+)-(*S,S*)- and (−)-(*R,R*)-enantiomers of *trans*-3,4-dihydroxy-3,4-dihydrobenzo[*c*]phenanthrene by rat and mouse liver microsomes and by a purified and reconstituted P-450 system. *J. Biol. Chem.* 261: 5404–5413.
183. Koreeda M, Moore PD, Wislocki PG, Levin W, Conney AH, Yagi H and Jerina DM (1978) Binding of benzo[*a*]pyene 7,8-diol 9,10-epoxides to DNA, RNA, and protein of mouse skin occurs with high stereoselectivity. *Science* 199: 778–781.
184. Pruess-Schwartz D, Sebti SM, Gilham PT and Baird WM (1984) Analysis of benzo[*a*]pyrene:DNA adducts formed in cells in culture by immobilized boronate chromatography. *Cancer Res.* 44: 4104–4110.
185. Pruess-Schwartz D, Baird WM, Yagi H, Jerina DM, Pigott MA and Dipple A (1987) Stereochemical specificity in the metabolic activation of benzo[*c*]phenanthrene to metabolites that covalently bind to DNA in rodent embryo cell cultures. *Cancer Res.* 47: 4032–4037.
186. Einholf HJ, Amin S, Yagi H, Jerina DM and Baird WM (1996) Benzo[*c*]phenanthrene is activated to DNA-binding diol epoxides in the human mammary carcinoma cell line MCF-7 but only limited activation occurs in mouse skin. *Carcinogenesis* 17: 2237–2244.
187. Einholf HJ, Story WT, Marcus CB, Larsen MC, Jefcoate CR, Greenlee WF, Yagi H, Jerina DM, Amin S, Park SS, Gelboin HV and Baird WM (1997) Role of cytochrome P450 enzyme induction in the metabolic activation of

benzo[*c*]phenanthrene in human cell lines and mouse epidermis. *Chem. Res. Toxicol.* 10: 609–617.

188. Melendez-Colon VJ, Luch A, Seidel A and Baird WM (2000) Formation of stable DNA adducts and apurinic sites upon metabolic activation of bay and fjord region polycyclic aromatic hydrocarbons in human cell cultures. *Chem. Res. Toxicol.* 13: 10–17.
189. Wood AW, Richard LC, Levin W, Thakker DR, Yagi H, Sayer JM, Jerina DM and Conney AH (1984) Mutagenicity of the enantiomers of the diastereomeric bay-region benzo[*c*]phenanthrene 3,4-diol-1,2-epoxides in bacterial and mammalian cells. *Cancer Res.* 44: 2320–2324.
190. Jerina DM, Sayer JM, Agarwal SK, Yagi H, Levin W, Wood AW, Conney AH, Preuss-Schwarz D, Baird WM, Pigott MA and Dipple A (1986) Reactivity and tumorigenicity of bay-region diol epoxides derived from polycyclic aromatic hydrocarbons. In: *Biological Reactive Intermediates III, Mechanisms of Action in Animal Models and Human Diseases* [Kocsis JJ, Jollow DJ, Witmer CM, Nelson JO and Snyder R (eds.)] pp 11–30, Plenum Press, New York.
191. Levin W, Chang RL, Wood AW, Thakker DR, Yagi H, Jerina DM and Conney AH (1986) Tumorigenicity of optical isomers of the diastereomeric bay-region 3,4-diol-1,2-epoxides of benzo[*c*]phenanthrene in murine tumor models. *Cancer Res.* 46: 2257–2261.
192. Levin W, Wood AW, Chang RL, Ittah Y, Croisy-Celcay M, Yagi H, Jerina DM and Conney AH (1980) Exceptionally high tumor-initiating activity of benzo[*c*]phenanthrene bay-region diol-epoxides on mouse skin. *Cancer Res.* 40: 3910–3914.
193. Hecht SS, Radock R, Amin S, Huie K, Melikian AA, Hoffmann D, Pataki J and Harvey RG (1985) Tumorigenicity of 5-methylchrysene dihydrodiols and dihydrodiol epoxides in newborn mice and on mouse skin. *Cancer Res.* 45: 1449–1452.
194. Hecht SS, Melikian AA and Amin S (1986) Methylchrysenes as probes for the mechanism of metabolic activation of carcinogenic methylated polynuclear aromatic hydrocarbons. *Acc. Chem. Res.* 19: 174–180.
195. Melikian AA, Amin S, Huie K, Hecht SS and Harvey RG (1988) Reactivity with DNA bases and mutagenicity toward *Salmonella typhimurium* of methylchrysene diol epoxide enantiomers. *Cancer Res.* 48: 1781–1787.
196. Moschel RC, Baird WM and Dipple A (1977) Metabolic activation of the carcinogen 7,12-dimethylbenz[*a*]anthracene for DNA binding. *Biochem. Biophys. Res. Commun.* 76: 1092–1098.

197. Bigger CA, Sawicki JT, Blake DM, Raymond LG and Dipple A (1983) Products of binding of 7,12-dimethylbenz[*a*]anthracene to DNA in mouse skin. *Cancer Res.* 43: 5647–5651.
198. Vericat JA, Cheng SC and Dipple A (1989) Absolute stereochemistry of the major 7,12-dimethylbenz[*a*]anthracene-DNA adducts formed in mouse cells. *Carcinogenesis* 10: 567–570.
199. Ralston SL, Lau HHS, Seidel A, Luch A, Platt KL and Baird WM (1994) Identification of dibenzo[*a,l*]pyrene-DNA adducts formed in cells in culture and in mouse skin. *Polycycl. Aromat. Compds.* 6: 199–206.
200. Ralston SL, Lau HHS, Seidel A, Luch A, Platt KL and Baird WM (1994) The potent carcinogen dibenzo[*a,l*]pyrene is metabolically activated to fjord-region 11,12-diol 13,14-epoxides in human mammary carcinoma MCF-7 cell cultures. *Cancer Res.* 54: 887–890.
201. Ralston SL, Seidel A, Luch A, Platt KL and Baird WM (1995) Stereoselective activation of dibenzo[*a,l*]pyrene to (−)-*anti*(11*R*,12*S*,13*S*,14*R*)- and (+)-*syn*(11*S*,12*R*,13*S*,14*R*)-11,12-diol-13,14-epoxides which bind extensively to deoxyadenosine residues of DNA in the human mammary carcinoma cell line MCF-7. *Carcinogenesis* 16: 2899–2907.
202. Meehan T and Straub K (1979) Double-stranded DNA stereoselectively binds benzo[*a*]pyrene diol epoxides. *Nature* 277: 410–412.
203. Reardon DB, Prakash AS, Hilton BD, Roman JM, Pataki J, Harvey RG and Dipple A (1987) Characterization of 5-methylchrysene-1,2-dihydrodiol-3,4-epoxide-DNA adducts. *Carcinogenesis* 8: 1317–1322.
204. Sawicki JT, Moschel RC and Dipple A (1983) Involvement of both *syn*- and *anti*-dihydrodiol epoxides in the binding of 7,12-dimethylbenz[*a*]anthracene to DNA in mouse embryo cell cultures. *Cancer Res.* 43: 3212–3218.
205. Cheng SC, Prakash AS, Pigott MA, Hilton BD, Roman JM, Lee H, Harvey RG and Dipple A (1988) Characterization of 7,12-dimethylbenz[*a*]-anthracene-adenine nucleoside adducts. *Chem. Res. Toxicol.* 1: 216–221.
206. Cavalieri EL, Rogan EG, Higginbotham S, Cremonesi P and Salmasi S (1989) Tumor initiating activity in mouse skin and carcinogenicity in rat mammary gland of dibenzo[*a*]pyrenes: the very potent environmental carcinogen dibenzo[*a,l*]pyrene. *J. Cancer Res. Clin. Oncol.* 115: 67–72.
207. Cavalieri EL and Rogan EG (1984) Role of radical cations in aromatic hydrocarbon carcinogenesis. *Environ. Health Perspect.* 64: 69–84.
208. Cavalieri EL and Rogan EG (1985) One-electron oxidation in aromatic hydrocarbon carcinogenesis. In: *Polycyclic Hydrocarbons and Carcinogene-*

sis [Harvey RG (ed.)] ACS Symposium Series 283, pp 289–305, American Chemical Society, Washington, D.C.

209. Cavalieri EL and Rogan EG (1995) Central role of radical cations in metabolic activation of aromatic hydrocarbon carcinogenesis. *Xenobiotica* 25: 677–688.
210. Cremonesi P, Cavalieri EL and Rogan EG (1989) One-electron oxidation of 6-substituted benzo[*a*]pyrenes by manganic acetate. *J. Org. Chem.* 54: 3561–3570.
211. Rogan EG, Katomski PA, Roth RW and Cavalieri EL (1979) Horseradish peroxidase/hydrogen peroxide-catalyzed binding of aromatic hydrocarbons to DNA. *J. Biol. Chem.* 254: 7055–7059.
212. Rogan EG, Hakam A and Cavalieri EL (1983) Structure elucidation of a 6-methylbenzo[*a*]pyrene-DNA adduct formed by horseradish peroxidase *in vitro* and mouse skin *in vivo*. *Chem.-Biol. Interact.* 47: 111–122.
213. Rogan EG, Cavalieri EL, Tibbels S, Cremonesi P, Warner C, Nagel D, Tomer K, Cerny R and Gross M (1988) Synthesis and identification of benzo[*a*]pyrene-guanine nucleoside adducts formed by electrochemical oxidation and horseradish peroxidase-catalyzed reaction of benzo[*a*]pyrene with DNA. *J. Am. Chem. Soc.* 110: 4023–4029.
214. Cavalieri EL, Rogan EG, Devanesan PD, Cremonesi P, Cerny RL, Gross ML and Bodell WJ (1990) Binding of benzo[*a*]pyrene to DNA by cytochrome P-450-catalyzed one-electron oxidation in rat liver microsomes and nuclei. *Biochemistry* 29: 4820–4827.
215. RamaKrishna NV, Devanesan PD, Rogan EG, Cavalieri EL, Jeong H, Jankowiak R and Small GJ (1992) Mechanism of metabolic activation of the potent carcinogen 7,12-dimethylbenz[*a*]anthracene. *Chem. Res. Toxicol.* 5: 220–226.
216. Devanesan PD, RamaKrishna NVS, Todorovic R, Rogan EG, Cavalieri EL, Jeong H, Jankowiak R and Small GJ (1992) Identification and quantitation of benzo[*a*]pyrene-DNA adducts formed by rat liver microsomes *in vitro*. *Chem. Res. Toxicol.* 5: 302–309.
217. Rogan EG, Devanesan PD, RamaKrishna NVS, Higginbotham S, Padmavathi NS, Chapman K, Cavalieri EL, Jeong H, Jankowiak R and Small GJ (1993) Identification and quantitation of benzo[*a*]pyrene-DNA adducts formed in mouse skin. *Chem. Res. Toxicol.* 6: 356–363.
218. Devanesan PD, RamaKrishna NVS, Padmavathi NS, Higginbotham S, Rogan EG, Cavalieri EL, Marsch GA, Jankowiak R and Small GJ (1993)

Identification and quantitation of 7,12-dimethylbenz[*a*]anthracene-DNA adducts formed in mouse skin. *Chem. Res. Toxicol.* 6: 364–371.

219. Chen L, Devanesan PD, Higginbotham S, Ariese F, Jankowiak R, Small GJ, Rogan EG and Cavalieri EL (1996) Expanded analysis of benzo[*a*]pyrene-DNA adducts formed *in vitro* and in mouse skin: their significance in tumor initiation. *Chem. Res. Toxicol.* 9: 897–903.
220. Hanson AA, Li KM, Lin CH, Jankowiak R, Small GJ, Rogan EG and Cavalieri EL (2000) Synthesis and structure determination of 6-methyl-benzo[*a*]pyrene-deoxyribonucleoside adducts and their identification and quantitation *in vitro* and in mouse skin. *Chem.-Biol. Interact.* 128: 65–90.
221. Cavalieri EL, Rogan EG, Roth RW, Saugier RK and Hakam A (1983) The relationship between ionization potential and horseradish peroxidase/hydrogen peroxide-catalyzed binding of aromatic hydrocarbons to DNA. *Chem.-Biol. Interact.* 47: 87–109.
222. Estabrook RW, Werringloer J, Capdevila J and Prough RA (1978) The role of cytochrome P-450 and the microsomal electron transport system: the oxidative metabolism of benzo[*a*]pyrene. In: *Polycyclic Hydrocarbons and Cancer* [Gelboin HV and Ts'o POP (eds.)] Volume I: Environment, Chemistry, and Metabolism, pp 285–319, Academic Press, New York.
223. Phillips DH and Sims P (1979) Polycyclic aromatic hydrocarbon metabolites: their reactions with nucleic acids. In: *Chemical Carcinogens and DNA* [Grover PL (ed.)] pp 29–57, CRC Press, Boca Raton, FL.
224. Grover PL (1986) Pathways involved in the metabolism and activation of polycyclic hydrocarbons. *Xenobiotica* 16: 915–931.
225. Holder G, Yagi H, Dansette P, Jerina DM, Levin W, Liu AYH and Conney AH (1974) Effects of inducers and epoxide hydrase on the metabolism of benzo[*a*]pyrene by liver microsomes and a reconstituted system: analysis by high-pressure liquid chromatography. *Proc. Natl. Acad. Sci. USA* 71: 4356–4360.
226. Yang SK, Roller PP and Gelboin HV (1977) Enzymatic mechanism of benzo[*a*]pyrene conversion to phenols and diols and an improved high-pressure liquid chromatographic separation of benzo[*a*]pyrene derivatives. *Biochemistry* 16: 3680–3687.
227. Gelboin HV (1980) Benzo[*a*]pyrene metabolism, activation and carcinogenesis: role and regulation of mixed-function oxidases and related enzymes. *Physiol. Rev.* 60: 1107–1166.
228. Nagata C, Tagashira Y and Kodama M (1974) Metabolic activation of benzo-[*a*]pyrene: significance of the free radical. In: *Chemical Carcinogenesis* [Ts'o POP and DiPaolo JA (eds.)] Part A, pp 87–111, Marcel Dekker, New York.

229. Lorentzen RJ, Caspary WJ, Lesko SA and Ts'o POP (1975) The autooxidation of 6-hydroxybenzo[*a*]pyrene and 6-oxobenzo[*a*]pyrene radical, reactive metabolites of benzo[*a*]pyrene. *Biochemistry* 14: 3970–3977.
230. Lesko SA, Caspary W, Lorentzen RJ and Ts'o POP (1975) Enzymic formation of 6-oxobenzo[*a*]pyrene radical in rat liver homogenates for carcinogenic benzo[*a*]pyrene. *Biochemistry* 14: 3978–3984.
231. Lesko SA, Lorentzen RJ and Ts'o POP (1978) Benzo[*a*]pyrene metabolism: one-electron pathway and the role of nuclear enzymes. In: *Polycyclic Hydrocarbons and Cancer* [Gelboin HV and Ts'o POP (eds.)] Volume I: Environment, Chemistry, and Metabolism, pp 261–269, Academic Press, New York.
232. Buhler DR, Unlü F, Thakker DR, Slaga TJ, Conney AH, Wood AW, Chang RL, Lewin W and Jerina DM (1983) Effect of 6-fluoro substituent on the metabolism and biological activity of benzo[*a*]pyrene. *Cancer Res.* 43: 1541–1549.
233. Cavalieri EL, Rogan EG, Cremonesi P and Devanesan PD (1988) Radical cations as precursors in the metabolic formation of quinones from benzo[*a*]pyrene and 6-fluorobenzo[*a*]pyrene: fluoro substitution as a probe for one-electron oxidation in aromatic substrates. *Biochem. Pharmacol.* 37: 2173–2182.
234. Cavalieri EL, Devanesan PD and Rogan EG (1988) Radical cations in the horseradish peroxidase- and prostaglandin H synthase-mediated metabolism and binding of benzo[*a*]pyrene to deoxyribonucleic acid. *Biochem. Pharmacol.* 37: 2183–2187.
235. Daly JW, Jerina DM and Witkop B (1972) Arene oxides and the NIH shift: the metabolism, toxicity and carcinogenicity of aromatic compounds. *Experientia* 28: 1129–1164.
236. Yang SK, Roller PP, Fu PP, Harvey RG and Gelboin HV (1977) Evidence for a 2,3-epoxide as an intermediate in the microsomal metabolism of benzo[*a*]pyrene to 3-hydroxybenzo[*a*]pyrene. *Biochem. Biophys. Res. Commun.* 77: 1176–1182.
237. Mulder PPJ, Devanesan P, van Alem K, Lodder G, Rogan EG and Cavalieri EL (2003) Fluorobenzo[*a*]pyrenes as probes of the mechanism of cytochrome P450-catalyzed oxygen transfer in aromatic oxygenations. *Free Radic. Biol. Med.* 34: 734–745.
238. Li KM, Todorovic R, Rogan EG, Cavalieri EL, Ariese F, Suh M, Jankowiak R and Small GJ (1995) Identification and quantitation of dibenzo[*a,l*]pyrene-DNA adducts formed by rat liver microsomes *in vitro*: preponderance of depurination adducts. *Biochemistry* 34: 8043–8049.

239. Todorovic R, Ariese F, Devanesan PD, Jankowiak R, Small GJ, Rogan EG and Cavalieri EL (1997) Determination of benzo[*a*]pyrene- and 7,12-dimethylbenz[*a*]anthracene-DNA adducts formed in rat mammary glands. *Chem. Res. Toxicol.* 10: 941–947.
240. Chakravarti D, Mailander P, Franzen J, Higginbotham S, Cavalieri EL and Rogan EG (1998) Detection of dibenzo[*a,l*]pyrene-induced H-*ras* codon 61 mutant gene in preneoplastic SENCAR mouse skin using a new PCR-RFLP method. *Oncogene* 16: 3203–3210.
241. Chakravarti D, Pelling JC, Cavalieri EL and Rogan EG (1995) Relating aromatic hydrocarbon-induced DNA adducts and c-Ha-*ras* mutations in mouse skin papillomas: the role of apurinic sites. *Proc. Natl. Acad. Sci. USA* 92: 10422–10426.
242. Cavalieri EL and Rogan EG (1996) The primary role of apurinic sites in tumor initiation. *Polycycl. Aromat. Compds.* 10: 251–258.
243. Cavalieri EL and Rogan EG (2002) A unified mechanism in the initiation of cancer. *Ann. N.Y. Acad. Sci.* 959: 341–354.
244. Chakravarti D, Mailander PC, Cavalieri EL and Rogan EG (2000) Evidence that error-prone DNA repair converts dibenzo[*a,l*]pyrene-induced depurinating lesions into mutations: formation, clonal proliferation and regression of initiated cells carrying H-*ras* oncogene mutations in early preneoplasia. *Mutat. Res.* 456: 17–32.
245. Lorentzen RJ and Ts'o POP (1977) Benzo[*a*]pyrenedione/benzo[*a*]pyrenediol oxidation-reduction couples and the generation of reactive reduced molecular oxygen. *Biochemistry* 16: 1467–1473.
246. Cerutti P, Shinohara K, Ide ML and Remsen J (1978) Formation and repair of benzo[*a*]pyrene-induced DNA damage in mammalian cells. In: *Polycyclic Hydrocarbons and Cancer* [Gelboin HV and Ts'o POP (eds.)] Volume II: Molecular and Cell Biology, pp 203–212, Academic Press, New York.
247. Chevion M (1988) A site-specific mechanism for free radical induced biological damage: the essential role of redox-active transition metals. *Free Radic. Biol. Med.* 5: 27–37.
248. Blakely WF, Fuciarelli AF, Wegher BJ and Dizdaroglu M (1990) Hydrogen peroxide-induced base damage in deoxyribonucleic acid. *Radiat. Res.* 121: 338–343.
249. Leadon SA, Stampfer MR and Bartley J (1988) Production of oxidative DNA damage during the metabolic activation of benzo[*a*]pyrene in human mammary epithelial cells correlates with cell killing. *Proc. Natl. Acad. Sci. USA* 85: 4365–4368.

250. Frenkel K, Wei L and Wei H (1995) 7,12-Dimethylbenz[*a*]anthracene induces oxidative DNA modification *in vivo*. *Free Radic. Biol. Med.* 19: 373–380.
251. Mauthe RJ, Cook VM, Coffing SL and Baird WM (1995) Exposure of mammalian cell cultures to benzo[*a*]pyrene and light results in oxidative DNA damage as measured by 8-hydroxydeoxyguanosine formation. *Carcinogenesis* 16: 133–137.
252. Kim KB and Lee BM (1997) Oxidative stress to DNA, protein, and antioxidant enzymes (superoxide dismutase and catalase) in rats treated with benzo[*a*]pyrene. *Cancer Lett.* 113: 205–212.
253. Joseph P and Jaiswal AK (1994) NAD(P)H:quinone oxidoreductase$_1$ (DT diaphorase) specifically prevents the formation of benzo[*a*]pyrene quinone-DNA adducts generated by cytochrome P4501A1 and P450 reductase. *Proc. Natl. Acad. Sci. USA* 91: 8413–8417.
254. Joseph P and Jaiswal AK (1998) NAD(P)H:quinone oxidoreductase 1 reduces the mutagenicity of DNA caused by NAD(P)H:P450 reductase-activated metabolites of benzo[*a*]pyrene quinones. *Br. J. Cancer* 77: 709–719.
255. Long II DJ, Waikel RL, Wang XJ, Perlaky L, Roop DR and Jaiswal AK (2000) NAD(P)H:quinone oxidoreductase 1 deficiency increases susceptibility to benzo[*a*]pyrene-induced mouse skin carcinogenesis. *Cancer Res.* 60: 5913–5915.
256. Long II DJ, Waikel RL, Wang XJ, Roop DR and Jaiswal AK (2001) NAD(P)H:quinone oxidoreductase 1 deficiency and increased susceptibility to 7,12-dimethylbenz[*a*]anthracene-induced carcinogenesis in mouse skin. *J. Natl. Cancer Inst.* 93: 1166–1170.
257. Joseph P, Long II DJ, Klein-Szanto AJ and Jaiswal AK (2000) Role of NAD(P)H:quinone oxidoreductase 1 (DT diaphorase) in protection against quinone toxicity. *Biochem. Pharmacol.* 60: 207–214.
258. Berenblum I and Schoental R (1943) The metabolism of B[*a*]P in mice and rats. I. The isolation of a hydroxy and a quinone derivative, and a consideration of their biological significance. *Cancer Res.* 3: 145–150.
259. Yang Y, Griffiths WJ, Nordling M, Nygren J, Möller L, Bergman J, Liepinsh E, Otting G, Gustafsson JA, Rafter J and Sjövall J (2000) Ring opening of benzo[*a*]pyrene in the germ-free rat is a novel pathway for formation of potentially genotoxic metabolites. *Biochemistry* 39: 15585–15591.
260. Nordling MM, Nygren J, Bergman J, Sundberg K and Rafter JJ (2002) Toxicological characterization of a novel *in vivo* benzo[*a*]pyrene metabolite, 7-oxo-benz[*d*]anthracene-3,4-dicarboxylic acid anhydride. *Chem. Res. Toxicol.* 15: 1274–1280.

261. Smithgall TE, Harvey RG and Penning TM (1986) Regio- and stereospecificity of homogenous 3α-hydroxysteroid-dihydrodiol dehydrogenase for *trans*-dihydrodiol metabolites of polycyclic aromatic hydrocarbons. *J. Biol. Chem.* 261: 6184–6191.
262. Smithgall TE, Harvey RG and Penning TM (1988) Spectroscopic identification of *ortho*-quinones as the products of polycyclic aromatic *trans*-dihydrodiol oxidation catalyzed by dihydrodiol dehydrogenase. A potential route of proximate carcinogen metabolism. *J. Biol. Chem.* 263: 1814–1820.
263. Glatt HR, Vogel K, Bentley P and Oesch F (1979) Reduction of benzo[*a*]-pyrene mutagenicity by dihydrodiol dehydrogenase. *Nature* 277: 319–320.
264. Vogel K, Bentley P, Platt KL and Oesch F (1980) Rat liver cytoplasmic dihydrodiol dehydrogenase. Purification to apparent homogeneity and properties. *J. Biol. Chem.* 255: 9621–9625.
265. Penning TM, Mukharji I, Barrows S and Talalay P (1984) Purification and properties of a 3α-hydroxysteroid dehydrogenase of rat liver cytosol and its inhibition by anti-inflammatory drugs. *Biochem. J.* 222: 601–611.
266. Penning TM, Burczynski ME, Hung CF, McCoull K, Palackal NT and Tsuruda LS (1999) Dihydrodiol dehydrogenases and polycyclic aromatic hydrocarbon activation: generation of reactive and redox active *o*-quinones. *Chem. Res. Toxicol.* 12: 1–18.
267. Hyndman D, Bauman DR, Heredia VV and Penning TM (2003) The aldo-keto reductase superfamily homepage (www.med.upenn.edu/akr). *Chem.-Biol. Interact.* 143–144: 621–631.
268. Burczynski ME, Harvey RG and Penning TM (1998) Expression and characterization of four recombinant human dihydrodiol dehydrogenase isoforms: oxidation of *trans*-7,8-dihydroxy-7,8-dihydrobenzo[*a*]pyrene to the activated *o*-quinone metabolite of benzo[*a*]pyrene-7,8-dione. *Biochemistry* 37: 6781–6790.
269. Bolton JL, Trush MA, Penning TM, Dryhurst G and Monks TJ (2000) Role of quinones in toxicology. *Chem. Res. Toxicol.* 13: 135–160.
270. Palackal NT, Lee SH, Harvey RG, Blair IA and Penning TM (2002) Activation of polycyclic aromatic hydrocarbon *trans*-dihydrodiol proximate carcinogens by human aldo-keto reductase (AKR1C) enzymes and their functional overexpression in human lung carcinoma (A549) cells. *J. Biol. Chem.* 277: 24799–24808.
271. Burczynski ME, Lin HK and Penning TM (1999) Isoform-specific induction of a human aldo-keto reductase by polycyclic aromatic hydrocarbons (PAHs), electrophiles, and oxidative stress: implications for the alternative pathway

of PAH activation catalyzed by human dihydrodiol dehydrogenase. *Cancer Res.* 59: 607–614.

272. Palackal NT, Burczynski ME, Harvey RG and Penning TM (2001) The ubiquitous aldehyde reductase (AKR1A1) oxidizes proximate carinogen *trans*-dihydrodiols to *o*-quinones: potential role in polycyclic aromatic hydrocarbon activation. *Biochemistry* 40: 10901–10910.
273. Palackal NT, Burczynski ME, Harvey RG and Penning TM (2001) Metabolic activation of polycyclic aromatic hydrocarbon *trans*-dihydrodiols by ubiquitously expressed aldehyde reductase (AKR1A1). *Chem.-Biol. Interact.* 130–132: 815–824.
274. Penning TM, Ohnishi ST, Ohnishi T and Harvey RG (1996) Generation of reactive oxygen species during the enzymatic oxidation of polycyclic aromatic hydrocarbon *trans*-dihydrodiols catalyzed by dihydrodiol dehydrogenase. *Chem. Res. Toxicol.* 9: 84–92.
275. Flowers-Geary L, Harvey RG and Penning TM (1992) Examination of polycyclic aromatic hydrocarbon *o*-quinones produced by dihydrodiol dehydrogenase as substrates for redox-cycling in rat liver. *Biochem. (Life Sci. Adv.)* 11: 49–58.
276. Shibutani S, Takeshita M and Grollman AP (1991) Insertion of specific bases during DNA synthesis past the oxidation-damaged base 8-oxodG. *Nature* 349: 431–434.
277. Cheng KC, Cahill DS, Kasai H, Nishimura S and Loeb LA (1992) 8-Hydroxyguanine, an abundant form of oxidative DNA damage, causes $G \rightarrow T$ and $A \rightarrow C$ substitutions. *J. Biol. Chem.* 267: 166–172.
278. Breen AP and Murphy JA (1995) Reactions of oxy radicals with DNA. *Free Radic. Biol. Med.* 18: 1033–1077.
279. Flowers L, Ohnishi ST and Penning TM (1997) DNA strand scission by polycyclic aromatic hydrocarbon *o*-quinones: role of reactive oxygen species, Cu(II)/Cu(I) redox cycling, and *o*-semiquinone anion radicals. *Biochemistry* 36: 8640–8648.
280. Yu D, Berlin JA, Penning TM and Field J (2002) Reactive oxygen species generated by PAH *o*-quinones cause change-in-function mutations in p53. *Chem. Res. Toxicol.* 15: 832–842.
281. Yamamoto K and Kawanishi S (1989) Hydroxyl free radical is not the main active species in site-specific DNA damage induced by copper (II) ion and hydrogen peroxide. *J. Biol. Chem.* 264: 15435–15440.
282. Flowers L, Bleczinski WF, Burczynski ME, Harvey RG and Penning TM (1996) Disposition and biological activity of benzo[*a*]pyrene-7,8-dione. A

genotoxic metabolite generated by dihydrodiol dehydrogenase. *Biochemistry* 35: 13664–13672.

283. Chaudhary AK, Nokubo M, Reddy GR, Yeola SN, Morrow JD, Blair IA and Marnett LJ (1994) Detection of endogenous malondialdehyde-deoxyguanosine adducts in human liver. *Science* 265: 1580–1582.
284. Plastaras JP, Dedon PC and Marnett LJ (2002) Effects of DNA structure on oxopropenylation by the endogenous mutagens malondialdehyde and base propenal. *Biochemistry* 41: 5033–5042.
285. Marnett LJ (2002) Oxy radicals, lipid peroxidation and DNA damage. *Toxicology* 181–182: 219–222.
286. Murty VS and Penning TM (1992) Characterization of mercapturic acid and glutathionyl conjugates of benzo[*a*]pyrene-7,8-dione by two-dimensional NMR. *Bioconjug. Chem.* 3: 218–224.
287. Shou M, Harvey RG and Penning TM (1993) Reactivity of benzo[*a*]pyrene-7,8-dione with DNA. Evidence for the formation of deoxyguanosine adducts. *Carcinogenesis* 14: 475–482.
288. McCoull KD, Rindgen D, Blair IA and Penning TM (1999) Synthesis and characterization of polycyclic aromatic hydrocarbon *o*-quinone depurinating N7-guanine adducts. *Chem. Res. Toxicol.* 12: 237–246.
289. Penning TM, Palackal NT, Blair IA and Harvey RG (2002) The aldo-keto reductases and polycyclic aromatic hydrocarbon activation. *Polycycl. Aromat. Compds.* 22: 791–800.
290. Flowers-Geary L, Bleczinski W, Harvey RG and Penning TM (1996) Cytotoxicity and mutagenicity of polycyclic aromatic hydrocarbon *ortho*-quinones produced by dihydrodiol dehydrogenase. *Chem.-Biol. Interact.* 99: 55–72.
291. Hecht SS, Amin S, Rivenson A and Hoffmann D (1979) Tumor initiating activity of 5,11-dimethylchrysene and the structural requirements favoring carcinogenicity of methylated polynuclear aromatic hydrocarbons. *Cancer Lett.* 8: 65–70.
292. Amin S, Misra B, Braley J and Hecht SS (1991) Comparative tumorigenicity in newborn mice of chrysene- and 5-alkylchrysene-1,2-diol-3,4-epoxides. *Cancer Lett.* 58: 115–118.
293. Geacintov NE, Cosman M, Hingerty BE, Amin S, Broyde S and Patel DJ (1997) NMR solution structures of stereoisomeric covalent polycyclic aromatic carcinogen-DNA adducts: principles, patterns, and diversity. *Chem. Res. Toxicol.* 10: 111–146.

294. Flesher JW and Sydnor KL (1971) Carcinogenicity of derivatives of 7,12-dimethylbenz[*a*]anthracene. *Cancer Res.* 31: 1951–1954.
295. Flesher JW and Sydnor KL (1973) Possible role of 6-hydroxymethylbenzo-[*a*]pyrene as a proximate carcinogen of benzo[*a*]pyrene and 6-methylbenzo[*a*]pyrene. *Int. J. Cancer* 11: 433–437.
296. Watabe T, Ishizuka T, Isobe M and Ozawa N (1982) A 7-hydroxymethyl sulfate ester as an active metabolite of 7,12-dimethylbenz[*a*]anthracene. *Science* 215: 403–405.
297. Miller EC and Miller JA (1986) Electrophilic sulfuric acid ester metabolites as ultimate carcinogens. *Adv. Exp. Med. Biol.* 197: 583–595.
298. Baird WM, Dipple A, Grover PL, Sims P and Brookes P (1973) Studies on the formation of hydrocarbon-deoxyribonucleoside products by the binding of derivatives of 7-methylbenz[*a*]anthracene to DNA in aqueous solution and in mouse embryo cells in culture. *Cancer Res.* 33: 2386–2392.
299. Rayman MP and Dipple A (1973) Structure and activity in chemical carcinogenesis. Comparison of the reactions of 7-bromomethylbenz[*a*]anthracene and 7-bromomethyl-12-methylbenz[*a*]anthracene with deoxyribonucleic acid *in vitro*. *Biochemistry* 12: 1202–1207.
300. Wettstein JG, Chien MT and Flesher JW (1979) DNA binding and mutagenicity of 7-substituted derivatives of benz[*a*]anthracene. *Drug Chem. Toxicol.* 2: 383–396.
301. Tay LK, Sydnor KL and Flesher JW (1979) Binding of 6-hydroxymethylbenzo[*a*]pyrene and 6-acetoxymethylbenzo[*a*]pyrene to DNA. *Chem.-Biol. Interact.* 25: 35–44.
302. Dipple A, Moschel RC and Hudgins WR (1982) Selectivity of alkylation and aralkylation of nucleic acid components. *Drug Metab. Rev.* 13: 249–268.
303. Rogan EG, Roth RW, Katomski-Beck PA, Laubscher JR and Cavalieri EL (1980) Nonenzymatic ATP-mediated binding of hydroxymethyl derivatives of aromatic hydrocarbons to DNA. *Chem.-Biol. Interact.* 31: 51–63.
304. Dipple A, Tomaszewski JE, Moschel RC, Bigger CA, Nebzydoski JA and Egan M (1979) Comparison of metabolism-mediated binding to DNA of 7-hydroxymethyl-12-methylbenz[*a*]anthracene and 7,12-dimethylbenz[*a*]anthracene. *Cancer Res.* 39: 1154–1158.
305. Watabe T (1983) Metabolic activation of 7,12-dimethylbenz[*a*]anthracene (DMBA) and 7-methylbenz[*a*]anthracene (7-MBA) by rat liver P-450 and sulfotransferase. *J. Toxicol. Sci.* 8: 119–131.

306. Watabe T, Fujieda T, Hiratsuka A, Ishizuka T, Hakamata Y and Ogura K (1985) The carcinogen, 7-hydroxymethyl-12-methylbenz[*a*]anthracene, is activated and covalently binds to DNA via a sulphate ester. *Biochem. Pharmacol.* 34: 3002–3005.
307. Watabe T, Hakamata Y, Hiratsuka A and Ogura K (1986) A 7-hydroxymethyl sulphate ester as an active metabolite of the carcinogen, 7-hydroxymethylbenz[*a*]anthracene. *Carcinogenesis* 7: 207–214.
308. Okuda H, Hiratsuka A, Nojima H and Watabe T (1986) A hydroxymethylsulfate ester as an active metabolite of the carcinogen, 5-hydroxymethylchrysene. *Biochem. Pharmacol.* 35: 535–538.
309. Watabe T, Ogura K, Okuda H and Hiratsuka A (1989) Hydroxymethylsulfate esters as reactive metabolites of the carcinogen, 7-methyl- and 7,12-dimethylbenz[*a*]anthracenes and 5-methylchrysene. In: *Xenobiotic Metabolism and Disposition* [Kato R, Estabrook RW and Cayen MN (eds.)] pp 393–400, Taylor and Francis, London.
310. Surh YJ, Liem A, Miller EC and Miller JA (1989) Metabolic activation of the carcinogen 6-hydroxymethylbenzo[*a*]pyrene: formation of an electrophilic sulfuric acid ester and benzylic DNA adducts in rat liver *in vivo* and reactions *in vitro*. *Carcinogenesis* 10: 1519–1528.
311. Surh YJ, Liem A, Miller EC and Miller JA (1990) The strong hepatocarcinogenicity of the electrophilic and mutagenic metabolite 6-sulfooxymethylbenzo[*a*]pyrene and its formation of benzylic DNA adducts in the livers of infant male B6C3F1 mice. *Biochem. Biophys. Res. Commun.* 172: 85–91.
312. Surh YJ, Liem A, Miller EC and Miller JA (1991) 7-Sulfooxymethyl-12-methylbenz[*a*]anthracene is an electrophilic mutagen, but does not appear to play a role in carcinogenesis by 7,12-dimethylbenz[*a*]anthracene or 7-hydroxymethylbenz[*a*]anthracene. *Carcinogenesis* 12: 339–347.
313. Flesher JW, Horn J and Lehner AF (1997) 6-Sulfooxymethylbenzo[*a*]pyrene is an ultimate electrophilic and carcinogenic form of the intermediary metabolite 6-hydroxymethylbenzo[*a*]pyrene. *Biochem. Biophys. Res. Commun.* 234: 554–558.
314. Dewhurst F, Kitchen DA and Calcutt G (1972) The carcinogenicity of some 6-substituted benzo[*a*]pyrene derivatives in mice. *Br. J. Cancer* 26: 506–508.
315. Cavalieri E, Roth R, Grandjean C, Althoff J, Patil K, Liakus S and Marsh S (1978) Carcinogenicity and metabolic profiles of 6-substituted benzo[*a*]pyrene derivatives on mouse skin. *Chem.-Biol. Interact.* 22: 53–67.

316. Sydnor KL, Bergo CH and Flesher JW (1980) Effect of various substituents in the 6-position on the relative carcinogenic activity of a series of benzo[*a*]pyrene derivatives. *Chem.-Biol. Interact.* 29: 159–167.
317. Flesher JW, Myers SR and Blake JW (1986) Bioalkylation of polynuclear aromatic hydrocarbons: a predictor of carcinogenic activity. In: *Polynuclear Aromatic Hydrocarbons: Chemistry, Characterization and Carcinogenesis* [Cooke M and Dennis AJ (eds.)] pp 271–284, Battelle Press, Columbus, OH.
318. Flesher JW, Stansbury KH and Sydnor KL (1982) S-adenosyl-L-methionine is a carbon donor in the conversion of benzo[*a*]pyrene to 6-hydroxymethylbenzo[*a*]pyrene by rat liver S-9. *Cancer Lett.* 16: 91–94.
319. Flesher JW, Myers SR and Stansbury KH (1990) The site of substitution of the methyl group in the bioalkylation of benzo[*a*]pyrene. *Carcinogenesis* 11: 493–496.
320. Flesher JW, Myers SR and Blake JW (1984) Biosynthesis of the potent carcinogen 7,12-dimethylbenz[*a*]anthracene. *Cancer Lett.* 24: 335–343.
321. Myers SR, Blake JW and Flesher JW (1988) Bioalkylation and biooxidation of anthracene, *in vitro* and *in vivo*. *Biochem. Biophys. Res. Commun.* 151: 1441–1445.
322. Flesher JW and Myers SR (1990) Bioalkylation of benz[*a*]anthracene as a biochemical probe for carcinogenic activity. Lack of bioalkylation in a series of six noncarcinogenic polynuclear aromatic hydrocarbons. *Drug Metab. Disp.* 18: 163–167.
323. Myers SR and Flesher JW (1991) Metabolism of chrysene, 5-methylchrysene, 6-methylchrysene and 5,6-dimethylchrysene in rat liver cytosol, *in vitro*, and in rat subcutaneous tissue, *in vivo*. *Chem-Biol. Interact.* 77: 203–221.
324. Myers SR and Flesher JW (1991) Bioalkylation of benz[*a*]anthracene, 7-methylbenz[*a*]anthracene, and 12-methylbenz[*a*]anthracene in rat lung cytosol preparations. *Biochem. Pharmacol.* 41: 1683–1689.
325. Stansbury KH, Flesher JW and Gupta RC (1994) Mechanism of aralkyl-DNA adduct formation from benzo[*a*]pyrene *in vivo*. *Chem. Res. Toxicol.* 7: 254–259.
326. Rice JE, Rivenson A, Braley J and LaVoie EJ (1987) Methylated derivatives of pyrene and fluorene: evaluation of genotoxicity in the hepatocyte/DNA repair test and tumorigenic activity in newborn mice. *J. Toxicol. Environ. Health* 21: 525–532.

327. Glatt HR, Werle-Schneider G, Enders N, Monnerjahn S, Pudil J, Czich A, Seidel A and Schwarz M (1994) 1-Hydroxymethylpyrene and its sulfuric acid ester: toxicological effects *in vitro* and *in vivo* and metabolic aspects. *Chem.-Biol. Interact.* 92: 305–319.
328. Engst W, Landsiedel R, Hermersdörfer H, Doehmer J and Glatt HR (1999) Benzylic hydroxylation of 1-methylpyrene and 1-ethylpyrene by human and rat cytochromes P450 individually expressed in V79 Chinese hamster cells. *Carcinogenesis* 20: 1777–1785.
329. Glatt HR (2000) Sulfotransferases in the bioactivation of xenobiotics. *Chem.-Biol. Interact.* 129: 141–170.
330. Glatt HR, Boeing H, Engelke CEH, Ma L, Kuhlow A, Pabel U, Pomplun D, Teubner W and Meinl W (2001) Human cytosolic sulphotransferases: genetics, characteristics, toxicological aspects. *Mutat. Res.* 482: 27–40.
331. Surh YJ, Blomquist JC, Liem A and Miller JA (1990) Metabolic activation of 9-hydroxymethyl-10-methylanthracene and 1-hydroxymethylpyrene to electrophilic, mutagenic and tumorigenic sulfuric acid esters by rat hepatic sulfotransferase activity. *Carcinogenesis* 11: 1451–1460.
332. Glatt HR, Henschler R, Phillips DH, Blake JW, Steinberg P, Seidel A and Oesch F (1990) Sulfotransferase-mediated chlorination of 1-hydroxymethylpyrene to a mutagen capable of penetrating indicator cells. *Environ. Health Perspect.* 88: 43–48.
333. Surh YJ, Blomquist JC and Miller JA (1991) Activation of 1-hydroxymethylpyrene to an electrophilic and mutagenic metabolite by rat hepatic sulfotransferase activity. *Adv. Exp. Med. Biol.* 283: 383–391.
334. Glatt HR, Bartsch I, Christoph S, Coughtrie MW, Falany CN, Hagen M, Landsiedel R, Pabel U, Phillips DH, Seidel A and Yamazoe Y (1998) Sulfotransferase-mediated activation of mutagens studied using heterologous expression systems. *Chem.-Biol. Interact.* 109: 195–219.
335. Ma L, Kuhlow A and Glatt HR (2002) Ethanol enhances the activation of 1-hydroxymethylpyrene to DNA adduct-forming species in the rat. *Polycycl. Aromat. Compds.* 22: 933–946.
336. Horn J, Flesher JW and Lehner AF (1996) 1-Sulfooxymethylpyrene is an electrophilic mutagen and ultimate carcinogen of 1-methyl- and 1-hydroxymethylpyrene. *Biochem. Biophys. Res. Commun.* 228: 105–109.
337. Nemoto N (1981) Glutathione, glucuronide, and sulfate transferase in polycyclic aromatic hydrocarbon metabolism. In: *Polycyclic Hydrocarbons and Cancer* [Gelboin HV and Ts'o POP (eds.)] Volume III, pp 213–258, Academic Press, New York.

338. Mulder GJ and Jakoby WB (1990) Sulfation. In: *Conjugation Reactions in Drug Metabolism: An Integrated Approach* [Mulder GJ (ed.)] pp 107–162, Taylor & Francis, London.
339. Mulder GJ, Coughtrie MWH and Burchell B (1990) Glucuronidation. In: *Conjugation Reactions in Drug Metabolism: An Integrated Approach* [Mulder GJ (ed.)] pp 51–106, Taylor & Francis, London.
340. Tephly TR and Burchell B (1990) UDP-glucuronosyltransferases: a family of detoxifying enzymes. *Trends Pharmacol. Sci.* 11: 276–279.
341. Bock KW (1991) Roles of UDP-glucuronosyltransferases in chemical carcinogenesis. *Crit. Rev. Biochem. Mol. Biol.* 26: 129–150.
342. Ritter JK (2000) Roles of glucuronidation and UDP-glucuronosyltransferases in xenobiotic bioactivation reactions. *Chem.-Biol. Interact.* 129: 171–193.
343. Nemoto N, Takyama S, Nagao M and Umezawa K (1978) Modification of the mutagenicity of benzo[*a*]pyrene on bacteria by substrates of enzymes producing water soluble conjugates. *Toxicol. Lett.* 2: 205–211.
344. Owens IS, Koteen GM and Legraverend C (1979) Mutagenesis of certain benzo[*a*]pyrene phenols *in vitro* following further metabolism by mouse liver. *Biochem. Pharmacol.* 28: 1615–1622.
345. Hu Z and Wells PG (1992) *In vitro* and *in vivo* biotransformation and covalent binding of benzo[*a*]pyrene in Gunn and RHA rats with a genetic deficiency in bilirubin uridine diphosphate-glucuronosyltransferase. *J. Pharmacol. Exp. Ther.* 263: 334–342.
346. Hu Z and Wells PG (1994) Modulation of benzo[*a*]pyrene bioactivation by glucuronidation in lymphocytes and hepatic microsomes from rats with a hereditary deficiency in bilirubin UDP-glucuronosyltransferase. *Toxicol. Appl. Pharmacol.* 127: 306–313.
347. Vienneau DS, DeBoni U and Wells PG (1995) Potential genoprotective role for UDP-glucuronosyltransferases in chemical carcinogenesis: initiation of micronuclei by benzo[*a*]pyrene and benzo[*e*]pyrene in UDP-glucuronosyltransferase-deficient cultured rat skin fibroblasts. *Cancer Res.* 55: 1045–1051.
348. Bock KW, von Clausbruch UC, Kaufmann R, Lilienblum W, Oesch F, Pfeil H and Platt KL (1980) Functional heterogeneity of UDP-glucuronosyltransferase in rat tissues. *Biochem. Pharmacol.* 29: 495–500.
349. Bock KW, Lipp HP and Bock-Hennig BS (1990) Induction of drug-metabolizing enzymes by xenobiotics. *Xenobiotica* 20: 1101–1111.

350. Bock KW, Gschaidmeier H, Seidel A, Baird S and Burchell B (1992) Mono- and diglucuronide formation from chrysene and benzo[*a*]pyrene phenols by 3-methylcholanthrene-inducible phenol UDP-glucuronosyltransferase (UGT1A1). *Mol. Pharmacol.* 42: 613–618.
351. Lilienblum W, Bock-Henning BS and Bock KW (1985) Protection against toxic redox cycles between benzo[*a*]pyrene-3,6-quinone and its quinol by 3-methylcholanthrene-inducible formation of the quinol mono- and diglucuronide. *Mol. Pharmacol.* 27: 451–458.
352. Gschaidmeier H, Seidel A, Burchell B and Bock KW (1995) Formation of mono- and diglucuronides and other glycosides of benzo[*a*]pyrene-3,6-quinol by V79 cell-expressed human phenol UDP-glucuronosyltransferases of the UGT1 gene complex. *Biochem. Pharmacol.* 49: 1601–1606.
353. Jin CJ, Miners JO, Burchell B and Mackenzie PI (1993) The glucuronidation of hydroxylated metabolites of benzo[*a*]pyrene and 2-acetylaminofluorene by cDNA-expressed human UDP-glucuronosyltransferases. *Carcinogenesis* 14: 2637–2639.
354. Nemoto N and Gelboin HV (1976) Enzymatic conjugation of benzo[*a*]pyrene oxides, phenols and dihydrodiols with UDP-glucuronic acid. *Biochem. Pharmacol.* 25: 1221–1226.
355. Grove AD, Kessler FK, Metz RP and Ritter JK (1997) Identification of a rat oltripaz-inducible UDP-glucuronosyltransferase (UGT1A7) with activity towards benzo[*a*]pyrene-7,8-dihydrodiol. *J. Biol. Chem.* 272: 1621–1627.
356. Fang JL, Beland FA, Doerge DR, Wiener D, Guillemette C, Marques MM and Lazarus P (2002) Characterization of benzo[*a*]pyrene-*trans*-7,8-dihydrodiol glucuronidation by human tissue microsomes and overexpressed UDP-glucuronosyltransferase enzymes. *Cancer Res.* 62: 1978–1986.
357. Boyland E and Sims P (1957) Metabolism of polycyclic compounds. 11. The conversion of naphthalene into 2-hydroxy-1-naphthyl sulphate in the rabbit. *Biochem. J.* 66: 38–40.
358. Sims P (1962) Metabolism of polycyclic compounds. 19. The metabolism of phenanthrene in rabbits and rats: phenols and sulphuric esters. *Biochem. J.* 84: 558–563.
359. Richter P, Kauffman FC and Zaleski J (1994) Predominance of glucuronidation over sulfation in metabolism of 1-hydroxybenzo[*a*]pyrene by isolated rat hepatocytes. *Toxicol. Lett.* 74: 79–90.
360. Surh YJ, Kwon H and Tannenbaum SR (1993) Sulfotransferase-mediated activation of 4-hydroxy- and 3,4-dihydroxy-3,4-dihydrocyclopenta[*c,d*]-pyrene, major metabolites of cyclopenta[*c,d*]pyrene. *Cancer Res.* 53: 1017–1022.

361. Surh YJ and Tannenbaum SR (1995) Sulfotransferase-mediated activation of 7,8,9,10-tetrahydro-7-ol, 7,8-dihydrodiol, and 7,8,9,10-tetraol derivatives of benzo[*a*]pyrene. *Chem. Res. Toxicol.* 8: 693–698.
362. Glatt HR, Pauly K, Frank H, Seidel A, Oesch F, Harvey RG and Werle-Schneider G (1994) Substrate-dependent sex differences in the activation of benzylic alcohols to mutagens by hepatic sulfotransferases of the rat. *Carcinogenesis* 15: 2605–2611.
363. Nemoto N, Takayma S and Gelboin HV (1977) Enzymatic conversion of benzo[*a*]pyrene phenols, dihydrodiols and quinones to sulfate conjugates. *Biochem. Pharmacol.* 26: 1825–1829.
364. Nemoto N, Takayama S and Gelboin HV (1978) Sulfate conjugation of benzo[*a*]pyrene metabolites and derivatives. *Chem.-Biol. Interact.* 23: 19–30.
365. Autrup H (1979) Separation of water-soluble metabolites of benzo[*a*]pyrene formed by cultured human colon. *Biochem. Pharmacol.* 28: 1727–1730.
366. Boyland E (1986) The metabolism of foreign compounds and the induction of cancer. *Xenobiotica* 16: 899–913.
367. Ramesh A, Inyang F, Hood DB, Archibong AE, Knuckles ME and Nyanda AM (2001) Metabolism, bioavailability, and toxicokinetics of benzo[*a*]pyrene in F-344 rats following oral administration. *Exp. Toxicol. Pathol.* 53: 275–290.
368. Coles B and Ketterer B (1990) The role of glutathione and glutathione transferases in chemical carcinogenesis. *Crit. Rev. Biochem. Mol. Biol.* 25: 47–70.
369. Armstrong RN (1991) Glutathione *S*-transferases: reaction mechanism, structure, and function. *Chem. Res. Toxicol.* 4: 131–140.
370. Bourne MC and Young L (1934) The metabolism of naphthalene in rabbits. *Biochem. J.* 28: 803–808.
371. Boyland E and Levi AA (1936) Metabolism of polycyclic compounds. III. Anthrylmercapturic acid. *Biochem. J.* 30: 1225–1227.
372. Booth J, Boyland E and Sims P (1961) An enzyme from rat liver catalysing conjugations with glutathione. *Biochem. J.* 79: 516–524.
373. Hayes JD and Strange RC (2000) Glutathione *S*-transferase polymorphisms and their biological consequences. *Pharmacology* 61: 154–166.
374. Grover PL (1974) K-region epoxides of polycyclic hydrocarbons: formation and further metabolism by rat-lung preparations. *Biochem. Pharmacol.* 23: 333–343.
375. Keysell GR, Booth J and Sims P (1975) Glutathione conjugates as metabolites of benz[*a*]anthracene. *Xenobiotica* 5: 439–448.

376. Nemoto N, Gelboin HV, Habig WH, Ketley JN and Jakoby WB (1975) K-region benzo[*a*]pyrene-4,5-oxide is conjugated by homogeneous glutathione *S*-transferases. *Nature* 255: 512.
377. Glatt HR, Friedberg T, Grover PL, Sims P and Oesch F (1983) Inactivation of a diol-epoxide and a K-region epoxide with high efficiency by glutathione transferase X. *Cancer Res.* 43: 5713–5717.
378. Jernström B, Martinez M, Meyer DJ and Ketterer B (1985) Glutathione conjugation of the carcinogenic and mutagenic electrophile (±)-7β-8α-dihydroxy-9α,10β-oxy-7,8,9,10-tetrahydrobenzo[*a*]pyrene catalyzed by purified rat liver glutathione transferases. *Carcinogenesis* 6: 85–89.
379. Robertson IGC and Jernström B (1986) The enzymatic conjugation of glutathione with bay-region diolepoxides of benzo[*a*]pyrene, benz[*a*]anthracene and chrysene. *Carcinogenesis* 7: 1633–1636.
380. Hu X, Xia H, Srivastava SK, Herzog C, Awasthi YC, Ji X, Zimniak P and Singh SV (1997) Activity of four allelic forms of glutathione *S*-transferase hGSTP1-1 for diol epoxides of polycyclic aromatic hydrocarbons. *Biochem. Biophys. Res. Commun.* 238: 397–402.
381. Hayakawa T, Udenfriend S, Yagi H and Jerina DM (1975) Substrate and inhibitors of hepatic glutathione *S*-epoxide transferase. *Arch. Biochem. Biophys.* 170: 438–451.
382. Nemoto N and Gelboin HV (1975) Assay and properties of glutathione-*S*-benzo[*a*]pyrene-4,5-oxide transferase. *Arch. Biochem. Biophys.* 170: 739–742.
383. Waterfall JF and Sims P (1972) Epoxy derivatives of aromatic polycyclic hydrocarbons. *Biochem. J.* 128: 265–277.
384. Jernström B, Funk M, Frank H, Mannervik B and Seidel A (1996) Glutathione *S*-transferase A1–1-catalysed conjugation of bay and fjord region diol epoxides of polycyclic aromatic hydrocarbons with glutathione. *Carcinogenesis* 17: 1491–1498.
385. Sundberg K, Widersten M, Seidel A, Mannervik B and Jernström B (2002) Glutathione conjugation of bay- and fjord-region diol epoxides of polycyclic aromatic hydrocarbons by glutathione transferases M1-1 and P1-1. *Chem. Res. Toxicol.* 10: 1221–1227.
386. Fields WR, Morrow CS, Doss AJ, Sundberg K, Jernström B and Townsend AJ (1998) Overexpression of stably transfected human glutathione *S*-transferase P1-1 protects against DNA damage by benzo[*a*]pyrene diol-epoxide in human T47D cells. *Mol. Pharmacol.* 54: 298–304.
387. Hu X, Herzog C, Zimniak P and Singh SV (1999) Differential protection against benzo[*a*]pyrene-7,8-dihydrodiol-9,10-epoxide-induced DNA

damage in HepG2 cells stably transfected with allelic variants of pi class human glutathione *S*-transferase. *Cancer Res.* 59: 2358–2362.

388. Sundberg K, Dreij K, Seidel A and Jernström B (2002) Glutathione conjugation and DNA adduct formation of dibenzo[*a,l*]pyrene and benzo[*a*]pyrene diol epoxides in V79 cells stably expressing different human glutathione transferases. *Chem. Res. Toxicol.* 15: 170–179.
389. Henderson CJ, Smith AG, Ure J, Brown K, Bacon EJ and Wolf CR (1998) Increased skin tumorigenesis in mice lacking pi class glutathione *S*-transferases. *Proc. Natl. Acad. Sci. USA* 95: 5275–5280.
390. Nesnow S, Ross JA, Mass MJ and Stoner GD (1998) Mechanistic relationship between DNA adducts, oncogene mutations, and lung tumorigenesis in strain A mice. *Exp. Lung Res.* 24: 395–305.
391. Ross JA and Nesnow S (1999) Polycyclic aromatic hydrocarbons: correlations between DNA adducts and *ras* oncogene mutations. *Mutat. Res.* 424: 155–166.
392. Denissenko MF, Pao A, Tang MS and Pfeifer GP (1996) Preferential formation of benzo[*a*]pyrene adducts at lung cancer mutational hotspots in *p53*. *Science* 274: 430–432.
393. Pfeifer GP and Denissenko MF (1998) Formation and repair of DNA lesions in the *p53* gene: relation to cancer mutations. *Environ. Mol. Mutagen.* 31: 197–205.
394. Colapietro AM, Goodell AL and Smart RC (1993) Characterization of benzo[*a*]pyrene-initiated mouse skin papillomas for Ha-*ras* mutations and protein kinase C levels. *Carcinogenesis* 14: 2289–2295.
395. Quintanilla M, Brown K, Ramsden M and Balmain A (1986) Carcinogen-specific mutation and amplification of Ha-*ras* during mouse skin carcinogenesis. *Nature* 322: 78–80.
396. Strauss BS (1991) The 'A rule' of mutagen specificity: a consequence of DNA polymerase bypass of non-instructional lesions? *Bioessays* 13: 79–84.
397. Autrup H and Seremet T (1986) Excretion of benzo[*a*]pyrene-Gua adduct in the urine of benzo[*a*]pyrene-treated rats. *Chem.-Biol. Interact.* 60: 217–226.
398. Rogan EG, RamaKrishna NV, Higginbotham S, Cavalieri EL, Jeong H, Jankowiak R and Small GJ (1990) Identification and quantitation of 7-(benzo[*a*]pyren-6-yl)guanine in the urine and feces of rats treated with benzo[*a*]pyrene. *Chem. Res. Toxicol.* 3: 441–444.
399. Roberts KP, Lin CH, Singhal M, Casale GP, Small GJ and Jankowiak R (2000) On-line identification of depurinating DNA adducts in human urine by capillary electrophoresis-fluorescence line narrowing spectroscopy. *Electrophoresis* 21: 799–806.

400. Casale GP, Singhal M, Bhattacharya S, RamaNathan R, Roberts KP, Barbacci DC, Zhao J, Jankowiak R, Gross ML, Cavalieri EL, Small GJ, Rennard SI, Mumford JL and Shen M (2001) Detection and quantification of depurinated benzo[*a*]pyrene-adducted DNA bases in the urine of cigarette smokers and women exposed to household coal smoke. *Chem. Res. Toxicol.* 14: 192–201.
401. Nakamura J and Swenberg JA (1999) Endogenous apurinic/apyrimidinic sites in genomic DNA of mammalian tissues. *Cancer Res.* 59: 2522–2526.
402. Atamna H, Cheung I and Ames BN (2000) A method for detecting abasic sites in living cells: age-dependent changes in base excision repair. *Proc. Natl. Acad. Sci. USA* 97: 686–691.
403. Braithwaite E, Wu X and Wang Z (1999) Repair of DNA lesions: mechanisms and relative repair efficiencies. *Mutat. Res.* 424: 207–219.
404. Lindahl T (2000) Suppression of spontaneous mutagenesis in human cells by DNA base excision-repair. *Mutat. Res.* 462: 129–135.
405. Braithwaite E, Wu X and Wang Z (1998) Repair of DNA lesions induced by polycyclic aromatic hydrocarbons in human cell-free extracts: involvement of two excision repair mechanisms *in vitro*. *Carcinogenesis* 19: 1239–1246.

3 Biomonitoring of Polycyclic Aromatic Hydrocarbons — Human Exposure

Albrecht Seidel

Biochemical Institute for Environmental Carcinogens,
Prof. Dr. Gernot Grimmer-Foundation, Grosshansdorf, Germany
E-mail: albrecht.seidel@biu-grimmer.de

3.1 Introduction

The carcinogenic activity of polycyclic aromatic hydrocarbons (PAHs) in laboratory animals and men is well documented.[1–4] The biological activity of PAHs can be deduced from a number of working place studies and, in particular, from effect balance analyses that have been conducted with some environmentally relevant matrices including automobile exhaust (Otto motor) condensate, diesel engine exhaust extract, tobacco smoke condensate (sidestream), flue gas of hard coal fired residence furnaces, and used engine oil. The latter have shown that the PAH containing fractions in these matrices are responsible for 70–90% of its carcinogenic activity.[5–12] Due to their formation during incomplete combustion processes of organic material, PAHs are widespread in the environment and are occasionally detected in extraordinary high concentrations at certain working places such as cokeries. Therefore, these compounds pose a substantial health risk for humans. Strikingly high incidences for lung and skin cancer observed for PAH-exposed individuals at working places have consequently led to periodical exposure measurements according to regulations. However, even workers at identical working places might be exposed to substantially different PAH levels due to differences in their breathing behaviour and the observance of personal precautions, e.g., usage of a breathing mask.[13,14]

In comparison to ambient air measurements of PAH levels at working places ('ambient monitoring'), determination of metabolites of incorporated PAHs ('biomonitoring') that are mainly formed in liver, lung, skin, and kidney by various enzymes (see Chapter 2) was proven to be a more exact measure of the actual and individual overall PAH body burden. Monitoring of this overall body burden must be differentiated from 'effect monitoring' that determines the interaction between reactive metabolites and endogenous macromolecules such as DNA and/or proteins (see Chapter 4). In the past, monitoring of the PAH body burden in humans was based on determination of urinary excreted 1-hydroxypyrene (1-OHP) and of its glucuronide (Figure 3.1). 1-OHP is the main metabolite of pyrene and a major constituent of PAH profiles detected in the environment and at working places. In more recent years, additional PAH metabolites such as phenols and dihydrodiols of phenanthrene, 3-hydroxybenzo[*a*]pyrene

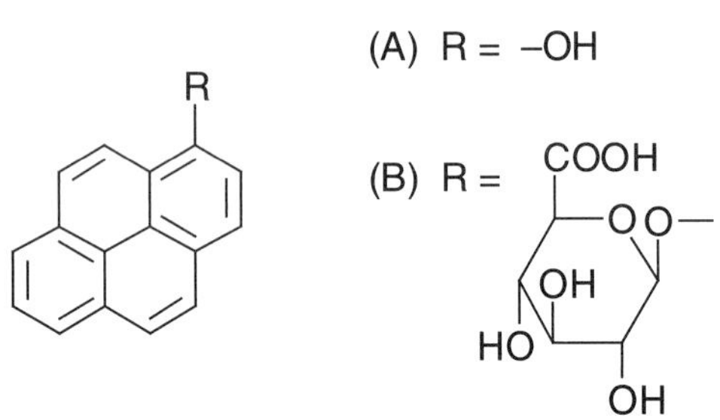

Figure 3.1: Structure of 1-hydroxypyrene (A) and its glucuronide (B).

(3-OH-B[*a*]P), and 3-hydroxybenz[*a*]anthracene (3-OH-B[*a*]A) were considered and integrated into biomonitoring studies.[15–26]

3.2 Studies at Work Places/Occupational Exposure to PAHs

In a number of studies significant increases in the rates of various types of cancer were observed and correlated with high levels of exposure to PAHs. For example, high PAH levels were found at working places of cokeries, aluminum, iron and steel plants, as well as industrial plants producing and/or manufacturing coal tar, pitch, creosote, rubber, mineral oil, soot, carbon black, fire-proof material and graphite electrodes (cf. Chapter 4). In addition, chimney-sweeps, asphalt and roofing workers are among occupational groups exposed to high levels of PAHs. Significant increased cancer incidences for the lung and other organs (scrotum,[27] bladder,[27] kidney[28] and colon[29]) were reported for employees of gasworks and cokeries.[27–34] Furthermore, increased rates of skin cancer (scrotum) and neoplasms in other organs were observed in workers with occupationally related continuous exposure to coal tar, coal tar pitch, and creosote (wood impregnation plant).[35] A causal relationship between PAH exposure and an increased cancer incidence can be assumed to be likely for the majority of these cases since the concentrations of PAHs in such sources are very high. For example, coal tar pitch contains 1–2% of benzo[*a*]pyrene (B[*a*]P) and consists of about 50% PAHs.[36] Increased cancer rates of the lung and other organs were also reported for employees of aluminum foundries (Soederberg technology)[37–40] and similar working places.[41] Increased skin, scrotum, and lung cancer incidences were documented for employees in mineral oil industries (using mostly cutting oil).[42–44] Moreover, workers producing soot and carbon black were found to be at significantly increased cancer risk for various organs.[45] Increased rates of bladder, stomach and lung cancer, as well as leukemia, were established for workers in the rubber industry.[44,46,47] In the case of chimney-sweeps, an unusually high incidence of scrotum cancer ('chimney-sweep cancer') had been already observed more than 200 years ago.[48] Later studies reported the frequent occurrence of cancerous alterations of the respiratory tract, esophagus and liver in the same occupational group.[49–52]

Despite the lack of early epidemiological studies on the health risk for road workers, this occupational group must be considered as being at risk in the past since some of the materials used were blended with coal tar pitch.[53,54] In the meantime, handling of this type of crude material (carbo bitumen) has been restricted or completely banned. Thus, an unusually high occupational PAH exposure is rather unlikely these days. However, an increased risk of lung cancer among asphalt workers has been suggested by epidemiological evidence based on large scale statistical analyses and the voluntary discontinuation of coal tar usage by the asphalt industry in Western Europe. The avoidance of coal tar during the past few decades provides an opportunity to answer the question of whether bitumen exposure *per se* poses a carcinogenic risk.[55] To address this question, a historical cohort of asphalt workers was assembled by the *International Agency of Research on Cancer* (IARC) in eight different countries (Denmark, Finland, France, Germany, the Netherlands, Norway, Sweden and Israel) in order to obtain diverse exposure profiles and a sufficient number of cases for the main health outcome of interest, i.e., the occurrence of lung cancer. As a result of these studies, the workers employed in road paving, asphalt mixing and other jobs entailing exposure to bitumen fume, were identified with a small increase in lung cancer mortality risk as compared to workers in ground and building construction. However, exposure assessment was limited and confounding from carcinogen exposure in other industries, tobacco smoking and other lifestyle factors could not be ruled out. Thus the results of these analyses did not allow to conclude on the presence or absence of a causal link between exposure to bitumen fumes and an increased risk of developing lung cancer.[56–58] In contrast to these uncertain associations, significantly increased rates of cancer of the lung[59] and of other organs (larynx, esophagus, stomach, bladder) clearly point to an occupationally related cancer risk for roofing workers.[60]

A tremendous number of studies on the exposure to PAHs at working places have been conducted. Some of them that included the measurement of B[*a*]P are compiled in Table 3.1. The data indicate that the concentrations are varying for comparable or even identical working places; but they

Table 3.1: Benzo[*a*]pyrene concentrations at different working places

Work place	*Country*	*Year*	B[*a*]P [*μg/m³*]	*Ref.*
Cokery				
Oven platform	Poland	1978	20–383	Braszczynska *et al.*[161]
modernized			0–6.8	Braszczynska *et al.*[161]
Oven platform	Sweden	1982	9.4–13.5	Lindstedt *et al.*[162]
driver of filling container			4.7–17	Lindstedt *et al.*[162]
Oven platform	Germany	1983	22.3–33.0	Blome[163]
driver of filling container			4.5	Blome[163]
ramp workers			1.33	Blome[163]
driver			0.16–0.93	Blome[163]
Oven platform	Germany	1992	10.6–15.8	Grimmer *et al.*[15]
driver of filling container			5.8–10.1	Grimmer *et al.*[15]
machinist			0.9–4.9	Grimmer *et al.*[15]
Aluminum plant (Soederberg technique)				
Various working places	Norway	1978	11.3–854	Bjørseth *et al.*[164]
Copper-, brass-, and zinc foundry	Germany	1983	<0.05–0.1	Blome[165]
Steel foundry			0.38–57.5*	
	Canada	1982		Verma *et al.*[166]
	Germany	1983		Blome[165]
		1986		Knecht *et al.*[167]
		1988		Coenen[168]
	Finland	1981		Schimberg[169]
Light metal foundry	Germany	1983	0.05–2.75	Blome[165]
Road paving				
Using carbo bitumen	Germany	1989	0.7–22.0	Knecht *et al.*[170]
Using petro bitumen	Germany	1990	0.02	Tobias *et al.*[171]
	Denmark	1989	4.0	Hansen[172]
Roofing	Germany	1983	14.0	Blome[165]
	USA	1982	0.4–11.0	Malaiyandi *et al.*[36] & Tharr[173]
Optical industry	Germany	1983	<0.05–19.7	Blome[165]
Tar refinery	Germany	1979	3.6	Schunk[174]
Bitumen production	Germany	1983	<0.5	Blome[165]

(continued)

Table 3.1: (continued)

Work place	*Country*	*Year*	B[*a*]P [*μg/m*^{3}]	*Ref.*
Brickyard	France	1987	3.4	Lesage *et al.*[175]
Rubber industry	USA	1980	0–32.3	Williams *et al.*[176]
	Finland	1982	<0.02–0.25	Enwald[177]
	Germany	1983	<0.05	Blome[165]

*Range of B[*a*]P concentration found in the five referenced steel foundry studies from Canada, Germany and Finland.

also show that major progress has been made to improve the situation at working places and to reduce PAH emissions by technical innovations. A good example are cokeries for which a maximum level of 5 μg B[*a*]P/m^3 at the oven platform is regulated by law. Similarly, limit values of 2 and 0.15 μg B[*a*]P/m^3 are effective at other work places in Germany and plants in France producing coal electrodes.[61,62] Regarding the effectiveness of human biomonitoring it is important to point out that PAH profiles are not only varying between different sources of emission, but also at the same work place in a time-depending manner. These issues clearly impede a general extrapolation from measured B[*a*]P concentrations to those of pyrene or phenanthrene, and *vice versa.*

3.3 Non-Occupational Exposure to PAHs

Non-occupational and background PAH exposure of the general population occurs through contaminated air, water, and diet (cf. Chapter 4). At present, limit values/concentrations of 1 and 10 ng/m^3 B[*a*]P are recommended or regulated in some countries, e.g., in Italy and Germany. Such high concentrations are only rarely detected in the atmosphere of rural regions or even in urban areas. However, elevated levels of PAHs may be present in areas with high traffic burden, in street tunnels or in industrial areas, e.g., close to cokeries (reviewed in 1998 IPCS Paper[3]).

Given a daily air consumption of 11 m^3 for a normal healthy adult person, the average uptake of B[*a*]P by inhalation can be calculated as about 15 ng. Based on a usual PAH profile detectable in the atmosphere of non-contaminated areas however,[63] the daily uptake of pyrene and phenanthrene is about 200 and 1000 ng, respectively. Since the level of B[*a*]P was found in the range of 0.1–1 ng/L for drinking water, the human body burden from this source is calculated to be 2–3 ng per day at maximum, but usually much less. From this source, the daily uptake of pyrene and phenanthrene is about 10 and 20 times higher, respectively, as compared to B[*a*]P. Diet is the major source of PAH uptake in the non-occupationally exposed general population. The concentration of B[*a*]P in meat and meat products in Germany is restricted to 1 μg/kg. In most of the common food sources tested, however, much lower levels of B[*a*]P were found. This is in contrast to sea food, including oysters, mussels and smoked fish, and to some vegetables such as green cabbage, where higher concentrations have been detected. Based on the limit value of 1 μg/kg, a maximal daily uptake of about 1 μg B[*a*]P from food sources can be estimated,[64] although higher values are likely for certain countries, such as the Netherlands.[65] The amounts of pyrene and phenanthrene incorporated via diet are usually 3–10 times higher as compared to B[*a*]P.

3.4 Metabolism and Excretion of PAHs

Intravitally, PAHs are enzymatically converted into arene oxides that either undergo spontaneous isomerization to phenols or hydrolysis to *trans*-dihydrodiols catalyzed by microsomal epoxide hydrolase (mEH). Dihydrodiols can be further oxidized to vicinal diol-epoxides (see Chapter 2). Both oxidation steps are regio- and stereospecifically catalyzed by cytochrome P450-dependent monooxygenases (CYP). From this superfamily of proteins, the enzymes CYP1A1, 1A2, 1B1 and 3A4 are predominantly involved in metabolic activation of PAHs. PAH diol-epoxides possessing a bay- and/or fjord-region were identified as ultimate carcinogens[66,67] that covalently bind to nucleophilic centers within DNA.[68,69] PAH-DNA adduct formation is considered as the initial step during cancerous transformation of a cell[69,70] (cf. Chapters 2 and 5). Alternatively, phenols and dihydrodiols may be further modified and subsequently excreted into the urine or feces as water soluble sulfates and/or glucuronide conjugates. These derivatives are enzymatically formed by conjugation with sulfuric acid or glucuronic acid (see Chapter 2). In contrast, arene oxides and diol-epoxides may be converted into glutathione (GSH) conjugates by glutathione *S*-transferases (GST). After enzymatic removal of two amino acid residues (glycine and glutaminic acid) from GSH conjugates and subsequent *N*-acetylation of the remaining cysteinyl adduct, the resulting derivatives, termed 'mercapturic acids', can be excreted into the urine. Formation of mercapturic acids was found to be a major detoxification pathway in rodents, but much less is known about its importance in man. For biomonitoring studies in humans however, this pathway needs to be considered due to the predominant elimination of low molecular weight metabolites (<475 Da) via urine. (The molecular weights of the mercapturic acid derivatives of phenanthrene, pyrene and B[*a*]P are 357, 381 and 431 Da, respectively.) In contrast, metabolites with molecular weights >475 Da are predominantly excreted via feces.[71] In additon to mercapturic acids, the excretion of other mercapto derivatives of PAH metabolites such as mercaptoacetic, mercaptolactic and mercaptopyruvic acids was detected in rodents.[72] However, at present it remains unclear of whether or not these pathways may also play an important role during PAH metabolism in humans.

Animal studies have shown that PAHs are quickly transported from the site of application (gastrointestinal tract, lung, skin) to other organs via blood and lymphatic vessels.[73] For instance, local administration of PAHs onto the skin of rodents was succeeded by PAH-DNA adduct formation in the lung tissue of these animals.[74–76] Excretion of PAHs (given by gavage or intraperitoneal) was found to be completed after 3 days.[77–80] Viau *et al.*[81] reported that 57% of a given amount of pyrene is excreted via urine and 18% via feces within 24 hours in rats. Bouchard *et al.*[82] estimated a half-life of about 5–7 hours for this species and compound. For humans, inhaled PAHs from cigarette smoke are excreted within about 24 hours [J. Jacob, personal communication]. This observation is supported by Brzeznicki *et al.*[83] who investigated PAH exposure in workers of an aluminum plant (half-life of 9.8 hours). A half-life of about 12 hours was reported by Viau *et al.*[84] for a group of volunteers that had received pyrene either by dermal application or through ingestion with foodstuff. In contrast, Buckley and Lioy, found a half-life of 4.4 hours after administration of pyrene with foodstuff.[85]

3.5 Biomonitoring of PAHs and Their Metabolites

3.5.1 Principal Considerations

Initial experiments that used the mutagenic potential of human urine as an indicator of PAH exposure produced inconsistent and unsatisfying results — due to a low sensitivity and poor specificity.[17,86–90] Methods were then developed to directly determine the concentration of excreted PAH metabolites in the urine. Compared to any extrapolations from working place ambient air measurements, determination of the urinary metabolites provides a much better estimate of the actual overall systemic dose of PAHs incorporated by a single person. Not only PAH uptake by inhalation, but also the uptake by ingestion and percutaneous absorption can be detected by this approach.[91] The following prerequisites must be met by methods of biomonitoring of PAH metabolites in order to provide sufficient information on the overall PAH body burden:

(i) Determination of analytes (metabolites) should be unequivocal, i.e., specific and reproducible;
(ii) Given a metabolite concentration of >1–10 ng/L, the sensitivity of the method should allow a practical sample size;
(iii) In order to keep the error range small, the metabolites used for analysis should be predominantly excreted via urine;
(iv) As much metabolites as possible should be included into the analysis of a particular PAH, e.g., all phenols and dihydrodiols in the case of phenanthrene. The ratios between individual metabolites are not constant, but rather depend on the enzymatic status of the subject under consideration;
(v) The PAH selected as an indicator compound should represent a major constituent of all different kinds of emissions. Ideally, several PAHs and their metabolites should be included into analysis since PAH profiles may differ substantially depending on the particular emission source.

3.5.2 1-Hydroxypyrene and Its Glucuronide

Since it fulfills at least conditions (i), (ii) and (v), 1-OHP appears to be a suitable biomarker that would allow an estimation of the overall body burden on PAHs. 1-OHP represents the main metabolite of pyrene in mammals,[92–97] although additional metabolites have been identified including *trans*-4,5-dihydroxy-4,5-dihydropyrene (i.e., the K-region dihydrodiol),[92–96] pyrene 1,6- and 1,8-quinone,[93] 1,6- and 1,8-dihydroxypyrene,[92,96] and two unspecified phenolic dihydrodiols.[96] Keimig and Morgan[95,98] developed a method to determine 1-OHP by HPLC after acidic cleavage of its conjugates and extraction with dichloro methane. This method was modified and significantly improved by Jongeneelen *et al.*,[18,99] who introduced an enzymatic cleavage of the conjugates by β-glucuronidase/sulfatase. After concentration on a SepPak-C_{18} cartridge, HPLC separation (water/methanol gradient) and fluorescence detection resulted in a recovery rate of 78% with variation coefficients of 2.6% ($n = 8$) for 60 μg/L 1-OHP. An improvement of the method finally led to recovery rates of 83–88% and a quantification limit of 0.2 μg/L for a sample size of 20 μL. Furthermore, the long-term stability of 1-OHP in stock solutions makes it suitable to serve as a reference compound (about 12 months for 39.6 μg 1-OHP/L solvent[100]). Later, Lintelmann and Boos[101,102] developed an automated HPLC using a two column-switching technique and the enrichment of 1-OHP on copper phthalocyanin trisulfonate-saturated porous glass. The detection limit for this method was determined as 0.01 pmol ±4.9% ($n = 7$).

Elevated excretion levels of 1-OHP were observed in the urine of workers occupationally exposed to PAHs (cokeries, road constructions, wood impregnation plants, aluminum smelter) and for patients treated with tar ointments. Selected data are summarized in Table 3.2 and reported values are given as μg/g creatinine and/or as μg/L urine. Occasionally, values are reported in μg/24 hour urine; for comparison purposes, these data were converted into μg/g creatinine based on the assumption of a daily excretion of 1.2 g creatinine for women and 1.8 g for men.

In one study, the urinary excretion of 1-OHP correlated with the lung cancer risk of workers occupationally exposed to PAHs.[103] A value of 2.3 μmol 1-OHP/mol creatinine (4.4 μg/g) was associated with a relative lung cancer risk of 1.3. This value was suggested as an occupational exposure limit (OEL) for coke oven workers.[103,104] For aluminum

Table 3.2: Levels of 1-hydroxypyrene in the urine of occupationally PAH-exposed individuals and control subjects

Exposure	*No. of individuals*	*μg 1-OHP/ g creatinine*	*μg 1-OHP/L*	*Ref.*
Non-exposed at working place				
General population (Germany)	8		0.3	Grimmer *et al.*[178]
General population (China)	70	0.8–4.6		Zhao *et al.*[179]
Non-smoker	10	0.03–0.21	9.4–13.5	Jacob *et al.*[19]
	11	<0.10–0.30	4.7–17	Angerer *et al.*[180]
	19	0.04		Granella *et al.*[181]
Smoker	11	<0.10–0.60	<0.10–0.80	Angerer *et al.*[180]
	9	0.06–0.6		Jacob *et al.*[19]
	22	0.08		Granella *et al.*[181]
Tar ointment-treated	8	0.94–5.81		Angerer *et al.*[180]
patients	25		7.6	Clonfero *et al.*[182]
PAH-exposed at working place				
Coal-electrodes	6	7.1–82.6	9.7–73.5	Angerer *et al.*[180]
production	4	18.1–29.6	20.1–50.4	Angerer *et al.*[180]
	14	2.2–125.0	3.2–160.8	Angerer *et al.*[180]
	13	0.58–16.8		Angerer *et al.*[180]
	67	0.2–326		Mannschreck *et al.*[20]
pre-shift	34	7.5		Bentsen-Farmen *et al.*[183]
post-shift	34	19.7		Bentsen-Farmen *et al.*[183]
Graphite oil treatment of glass forms	10	0.1–3.8	<0.1–4.2	Angerer *et al.*[180]
Aluminum smelter	37	0.7–126.2	0.2–99.2	Angerer *et al.*[180]
	5	2.3–17		Vu-Duc *et al.*[105]
Steel plant (China)	12	3.5		Zhao *et al.*[179]
Tar impregnation	3	7.1–41.2	12.7–77.1	Angerer *et al.*[180]
	3	1.3–38.5		Jongeneelen *et al.*[99]
	3	1.0–9.1		Jongeneelen[103]
	1	80.9–158		Jongeneelen *et al.*[184]
	3	8.2–134.9		Jongeneelen[103]
	6	4.8–146.5		Jongeneelen[103]
	9	1.2–55.9		Jongeneelen[103]
Meat smoking	13	<0.1–1.1	<0.1–1.1	Angerer *et al.*[180]

(continued)

Table 3.2: (continued)

Exposure	*No. of individuals*	*μg 1-OHP/ g creatinine*	*μg 1-OHP/L*	*Ref.*
Incineration	53	<0.1–0.8	<0.1–1.3	Angerer *et al.*[180]
Sleeper plant	14	0.4–8.7		Jongeneelen[103]
Coal tar distillation	4	2.3–25.1		Jongeneelen *et al.*[184]
Road paving	31	0.8–16.4		Jongeneelen *et al.*[185]
	4		2.6	Grimmer *et al.*[178]
using petro bitumen				
(shift)	13		1.0–3.5	Knecht *et al.*[186]
(post-shift)	20		0.4–2.4	Knecht *et al.*[186]
using carbo bitumen				
(pre-shift)	15		0.4–10.9	Knecht *et al.*[186]
(shift)	5		1.3–42.4	Knecht *et al.*[186]
(post-shift)	12		5.3–92.1	Knecht *et al.*[186]
Cokery				
(China)	15	36.3		Zhao *et al.*[179]
	31	15.9		Zhao *et al.*[179]
(Sweden) before oven renovation	10		3.8–88	Levin *et al.*[187]
after oven renovation	10		0.7–17	Levin *et al.*[187]
(Germany) oven platform	24	3.3–79.4		Strunk *et al.*[188]
(Germany) oven platform	8		14.1–118.5	Grimmer *et al.*[15]
driver	4		8.2–21.0	Grimmer *et al.*[15]
machinist	4		2.0–6.8	Grimmer *et al.*[15]
Car repair shop				
non-smoker	40	0.07		Granella *et al.*[181]
smoker	25	0.13		Granella *et al.*[181]

industry workers, an OEL of 4.9 μmol 1-OHP/mol creatinine (9.4 μg/g) was defined.[104] These values are part of a benchmark guideline to use 1-OHP as a biomarker for PAH exposures at working places,[104] and they are based on epidemiological data correlating B[*a*]P exposure with cancer risk. However, such a concept implies a constant ratio between pyrene and B[*a*]P; an assumption that is realized for identical working places at best. Even in those cases, major differences in the PAH profiles are common.[15] The relevance of 1-OHP as a biomarker for PAH-exposed workers in the aluminum industry was critically discussed by Vu-Duc and Lafontaine.[105] In a study with five volunteers who had consumed similar amounts of identical food for 5 consecutive days, Viau *et al.* recently demonstrated that urinary excretion of 1-OHP did not correlate with the ingested dose of pyrene under normal feeding conditions.[106]

Since 1-OHP glucuronide (1-OHPG) is the main metabolically formed conjugate from pyrene, Strickland *et al.*[107,108] developed an analytical method based on the combination of immuno-affinity chromatography (IAC) and HPLC. The authors found a 5-fold higher fluorescence yield for 1-OHPG compared to 1-OHP, whereas Singh *et al.*[109] reported only a 3-fold higher value. Despite these differences, the method has been validated and was found to produce similar results as compared to other methods, e.g., conventional HPLC[109] or IAC in combination with synchronous fluorescence-spectroscopy.[110] In a subsequent study, Strickland *et al.* measured 1-OHPG in the urine of occupationally PAH-exposed workers. They found 2.16 pmol 1-OHPG/mL urine as compared to 0.38 pmol/mL for low or non-exposed control subjects; the levels were significantly increased in smokers (1.82 pmol/mL) compared to non-smokers (0.75 pmol/mL).[108,111–113] Kang *et al.*[114] investigated the inter-individual variation of the 1-OHPG excretion after consumption of charbroiled beef. They attempted to correlate the corresponding 1-OHPG values with the level of PAH-DNA adducts detected in white blood cells by an enzyme-linked immunosorbent assay (cf. Table 4.1 in Chapter 4). However, only 40% of the investigated subjects were found with significantly increased DNA adduct levels. The urinary excretion levels of 1-OHPG returned to control values 24–75 hours after ingestion. These results were confirmed in a subsequent study.[115]

3.5.3 Enzyme Polymorphisms and Excretion Levels of 1-Hydroxypyrene

In principle, the inter-individual differences in the excretion levels of 1-OHP may be a consequence of genetic polymorphisms of the enzymes involved in metabolism of pyrene. Genetic variants of CYP (phase-I) and GST enzymes (phase-II) that contribute to PAH metabolism were investigated and characterized[116,117] (cf. Chapters 2 and 10). Wu *et al.*[118] investigated the effect of the *CYP1A1 Msp*I genotype on the excretion of 1-OHP in coke oven workers from Taiwan. In this study, a 2-fold increase in 1-OHP concentrations was found in individuals homozygous for this allele in comparison to controls that were heterozygous or that carried only wild-type/reference alleles ($P = 0.04$). In addition, decreasing urinary 1-OHP levels were found in the following order: topside oven workers/homozygous variant trait > topside oven workers/heterozygous variant trait > sideoven workers/homozygous variant trait > and sideoven workers/heterozygous trait ($P < 0.001$).

Similar results were reported in a study from Merlo *et al.*[119] that examined traffic police officers exposed to ambient air PAHs in the center of a city. Officers carrying the heterozygous variant of the *CYP1A1 Msp*I genotype and consuming less than 15 cigarettes per day were detected with higher concentrations of urinary 1-OHP as compared to individuals homozygous for the wild-type/reference allele. In contrast, no significant genetic influence was established for non-smoking police men or persons consuming more than 15 cigarettes per day.[119] In a group of occupationally non-exposed Japanese, Hawaiian or Caucasian subjects, Nerurkar *et al.*[120] found 2-fold higher urinary 1-OHP levels in smokers with heterozygous *CYP1A1 Msp*I genotype compared to smokers belonging to the reference group ($P = 0.02$). Similar results were obtained in the same study for the *CYP1A1* Ile^{462}Val genotype[120] (cf. Chapter 10). Pan *et al.*[121] however, did not find any effect of the *CYP1A1* Ile^{462}Val genotype on urinary 1-OHP excretion levels in coke oven workers from China. Alexandrie *et al.*[122] investigated workers in an aluminum smelter and reported a non-significant increase in the 1-OHP levels for individuals carrying the *CYP1A1* Ile^{462}Val genotype. In this study the highest concentrations of 1-OHP were detected in subjects with a *GSTM1* null genotype. Recently, Schocket *et al.*[123] reported that there is no influence of the *CYP1A1* Ile^{462}Val genotype on 1-OHP excretion among employees in primary aluminum production. In addition, further *CYP* polymorphisms such as *CYP1A1 Msp*I, *CYP1B1* Leu^{432}Val, *CYP2C9* Arg^{144}Cys and *CYP2C9* Ile^{359}Leu did not have any influence on the 1-OHP excretion levels in the subjects under consideration[123] (cf. Chapter 10). Alexandrie *et al.*[122] pointed out that 1-OHP excretion may not only be determined by extrahepatic CYP1A1 activity in the lung, but also by other CYP enzymes, particularly those expressed in the liver. This is supported by a study of Nan *et al.*[124] with coke oven workers in Korea. Since the *CYP1A1* $Ile^{462}Val$ polymorphism did not show any influence in this study, the authors suggested that this result may be due to a lower expression level of CYP1A1 in the Asian population as compared to Caucasians. Interestingly, a higher excretion level of 1-OHP was detected in coke oven workers carrying the CYP2E1 genotype c1/c2 or c2/c2.[124] Together, the available data indicate that individuals with certain polymorphic variants of *CYP1A1* and *CYP2E1* have a higher capacity to metabolically convert pyrene into 1-OHP.

The urinary 1-OHP concentration is determined via its glucuronide and sulfate conjugates. Therefore, the relationship between *GST* genotypes and 1-OHP levels may be only of indirect nature. Hong *et al.*[113] investigated the influence of *GSTM1* and *GSTT1* polymorphisms on the urinary excretion of 1-OHPG in smokers. Smokers carrying the *GSTM1* null genotype were found with increased excretion levels of 1-OHPG compared to control individuals. Conversely, the level of 1-OHPG was higher in *GSTT1*-positive smokers compared to *GSTT1*-deficient individuals.[113] Among coke oven workers in Italy, Brescia *et al.* reported an increase in 1-OHP excretion levels associated with *GSTM1* null genotype.[125] This observation is in accordance with a previous study from Alexandrie *et al.*[122] on workers in the primary aluminum production. In contrast, Schoket *et al.*[123] detected reduced 1-OHP excretion levels in *GSTM1*-deficient individuals working in an aluminum smelter. The authors also suggested an interaction between GSTM1 and GSTP1: subjects deficient for GSTM1 were measured with significantly reduced 1-OHP excretion levels when compared to GSTM1-positive individuals who simultaneously carried the *GSTP1* Ile^{105}/Ile^{105} genotype. Other studies on traffic police officers in Italy[119] or coke oven workers in Sweden[126] and China[127] report that polymorphic forms of GSTM1, GSTP1 or GSTT1 showed little, if any, influence on the 1-OHP excretion levels. Accordingly, Nan *et al.*[124] found no influence of *GSTM1* genotype on urinary 1-OHP excretion levels in Korean coke oven workers. For GSTT1-positive individuals, however, a higher urinary concentration of 1-OHP was reported in comparison to carriers of the *GSTT1* null genotype, although this increase was not statistically significant. These results are in line with the data collected from US-soldiers in Kuwait.[128] In addition, and based on the investigation of PAH-exposed coke oven workers in the Netherlands, van Delft *et al.* recently concluded that both the GSTM1 and GSTT1 status would have no influence on the levels of 1-OHP in the urine.[129]

3.5.4 Phenanthrene Metabolites

Phenanthrene metabolites were used in numerous studies as biomarkers for PAH exposure. This strategy fulfills all prerequisites listed above. Phenanthrene is metabolized to five isomeric phenols (1-, 2-, 3-, 4- and 9-hydroxyphenanthrene) and three *trans*-dihydrodiols (1,2-, 3,4- and 9,10-dihydroxyphenanthrene; see Figure 3.2), all of which are predominantly excreted as sulfate and glucuronide conjugates.[80,130–136] The isomeric phenanthrene phenols can be determined after enzymatic cleavage of their conjugates and subsequent derivatization by means of GC/MS-MS,[15,137,138]

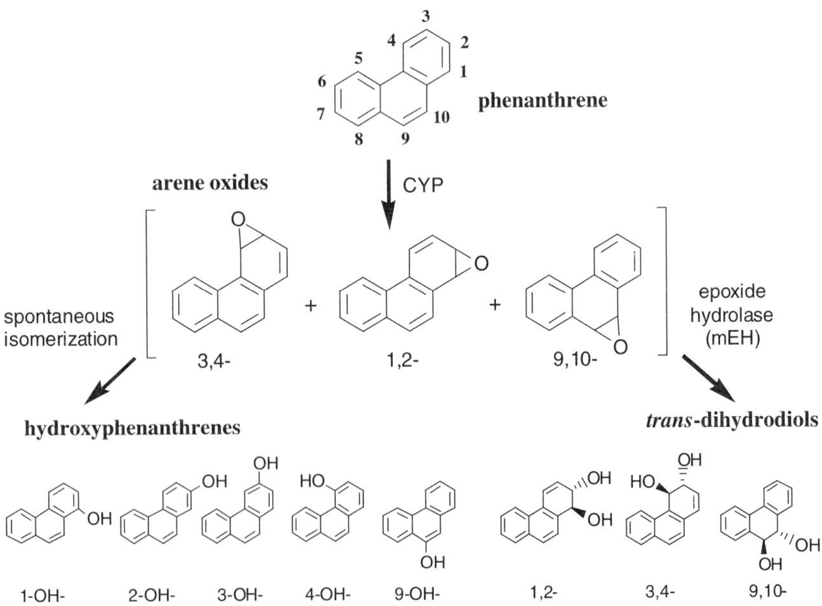

Figure 3.2: Metabolism of phenanthrene to phenols and *trans*-dihydrodiols.

or — after HPLC separation — directly by their fluorescence spectra.[139–142] Prior to HPLC-based determination, phenanthrene phenols can be concentrated on a silica gel modified with copper phthalocyanine trisulfonic acid,[140] or by means of sol-gel glass immunosorption.[141] The complete separation of all five isomeric phenanthrene phenols cannot sufficiently be achieved by HPLC alone;[140,141] however, it enables simultaneous determination of other PAH phenols including 1-OHP, 3-OH-B[*a*]A and 3-OH-B[*a*]P.[24,143] New capillary electrophoresis techniques were developed in order to quantify various PAH phenols.[26,144–146] Based on γ-cyclodextrin-modified micellar electrokinetic chromatography coupled with laser-induced fluorescence detection (CD-MEKC), simultaneous determination of the following twelve PAH phenols has been reported: 1-OHP, 1- and 2-naphthol, 1-hydroxyphenanthrene, 1- and 3-hydroxychrysene, 1-hydroxybenzo[*b*]fluoranthene, 1-, 3-, 7- and 9-OH-B[*a*]P and 2-hydroxyindeno[1,2,3-*cd*]pyrene.[26,146] The levels of 3-OH-B[*a*]A and 3-OH-B[*a*]P found in biological samples,[24,143] however, are only of limited toxicological importance. Both corresponding parent PAHs are also converted into mutagenic and carcinogenic metabolites (bay-region diol-epoxides) that are not detectable by this method (cf. below, and Chapter 2).

As a consequence of their poor intrinsic fluorescence properties and the low sensitivity of the method, phenanthrene dihydrodiols cannot sufficiently be determined by HPLC/fluorescence detection.[143] However, indirect methods have been developed to allow the determination of these metabolites in the urine of PAH-exposed subjects.[15,19] In principle, phenanthrene can be metabolically converted at three different molecule regions into *trans*-dihydrodiols (Figure 3.2). Compared to pyrene, these conversions occur to a much higher extent. The urinary phenanthrene metabolite profile therefore provides some additional information regarding the balance of contributing enzymes. Since CYP enzymes oxidize phenanthrene with different regioselectivity,[147,148] the induction status and the ratios of single enzyme forms contributing to the levels of certain *trans*-dihydrodiols can be monitored via urinary dihydrodiol profiles. For instance, induction of CYP1A2 in cigarette smokers has been demonstrated based on changes in the ratio of oxidized metabolites of phenanthrene at its 3,4- *versus* 1,2-position.[19]

A small increase in the total level of urinary phenanthrene metabolites could be observed in smokers compared to non-smokers.[149,150] In contrast, passive smokers did not show any comparable effect. Consumption of PAH-rich diet may cause an increase in the urinary level of phenanthrene phenols. However, an 8-fold higher concentration of phenanthrene in the diet (from 0.5 to 4.4 μg/kg) only led to a 2-fold increase in the excretion level of the corresponding phenols (from 0.76 to 1.46 nmol/mmol creatinine).[150] This finding is supported by results of Buckley and Lioy[85] who detected a 4–12-fold higher level of 1-OHP in the urine of volunteers who had eaten charbroiled beef with 250-fold higher concentrations of B[*a*]P compared to control food. It remains unclear however, whether the concentration of pyrene has increased in the charbroiled beef to a similar extent as B[*a*]P. A compilation of the excretion levels of phenanthrene metabolites in the urine of various PAH-exposed subjects can be found in Table 3.3.

3.5.5 Benzo[*a*]pyrene Metabolites

It has repeatedly been pointed out that the urine levels of phenanthrene and pyrene metabolites would not properly reflect the body burden on carcinogenic PAHs. Both compounds are biologically inactive and therefore not useful to serve as surrogates for carcinogenic PAHs such as B[*a*]P and dibenzo[*a,l*]pyrene (DB[*a,l*]P) (cf. Chapter 2). Accordingly, attempts were undertaken to establish analytical detection methods for metabolites of toxicologically more relevant PAHs such as B[*a*]A or B[*a*]P. The principle problem still persists however, that metabolites of higher molecular weight PAHs are predominantly excreted via feces rather than in urine.[151] Since urinary concentrations of these PAHs are therefore very low, the uncertainty in analysis is high and, as a consequence, the conclusions drawn are likely to be error-prone.

Becher and Bjørseth[21] described a method for measurement of the urinary levels of 3-OH-B[*a*]P. They reported concentrations of 0.12 μg/L in the urine of PAH-exposed workers. However, this procedure could not

Table 3.3: Excretion of phenanthrene metabolites in the urine of occupationally PAH-exposed individuals and control subjects

Work place	*No. of individuals*	*μg phenanthrols/ g creatinine*	*μg phenanthrols/ L*	*Ref.*
Non exposed				
General population	8		3.5	Grimmer *et al.*[178]
PAH-poor diet	8	1.3		Martin *et al.*[150]
PAH-rich diet	8	2.5		Martin *et al.*[150]
Non-smoker	6	1.5		Martin *et al.*[150]
Non-smoker	10	1.3		Jacob *et al.*[19]
Smoker	9	1.3		Martin *et al.*[150]
Smoker	6	2.4		Jacob *et al.*[19]
Smoker (lung cancer)	10	1.6		
PAH-exposed				
Road construction	4		34.9	Grimmer *et al.*[178]
pre-shift	4	6.9–25.7		Martin *et al.*[150]
post-shift	5	5.1–286		Martin *et al.*[150]
Wood impregnation plant				
pre-shift	1	44.5		Martin *et al.*[150]
post-shift	1	482		Martin *et al.*[150]
Cokery	24		70.3	Grimmer *et al.*[178]
Cokery	8	12		Strunk *et al.*[188]
oven platform	4		103–686	Grimmer *et al.*[15]
driver	4		45–88	Grimmer *et al.*[15]
machinist	4		7–22	Grimmer *et al.*[15]

be reproduced in other laboratories and, therefore, the method does not prevail in routine biomonitoring. Two years later, Jongeneelen *et al.*[17] described a HPLC method to determine 3-OH-B[*a*]P with a detection limit of 1 ng/24 hour urine. In a follow-up publication from the same group, a detection limit of 1 μg/L (4 nmol/L) was reported.[18] Nevertheless, the concentrations of 3-OH-B[*a*]P in the urine of workers in a coal tar distillation plant did not reach this threshold of detection, thus the method appeared not to be sensitive enough for human biomonitoring studies. Bouchard *et al.*[16] improved the sensitivity and reproducibility of the HPLC method by adding ascorbic acid to the solvents. Recently, Gündel and Angerer[22] reported on a HPLC method with fluorescence detection

Table 3.4: Excretion of 3-hydroxybenzo[*a*]pyrene (3-OH-B[*a*]P) in the urine of occupationally PAH-exposed individuals and control subjects

Work place	*No. of Subjects*	*ng 3-OH-B[a]P/L*	*Ref.*
Non exposed	48	6*	Grimmer *et al.*[178]
Cokeries	40	370	Grimmer *et al.*[178]
Road paving	10	19	Grimmer *et al.*[178]
Fire-proof material production	19	5–356	Gündel *et al.*[22]

* detection threshold: 1 ng/L

using a pre-column containing silica gel modified with copper phthalocyanine — a technique originally developed by Lintelmann and Boos[101,102] (cf. above). They reported detection thresholds of 6 ng/L and 8 ng/L for 3-OH-B[*a*]P and 3-OH-B[*a*]A, respectively, both of which could be analyzed simultaneously. Using this method, in the post-shift urine of workers ($n = 19$) from a fire-proof material producing plant, concentrations of 3–198 ng 3-OH-B[*a*]P and 15–1871 ng 3-OH-B[*a*]A/g creatinine have been found. Selected data of 3-OH-B[*a*]P concentrations measured in the urine of PAH-exposed workers are collected in Table 3.4.

Grimmer *et al.*[23] correlated the amount of B[*a*]P inhaled within an eight hour working shift with the excretion levels of its phenolic metabolites (sum of 3-, 7- and 9-phenols). The amount of 17.71 μg B[*a*]P incorporated through inhalation corresponded to an amount of 0.82 μg phenols excreted into urine. Unfortunately, the detection threshold levels for 3-OH-B[*a*]P using conventional fluorescence detectors is too high to allow an accurate measure of this analyte in the urine of occupationally non-exposed subjects.[143] More sensitive methods have therefore been established, e.g., laser-induced fluorescence detection coupled with HPLC, and γ-cyclodextrin-modified micellar electrokinetic chromatography in combination with the aforementioned method. Both methods allow the detection of very low concentrations in the range of 0.5-8 ng/L.[25,26] Applying these methods, coke oven workers were found with significantly increased urinary 1-OHP levels compared to non-exposed control subjects. In most of the collected samples however, a corresponding increase in the urinary levels of 3-OH-B[*a*]P could not be observed.[25] The determination of 1-OHP as a metabolic product of the non-carcinogenic PAH pyrene may well provide some clues on the PAH body burden of exposed subjects. Analysis of a derivative generated during metabolic activation of B[*a*]P towards its ultimate carcinogenic metabolites, i.e., bay-region B[*a*]P-7,8-diol-9,10-epoxides (B[*a*]PDE; cf. Figure 2.5 in Chapter 2), would certainly be the more appropiate parameter in monitoring exposed individuals with respect to any possible adverse health effects.

As a measure of the metabolic pathway that generates ultimate carcinogenic B[*a*]PDE isomers, Simpson *et al.*[152] developed a highly sensitive method for determination of the corresponding hydrolysis products, the B[*a*]P-7,8,9,10-tetraols (7,8,9,10-tetrahydroxy-7,8,9,10-tetrahydro-B[*a*]P isomers), in human urine. Herein, tetraols are concentrated by reverse phase and phenylboronic acid solid phase extraction, then subjected to chemical permethylation and finally quantified by GC/NICI-MS (selected ion mode). With this procedure a detection threshold of about 1 fmol of permethylated B[*a*]P-7,8,9,10-tetraols (signal to noise ratio ~3) could be established. The B[*a*]P-7,8,9,10-tetraol concentrations measured ranged from non-detectable to 0.2 fmol/μg creatinine for smokers, 0.07–0.92 fmol/μg creatinine for steel plant workers, and 0.7–19 fmol/μg creatinine for psoriasis patients who had received a medical treatment with coal tar ointments.[152] Bowman *et al.*[153] previously determined B[*a*]P-7,8,9,10-tetraols in the same group of psoriasis patients by IAC and synchronous fluorescence spectrophotometry according to the method described in Weston *et al.*[154] In this study, concentrations of 150 fmol/mL urine (mean value) were reported, whereas Simpson *et al.*[152] only found mean values of about 16 fmol/mL. More recently, B[*a*]P-7,8,9,10-tetraol and 1-OHP concentrations measured in the urine of coke oven workers at a steel plant in Taiwan were reported at 0.4 and 9.7 μmol/mol creatinine, respectively.[155] These levels were significantly higher than those in control subjects (0.03 and 0.4 μmol/mol creatinine, respectively). In addition, urinary B[*a*]P-7,8,9,10-tetraol concentrations could be positively correlated with urinary 1-OHP concentrations, suggesting that both together urinary 1-OHP and B[*a*]P-7,8,9,10-tetraol levels might serve as a combined biomarker for the assessment of the incorporation of carcinogenic PAHs. The obvious question of whether or not the determination of B[*a*]P-7,8,9,10-tetraols will be appropriate as a routine assay for PAH biomonitoring must presently remain unanswered in view of the very low levels of these compounds found even in occupationally (highly) exposed individuals.

For occupationally non-exposed individuals and, in particular for non-smokers, the diet must be considered as the main source of the daily uptake of PAHs[156] (cf. above and Chapter 4). Open fire cooking of food leads to production of PAHs.[157–159] In addition, non-cooked or somehow temperature-treated diet can be significantly contaminated with PAHs, thus reflecting the general occurrence of these pollutants in the environment as a consequence of incomplete combustion processes.[156] The weekly ingested amounts of B[*a*]P from a charbroiled or smoked diet were estimated at 0.01–4.0 μg/person.[64,157] Although no convincing epidemiological data exist, it appears reasonable to assume that PAHs may also play an important role in the etiology of colorectal cancers. This assumption is supported by findings of Alexandrov *et al.*,[160] who developed a fluorimetric assay and reported B[*a*]PDE-DNA adduct concentrations of 0.2–1.0 adducts/10^8 nucleotides in the colon mucosa of smokers and non-smokers.

3.6 Summary

The determination of PAH metabolites in human urine is the method of choice for the detection of recent PAH exposures, particularly when different absorption routes are combined (e.g., inhalation, percutaneous resorption, ingestion). The urinary 1-OHP level serves as a sensitive and correlative biomarker in the biomonitoring of PAH exposure on a collective basis. At present, 1-OHP is the most frequently applied biomarker in health care-related occupational medicine. However, it needs to be considered that (i) the PAH profiles may vary substantially between different working places; (ii) large inter-individual variations in urinary 1-OHP levels, occasionally observed in individuals with comparable exposures, may be due to inter-individual differences in bioavailability, polymorphic enzyme activity, and enterohepatic metabolite cycling *versus* urinary excretion. According to these limitations, an individual health risk cannot be deduced from urinary 1-OHP levels. Several studies suggested that the value of urinary biomonitoring can be improved and further strengthened by determination of metabolites from phenanthrene, i.e., five phenols and three dihydrodiols (Figure 3.2). The phenanthrene metabolite profile not only depends on the degree of exposure, but it also provides additional information on the individual status of enzymes involved in PAH metabolism (e.g., CYP and mEH).

3.7 Conclusions

Since biomonitoring of 1-OHP and/or phenanthrene metabolites does not reflect the internal body burden of carcinogenic PAHs accurately, there is a great need for practicable and reliable methods allowing the determination of effect markers of carcinogenic PAHs such as B[*a*]P and DB[*a,l*]P. New strategies such as determination of the hydrolysis products of bay- and fjord-region diol-epoxides, e.g., B[*a*]P-7,8,9,10-tetraols and DB[*a,l*]P-11,12,13,14-tetraols, though present only in very small concentrations in the urine, may indicate the future direction of human biomonitoring. At present, these methods appear to be too laborious for routine application in occupational biomonitoring studies.

Acknowledgment

The author is grateful to Prof. Dr. Jürgen Jacob for his help and critical reading of the manuscript.

References

1. IARC (1973) *Monographs on the Evaluation of Carcinogenic Risk of the Chemical to Man.* Volume 3: Certain Polycyclic Aromatic Hydrocarbons and Heterocyclic Compounds. International Agency for Research on Cancer, Lyon, France.
2. IARC (1983) *Monographs on the Evaluation of Carcinogenic Risk of the Chemical to Humans.* Volume 32: Polynuclear Aromatic Compounds, Part 1, Chemical, Environment and Experimental Data. International Agency for Research on Cancer, Lyon, France.
3. IPCS Environmental Health Criteria 202 (1998) *Selected Non-Heterocyclic Polycyclic Aromatic Hydrocarbons.* WHO, Geneva.

4. Jacob J (1996) The significance of polycyclic aromatic hydrocarbons as environmental carcinogens. *Pure Appl. Chem.* 68: 301–308.
5. Grimmer G, Brune H, Deutsch-Wenzel R, Dettbarn G, Jacob J, Naujack KW, Mohr U and Ernst H (1987) Contribution of polycyclic aromatic hydrocarbons and nitro-derivatives to the carcinogenic impact of diesel engine exhaust condensate evaluated by implantation into the lungs of rats. *Cancer Lett.* 37: 173–180.
6. Grimmer G, Brune H, Deutsch-Wenzel R, Naujack KW, Misfeld J and Timm J (1983) On the contribution of polycyclic aromatic hydrocarbons to the carcinogenic impact of automobile exhaust condensate evaluated by local application onto mouse skin. *Cancer Lett.* 21: 105–113.
7. Grimmer G, Brune H, Deutsch-Wenzel R, Dettbarn G and Misfeld J (1984) Contribution of polycyclic aromatic hydrocarbons to the carcinogenic impact of gasoline engine exhaust condensate evaluated by implantation into the lung of rats. *J. Natl. Cancer Inst.* 72: 733–739.
8. Grimmer G, Brune H, Deutsch-Wenzel R, Dettbarn G, Misfeld J, Abel U and Timm J (1984) The contribution of polycyclic aromatic hydrocarbons to the carcinogenic impact of emission condensate from coal-fired residential furnaces evaluated by topical application to the skin of mice. *Cancer Lett.* 23: 167–176.
9. Grimmer G, Brune H, Deutsch-Wenzel R, Dettbarn G, Misfeld J, Abel U and Timm J (1985) The contribution of polycyclic aromatic hydrocarbon fractions with different boiling ranges to the carcinogenic impact of emission condensate from coal-fired residential furnaces as evaluated by topical application to the skin of mice. *Cancer Lett.* 28: 203–211.
10. Grimmer G, Brune H, Deutsch-Wenzel R, Dettbarn G and Misfeld J (1987) Contribution of polycyclic aromatic hydrocarbons and polar polycyclic aromatic compounds to the carcinogenic impact of flue gas condensate from coal-fired residential furnaces evaluated by implantation into the rat lung. *J. Natl. Cancer Inst.* 78: 935–941.
11. Grimmer G, Brune H, Dettbarn G, Naujack KW, Mohr U and Wenzel-Hartung R (1988) Contribution of polycyclic aromatic compounds to the carcinogenicity of sidestream smoke of cigarettes evaluated by implantation into the lungs of rats. *Cancer Lett.* 43: 173–177.
12. Grimmer G, Dettbarn G, Brune H, Deutsch-Wenzel R and Misfeld J (1982) Quantification of the carcinogenic effect of polycyclic aromatic hydrocarbons in used engine oil by topical application onto the skin of mice. *Int. Arch. Occup. Environ. Health* 50: 95–100.

13. Grimmer G, Jacob J, Dettbarn G and Naujack KW (1997) Determination of urinary metabolites of polycyclic aromatic hydrocarbons (PAH) for the risk assessment of PAH-exposed workers. *Int. Arch. Occup. Environ. Health* 69: 231 239.
14. Seidel A, Dahmann D, Krekeler H and Jacob J (2002) Biomonitoring of polycyclic aromatic compounds in the urine of mining workers occupationally exposed to diesel-exhaust. *Int. J. Hyg. Environment. Health* 204: 333–338.
15. Grimmer G, Dettbarn G and Jacob J (1993) Metabolism and excretion of polycyclic aromatic hydrocarbons in rat and in human. *Int. Arch. Occup. Environ. Health* 65: 189–199.
16. Bouchard M, Dodd C and Viau C (1994) Improved procedure for the high-performance liquid chromatographic determination of monohydroxylated PAH metabolites in urine. *J. Anal. Toxicol.* 18: 261–264.
17. Jongeneelen FJ, Leijdekkers CM, Bos RP, Theuws JLG and Henderson PT (1985) Excretion of 3-hydroxybenzo[*a*]pyrene and mutagenicity in rat urine after exposure to benzo[*a*]pyrene. *J. Appl. Toxicol.* 5: 277–282.
18. Jongeneelen FJ, Anzion RBM and Henderson PT (1987) Determination of hydroxylated metabolites of polycyclic aromatic hydrocarbons in urine. *J. Chromat.* 413: 227–232.
19. Jacob J, Grimmer G and Dettbarn G (1999) Profile of urinary phenanthrene metabolites in smokers and non-smokers. *Biomarkers* 4: 319–327.
20. Mannschreck C, Gündel J and Angerer J (1996) Occupational exposure to PAHs — biological monitoring of hydroxylated metabolites. *Polycycl. Aromat. Compds.* 11: 11–18.
21. Becher G and Bjørseth A (1983) Determination of exposure to polycyclic aromatic hydrocarbons by analysis of human urine. *Cancer Lett.* 17: 301–311.
22. Gündel J and Angerer J (2000) High-performance liquid chromatographic method with fluorescence detection for the determination of 3-hydroxybenzo[*a*]pyrene and 3-hydroxybenz[*a*]anthracene in the urine of polycyclic aromatic hydrocarbon-exposed workers. *J. Chromatogr. B* 738: 47–55.
23. Grimmer G, Naujack KW and Dettbarn G (1993) Determination of urinary metabolites of polycyclic aromatic hydrocarbons among coke oven workers. In: *INRS, Maitriser le risque au poste de travail.* Institut National de Recherche et de Securite, pp 227–233, Presses Universitaires de Nancy, France.
24. Gündel J, Schaller KH and Angerer J (2000) Occupational exposure to polycyclic aromatic hydrocarbons in a fireproof stone producing plant: biological

monitoring of 1-hydroxypyrene, 1-, 2-, 3- and 4-hydroxyphenanthrene, 3-hydroxybenz[*a*]anthracene and 3-hydroxybenzo[*a*]pyrene. *Int. Arch. Occup. Environ. Health* 73: 270–274.
25. Ariese F, Verkaik M, Hoornweg GP, van de Nesse RJ, Jukema-Leenstra SR, Hofstraat JW, Gooijer C and Velthorst NH (1994) Trace analysis of 3-hydroxybenzo[*a*]pyrene in urine for the biomonitoring of human exposure to polycyclic aromatic hydrocarbons. *J. Anal. Toxicol.* 18: 195–204.
26. Smith CJ, Grainger J and Patterson DG (1998) Separation of polycyclic aromatic hydrocarbon metabolites by γ-cyclodextrin-modified micellar electrokinetic chromatography with laser-induced fluorescence detection. *J. Chromatogr. A* 803: 241–245.
27. Doll R, Fisher REW, Gammo EJ, Gunn W, Hughes GO, Tyrer FH and Wilson W (1972) Mortality of gasworkers — final report of a prospective study. *Br. J. Ind. Med.* 29: 394–406.
28. Redmond CK, Strobino BR and Cypress RH (1976) Cancer experience among coke by-product workers. *Ann. N.Y. Acad. Sci.* 217: 102–115.
29. Redmond CK, Wieand HS, Rockette HE, Sass R and Winberg G (1981) *Long-Term Mortality Experience of Steelworkers.* DHHS [NIOSH] Publication No. 81–120, National Institute for Occupational Safety & Health, Cincinnati, OH, USA.
30. Lloyd JW (1971) Long-term mortality of steel-workers. Respiratory cancer in coke plant workers. *J. Occup. Med.* 13: 53–68.
31. Mazumbar S, Redmond C, Sollecito W and Sussman N (1975) An epidemiological study of exposure to coal tar pitch volatiles among coke oven workers. *J. Air Poll. Contr. Ass.* 25: 382–389.
32. Manz A, Berger J and Waltsgott H (1983) Zur Frage des Berufskrebses bei Beschäftigten der Gasindustrie — Kohortenstudie. Bundesanstalt für Arbeitsschutz und Unfallforschung, Wirtschaftsverlag, Bremerhaven.
33. Chau N, Bertrand JP, Mur JM, Figueredo A, Patris A, Moulin JJ and Pham QT (1993) Mortality in retired coke oven plant workers. *Br. J. Ind. Med.* 50: 127–135.
34. Constantino JP, Redmond CK and Bearden A (1995) Occupationally related cancer risk among coke oven workers: 30 years of follow-up. *J. Occup. Environ. Med.* 37: 597–604.
35. Karlehagen S, Andersen A and Ohlson CG (1992) Cancer incidence among creosote-exposed workers. *Scand. J. Work Environ. Health* 18: 26–29.
36. Malaiyandi M, Benedek A, Holko AP and Bancsi JJ (1982) Measurement of potentially hazardous polynuclear aromatic hydrocarbons from occupational

exposure during roofing and paving operations. In: *Polynuclear Aromatic Hydrocarbons: Physical and Biological Chemistry* [Cooke M, Dennis AJ and Fisher GL (eds.)] 6th International Symposium on Polycyclic Aromatic Compounds, pp 471–489, Battclle-Prcss, Columbus, OII, USA.

37. Rockette HE and Arena VC (1983) Mortality studies of aluminium reduction plant workers: potroom and carbon department. *J. Occup. Med.* 25: 549–557.
38. Gibbs GW (1985) Mortality of aluminium reduction plant workers, 1950 through 1977. *J. Occup. Med.* 27: 761–770.
39. Silverstein M, Meizlish N, Park R and Mirer F (1985) Mortality among workers exposed to coal tar pitch volatiles and welding emissions: an exercise in epidemiologic triage. *Am. J. Public Health* 75: 1283–1287.
40. Armstrong B, Tremblay C, Baris D and Theriault G (1985) Lung cancer mortality and polynuclear aromatic hydrocarbons: a case-cohort study of aluminium production workers in Arvida, Quebec, Canada. *Am. J. Epidemiol.* 139: 250–262.
41. Tola S, Koskela RS, Hernberg S and Järvinen E (1979) Lung cancer mortality among iron foundry workers. *J. Occup. Med.* 21: 753–760.
42. Thony C, Thony J, Lafontaine M and Limasset JC (1975) Occurrence of carcinogenic polycyclic aromatic hydrocarbons in some cutting oils. Study of the corresponding risk. *Arch. Malad. Prof. Mèd. Trav. Sécurité Soc.* 36: 37–51.
43. Bingham E, Trosset RP and Warshawsky D (1980) Carcinogenic potential of petroleum hydrocarbons. A critical review of the literature. *J. Environ. Pathol. Toxicol.* 3: 483–563.
44. IARC (1984) *Monographs on the Evaluation of the Carcinogenic Risk of Chemicals to Humans.* Volume 33: Polynuclear Aromatic Compounds, Part 2, Carbon Blacks, Mineral Oils and Some Nitroarenes. International Agency for Research on Cancer, Lyon, France.
45. Troitskaya NA, Velichkovsky BT, Kogan FM and Kuzminykh SI (1980) On carcinogenic hazards in the carbon black industry. *Vopr. Onkol.* 26: 63–67.
46. Blum S, Arp EW Jr, Smith AH and Tyroler HA (1979) Stomach cancer among rubber workers: an epidemiologic investigation. In: *Dusts and Disease* [Lemen R and Dement JM (eds.)] Proceedings of the Conference on Occupational Exposures to Fibrous and Particulate Dust and Their Extension into the Environment, pp 325–334, Pathotox Publishers, Park Forest South, IL, USA.

47. Parkes HG, Veys CA, Waterhouse JAH, Cook D and Peters AT (1982) Cancer mortality in the British rubber industry. *Br. J. Ind. Med.* 39: 209–220.
48. Pott P (1775) *Chirurgical Observations Relative to the Cataract, The Polypus of the Nose, The Cancer of the Scrotum, The Different Kinds of Ruptures, and The Mortification of the Toes and Feet.* Hawes, Clarke & Collins, London.
49. Hogstedt C, Andersson K, Frenning B and Gustavsson A (1982) A cohort study on mortality among long-time employed Swedish chimney sweeps. *Scand. J. Work Environ. Health* 8, Suppl 1: 72–78.
50. Cadez E (1983) Ergebnisse langjähriger Untersuchungen über den Einfluß von Schadstoffen auf die Gesundheit von Schornsteinfegern. *Staub Reinhalt. Luft* 43: 116–117.
51. Gustavsson P, Gustavsson A and Hogstedt C (1987) Excess mortality among Swedish chimney sweeps. *Br. J. Ind. Med.* 44: 738–743.
52. Gustavsson P and Gustavsson A (1988) Excess of cancer in Swedish chimney sweeps. *Br. J. Ind. Med.* 45: 777–781.
53. Hansen ES (1989) Cancer incidence in an occupational cohort exposed to bitumen fumes. *Scand. J. Work. Environ. Health* 15: 101–105.
54. Jöckel KH, Ahrens W, Wichmann HE, Becher H, Bolm-Audorff U, Jahn I, Molik B, Greiser E and Timm J (1992) Occupational and environmental hazards associated with lung cancer. *Int. J. Epidemiol.* 21: 202–213.
55. Partanen TJ, Boffetta P, Heikkilä PR, Frentzel-Beyme RR, Heederik D, Hours M, Järvholm BG, Kauppinen TP, Kromhout H, Langård S, Svane OB, Bernstein M, Bertazzi PA, Kogevinas M, Neuberger MA, Pannett B and Sunyer J (1995) Cancer risk for European asphalt workers. *Scand. J. Work Environ. Health* 21: 252–258.
56. Boffetta P, Burstyn I, Partanen T, Kromhout H, Svane O, Langård S, Järvholm B, Frentzel-Beyme R, Kauppinen T, Stücker I, Shaham J, Heederik D, Ahrens W, Bergdahl I, Cenée S, Ferro G, Heikkilä P, Hooiveld M, Johansen C, Randem B and Schill W (2001) *IARC Epidemiological Study of Cancer Mortality Among European Asphaltworkers.* IARC Internal Report No. 01/003, International Agency for Research on Cancer, Lyon, France.
57. Boffetta P, Burstyn I, Partanen T, Kromhout H, Svane O, Langård S, Järvholm B, Frentzel-Beyme R, Kauppinen T, Stücker I, Shaham J, Heederik D, Ahrens W, Bergdahl I, Cenée S, Ferro G, Heikkilä P, Hooiveld M, Johansen C, Randem BG and Schill W (2003) Cancer mortality among European asphalt workers: an international epidemiological study. I. Results of the analysis based on job titles. *Am. J. Ind. Med.* 43: 18–27.

58. Boffetta P, Burstyn I, Partanen T, Kromhout H, Svane O, Langård S, Järvholm B, Frentzel-Beyme R, Kauppinen T, Stücker I, Shaham J, Heederik D, Ahrens W, Bergdahl I, Cenée S, Ferro G, Heikkilä P, Hooiveld M, Johansen C, Randem BG and Schill W (2003) Cancer mortality among European asphalt workers: an international epidemiological study. II. Exposure to bitumen fume and other agents. *Am. J. Ind. Med.* 43: 28–39.
59. Milham S (1982) *Occupational Mortality in Washington State 1950–1979.* Prepared under contract No. 210-80-0088, National Institute for Occupational Safety and Health (NIOSH), Cincinnati, OH, USA.
60. Hammond EC, Selikof IJ, Lawther PL and Seidman H (1976) Inhalation of benzpyrene and cancer in man. *Ann. N.Y. Acad. Sci.* 271: 116–124.
61. German Federal Department for Worker Safety (1989) 23. TRK-Wert for benzo[*a*]pyrene. In: *Bundesarbeitsblatt 10*, p 58, Dortmund, Germany.
62. Lafontaine M, Attenont H, Hubert G, Taiclet A and Truy S (1990) Emission of polycyclic aromatic hydrocarbons in foundries. *Cah. Notes Doc.* 141: 799–807.
63. Grimmer G, Jacob J and Naujack KW (1997) Atmospheric emission of polycyclic aromatic hydrocarbons in sampling areas of the German environmental specimen bank. Method for the precise measurement of gaseous and particle-associated polycyclic aromatic hydrocarbons in the sub-nanogram range using deuterated internal standards. *Chemosphere* 34: 2213–2226.
64. Jacob J and Grimmer G (1996) Metabolism and excretion of polycyclic aromatic hydrocarbons in rat and in human. *Centr. Eur. J. Publ. Health* 4: 33–39.
65. Vorhoeve P and Compaan H (1996) *Polycyclische aromaten in het Nederlandse milieu.* TNO-MT Report. Nr. CL 80/37-122, Delft, The Netherlands.
66. Thakker DR, Levin W, Wood AW, Conney AH, Yagi H and Jerina DM (1988) Stereoselective biotransformation of polycyclic aromatic hydrocarbons to ultimate carcinogens. In: *Drug Stereochemistry — Analytical Methods and Pharmacology* [Wainer IW and Drayer DE (eds.)] pp 271–296, Marcel Dekker, New York.
67. Phillips DH and Grover PL (1994) Polycyclic hydrocarbon activation: bay regions and beyond. *Drug Metab. Rev.* 26: 443–467.
68. Szeliga J and Dipple A (1998) DNA adduct formation by polycyclic aromatic hydrocarbon dihydrodiol epoxides. *Chem. Res. Toxicol.* 11: 1–11.
69. Baird WM and Ralston SL (1997) Carcinogenic polycyclic aromatic hydrocarbons. In: *Comprehensive Toxicology* [Sipes IG, McQueen CA and

Gandolfi AJ (eds.)] Volume 12: Chemical Carcinogens and Anticarcinogens, pp 171–200, Pergamon Press, Oxford, UK.

70. Ross JA and Nesnow S (1999) Polycyclic aromatic hydrocarbons: correlations between DNA adducts and *ras* oncogene mutations. *Mutat. Res.* 424: 155–166.
71. Nau H (1994) Toxikokinetik. In: *Lehrbuch der Toxikologie* [Marquardt H and Schäfer SG (eds.)] p 49, BI-Wissenschaftsverlag, Mannheim, Leipzig, Wien, Zürich.
72. Horning MG, Sheng LS, Nowlin JG, Lertratanangkoon K and Horning EC (1987) Analytical methods for the study of urinary thioether metabolites in the rat and guinea pig. *J. Chromatogr.* 399: 303–319.
73. Mitchell CE (1982) Distribution and retention of benzo[*a*]pyrene in rats after inhalation. *Toxicol. Lett.* 11: 35–42.
74. Schoket B, Hewer A, Grover PL and Phillips DH (1988) Covalent binding of components of coal-tar, creosote and bitumen to the DNA of the skin and lungs of mice following topical application. *Carcinogenesis* 9: 1253–1258.
75. Randerath E, Mittal D and Randerath K (1988) Tissue distribution of covalent DNA damage in mice treated dermally with cigarette 'tar': preference for lung and heart DNA. *Carcinogenesis* 9: 75–80.
76. Carmichael PL, Jacob J, Grimmer G and Phillips DH (1990) Analysis of the polycyclic aromatic hydrocarbon content of petrol and diesel engine lubricating oils and determination of DNA adducts in topically treated mice by ^{32}P-postlabelling. *Carcinogenesis* 11: 2025–2032.
77. Aitio A (1974) Different elimination and effect on mixed function oxidase of 20-methylcholanthrene after intragastric and intraperitoneal administration. *Res. Commun. Chem. Pathol. Pharmacol.* 9: 701–710.
78. Grimmer G, Brune H, Dettbarn G, Heinrich U, Jacob J, Mohtashamipur E, Norpoth K, Pott F and Wenzel-Hartung R (1988) Urinary and faecal excretion of chrysene and chrysene metabolites by rat after oral, intraperitoneal, intratracheal or intrapulmonary application. *Arch. Toxicol.* 62: 401–405.
79. Jacob J, Brune H, Dettbarn G, Grimmer G, Heinrich U, Mohtashamipur E, Norpoth K, Pott F and Wenzel-Hartung R (1989) Urinary and faecal excretion of pyrene and hydroxypyrene by rats after oral, intraperitoneal, intratracheal or intrapulmonary application. *Cancer Lett.* 46: 15–20.
80. Grimmer G, Brune H, Dettbarn G, Jacob J, Mohtashamipur E, Norpoth K and Wenzel-Hartung R (1991) Urinary and faecal excretion of phenanthrene and phenanthrols by rats following oral, intraperitoneal, or intrapulmonary application. *Polycycl. Aromat. Compds.* 2: 39–47.

81. Viau C, Bouchard M, Carrier G, Brunet R and Krishnan K (1999) The toxicokinetics of pyrene and its metabolites in rats. *Toxicol. Lett.* 108: 201–207.
82. Bouchard M, Krishnan K and Viau C (1998) Kinetics of tissue distribution and elimination of pyrene and 1-hydroxypyrene following intravenous administration of [^{14}C]pyrene in rats. *Toxicol. Sci.* 46: 11–20.
83. Brzeznicki S, Jakubowski M and Czerski B (1997) Elimination of 1-hydroxypyrene after human volunteer exposure to polycyclic aromatic hydrocarbons. *Int. Arch. Occup. Environ. Health* 70: 257–260.
84. Viau C, Carrier G, Vyskocil A and Dodd C (1995) Urinary excretion kinetics of 1-hydroxypyrene in volunteers exposed to pyrene by the oral and dermal route. *Sci. Total. Environ.* 163: 179–186.
85. Buckley TJ and Lioy PJ (1992) An examination of the time course from human dietary exposure to polycyclic aromatic hydrocarbons to urinary elimination of 1-hydroxypyrene. *Br. J. Ind. Med.* 49: 113–124.
86. Möller M and Dybing E (1980) Mutagenicity studies with urine concentrates in the occupational environment. *J. Work. Environ. Health* 6: 216–220.
87. Venier P, Clonfero E, Cottica D, Gava C, Zordan M, Bozzoli L and Levis AG (1985) Mutagenic activity and polycyclic aromatic hydrocarbon levels in urine of workers exposed to coal tar pitch volatiles in an anode plant. *Carcinogenesis* 6: 749–752.
88. Kriebel D, Commoner B, Bollinger D, Bronsdon A, Gold J and Henry J (1983) Detection of occupational exposure to genotoxic agents with urinary mutagen assay. *Mutat. Res.* 108: 67–79.
89. Bos RP, Kromhout H, Ikink H, de Haan W, Koppejan J and Theuws JLG (1989) Mutagens in urine of non-smoking and smoking workers in an aircraft tyre retreading plant. Skin exposure as a causal factor? *Mutat. Res.* 223: 41–48.
90. Recio L, Enoch H, Hannah MA and Hill RH (1984) Application of urine mutagenicity to monitor coal liquifaction workers. *Mutat. Res.* 136: 210–217.
91. Lauwerys RR and Hoet P (1993) *Industrial Chemical Exposure. Guidelines for Biological Biomonitoring.* 2nd Edition. Lewis Publishers, CRC Press, Boca Raton, FL, USA.
92. Boyland E and Sims P (1964) Metabolism of polycyclic compounds. 23. The metabolism of pyrene in rats and rabbits. *Biochem. J.* 90: 391–398.
93. Harper KH (1957) The metabolism of pyrene. *Br. J. Cancer* 11: 499–507.
94. Harper KH (1958) The intermediary metabolism of pyrene. *Br. J. Cancer* 12: 116–120.

95. Keimig SD, Kirby KW, Morgan DP, Kaiser JE and Hubert TD (1983) Identification of 1-hydroxypyrene as a major metabolite of pyrene in pig urine. *Xenobiotica* 13: 415–420.
96. Jacob J, Grimmer G, Raab G and Schmoldt A (1982) The metabolism of pyrene by rat liver microsomes and the influence of various monooxygenase inducers. *Xenobiotica* 12: 45–53.
97. Grimmer G, Dettbarn G, Naujack KW and Jacob J (1994) Relationship between inhaled PAH and urinary excretion of phenanthrene, pyrene and benzo[*a*]pyrene metabolites in coke plant workers. *Polycycl. Aromat. Compds.* 5: 269–277.
98. Keimig SD and Morgan DP (1983) Establishing the dose-detection threshold of index polycyclic aromatic hydrocarbon compounds. In: *Polycyclic Aromatic Hydrocarbons* [Cooke M and Dennis AJ (eds.)] 7th International Symposium on Polycyclic Aromatic Compounds, pp 663–677, Battelle-Press, Columbus, OH, USA.
99. Jongeneelen FJ, Anzion RBM, Leijdekkers CM, Bos RP and Henderson PT (1985) 1-Hydroxypyrene in human urine after exposure to coal tar and coal tar derived products. *Int. Arch. Occup. Environ. Health* 57: 47–55.
100. Jongeneelen FJ and Anzion RBM (1991) 1-Hydroxypyrene In: *Analyses of Hazardous Substances in Biological Materials* [Angerer J and Schaller KH (eds.)] Volume 3, pp 151–169, Wiley-VCH, Weinheim, Germany.
101. Lintelmann J and Boos KS (1990) Determination of the PAH metabolite 1-hydroxypyrene by a HPLC-column switching technique. In: *InCom' 90. Instrumentalized Analytical Chemistry and Computer Technology.* [Günther W, Matthes JP and Perkampus HH (eds.)] pp 128–129, GIT-Verlag, Darmstadt, Germany.
102. Boos KS, Lintelmann J and Kettrup A (1992) Coupled-column high-performance liquid chromatographic method for the determination of 1-hydroxypyrene in urine of subjects exposed to polycyclic aromatic hydrocarbons. *J. Chromat.* 600: 189–194.
103. Jongeneelen FJ (1992) Biological exposure limit for occupational exposure to coal tar pitch volatiles at coke-ovens. *Int. Arch. Occup. Environ. Health* 63: 511–516.
104. Jongeneelen FJ (2001) Benchmark guideline for urinary 1-hydroxypyrene as biomarker of occupational exposure to polycyclic aromatic hydrocarbons. *Ann. Occup. Hyg.* 45: 3–13.
105. Vu-Duc T and Lafontaine M (1996) 1-Hydroxypyrene in human urines as biomarker of exposure to PAH in work related processes in an aluminium production plant. *Polycycl. Aromat. Compds.* 11: 1–9.

106. Viau C, Diakite A, Zuzgyte A, Tuchweber B, Blais C, Bouchard M and Vyscozil A (2002) Is 1-hydroxypyrene a reliable bioindicator of measured dietary polycyclic aromatic hydrocarbon under normal conditions? *J. Chromatogr. B, Analyt. Technol. Biomed. Life Sci.* 778: 165–177.
107. Strickland PT, Kang D, Bowman ED, Fitzwilliam A, Downing TE, Rothman N, Groopman JD and Weston A (1994) Identification of 1-hydroxypyrene glucuronide as a major pyrene metabolite in human urine by synchronous fluorescence spectroscopy and gas chromatography-mass spectrometry. *Carcinogenesis* 15: 483–487.
108. Strickland P and Kang D (1999) Urinary 1-hydroxypyrene and other PAH metabolites as biomarkers of exposure to environmental PAH in air particulate matter. *Toxicol. Lett.* 108: 191–199.
109. Singh R, Tucek M, Maxa K, Tenglerova J and Weyand EH (1995) A rapid and simple method for the analysis of 1-hydroxypyrene glucuronide: a potential biomarker for polycyclic aromatic hydrocarbon exposure. *Carcinogenesis* 16: 2909–2915.
110. Lee CK, Cho SH, Kang JW, Lee SJ, Lu YS, Sung J, Strickland PT and Kang D (1999) Comparison of three analytical methods for 1-hydroxypyrene glucuronide in urine after non-occupational exposure to polycyclic aromatic hydrocarbons. *Toxicol. Lett.* 108: 209–215.
111. Kang D, Rothman N, Cho SH, Lim HS, Kwon HJ, Kim SM, Schwartz B and Strickland PT (1995) Association of exposure to polycyclic aromatic hydrocarbons (estimated from job category) with concentration of 1-hydroxypyrene glucuronide in urine from workers at a steel plant. *Occup. Environ. Med.* 52: 593–599.
112. Strickland P, Kang D and Sithisarankul P (1996) Polycyclic aromatic hydrocarbon metabolites in urine as biomarkers of exposure and effect. *Environ. Health Persp.* 104, Suppl 5: 927–932.
113. Hong YC, Leem JH, Park HS, Lee KH, Lee SJ, Lee CK and Kang D (1999) Variations in urinary 1-hydroxypyrene glucuronide in relation to smoking and the modification effects of GSTM1 and GSTT1. *Toxicol. Lett.* 108: 217–223.
114. Kang DH, Rothman N, Poirier MC, Greenberg A, Hsu CH, Schwartz BS, Baser ME, Groopman JD, Weston A and Strickland PT (1995) Interindividual differences in the concentration of 1-hydroxypyrene-glucuronide in urine and polycyclic aromatic hydrocarbon-DNA adducts in peripheral white blood cells after charbroiled beef consumption. *Carcinogenesis* 16: 1079–1085.

115. Sithisarankul P, Vineis P, Kang D, Rothman N, Caporaso N and Strickland P (1997) Association of 1-hydroxypyrene-glucuronide in human urine with cigarette smoking and broiling or roasted meat consumption. *Biomarkers* 2: 217–221.
116. Bartsch H, Nair U, Risch A, Rojas M, Wikman H and Alexandrov K (2000) Genetic polymorphism of *CYP* genes, alone or in combination, as a risk modifier of tobacco-related cancers. *Cancer Epidemiol. Biomarkers Prev.* 9: 3–28.
117. Autrup H (2000) Genetic polymorphisms in human xenobiotica metabolizing enzymes as susceptibility factors in toxic response. *Mutat. Res.* 464: 65–76.
118. Wu MT, Huang SL, Ho CK, Yeh YF and Christiani DC (1998) Cytochrome *P4501A1 Msp*I polymorphism and urinary 1-hydroxypyrene concentrations in coke-oven workers. *Cancer Epidemiol. Biomarkers Prev.* 7: 823–829.
119. Merlo F, Andreassen A, Weston A, Pan CF, Haugen A, Valerio F, Reggiardo G, Fontana V, Garte S, Puntoni R and Abbondandolo A (1998) Urinary excetion of 1-hydroxypyrene as a marker of exposure to urban air level of polycyclic aromatic hydrocarbons. *Cancer Epidemiol. Biomarkers Prev.* 7: 147–155.
120. Nerurkar PV, Okinaka L, Aoki C, Seifried A, Lum-Jones A, Wilkens LR and LeMarchand L (2000) CYP1A1, GSTM1, and GSTP1 genetic polymorphisms and urinary 1-hydroxypyrene excretion in non-occupationally exposed individuals. *Cancer Epidemiol. Biomarkers Prev.* 9: 1119–1122.
121. Pan G, Hanaoka T, Yamano Y, Hara K, Ichiba M, Wang Y, Zhang J, Feng Y, Shujuan Z, Guan D, Gao G, Liu N and Takahashi K (1998) A study of multiple biomarkers in coke oven workers — a cross-sectional study in China. *Carcinogenesis* 19: 1963–1968.
122. Alexandrie AK, Warholm M, Carstensen U, Axmon A, Hagmar L, Levin JO, Östman C and Rannug A (2000) CYP1A1 and GSTM1 polymorphisms affect urinary 1-hydroxypyrene levels after PAH exposure. *Carcinogenesis* 21: 669–676.
123. Schoket B, Papp G, Lévay K, Mracková G, Kadlubar FF and Vincze I (2001) Impact of metabolic genotypes on levels of biomarkers of genotoxic exposure. *Mutat. Res.* 482: 57–69.
124. Nan HM, Kim H, Lim HS, Choi JK, Kawamoto T, Kang JW, Lee CH, Kim YD and Kwon EH (2001) Effects of occupation, lifestyle and genetic polymorphisms of CYP1A1, CYP2E1, GSTM1 and GSTT1 on urinary 1-hydroxypyrene and 2-naphthol concentrations. *Carcinogenesis,* 22: 787–793.
125. Brescia G, Celotti L, Clonfero E, Neumann HG, Forni A, Foá V, Pisoni M, Ferri GM and Assenato G (1999) The influence of cytochrome P450 1A1

and glutathione *S*-transferase M1 genotype on biomarker levels in coke-oven workers. *Arch. Toxicol.* 73: 431–439.

126. Øvrebø S, Ryberg D, Haugen A and Leira HL (1998) Glutathione *S*-transferase M1 and P1 genotypes and urinary excretion of 1-hydroxypyrene in coke-oven workers. *Sci. Total Environ.* 220: 25–31.
127. Zhang J, Ichiba M, Zhang S, Hanaoka T, Pan G, Yamano Y, Takahashi K and Tomokuni K (2001) Urinary 1-hydroxypyrene in coke oven workers relative to exposure, alcohol consumption, and metabolic enzymes. *Occup. Environ. Med.* 58: 716–721.
128. Poirier MC, Weston A, Schoket B, Shamkhani H, Pan CF, McDiarmid MA, Scott BG, Deeter DP, Heller JM, Jacobsen-Kram D and Rothman N (1998) Biomonitoring of United States army soldiers serving in Kuwait in 1991. *Cancer Epidemiol. Biomarkers Prev.* 7: 545–551.
129. Van Delft JHM, Steenwinkel MJST, van Asten JG, de Vogel N, Bruijntjes-Rozier TCDM, Schouten T, Cramers P, Maas L, van Herwijnen MH, van Schooten FJ and Hopmans MJ (2001) Biological monitoring the exposure to polycyclic aromatic hydrocarbons of coke-oven workers in relation to smoking and genetic polymorphisms for *GSTM1* and *GSTT1*. *Ann. Occup. Hyg.* 45: 395–408.
130. Boyland E and Wolf G (1950) Metabolism of polycyclic compounds. 6. Conversion of phenanthrene into dihydroxydihydrophenanthrene. *Biochem. J.* 47: 64–69.
131. Boyland E and Sims P (1962) Metabolism of polycyclic compounds. 20. The metabolism of phenanthrene in rabbits and rats: mercapturic acids and related compounds. *Biochem. J.* 84: 564–570.
132. Boyland E and Sims P (1962) Metabolism of polycyclic compounds. 21. The metabolism of phenanthrene in rabbits and rats: dihydrodihydroxy compounds and related glucosiduronic acids. *Biochem. J.* 84: 571–582.
133. Boyland E and Sims P (1962) Metabolism of polycyclic compounds. 22. The metabolism of (±)-*trans*-9,10-dihydro-9,10-dihydroxyphenanthrene in rats. *Biochem. J.* 84: 583–586.
134. Sims P (1962) Metabolism of polycyclic compounds. 19. The metabolism of phenanthrene in rabbits and rats: phenols and sulphuric esters. *Biochem. J.* 84: 558–563.
135. Lertratanangkoon K, Horning MG, Middleditch BS, Tsang WS and Griffin GW (1982) Urinary bivalent sulfur metabolites of phenanthrene excreted by the rat and guinea pig. *Drug Metab. Dispos.* 10: 614–623.

136. Struble CB, Larsen GL, Feil VJ and Bakke JE (1986) Metabolism of 9-hydroxy-9,10-dihydro-10-cysteinyl-phenanthrene in rats. In: *Polynuclear Aromatic Hydrocarbons: Chemistry, Characterization and Carcinogenesis* [Cooke M and Dennis AJ (eds.)] 9th International Symposium on Polycyclic Aromatic Compounds, pp 893–900, Battelle-Press, Columbus, OH, USA.
137. Gmeiner G, Krassnig C, Schmidt E and Tausch H (1998) Fast screening method for the profile analysis of polycyclic aromatic hydrocarbon metabolites in urine using derivatisation-solid-phase microextraction. *J. Chromatogr. B* 705: 132–138.
138. Elovaara E, Väänänen V and Mikkola J (2003) Simultaneous analysis of naphthols, phenanthrols, and 1-hydroxypyrene in urine as biomarkers of polycyclic aromatic hydrocarbon exposure: intraindividual variance in the urinary metabolite excretion profiles caused by intervention with β-naphthoflavone induction in the rat. *Arch. Toxicol.* 77: 183–193.
139. Angerer J, Mannschreck C and Gündel J (1997) Biological monitoring and biochemical effect monitoring of exposure to polycyclic aromatic hydrocarbons. *Int. Arch. Occup. Environ. Health* 70: 365–377.
140. Lintelmann J (1999) PAH metabolites (1-hydroxyphenanthrene, 4-hydroxyphenanthrene, 9-hydroxyphenanthrene, 1-hydroxypyrene). In: *Analyses of Hazardous Substances in Biological Materials* [Angerer J and Schaller KH (eds.)] Volume 6, pp 163–185, Deutsche Forschungsgemeinschaft (DFG), Wiley-VCH, New York.
141. Schedl M, Wilharm G, Achatz S, Kettrup A, Niessner R and Knopp D (2001) Monitoring polycyclic aromatic hydrocarbons in human urine: extraction and purification with a sol-gel glass immunosorbent. *Anal. Chem.* 73: 5669–5676.
142. Smith CJ, Walcott CJ, Huang W, Maggio V, Grainger J and Patterson DG (2002) Determination of selected monohydroxy metabolites of 2-, 3- and 4-ring polycyclic aromatic hydrocarbons in urine by solid-phase microextraction and isotope dilution gas chromatography-mass spectrometry. *J. Chromatogr. B, Anal. Technol. Biomed. Life Sci.* 778: 157–164.
143. Hollender J, Koch B and Dott W (2000) Biomonitoring of environmental polycyclic aromatic hydrocarbon exposure by simultaneous measurement of urinary phenanthrene, pyrene and benzo[*a*]pyrene hydroxides. *J. Chromatogr. B* 739: 225–229.
144. Xu X and Hurtubise RJ (1999) Influence of organic solvents in the capillary zone electrophoresis of polycyclic aromatic hydrocarbon metabolites. *J. Chromatogr. A* 829: 289–299.

145. Krismann U and Kleiböhmer W (1997) Separation of hydroxylated polycyclic aromatic hydrocarbons by micellar electrokinetic capillary chromatography. *J. Chromatogr. A* 774: 193–201.
146. Kuijt J, García Ruiz C, Stroomberg GJ, Marina ML, Ariese F, Brinkman UAT and Gooijer C (2001) Laser-induced fluorescence detection at 266 nm in capillary electrophoresis. Polycyclic aromatic hydrocarbon metabolites in biota. *J. Chromatogr. A* 907: 291–299.
147. Jacob J, Raab G, Soballa V, Schmalix W, Grimmer G, Greim H, Doehmer J and Seidel A (1996) Cytochrome P450-mediated activation of phenanthrene in genetically engineered V79 Chinese hamster cells. *Environ. Toxicol. Pharmacol.* 1: 1–11.
148. Jacob J, Doehmer J, Grimmer G, Soballa V, Raab G, Seidel A and Greim H (1996) Metabolism of phenanthrene, benz[*a*]anthracene, benzo[*a*]pyrene and benzo[*c*]phenanthrene by eight cDNA-expressed human and rat cytochromes. *Polycycl. Aromat. Compds.* 10: 1–9.
149. Hoepfner I, Dettbarn G, Scherer G, Grimmer G and Adlkofer F (1987) Hydroxyphenanthrenes in the urine of non-smokers. *Toxicol. Lett.* 35: 67–71.
150. Martin F, Hoepfner I, Scherer G, Adlkofer F, Dettbarn G and Grimmer G (1989) Urinary excretion of hydroxyphenanthrenes after intake of polycyclic aromatic hydrocarbons. *Environ. Intern.* 15: 41–47.
151. Jacob J, Brune H, Grimmer G, Heinrich U, Mohtashamipur E, Norpoth K, Pott F and Wenzel-Hartung R (1990) Urinary and faecal excretion of metabolites after various modes of administration of polycyclic aromatic hydrocarbons (PAH) to rats. In: *Environmental Hygiene* [Seemayer NH and Hadnagy W (eds.)] Volume II, pp 87–90, Springer-Verlag, New York.
152. Simpson CD, Wu MT, Christiani DC, Santella RM, Carmella SG and Hecht SS (2000) Determination of *r*-7,*t*-8,9,*c*-10-tetrahydroxy-7,8,9,10-tetrahydrobenzo[*a*]pyrene in human urine by gas chromatography/negative ion chemical ionization/mass spectrometry. *Chem. Res. Toxicol.* 13: 271–280.
153. Bowman ED, Rothman N, Hackl C, Santella RM and Weston A (1997) Interindividual variation in the levels of certain urinary polycyclic aromatic hydrocarbon metabolites following medicinal exposure to coal tar ointment. *Biomarkers* 2: 321–327.
154. Weston A, Bowman ED, Carr P, Rothman N and Strickland PT (1993) Detection of metabolites of polycyclic aromatic hydrocarbons in human urine. *Carcinogenesis* 14: 1053–1055.

155. Wu MT, Simpson CD, Christiani DC and Hecht SS (2002) Relationship of exposure to coke-oven emissions and urinary metabolites of benzo[*a*]pyrene and pyrene in coke-oven workers. *Cancer Epidemiol. Biomarkers Prev.* 11: 311–314.
156. Phillips DH (1999) Polycyclic aromatic hydrocarbons in the diet. *Mutat. Res.* 443: 139–147.
157. Lioy SH, Waldman JM, Greenberg A, Harkov R and Pietarinen C (1988) The Total Human Environmental Exposure Study (THEES) to benzo[*a*]pyrene: comparison of the inhalation and food pathways. *Arch. Environ. Health* 43: 304–312.
158. Waldman JM, Lioy PL, Greenberg A and Butler JP (1991) Analysis of human exposure to benzo[*a*]pyrene via inhalation and food ingestion in the total human environmental exposure study (THEES). *J. Exp. Anal. Environ. Epidem.* 1: 193–225.
159. Dipple A and Bigger CAH (1991) Mechanism of action of food-associated polycyclic aromatic hydrocarbon carcinogens. *Mutat. Res.* 259: 263–276.
160. Alexandrov K, Rojas M, Kadlubar FF, Lang NP and Bartsch H (1996) Evidence of *anti*-benzo[*a*]pyrene diolepoxide-DNA adduct formation in human colon mucosa. *Carcinogenesis* 17: 2081–2083.
161. Braszczynska Z, Lindscheid D and Osinska R (1978) Evaluation of the influence of new technologies in coke industry on aromatic polycyclic hydrocarbons concentrations in ambient air. *Med. Pr.* 29: 357–364.
162. Lindstedt G and Sollenberg J (1982) Polycyclic aromatic hydrocarbons in the occupational environment with special reference to benzo[*a*]pyrene measurements in Swedish industry. *Scand. J. Work Environ. Health* 8: 1–19.
163. Blome H (1981) Messungen polycyclischer aromatischer Kohlenwasserstoffe an Arbeitsplätzen — Beurteilung der Ergebnisse. *Staub Reinhalt. Luft* 41: 225–229.
164. Bjørseth A, Bjørseth O and Fjeldstad PE (1978) *Characterization of PAH in Work Atmospheres.* Report No. 6, Royal Norwegian Council for Scientific and Industrial Research, Oslo, Norway.
165. Blome H (1983) *Polyzyklische Aromatische Kohlenwasserstoffe (PAH) am Arbeitsplatz.* BIA-Report 3/83, Berufsgenossenschaftliches Institut für Arbeitsschutz, Sankt Augustin, Druck Center Meckenheim, Germany.
166. Verma DK, Muir DCF, Cunliffe S, Julian JA, Vogt JH, Rosenfeld J and Chovli A (1982) Polycyclic aromatic hydrocarbons in Ontario foundry emissions. *Ann. Occup. Hyg.* 25: 17–25.

167. Knecht U, Elliehausen H and Woitowitz HJ (1986) Gaseous and adsorbed PAH in an iron foundry. *Br. J. Ind. Med.* 43: 834–838.
168. Coenen W (1988) Expositionssituation in Gießereien. *Staub Reinhalt. Luft* 48: 189–195.
169. Schimberg RW (1981) Industrial hygienic measurements of polycyclic aromatic hydrocarbons in foundries. In: *Chemical Analysis and Biological Fate: Polynuclear Aromatic Hydrocarbons* [Cooke M and Dennis AJ (eds.)] 5th International Symposium on Polycyclic Aromatic Compounds, pp 755–762, Battelle-Press, Columbus, OH, USA.
170. Knecht U, Bolm-Audorff U and Woitowitz HJ (1989) Atmospheric concentrations of polycyclic aromatic hydrocarbons during chimney sweeping. *Br. J. Ind. Med.* 46: 479–482.
171. Tobias S, Knecht U and Woitowitz HJ (1990) Verringerung der Krebsgefährdung durch Verwendung von Bitumen im Schwarzdeckenbau. In: *Verhandlungen der Deutschen Gesellschaft für Arbeitsmedizin* [Meyer-Falke A and Jansen G (eds.)] pp 239–241, Gentner-Verlag, Stuttgart.
172. Hansen ES (1989) Cancer incidence in an occupational cohort exposed to bitumen fumes. *J. Work. Environ. Health* 15: 101–105.
173. Tharr DG (1982) *Health Hazard Evaluation Report: Roofing Sites, Rochester and Buffalo, NY.* US Department of Health and Human Services, Public Health Service, Centers for Disease Control, National Institute for Occupational Safety and Health, NIOSH Report No. HETA 81-432-1105, NTIS No. PB-84-141-860/A02, Cincinnati, OH, USA.
174. Schunk W (1979) Zur Frage der Beziehung zwischen der Exposition gegenüber polycyclischen Kohlenwasserstoffen und der Häufigkeit von Krebserkrankungen in zwei Chemiebetrieben Thüringens. *Z. Ärztl. Fortb.* 7384–7388.
175. Lesage J, Perrault G and Durand P (1987) Evaluation of worker exposure to polycyclic aromatic hydrocarbons. *Am. Ind. Hyg. Assoc. J.* 48: 753–759.
176. Williams TM, Harris RL, Arp EW, Symons MJ and van Ert MD (1980) Worker exposure to chemical agents in the manufacture of rubber tires and tubes. *Am. Ind. Hyg. Assoc. J.* 41: 204–211.
177. Enwald E (1982) *How Safe Is It to Work With Rubber?* Proceedings of the Scandinavian Rubber Conference, Helsinki, Finland.
178. Grimmer G, Dettbarn G, Naujack KW and Jacob J (1991) Excretion of hydroxy derivatives of polycyclic aromatic hydrocarbons of the masses 178, 202, 228 and 252 in the urine of coke and road workers. *Int. J. Environ. Anal. Chem.* 43: 177–186.

179. Zhao ZH, Quan WY and Tian DH (1990) Urinary 1-hydroxypyrene as an indicator of human exposure to ambient polycyclic aromatic hydrocarbons in coal-burning environment. *Sci. Total Environ.* 92: 145–154.
180. Angerer J, Schaller KH and Lehnert G (1993) Biomonitoring of occupational exposures towards polycyclic aromatic hydrocarbons — HPLC determinations of the metabolite 1-hydroxypyrene in urine. In: *INRS, Maitriser le risque au poste de travail.* Institut National de Recherche et de Securite, pp 163–171, Presses Universitaires de Nancy, France.
181. Granella M and Clonfero E (1993) Urinary excretion of 1-pyrenol in automotive repair workers. *Int. Arch. Occup. Environ. Health* 65: 241–245.
182. Clonfero E, Zordan M, Venier P, Paleologo M, Levis AG, Cottica D, Bozzoli L, Jongeneelen FJ, Bos RP and Anzion RBM (1989) Biological monitoring of human exposure to coal tar. Urinary excretion of total PAH, 1-hydroxypyrene and mutagens in psoriatic patients. *Int. Arch. Occup. Environ. Health* 61: 363–369.
183. Bentsen-Farmen RK, Botnen IV, Notø H, Jacob J and Øvrebø S (1999) Detection of polycyclic aromatic hydrocarbon metabolites by high-pressure liquid chromatography after purification on immunoaffinity columns in urine from occupationally exposed workers. *Int. Arch. Occup. Environ. Health* 72: 161–168.
184. Jongeneelen FJ, Bos RP, Anzion RBM, Theuws JLG and Henderson PT (1986) Biological monitoring of polycyclic aromatic hydrocarbons: metabolites in urine. *J. Work Environ. Health* 12: 137–143.
185. Jongeneelen FJ, Scheepers PTJ, Groenedijk A, van Aerts LA, Anzion RBM, Bos RP and Veenstra SJ (1988) Airborne concentrations, skin contamination, urinary metabolite excretion of polycyclic aromatic hydrocarbons among paving workers exposed to coal tar-derived road tars. *Am. Ind. Hyg. Assoc. J.* 49: 600–607.
186. Knecht U and Woitowitz HJ (1989) Risk of cancer from the use of tar bitumen in road works. *Br. J. Ind. Med.* 46: 24–30.
187. Levin JO, Rhén M and Sikström E (1995) Occupational PAH exposure: urinary 1-hydroxypyrene levels of coke oven workers, aluminium smelter pot-room workers, road pavers, and occupationally non-exposed persons in Sweden. *Sci. Total Environ.* 163: 169–177.
188. Strunk P, Ortlepp K, Heinz H, Rossbach B and Angerer J (2002) Ambient and biological monitoring of coke plant workers — determination of exposure to polycyclic aromatic hydrocarbons. *Int. Arch. Occup. Environ. Health* 75: 354–358.

4 Macromolecular Adducts as Biomarkers of Human Exposure to Polycyclic Aromatic Hydrocarbons

David H. Phillips

Institute of Cancer Research, Sutton, Surrey SM2 5NG, UK
E-mail: david.phillips@icr.ac.uk

4.1 Introduction

Since the discovery seventy years ago that members of the class of polycyclic aromatic hydrocarbons (PAHs) were the carcinogenic components of coal tar, it has become clear that these products of incomplete combustion of fossil fuels, vegetation and indeed, any carbonaceous material, are ubiquitous.[1] The demonstration in the 1960s of a correlation between carcinogenic potency of PAHs and extent of covalent modification of DNA in target organs was a fundamental discovery that identified DNA as the critical cellular target in carcinogenesis,[2] and the demonstration a decade later that a bay-region diol-epoxide was the ultimately reactive metabolite of benzo[*a*]pyrene (B[*a*]P)[3] paved the way to understanding the mechanism of action of this class of carcinogens. It is now clear that carcinogenic PAHs undergo metabolic activation and that DNA adduct formation is an essential step by which they and other genotoxic carcinogens exert their biological effects (cf. Chapters 2, 5 and 6). Several methods have been developed in the last 20 years that enable DNA adduct formation to be detected and quantitated in human subjects. Monitoring human exposure to carcinogens by means of DNA adduct formation provides an integrated measurement of carcinogen intake, metabolic activation and delivery to the target macromolecule in the target tissue. Although proteins are not the primary target for carcinogenesis, PAH modification of proteins, typically haemoglobin in red blood cells and albumin in serum, is also a useful way of monitoring human exposure to the compounds.

Human exposure to PAHs can be broadly considered under several headings: tobacco smoking, occupational, environmental and diet. The literature on smoking-related DNA and protein adducts, including those formed by PAHs, in human tissues is vast and has been reviewed recently;[4] therefore it will not be considered in detail here, except to mention where smoking may confound the attribution of PAH-DNA and PAH-protein adduct formation to other sources of exposure to the compounds.

4.2 Methods of Detection

A number of methods with diverse approaches to the detection of PAH-DNA adducts are available. All of these have been developed within the last 20 years, prior to which it was not feasible to monitor human exposure to PAHs or other environmental carcinogens by measuring DNA adducts in human tissues. However, since the development of these methods, many studies have been carried out to validate them in human populations and to use them to determine the effects of occupational and environmental exposure. Similarly, sensitive methods for the detection of PAH-protein adducts have been developed. The principles and applications of the DNA and protein adduct methods have been described extensively and the interested reader is referred to published reviews for a more complete treatment of their principles and uses.[5–13]

Because PAHs are highly fluorescent molecules that give rise to fluorescent DNA adducts, detection of specific fluorescent signals in DNA is a sensitive means of detecting PAH-DNA adducts. One method used, synchronous fluorescence spectroscopy (SFS) generally requires larger DNA samples than alternative methods. A combination of HPLC separation of adducts or hydrolysis products, combined with fluorescence detection, has been developed as a specific method for the detection of B[*a*]P-7,8-diol-9,10-epoxide (B[*a*]PDE)-DNA adducts (cf. Figure 2.7 in Chapter 2, and Chapter 3).

The production of antibodies to B[*a*]PDE-DNA adducts has made possible the development of several sensitive immunochemical techniques; these include enzyme-linked immunosorbent assay (ELISA), ultra sensitive enzyme radioimmunoassay (USERIA) and dissociation enhanced lanthanide fluoroimmunoassay (DELFIA). Because antibodies raised against B[*a*]PDE-DNA adducts show extensive cross-reaction with other PAH-DNA adducts, they can be used effectively as class-specific antibodies to detect PAH-DNA adducts in general in human tissues. Also, because the presence of PAH-DNA adducts in human tissues can elicit an immunological response in the host, detection of antibodies to B[*a*]PDE-DNA in human sera can be used as a biomarker of exposure to PAHs.

The ultrasensitive ^{32}P-postlabelling assay has found widespread application in human biomonitoring studies. In this procedure, DNA is digested and the resultant carcinogen-modified nucleotides are radiolabelled with [^{32}P]orthophosphate. Although it is not specific for PAHs, under certain digestion, labelling and chromatographic conditions, the assay is selective for bulky aromatic and/or hydrophobic adducts whose characteristics match those of PAH-DNA adducts. Nevertheless, it should be remembered that the true nature of the adducts detected in human tissues by the ^{32}P-postlabelling assay is, in most instances, not fully characterized.

Finally, PAH-protein adduct detection provides an alternative to DNA adduct detection.[14] Methods used for PAHs, bound either to albumin or haemoglobin, have either used antibodies or mass spectrometric methods to detect the bound or hydrolyzed products.

4.3 Occupational Exposure to PAHs

4.3.1 Iron and Steel Production

In an early ^{32}P-postlabelling analysis of Finnish iron foundry workers, various aromatic DNA adducts were detected in white blood cell DNA.[15] For various categories of exposure, the frequency of positive samples was 3/4 from the high-exposure group, 8/10 from medium exposure, 4/18 from low exposure and 1/9 from unexposed controls. In a subsequent study of DNA isolated from the white blood cells of 53 iron foundry workers, analyses were carried out by ^{32}P-postlabelling in 3 different laboratories.[16] In all 3 laboratories, levels were found to be significantly higher in the exposed workers than in a small group ($n = 6$) of controls, and the correlation coefficients between laboratories were all statistically highly significant. In experimental studies, *in vitro* metabolic activation of foundry filter extracts in the presence of DNA led to detection by ^{32}P-postlabelling of a series of adducts, one of which had the chromatographic characteristics of the major B[*a*]PDE-DNA adduct.[17] Another study of Finnish iron foundry workers ($n = 35$) and controls ($n = 10$) using ELISA also found a significant relationship between exposure to B[*a*]P and DNA adduct levels, after adjustment for smoking.[18]

More detailed analysis of workers by both ^{32}P-postlabelling and immunoassay has identified particular job categories as being particularly associated with high levels of DNA adducts.[19] Jobs of men with high adduct levels included sand preparation, moulding, shake-out and transport; pattern making, melting and fettling were 'low adduct' jobs. In another study the upward trend in adduct levels, measured by ELISA, with increasing PAH exposure was of borderline significance ($p = 0.06$) but there was a highly significant correlation between adduct levels and mutation frequencies at the *hprt* locus[20] (cf. Chapter 8). In a follow-up study, analysis of the same workers ($n = 24$) one year later revealed a significant drop in DNA adducts, which resulted in a significant correlation with exposure levels ($p = 0.05$) on the one hand, but in a lack of correlation with *hprt* mutation frequencies on the other; only when the analysis was restricted to those workers with detectable levels of PAH-DNA adducts ($n = 17$) was the correlation significant ($p = 0.005$).[21] The overall conclusions from the study of this group of Finnish workers for several years is that DNA adducts (on some occasions measured by ^{32}P-postlabelling, on others by ELISA or by both methods) can be detected in white blood cells in individuals whose occupational exposure is to 5 ng/m^3 B[*a*]P or higher;[22] polymorphisms in glutathione *S*-transferase (i.e., *GSTM1*) and cytochrome P450 genes (i.e., *CYP1A1*) do not appear to have a strong influence on adduct levels (cf. Chapters 2, 3 and 10).

Protein adducts have also been analyzed in iron foundry workers. However, it was found in one study that B[*a*]PDE-albumin adducts were similar in foundry workers (n = 70) and age-matched controls (n = 68), and there was no correlation between adducts and urinary levels of 1-hydroxypyrene (1-OHP; cf. Chapter 3).[23] In another study, PAH-albumin adducts were at higher levels in foundry workers than in a control group, but the number of subjects was small (13 workers, 12 controls) and the difference was of borderline significance.[24] However, adduct levels in the foundry workers were significantly higher after 6 weeks at work than they were immediately after 4 weeks on vacation ($p < 0.05$).

4.3.2 Aluminum Production

In the manufacture of aluminum, an electrolytic process, PAHs are produced from the cell anodes, which are made of ground coke and coal tar pitch. Aluminum production workers are therefore potentially exposed to PAHs by inhalation and skin absorption, and they have an elevated risk of lung and bladder cancer (cf. Chapter 3).

A number of studies have measured PAH-DNA adducts in aluminum production workers. Using the relatively insensitive method of SFS, only one of 30 workers tested positive for B[*a*]PDE-DNA adducts in white blood cells.[25] In a ^{32}P-postlabelling study of workers at two Hungarian plants, significantly higher levels of adducts were found in workers at one of the plants, but not at the other, compared with a control group.[26] However, a subsequent investigation of workers from these plants by both ^{32}P-postlabelling and ELISA found significantly elevated levels of adducts in the samples from both plants, by both analytical methods.[27] In a study of aluminum production workers by Dutch investigators,[28] PAH-DNA adducts in white blood cells detected by ^{32}P-postlabelling correlated well with estimated PAH exposure levels for the different job categories, but there was considerable inter-individual variation within job categories. However in another study using both USERIA and ^{32}P-postlabelling, no differences were noted in adduct levels between workers and controls by the former assay, while with the latter test mean levels were only slightly elevated in workers and the difference was not statistically significant.[29] When a group of 105 aluminum workers were monitored for the presence of antibodies to B[*a*]PDE-DNA adducts in their blood, 14 tested positive, compared with one out of 60 office workers.[30] The aluminum workers comprised 15 personnel not exposed to PAHs (2 positive), 15 with low exposure (1 positive), 54 with medium exposure (9 positive) and 21 with high exposure (2 positive). Thus this study did not show a strong relationship between PAH exposure and positivity for PAH-DNA antibodies in blood.

More recent studies have looked at the influence of genetic polymorphisms on PAH-DNA adduct levels in aluminum production workers (cf. Chapters 3 and 10). In the Hungarian workers, a significant higher level of adducts in individuals with *GSTM1* null genotype was observed with ELISA analysis, but not ^{32}P-postlabelling.[31] However, stronger correlations were observed with combinations of *GSTM1* and *GSTP1* genotypes. In a study of 98 potroom workers and 55 unexposed controls that analyzed lymphocyte DNA by ^{32}P-postlabelling, no increase in adduct levels was observed in the workers relative to the controls.[32] Subsequently, however, some associations between specific DNA adduct peaks detected by HPLC-^{32}P-postlabelling and polymorphisms in the gene of NAD(P)H-dependent quinone oxidoreductase 1 (NQO1 = DT-diaphorase) and *GSTM1* were observed[33] (cf. Chapters 2 and 10).

4.3.3 Coke Ovens and Graphite Electrode Manufacture

The process of turning coal into coke, carried out on a large scale to provide fuel for steel making, produces large quantities of PAHs and puts the workforce at risk from exposure. There have been a large number of studies of coke oven workers and some that have included workers in carbon electrodes, involving monitoring blood cell DNA for the presence of aromatic DNA adducts. The studies and their main findings are summarized in Table 4.1.

Table 4.1: Studies of DNA adducts in coke oven and carbon anode productions workers

Study	*Source of DNA*	*Method of analysis*	*Workplace*	*Subjects*	*Outcome*
Harris *et al.*[34]	Monocytes	USERIA, SFS, antibodies to B[*a*]PDE-DNA adducts in serum by ELISA	Coke oven	41 exposed workers 9 controls	Of 27 tested by USERIA, 18 had detectable adducts. Using SFS, 10 showed no detectable signal. 11/41 were positive for antibodies.
Haugen *et al.*[35]	Lymphocytes	USERIA, SFS, antibodies to B[*a*]PDE-DNA adducts in serum by ELISA	Coke oven	38 exposed workers	Exposure to 7–9 μg/m^3 B[*a*]P. 4/38 positive by SFS; 13/38 positive by USERIA; 12/38 tested positive for antibodies; 9 months later 8 lost titers, 9 gained titers and 21 remained unchanged.
Pan *et al.*[36]	WBC	^{32}P-postlabelling	Coke oven (China)	75 exposed workers 24 controls	DNA adduct levels did not correlate with PAH exposure or urinary 1-OHP. However, adduct levels correlated with 1-OHP (cf. Chapter 3) and alcohol consumption in workers with *CYP1A1* Ile/Val or Val/Val polymorphisms. No influence of *GSTM1* (cf. Chapter 10).
Hemminki *et al.*[37]	WBC	^{32}P-postlabelling, ELISA	Coke oven (Poland)	63 exposed workers 19 local controls 15 rural controls	Adduct levels in exposed workers significantly higher than in the rural controls. Adduct levels in local controls also elevated relative to rural controls.

(continued)

Table 4.1: (continued)

Study	*Source of DNA*	*Method of analysis*	*Workplace*	*Subjects*	*Outcome*
Hemminki *et al.*[38]	WBC	^{32}P-postlabelling	Coke oven (Poland)	24 battery workers 37 non-battery workers	Heavily exposed battery workers (exposure up to 90 μg/m^3 B[*a*]P) had significantly higher adduct levels than other workers (exposure levels below 1 μg/m^3 B[*a*]P).
van Schooten *et al.*[39]	WBC	ELISA	Coke oven	56 exposed workers 44 controls	Exposure up to 7.8 μg/m^3 B[*a*]P. 47% of exposed workers had detectable adducts, compared with 30% of controls. Smokers had significantly higher adduct levels than non-smokers. Adduct levels did not correlate with PAH in air or 1-OHP in urine.
Assennato *et al.*[40]	WBC	DELFIA	Coke oven	69 exposed workers	Exposure up to 12 μg/m^3 B[*a*]P. PAH-DNA adducts below limit of detection in 33 workers. A subgroup of door operators ($n = 11$) with high exposure had the highest mean value and highest proportion of positive samples.
Øvrebø *et al.*[41]	WBC	^{32}P-postlabelling, USERIA	Coke oven	23 low exposure 26 intermediate exposure 18 high exposure 13 controls	Both methods produced a positive correlation coefficient between exposure and adduct levels ($p = 0.0145$ for USERIA, $p = 0.0594$ for ^{32}P-postlabelling), but there was no significant correlation between adduct levels measured by the two methods. Smokers had higher levels than non-smokers by ^{32}P-postlabelling.

Øvrebø *et al.*[42]	—	Antibodies to B[*a*]PDE-DNA adducts in blood plasma by ELISA	Coke oven	23 low exposure 26 intermediate exposure 18 high exposure 13 controls (as in Ref. 41)	No significant association between antibody concentrations and PAH exposure, duration of work, smoking or age.
Rojas *et al.*[43]	Lymphocytes and monocytes	HPLC/fluorescence	Coke oven	39 exposed workers 39 controls	51% of workers tested positive, compared with 18% of controls. Mean level in workers 8-fold higher than in controls, but with large inter-individual variation. Among the workers, smokers had 3.5-fold higher adduct levels than non-smokers (not significant; small numbers).
Schell *et al.*[44]	WBC, lymphocytes	^{32}P-postlabelling	Coke oven	For WBC: 34 exposed workers 32 controls. For lymphocytes: 23 exposed workers 23 controls	Adduct levels were significantly higher in the workers' WBC compared to controls, but not in the study in which lymphocytes were investigated. In the latter, a weak trend for elevated adducts in smokers was noted.
Popp *et al.*[45]	Lymphocytes	^{32}P-postlabelling	Coke oven (Germany)	23 exposed workers and matched controls	Exposure up to 2.25 µg/m^3 B[*a*]P. DNA adducts were not significantly higher in workers than in controls, although adducts increased relative to B[*a*]P exposure in workers (marginally significant).

(continued)

Table 4.1: (continued)

Study	*Source of DNA*	*Method of analysis*	*Workplace*	*Subjects*	*Outcome*
Baan *et al.*[46]	Lymphocytes	^{32}P-postlabelling	Carbon electrode production	19 low exposure 19 intermediate exposure 17 high exposure	No significant difference between adduct levels in the 3 groups, although levels significantly higher in smokers than in non-smokers.
Lewtas *et al.*[47]	WBC	^{32}P-postlabelling	Coke oven (Czech Republic)	76 workers in 10 job categories	DNA adduct levels were significantly correlated with exposure to PAH ($r = 0.53$, $p < 0.001$), but not if the 2 most highly exposed individuals were omitted from the analysis. At high levels of exposure, the relationship with DNA adducts became non-linear. No effect of smoking on adduct levels was observed.
Kuljukka *et al.*[48]	WBC	^{32}P-postlabelling	Cokery workers in oil shale processing plant (Estonia)	49 exposed workers 10 controls	Exposure mean 10.4 µg/m^3 B[*a*]P. Adducts in workers were not significantly different from controls ($p = 0.098$). Smokers had higher levels than non-smokers ($p = 0.002$). Adduct levels correlated with urinary 1-OHP levels.
van Delft *et al.*[49]	Lymphocytes	^{32}P-postlabelling	Carbon anode production (Nether-lands)	14 low exposure 15 intermediate exposure 13 high exposure	Exposure to up to 5 µg/m^3 B[*a*]P. No difference in adduct levels between groups, although urinary 1-OHP levels were significantly elevated in the intermediate and high exposure groups.

Binkova *et al.*[50]	WBC, lymphocytes	^{32}P-postlabelling	Coke oven (Czech Republic, Slovakia)	68 exposed workers 56 controls	Exposure to up to 62 μg/m^3 B[*a*]P. Adduct levels significantly elevated in workers compared with controls in both WBC and lymphocytes. Smokers had significantly higher adduct levels in lymphocytes. No effect of *GSTM1* genotype.
Viezzer *et al.*[51]	Mononuclear cells	^{32}P-postlabelling	Coke oven (Italy)	70 exposed workers 17 controls	*GSTM1* null genotype increased DNA adduct levels in smoking workers with high PAH exposure. *GSTT1* positive individuals had higher adduct levels than *GSTT1* null ($p = 0.04$).
Pavanello *et al.*[52]	Mononuclear cells	HPLC/fluorescence	Coke oven	15 exposed workers 34 controls	The percentage of subjects whose adduct levels exceeded the 95 percentile control subject (46.7%) was significant (χ^2 test, $p < 0.01$).
Arnould *et al.*[53]	WBC	^{32}P-postlabelling, ELISA	Carbon electrode plant	17 exposed workers 10 controls	Exposure below 1 μg/m^3 B[*a*]P. Significantly higher levels of adducts in workers than in controls, but adduct levels did not correlate with airborne B[*a*]P concentrations.
Arnould *et al.*[54]	WBC	ELISA	Coke oven (Poland)	58 exposed workers 10 office workers	Exposure up to 45 μg/m^3 B[*a*]P. Adduct levels significantly elevated in workers ($p < 0.001$). Overall no difference between smokers and non-smokers, but significant difference when matched for PAH exposure. Adduct levels correlated with urinary 1-OHP levels.

(continued)

between the urban and suburban vendors were significant for the non-smokers only, and not for the smokers or for smokers and non-smokers combined.[74]

4.4 Environmental Exposure to PAHs

In the original study of DNA adduct levels of coke oven workers in Upper Silesia, Poland,[37] two control groups, local residents and rural residents from another part of the country, were included. An unexpected finding was that adduct levels in the local controls were nearly as high as in the coke oven workers. This was the first evidence of environmental pollution giving rise to DNA adducts in human populations. Subsequently, there have been several additional studies of the use of DNA adducts as biomarkers of exposure to environmental carcinogens (in particular, to PAHs) in this and other geographical regions of interest.

Significant seasonal variation in adduct levels, consistent with recorded fluctuations in air pollution levels, have been found in the white blood cells of the residents of Upper Silesia;[75,76] winter samples had 7-fold higher levels of adducts, determined by ELISA, than summer samples ($p = 0.002$) and although the difference was numerically smaller (2.5-fold) using ^{32}P-postlabelling, the results were still highly significant ($p < 0.001$).[75] Follow-up of this population has confirmed these findings, demonstrating elevated levels of PAH-DNA adducts in winter samples of oral mucosa cells (by immunohistochemical staining), as well as significant differences from controls for a number of other biomarkers, including sister chromatid exchanges, chromosomal aberrations and bleomycin sensitivity of lymphocytes;[77] however, no influence of *GSTM1* or *CYP1A1* genotypes on any of these biomarkers was found.

Another area of Poland subject to high atmospheric pollution is the city of Krakow and its environs. Here, a study of mothers and their new-born babies has revealed that air pollution levels correlated with PAH-DNA adduct levels, determined by ELISA, in the white blood cells of both mothers and infants.[78] There was also a correlation between high adduct levels in umbilical cord blood and reduced birth weight, length and head circumference.[79] Maternal smoking habits did not show any significant effect on placental DNA adduct levels.[80]

Another area of Central Europe with high levels of air pollution is Northern Bohemia in the Czech Republic, and a number of studies have been conducted centred on the city of Teplice.[81] In particular, DNA isolated from white blood cells and analyzed by ^{32}P-postlabelling was found to have adduct levels significantly associated with PAH exposure levels in a group of non-smoking women with outdoor occupations.[82] Winter samples of placentas from Teplice showed significantly higher DNA adduct levels than from a less industrialized area of Southern Bohemia, with higher adduct levels in the *GSTM1* null individuals.[83,84]

Seasonal differences in PAH-haemoglobin adducts were also investigated in a group of office workers in Milan.[85] Although adduct levels were higher in winter than in summer, only a small proportion of the subjects had detectable adducts (16% and 11%, respectively, among 65 subjects) and the difference was not statistically significant.

In a study that compared an urban group of Danish students (n = 74), a rural group of Danish agricultural students (n = 29) and a group of urban dwellers in Athens (n = 17), significantly different DNA adduct levels, determined by ^{32}P-postlabelling, were found in the order Athens > urban Denmark > rural Denmark.[86] In contrast, there were no significant differences in PAH-protein adduct levels between the two Danish groups (the Greek samples were not analyzed), determined by ELISA. Thus, there was no correlation between protein and DNA adducts, and *GSTM1* genotype did not influence DNA adduct levels.

The high lung cancer rate in the Xuan Wei province, China, particularly for women, is associated with the domestic use of smoky coal for cooking in unventilated houses. PAH-DNA adducts, measured by ELISA, were detectable in a high percentage of placentas and white blood cells obtained from women who lived in such houses, when compared with women from Beijing.[87] However, there was no dose response between air concentrations of B[*a*]P and adduct levels in the placenta. In contrast, DNA adducts were not detectable by ^{32}P-postlabelling in placentas or white blood cells from women exposed to smoky coal, although they were detectable in bronchoalveolar lavage cells, where they were at a 4-fold higher level than in control subjects.[88] Depurinated B[*a*]P-adducted bases were also detected in the urine of 3 out of 7 women exposed to coal smoke, but were not detectable in any samples from 13 control subjects[89] (cf. Chapter 2). In another ^{32}P-postlabelling study, exposure to domestic wood-burning stoves in the USA did not result in exposure-related DNA adducts in either placenta or white blood cells of a small group of women.[90]

In a study of 1345 individuals in Italy, comprising an urban area of the city of Pisa and the suburban zone of Cascina, it was found that positivity for detectable B[*a*]PDE-DNA adduct antibodies was 26% for the former and 17.9% for the latter (overall prevalence 21%); a difference that was statistically significant.[91]

Detection of PAH-like DNA adducts in human tissues as a means of monitoring for exposure to environmental tobacco smoke (in non-smokers) has, in most cases, not revealed differences in peripheral blood cells between exposed and unexposed individuals.[4] However, in a controlled experiment in which 15 non-smokers were monitored before and after spending 3 hours in a smoky atmosphere, exposure resulted in the formation of specific adducts in the induced sputum from some individuals, but no induction of adducts in peripheral lymphocytes.[92]

4.5 Coal Tar Therapy

The use of coal tar-based ointments to treat a variety of skin conditions, including psoriasis and eczema, is widespread. The presence of carcinogenic PAHs in these formulations therefore poses a potential genotoxic hazard to users.

An analysis by ^{32}P-postlabelling of DNA isolated from skin biopsies of psoriasis patients treated with coal tar preparations revealed the formation of aromatic DNA adducts.[93] Similar patterns of adducts were formed when explant samples of human skin were treated topically with coal tar in short-term organ culture, and also in the skin of mice treated *in vivo* with the agent. In another study, a similar complex pattern of adducts was observed in the treated skin of psoriasis patients, while immunofluorescence techniques revealed specific nuclear staining in the epidermal cells of all biopsies from treated patients but not from control biopsies obtained from untreated individuals.[94] Subsequent studies have examined blood cells of treated psoriasis patients for the presence of associated DNA and protein adducts. In one study that used ELISA, adduct levels were significantly higher in white blood cell DNA during treatment than they were 2–5 months later.[95] Another study using this technique observed a non-significant elevation in treated patients compared with untreated controls but no difference in PAH-albumin adduct levels.[96] One study that looked at adducts in skin and blood of coal tar-treated eczema patients found, using ^{32}P-postlabelling analysis, that aromatic DNA adduct levels in skin were significantly higher during and 1 week after treatment than before, and the same was true of adduct levels in lymphocytes, monocytes and granulocytes, albeit with lower levels of adduct formation than in skin.[97] In another study of 26 psoriatic patients, there was no elevation in adduct levels in DNA from lymphocytes and monocytes 3 days after clinical coal tar treatment, compared to 34 controls, measured by HPLC/fluorescence of B[*a*]PDE-DNA adducts.[52]

Taken together, these studies show that medicinal use of coal tar results in PAH-DNA adduct formation at the site of application and also, to a lesser extent, in white blood cells.

4.6 Diet

Many foods contain PAHs[98] and diet is considered to be the major source of human exposure for non-smokers who are not occupationally exposed to PAHs, or who are inhabitants of regions where environmental or atmospheric contamination with PAHs is not abnormally high.

When peripheral lymphocytes of city fire fighters and matched controls were analyzed for PAH-DNA adducts by ELISA, differences in adduct levels were observed only after adjustment for charcoal-grilled food consumption.[99] Individuals who ate such food up to three times in the month prior to blood collection were less likely to have detectable adducts, while those who ate charcoal-grilled food more than 3 times in this period had elevated levels. A possible explanation for this is that light consumption induces PAH-detoxifying enzymes, while heavier consumption saturates the detoxifying enzymes and leads to a higher concentration of DNA-damaging metabolites (cf. Chapter 2).

Similar findings have come from a study of woodland fire fighters, in which PAH-DNA adducts in white blood cells, measured by ELISA, were not correlated with recent fire fighting activity but were related to recent consumption of charcoal-grilled food.[100] Subsequently, no significant influence of *GSTM1* or *CYP1A1* genetic polymorphisms on DNA adduct levels was found in these subjects.[101]

Another study involved the monitoring of US army personnel involved in putting out oilfield fires in Kuwait in the aftermath of the Gulf War.[102] DNA adduct levels in the soldiers' white blood cells were found by DELFIA to be lower during the fire-fighting activity than they were before and after deployment. Although the study was not able to ascertain the precise reasons, a likely explanation was a reduction in consumption of charcoal-grilled food by the soldiers while on active duty in Kuwait. Corroborating evidence that human exposure to PAHs from these fires was not biologically significant has come from a study of placental DNA from Kuwaiti women exposed to the air pollution during pregnancy.[103] No difference was found between the adduct levels in these samples ($n = 40$) and the levels in a control group of placental DNA samples from UK mothers ($n = 24$).

In a controlled experiment, 4 volunteers ate a diet high in charcoal-grilled beef for 7 days, having avoided such foods for the previous 30 days.[104] In two cases, PAH-DNA adducts in peripheral white blood cells increased to 3 and 6 times baseline levels, respectively, while in the other two individuals no rise in levels was observed. In a subsequent study of 10 volunteers, DNA adduct levels increased in 4 of the subjects.[105] These results are similar to those found by van Maanen *et al.*,[106] who measured DNA adducts in mononuclear cells by ^{32}P-postlabelling of 21 individuals during periods of consumption of, and abstinence from, barbecued meat. Meat consumption was for 5 consecutive days, during which blood samples were collected on days 3 and 5. Twelve out of 21 subjects had detectable levels of DNA adducts on day 3, but on day 5 only one sample was positive. Six subjects gave positive samples 5 days prior to charcoal-grilled meat consumption, while in the days after meat consumption, 3 samples were positive on day 3, and 2 were positive on day 7. This study shows considerable inter-individual variation in adduct formation, although the rise in the number of samples positive on day 3 of the consumption period is statistically significant.

4.7 Discussion and Summary

It is apparent from the foregoing descriptions that there have been many studies published in the last 20 years that have demonstrated human exposure to PAHs results in the formation of DNA adducts. As these lesions are promutagenic, and DNA adduct formation is a critical early step in the carcinogenic process, this renders PAHs clearly associated with cancer causation in humans — a conclusion supported by a wealth of epidemiological evidence.

Diverse methods have been shown to be effective for measuring PAH-DNA adducts in human tissues, including ^{32}P-postlabelling, immunochemical assays and fluorescence spectroscopic methods. These methods have the sufficient sensitivity and selectivity to be applied to large populations, although they can be time-consuming and expensive techniques. Despite the fact that the assays have very different detection endpoints, in those studies in which the different assays have been compared, results are quantitatively similar.

Not all studies have shown a clear association between PAH exposure and DNA adduct formation; in some cases, tobacco smoking can be a confounding source of the same class of adducts and in others, diet may be a significant source of the compounds. Thus, the ubiquitous nature of PAHs may interfere with investigations designed to associate PAH exposure, and PAH-DNA adduct formation, with a particular source or to detect differences between exposed and unexposed populations. Nevertheless, when Peluso *et al.*[107] carried out a meta-analysis of 13 studies of DNA adduct formation on occupational cohorts using ^{32}P-postlabelling analysis, they found that the association between levels of adducts and air pollution exposure was significant in both heavily exposed workers (e.g., in coke ovens and foundries) and in less exposed urban workers (e.g., bus drivers and traffic police). Seven of the studies out of this meta-analysis made measurements of B[*a*]P levels and these found a linear correlation with adduct levels at low dose which became sublinear at high dose, demonstrating evidence for saturation.

In conclusion, it is now recognized that PAHs are ubiquitous in the environment and biomonitoring studies provide evidence of DNA damage, in the form of covalent DNA adducts, in many exposure scenarios. This applies to occupational exposures in many industries where workers are documented to be at elevated risk of respiratory cancers, and it is a reasonable assumption that this is due to exposure to PAHs (cf. Chapter 3). In general populations where PAH pollution is particularly high, DNA adduct levels are also elevated. In some cases it is known, and in others it is a reasonable assumption, that humans in these environments have increased cancer risk as a result. Tobacco smoking is a causative factor in cancer in many organs,[108] and smokers have elevated levels of DNA adducts in many tissues.[4] While tobacco smoke is a very complex mixture of carcinogens, there is extensive evidence that PAHs are major contributors to the total burden of DNA adduct formation inflicted on human tissues by tobacco smoking. Taken together with the other sources of PAHs described in this chapter, and the evidence for DNA adduct formation in these exposure circumstances, it is clear that the widespread occurrence of PAH-DNA adducts in human tissues emphasizes the importance of this class of compounds in the overall burden of human cancer.

Acknowledgment

Research in the author's laboratory has been funded principally by Cancer Research UK.

References

1. Phillips DH (1983) Fifty years of benzo[*a*]pyrene. *Nature* 303: 468–472.
2. Brookes P and Lawley PD (1964) Evidence for the binding of polynuclear aromatic hydrocarbons to the nucleic acids of mouse skin: relation between carcinogenic power of hydrocarbons and their binding to DNA. *Nature* 202: 781–784.

3. Sims P, Grover PL, Swaisland A, Pal K and Hewer A (1974) Metabolic activation of benzo[*a*]pyrene proceeds by a diol-epoxide. *Nature* 252: 326–328.
4. Phillips DH (2002) Smoking-related DNA and protein adducts in human tissues. *Carcinogenesis* 23: 1979–2004.
5. Dell'Omo M and Lauwerys RR (1993) Adducts to macromolecules in the biological monitoring of workers exposed to polycyclic aromatic hydrocarbons. *Crit. Rev. Toxicol.* 23: 111–126.
6. Strickland PT, Routledge MN and Dipple A (1993) Methodologies for measuring carcinogen adducts in humans. *Cancer Epidemiol. Biomarkers Prev.* 2: 607–619.
7. Poirier MC and Weston A (1996) Human DNA adduct measurements: state of the art. *Environ. Health Perspect.* 104, Suppl 5: 883–893.
8. Angerer J, Mannschreck C and Gündel J (1997) Biological monitoring and biochemical effect monitoring of exposure to polycyclic aromatic hydrocarbons. *Int. Arch. Occup. Environ. Health* 70: 365–377.
9. Jongeneelen FJ (1997) Methods for routine biological monitoring of carcinogenic PAH-mixtures. *Sci. Total. Environ.* 199: 141–149.
10. Kriek E, Rojas M, Alexandrov K and Bartsch H (1998) Polycyclic aromatic hydrocarbon-DNA adducts in humans: relevance as biomarkers for exposure and cancer risk. *Mutat. Res.* 400: 215–231.
11. Wild CP and Pisani P (1998) Carcinogen DNA and protein adducts as biomarkers of human exposure in environmental cancer epidemiology. *Cancer Detect. Prev.* 22: 273–283.
12. Poirier MC, Santella RM and Weston A (2000) Carcinogen macromolecular adducts and their measurement. *Carcinogenesis* 21: 353–359.
13. De Kok TM, Moonen HJ, van Delft J and van Schooten FJ (2002) Methodologies for bulky DNA adduct analysis and biomonitoring of environmental and occupational exposures. *J. Chromatogr. B Anal. Technol. Biomed. Life Sci.* 778: 345–355.
14. Phillips DH and Farmer PB (1995) Protein and DNA adducts as biomarkers of exposure to environmental mutagens. In: *Environmental Mutagenesis* [Phillips DH and Venitt S (eds.)] pp 367–395, Bios, Oxford, UK.
15. Phillips DH, Hemminki K, Alhonen A, Hewer A and Grover PL (1988) Monitoring occupational exposure to carcinogens: detection by ^{32}P-postlabelling of aromatic DNA adducts in white blood cells from iron foundry workers. *Mutat. Res.* 204: 531–541.

16. Savela K, Hemminki K, Hewer A, Phillips DH, Putman KL and Randerath K (1989) Interlaboratory comparison of the ^{32}P-postlabelling assay for aromatic DNA adducts in white blood cells of iron foundry workers. *Mutat. Res.* 224: 485–492.
17. Savela K, Kohan MJ, Walsh D, Perera FP, Hemminki K and Lewtas J (1996) *In vitro* characterization of DNA adducts formed by foundry air particulate matter. *Environ. Health Perspect.* 104, Suppl 3: 687–690.
18. Perera FP, Hemminki K, Young TL, Brenner D, Kelly G and Santella RM (1988) Detection of polycyclic aromatic hydrocarbon-DNA adducts in white blood cells of foundry workers. *Cancer Res.* 48: 2288–2291.
19. Hemminki K, Randerath K, Reddy MV, Putman KL, Santella RM, Perera FP, Young TL, Phillips DH, Hewer A and Savela K (1990) Postlabeling and immunoassay analysis of polycyclic aromatic hydrocarbons-adducts of deoxyribonucleic acid in white blood cells of foundry workers. *Scand. J. Work Environ. Health* 16: 158–162.
20. Perera FP, Tang DL, O'Neill JP, Bigbee WL, Albertini RJ, Santella R, Ottman R, Tsai WY, Dickey C, Mooney LA, Savela K and Hemminki K (1993) HPRT and glycophorin A mutations in foundry workers: relationship to PAH exposure and to PAH-DNA adducts. *Carcinogenesis* 14: 969–973.
21. Perera FP, Dickey C, Santella R, O'Neill JP, Albertini RJ, Ottman R, Tsai WY, Mooney LA, Savela K and Hemminki K (1994) Carcinogen-DNA adducts and gene mutation in foundry workers with low-level exposure to polycyclic aromatic hydrocarbons. *Carcinogenesis* 15: 2905–2910.
22. Hemminki K, Dickey C, Karlsson S, Bell D, Hsu Y, Tsai WY, Mooney LA, Savela K and Perera FP (1997) Aromatic DNA adducts in foundry workers in relation to exposure, life style and CYP1A1 and glutathione transferase M1 genotype. *Carcinogenesis* 18: 345–350.
23. Omland O, Sherson D, Hansen AM, Sigsgaard T, Autrup H and Overgaard E (1994) Exposure of iron foundry workers to polycyclic aromatic hydrocarbons: benzo[*a*]pyrene-albumin adducts and 1-hydroxypyrene as biomarkers for exposure. *Occup. Environ. Med.* 51: 513–518.
24. Lee BM, Yin BY, Herbert R, Hemminki K, Perera FP and Santella RM (1991) Immunologic measurement of polycyclic aromatic hydrocarbon-albumin adducts in foundry workers and roofers. *Scand. J. Work Environ. Health* 17: 190–194.
25. Vahakangas K, Trivers G, Rowe M and Harris CC (1985) Benzo[*a*]pyrene diolepoxide-DNA adducts detected by synchronous fluorescence spectrophotometry. *Environ. Health Perspect.* 62: 101–104.

26. Schoket B, Phillips DH, Hewer A and Vincze I (1991) ^{32}P-postlabelling detection of aromatic DNA adducts in peripheral blood lymphocytes from aluminium production plant workers. *Mutat. Res.* 260: 89–98.
27. Schoket B, Phillips DH, Poirier MC and Vincze I (1993) DNA adducts in peripheral blood lymphocytes from aluminum production plant workers determined by ^{32}P-postlabeling and enzyme-linked immunosorbent assay. *Environ. Health Perspect.* 99: 307–309.
28. Van Schooten FJ, Jongeneelen FJ, Hillebrand MJ, van Leeuwen FE, de Looff AJ, Dijkmans AP, van Rooij JG, den Engelse L and Kriek E (1995) Polycyclic aromatic hydrocarbon-DNA adducts in white blood cell DNA and 1-hydroxypyrene in the urine from aluminum workers: relation with job category and synergistic effect of smoking. *Cancer Epidemiol. Biomarkers Prev.* 4: 69–77.
29. Ovrebo S, Haugen A, Hemminki K, Szyfter K, Drablos PA and Skogland M (1995) Studies of biomarkers in aluminum workers occupationally exposed to polycyclic aromatic hydrocarbons. *Cancer Detect. Prev.* 19: 258–267.
30. Galati R, Zijno A, Crebelli R, Falasca G, Tomei F, Iecher F, Carta P and Verdina A (2001) Detection of antibodies to the benzo[*a*]pyrene diol epoxide-DNA adducts in sera from individuals exposed to low doses of polycyclic aromatic hydrocarbons. *J. Exp. Clin. Cancer Res.* 20: 359–364.
31. Schoket B, Papp G, Levay K, Mrackova G, Kadlubar FF and Vincze I (2001) Impact of metabolic genotypes on levels of biomarkers of genotoxic exposure. *Mutat. Res.* 482: 57–69.
32. Carstensen U, Yang K, Levin JO, Ostman C, Nilsson T, Hemminki K and Hagmar L (1999) Genotoxic exposures of potroom workers. *Scand. J. Work Environ. Health* 25: 24–32.
33. Tuominen R, Baranczewski P, Warholm M, Hagmar L, Moller L and Rannug A (2002) Susceptibility factors and DNA adducts in peripheral blood mononuclear cells of aluminium smelter workers exposed to polycyclic aromatic hydrocarbons. *Arch. Toxicol.* 76: 178–186.
34. Harris CC, Vahakangas K, Newman MJ, Trivers GE, Shamsuddin A, Sinopoli N, Mann DL and Wright WE (1985) Detection of benzo[*a*]pyrene diol epoxide-DNA adducts in peripheral blood lymphocytes and antibodies to the adducts in serum from coke oven workers. *Proc. Natl. Acad. Sci. USA* 82: 6672–6676.
35. Haugen A, Becher G, Benestad C, Vahakangas K, Trivers GE, Newman MJ and Harris CC (1986) Determination of polycyclic aromatic hydrocarbons in the urine, benzo[*a*]pyrene diol epoxide-DNA adducts in lymphocyte DNA,

and antibodies to the adducts in sera from coke oven workers exposed to measured amounts of polycyclic aromatic hydrocarbons in the work atmosphere. *Cancer Res.* 46: 4178–4183.
36. Pan G, Hanaoka T, Yamano Y, Hara K, Ichiba M, Wang Y, Zhang J, Feng Y, Shujuan Z, Guan D, Gao G, Liu N and Takahashi K (1998) A study of multiple biomarkers in coke oven workers — a cross-sectional study in China. *Carcinogenesis* 19: 1963–1968.
37. Hemminki K, Grzybowska E, Chorazy M, Twardowska-Saucha K, Sroczynski JW, Putman KL, Randerath K, Phillips DH, Hewer A, Santella RM, Young TL and Perera FP (1990) DNA adducts in human environmentally exposed to aromatic compounds in an industrial area of Poland. *Carcinogenesis* 11: 1229–1231.
38. Hemminki K, Grzybowska E, Chorazy M, Twardowska-Saucha K, Sroczynski JW, Putman KL, Randerath K, Phillips DH and Hewer A (1990) Aromatic DNA adducts in white blood cells of coke workers. *Int. Arch. Occup. Environ. Health* 62: 467–470.
39. Van Schooten FJ, van Leeuwen FE, Hillebrand MJ, de Rijke ME, Hart AA, van Veen HG, Oosterink S and Kriek E (1990) Determination of benzo[*a*]pyrene diol epoxide-DNA adducts in white blood cell DNA from coke-oven workers: the impact of smoking. *J. Natl. Cancer Inst.* 82: 927–933.
40. Assennato G, Ferri GM, Foa V, Strickland P, Poirier M, Pozzoli L and Cottica D (1993) Correlation between PAH airborne concentration and PAH-DNA adducts levels in coke-oven workers. *Int. Arch. Occup. Environ. Health* 65: S143–S145.
41. Ovrebo S, Haugen A, Phillips DH and Hewer A (1992) Detection of polycyclic aromatic hydrocarbon-DNA adducts in white blood cells from coke oven workers: correlation with job categories. *Cancer Res.* 52: 1510–1514.
42. Ovrebo S, Haugen A, Farmer PB and Anderson D (1995) Evaluation of biomarkers in plasma, blood, and urine samples from coke oven workers: significance of exposure to polycyclic aromatic hydrocarbons. *Occup. Environ. Med.* 52: 750–756.
43. Rojas M, Alexandrov K, Auburtin G, Wastiaux-Denamur A, Mayer L, Mahieu B, Sebastien P and Bartsch H (1995) *Anti*-benzo[*a*]pyrene diolepoxide-DNA adduct levels in peripheral mononuclear cells from coke oven workers and the enhancing effect of smoking. *Carcinogenesis* 16: 1373–1376.
44. Schell C, Popp W, Kraus R, Vahrenholz C and Norpoth K (1995) ^{32}P-postlabeling analysis of DNA adducts in different populations. *Toxicol. Lett.* 77: 299–307.

45. Popp W, Vahrenholz C, Schell C, Grimmer G, Dettbarn G, Kraus R, Brauksiepe A, Schmeling B, Gutzeit T, von Bülow J and Norpoth K (1997) DNA single strand breakage, DNA adducts, and sister chromatid exchange in lymphocytes and phenanthrene and pyrene metabolites in urine of coke oven workers. *Occup. Environ. Med.* 54: 176–183.
46. Baan RA, Steenwinkel MJ, van Asten S, Roggeband R and van Delft JH (1997) The use of benzo[*a*]pyrene diolepoxide-modified DNA standards for adduct quantification in ^{32}P-postlabelling to assess exposure to polycyclic aromatic hydrocarbons: application in a biomonitoring study. *Mutat. Res.* 378: 41–50.
47. Lewtas J, Walsh D, Williams R and Dobias L (1997) Air pollution exposure — DNA adduct dosimetry in humans and rodents: evidence for non-linearity at high doses. *Mutat. Res.* 378: 51–63.
48. Kuljukka T, Savela K, Vaaranrinta R, Mutanen P, Veidebaum T, Sorsa M and Peltonen K (1998) Low response in white blood cell DNA adducts among workers in a highly polluted cokery environment. *J. Occup. Environ. Med.* 40: 529–537.
49. Van Delft JH, Steenwinkel MJ, van Asten JG, van Es J, Kraak A and Baan RA (1998) Monitoring of occupational exposure to polycyclic aromatic hydrocarbons in a carbon-electrode manufacturing plant. *Ann. Occup. Hyg.* 42: 105–114.
50. Binkova B, Topinka J, Mrackova G, Gajdosova D, Vidova P, Stavkova Z, Peterka V, Pilcik T, Rimar V, Dobias L, Farmer PB and Sram RJ (1998) Coke oven workers study: the effect of exposure and GSTM1 and NAT2 genotypes on DNA adduct levels in white blood cells and lymphocytes as determined by ^{32}P-postlabelling. *Mutat. Res.* 416: 67–84.
51. Viezzer C, Norppa H, Clonfero E, Gabbani G, Mastrangelo G, Hirvonen A and Celotti L (1999) Influence of GSTM1, GSTT1, GSTP1, and EPHX gene polymorphisms on DNA adduct level and HPRT mutant frequency in coke-oven workers. *Mutat. Res.* 431: 259–269.
52. Pavanello S, Favretto D, Brugnone F, Mastrangelo G, Dal Pra G and Clonfero E (1999) HPLC/fluorescence determination of *anti*-BPDE-DNA adducts in mononuclear white blood cells from PAH-exposed humans. *Carcinogenesis* 20: 431–435.
53. Arnould JP, Pfohl-Leszkowicz A, Bach V, Libert JP and Belegaud J (1999) Biological monitoring exposure of workers from plant producing carbon electrodes: quantification of benzo[*a*]pyrene DNA-adducts in leukocytes, by a ^{32}P-postlabelling method and an immunoassay. *Hum. Exp. Toxicol.* 18: 314–321.

54. Arnould JP, Kubiak R, Belowski J, Belegaud J and Szczeklik J (2000) Detection of benzo[*a*]pyrene-DNA adducts in leukocytes of coke oven workers. *Pathol. Biol.* 48: 548–553.
55. Rojas M, Cascorbi I, Alexandrov K, Kriek E, Auburtin G, Mayer L, Kopp-Schneider A, Roots I and Bartsch H (2000) Modulation of benzo[*a*]pyrene diolepoxide-DNA adduct levels in human white blood cells by CYP1A1, GSTM1 and GSTT1 polymorphism. *Carcinogenesis* 21: 35–41.
56. Van Delft JH, Steenwinkel MS, van Asten JG, de Vogel N, Bruijntjes-Rozier TC, Schouten T, Cramers P, Maas L, van Herwijnen MH, van Schooten F and Hopmans PM (2001) Biological monitoring the exposure to polycyclic aromatic hydrocarbons of coke oven workers in relation to smoking and genetic polymorphisms for GSTM1 and GSTT1. *Ann. Occup. Hyg.* 45: 395–408.
57. Teixeira JP, Gaspar J, Martinho G, Silva S, Rodrigues S, Mayan O, Martin E, Farmer PB and Rueff J (2002) Aromatic DNA adduct levels in coke oven workers: correlation with polymorphisms in genes GSTP1, GSTM1, GSTT1 and CYP1A1. *Mutat. Res.* 517: 147–155.
58. Ferreira MF Jr, Tas S, dell'Omo M, Goormans G, Buchet JP and Lauwerys R (1994) Determinants of benzo[*a*]pyrenediol epoxide adducts to haemoglobin in workers exposed to polycyclic aromatic hydrocarbons. *Occup. Environ. Med.* 51: 451–455.
59. Tas S, Buchet JP and Lauwerys R (1994) Determinants of benzo[*a*]pyrene diol epoxide adducts to albumin in workers exposed to polycyclic aromatic hydrocarbons. *Int. Arch. Occup. Environ. Health* 66: 343–348.
60. Kure EH, Andreassen A, Ovrebo S, Grzybowska E, Fiala Z, Strozyk M, Chorazy M and Haugen A (1997) Benzo[*a*]pyrene-albumin adducts in humans exposed to polycyclic aromatic hydrocarbons in an industrial area of Poland. *Occup. Environ. Med.* 54: 662–666.
61. Herbert R, Marcus M, Wolff MS, Perera FP, Andrews L, Godbold JH, Rivera M, Stefanidis M, Lu XQ, Landrigan PJ and Santella RM (1990) Detection of adducts of deoxyribonucleic acid in white blood cells of roofers by ^{32}P-postlabeling. Relationship of adduct levels to measures of exposure to polycyclic aromatic hydrocarbons. *Scand. J. Work Environ. Health* 16: 135–143.
62. Ichiba M, Hagmar L, Rannug A, Hogstedt B, Alexandrie AK, Carstensen U and Hemminki K (1994) Aromatic DNA adducts, micronuclei and genetic polymorphism for CYP1A1 and GST1 in chimney sweeps. *Carcinogenesis* 15: 1347–1352.

63. Lee J, Kang D, Lee KH, Ichiba M, Zhang J, Tomokuni K, Hwang ES, Park CG, Ha M, Kim S, Han SB, Choi JW, Lee E, Jang JY, Strickland PT, Hirvonen A and Cho SH (2002) Influence of GSTM1 genotype on association between aromatic DNA adducts and urinary PAH metabolites in incineration workers. *Mutat. Res.* 514: 213–221.
64. Abdel-Rahman AG, Allam MF, Mansour MT and Mustafa MA (2001) PAH-DNA adducts in a petrol refinery in Egypt. *Eur. J. Cancer Prev.* 10: 469–472.
65. Scheepers PT, Coggon D, Knudsen LE, Anzion R, Autrup H, Bogovski S, Bos RP, Dahmann D, Farmer P, Martin EA, Micka V, Muzyka V, Neumann HG, Poole J, Schmidt-Ott A, Seiler F, Volf J and Zwirner-Baier I (2002) BIOMarkers for occupational diesel exhaust exposure monitoring (BIOMODEM) — a study in underground mining. *Toxicol. Lett.* 134: 305–317.
66. Ruchirawa M, Mahidol C, Tangjarukij C, Pui-ock S, Jensen O, Kampeera-wipakorn O, Tuntaviroon J, Aramphongphan A and Autrup H (2002) Exposure to genotoxins present in ambient air in Bangkok, Thailand — particle associated polycyclic aromatic hydrocarbons and biomarkers. *Sci. Total. Environ.* 287: 121–132.
67. Peluso M, Merlo F, Munnia A, Valerio F, Perrotta A, Puntoni R and Parodi S (1998) ^{32}P-postlabeling detection of aromatic adducts in the white blood cell DNA of nonsmoking police officers. *Cancer Epidemiol. Biomarkers Prev.* 7: 3–11.
68. Hemminki K, Soderling J, Ericson P, Norbeck HE and Segerback D (1994) DNA adducts among personnel servicing and loading diesel vehicles. *Carcinogenesis* 15: 767–769.
69. Hemminki K, Zhang LF, Kruger J, Autrup H, Tornqvist M and Norbeck HE (1994) Exposure of bus and taxi drivers to urban air pollutants as measured by DNA and protein adducts. *Toxicol. Lett.* 72: 171–174.
70. Nielsen PS, Andreassen A, Farmer PB, Ovrebo S and Autrup H (1996) Biomonitoring of diesel exhaust-exposed workers. DNA and hemoglobin adducts and urinary 1-hydroxypyrene as markers of exposure. *Toxicol. Lett.* 86: 27–37.
71. Nielsen PS, de Pater N, Okkels H and Autrup H (1996) Environmental air pollution and DNA adducts in Copenhagen bus drivers — effect of GSTM1 and NAT2 genotypes on adduct levels. *Carcinogenesis* 17: 1021–1027.
72. Palli D, Russo A, Masala G, Saieva C, Guarrera S, Carturan S, Munnia A, Matullo G and Peluso M (2001) DNA adduct levels and DNA repair polymorphisms in traffic-exposed workers and a general population sample. *Int. J. Cancer* 94: 121–127.

73. Yang K, Airoldi L, Pastorelli R, Restano J, Guanci M and Hemminki K (1996) Aromatic DNA adducts in lymphocytes of humans working at high and low traffic density areas. *Chem. Biol. Interact.* 101: 127–136.
74. Pastorelli R, Restano J, Guanci M, Maramonte M, Magagnotti C, Allevi R, Lauri D, Fanelli R and Airoldi L (1996) Hemoglobin adducts of benzo[*a*]pyrene diolepoxide in newspaper vendors: association with traffic exhaust. *Carcinogenesis* 17: 2389–2394.
75. Perera FP, Hemminki K, Gryzbowska E, Motykiewicz G, Michalska J, Santella RM, Young TL, Dickey C, Brandt-Rauf P, De Vivo I, Blaner W, Tsai WY and Chorazy M (1992) Molecular and genetic damage in humans from environmental pollution in Poland. *Nature* 360: 256–258.
76. Perera F, Brenner D, Jeffrey A, Mayer J, Tang D, Warburton D, Young TI, Wazneh L, Latriano L, Motykiewicz G, Grzybowska E, Chorazy M, Hemminki K and Santella R (1992) DNA adducts and related biomarkers in populations exposed to environmental carcinogens. *Environ. Health Perspect.* 98: 133–137.
77. Motykiewicz G, Michalska J, Pendzich J, Malusecka E, Strozyk M, Kalinowska E, Butkiewicz D, Mielzynska D, Midro A, Santella RM and Chorazy M (1998) A molecular epidemiology study in women from Upper Silesia, Poland. *Toxicol. Lett.* 96–97: 195–202.
78. Whyatt RM, Santella RM, Jedrychowski W, Garte SJ, Bell DA, Ottman R, Gladek-Yarborough A, Cosma G, Young TL, Cooper TB, Randall MC, Manchester DK and Perera FP (1998) Relationship between ambient air pollution and DNA damage in Polish mothers and newborns. *Environ. Health Perspect.* 106, Suppl 3: 821–826.
79. Perera FP, Whyatt RM, Jedrychowski W, Rauh V, Manchester D, Santella RM and Ottman R (1998) Recent developments in molecular epidemiology: a study of the effects of environmental polycyclic aromatic hydrocarbons on birth outcomes in Poland. *Am. J. Epidemiol.* 147: 309–314.
80. Whyatt RM, Bell DA, Jedrychowski W, Santella RM, Garte SJ, Cosma G, Manchester DK, Young TL, Cooper TB, Ottman R and Perera FP (1998) Polycyclic aromatic hydrocarbon-DNA adducts in human placenta and modulation by CYP1A1 induction and genotype. *Carcinogenesis* 19: 1389–1392.
81. Sram RJ (2001) *Teplice Program: Impact of Air Pollution on Human Health.* Academia, Prague.

82. Binkova B, Lewtas J, Miskova I, Lenicek J and Sram R (1995) DNA adducts and personal air monitoring of carcinogenic polycyclic aromatic hydrocarbons in an environmentally exposed population. *Carcinogenesis* 16: 1037–1046.
83. Topinka J, Binkova B, Mrackova G, Stavkova Z, Benes I, Dejmek J, Lenicek J and Sram RJ (1997) DNA adducts in human placenta as related to air pollution and to GSTM1 genotype. *Mutat. Res.* 390: 59–68.
84. Topinka J, Binkova B, Mrackova G, Stavkova Z, Peterka V, Benes I, Dejmek J, Lenicek J, Pilcik T and Sram RJ (1997) Influence of GSTM1 and NAT2 genotypes on placental DNA adducts in an environmentally exposed population. *Environ. Mol. Mutagen.* 30: 184–195.
85. Pastorelli R, Guanci M, Restano J, Berri A, Micoli G, Minoia C, Alcini D, Carrer P, Negri E, La Vecchia C, Fanelli R and Airoldi L (1999) Seasonal effect on airborne pyrene, urinary 1-hydroxypyrene, and benzo[*a*]pyrene diol epoxide-hemoglobin adducts in the general population. *Cancer Epidemiol. Biomarkers Prev.* 8: 561–565.
86. Nielsen PS, Okkels H, Sigsgaard T, Kyrtopoulos S and Autrup H (1996) Exposure to urban and rural air pollution: DNA and protein adducts and effect of glutathione-*S*-transferase genotype on adduct levels. *Int. Arch. Occup. Environ. Health* 68: 170–176.
87. Mumford JL, Lee X, Lewtas J, Young TL and Santella RM (1993) DNA adducts as biomarkers for assessing exposure to polycyclic aromatic hydrocarbons in tissues from Xuan Wei women with high exposure to coal combustion emissions and high lung cancer mortality. *Environ. Health Perspect.* 99: 83–87.
88. Gallagher J, Mumford J, Li X, Shank T, Manchester D and Lewtas J (1993) DNA adduct profiles and levels in placenta, blood and lung in relation to cigarette smoking and smoky coal emissions. In: *Postlabelling Methods for Detection of DNA Adducts* [Phillips DH, Castegnaro M, and Bartsch H (eds.)] IARC Scientific Publication No. 124, pp 283–292, International Agency for Research on Cancer, Lyon, France.
89. Casale GP, Singhal M, Bhattacharya S, RamaNathan R, Roberts KP, Barbacci DC, Zhao J, Jankowiak R, Gross ML, Cavalieri EL, Small GJ, Rennard SI, Mumford JL and Shen M (2001) Detection and quantification of depurinated benzo[*a*]pyrene-adducted DNA bases in the urine of cigarette smokers and women exposed to household coal smoke. *Chem. Res. Toxicol.* 14: 192–201.

90. Reddy MV, Kenny PC and Randerath K (1990) ^{32}P-assay of DNA adducts in white blood cells and placentas of pregnant women: lack of residential wood combustion-related adducts but presence of tissue-specific endogenous adducts. *Teratog. Carcinog. Mutag.* 10: 373–384.
91. Petruzzelli S, Celi A, Pulera N, Baliva F, Viegi G, Carrozzi L, Ciacchini G, Bottai M, Di Pede F, Paoletti P and Giuntini C (1998) Serum antibodies to benzo[*a*]pyrene diol epoxide-DNA adducts in the general population: effects of air pollution, tobacco smoking, and family history of lung diseases. *Cancer Res.* 58: 4122–4126.
92. Besaratinia A, Maas LM, Brouwer EM, Moonen EJ, De Kok TM, Wesseling GJ, Loft S, Kleinjans JC and van Schooten FJ (2002) A molecular dosimetry approach to assess human exposure to environmental tobacco smoke in pubs. *Carcinogenesis* 23: 1171–1176.
93. Schoket B, Horkay I, Kosa A, Paldeak L, Hewer A, Grover PL and Phillips DH (1990) Formation of DNA adducts in the skin of psoriasis patients, in human skin in organ culture, and in mouse skin and lung following topical application of coal-tar and juniper tar. *J. Invest. Dermatol.* 94: 241–246.
94. Zhang YJ, Li Y, DeLeo VA and Santella RM (1990) Detection of DNA adducts in skin biopsies of coal tar-treated psoriasis patients: immunofluorescence and ^{32}P postlabeling. *Skin Pharmacol.* 3: 171–179.
95. Paleologo M, van Schooten FJ, Pavanello S, Kriek E, Zordan M, Clonfero E, Bezze C and Levis AG (1992) Detection of benzo[*a*]pyrene-diol-epoxide-DNA adducts in white blood cells of psoriatic patients treated with coal tar. *Mutat. Res.* 281: 11–16.
96. Santella RM, Perera FP, Young TL, Zhang YJ, Chiamprasert S, Tang D, Wang LW, Beachman A, Lin JH and DeLeo VA (1995) Polycyclic aromatic hydrocarbon-DNA and protein adducts in coal tar treated patients and controls and their relationship to glutathione *S*-transferase genotype. *Mutat. Res.* 334: 117–124.
97. Godschalk RW, Ostertag JU, Moonen EJ, Neumann HA, Kleinjans JC and van Schooten FJ (1998) Aromatic DNA adducts in human white blood cells and skin after dermal application of coal tar. *Cancer Epidemiol. Biomarkers Prev.* 7: 767–773.
98. Phillips DH (1999) Polycyclic aromatic hydrocarbons in the diet. *Mutat. Res.* 443: 139–147.
99. Liou SH, Jacobson-Kram D, Poirier MC, Nguyen D, Strickland PT and Tockman MS (1989) Biological monitoring of fire fighters: sister chromatid

exchange and polycyclic aromatic hydrocarbon-DNA adducts in peripheral blood cells. *Cancer Res.* 49: 4929–4935.

100. Rothman N, Correa-Villasenor A, Ford DP, Poirier MC, Haas R, Hansen JA, O'Toole T and Strickland PT (1993) Contribution of occupation and diet to white blood cell polycyclic aromatic hydrocarbon-DNA adducts in wildland firefighters. *Cancer Epidemiol. Biomarkers Prev.* 2: 341–347.
101. Rothman N, Shields PG, Poirier MC, Harrington AM, Ford DP and Strickland PT (1995) The impact of glutathione *S*-transferase M1 and cytochrome P450 1A1 genotypes on white-blood-cell polycyclic aromatic hydrocarbon-DNA adduct levels in humans. *Mol. Carcinog.* 14: 63–68.
102. Poirier MC, Weston A, Schoket B, Shamkhani H, Pan CF, McDiarmid MA, Scott BG, Deeter DP, Heller JM, Jacobson-Kram D and Rothman N (1998) Biomonitoring of United States Army soldiers serving in Kuwait in 1991. *Cancer Epidemiol. Biomarkers Prev.* 7: 545–551.
103. Marafie EM, Marafie I, Emery SJ, Waters R and Jones NJ (2000) Biomonitoring the human population exposed to pollution from the oil fires in Kuwait: analysis of placental tissue using ^{32}P-postlabeling. *Environ. Mol. Mutagen.* 36: 274–282.
104. Rothman N, Poirier MC, Baser ME, Hansen JA, Gentile C, Bowman ED and Strickland PT (1990) Formation of polycyclic aromatic hydrocarbon-DNA adducts in peripheral white blood cells during consumption of charcoal-broiled beef. *Carcinogenesis* 11: 1241–1243.
105. Kang DH, Rothman N, Poirier MC, Greenberg A, Hsu CH, Schwartz BS, Baser ME, Groopman JD, Weston A and Strickland PT (1995) Interindividual differences in the concentration of 1-hydroxypyrene-glucuronide in urine and polycyclic aromatic hydrocarbon-DNA adducts in peripheral white blood cells after charbroiled beef consumption. *Carcinogenesis* 16: 1079–1085.
106. Van Maanen JM, Moonen EJ, Maas LM, Kleinjans JC and van Schooten FJ (1994) Formation of aromatic DNA adducts in white blood cells in relation to urinary excretion of 1-hydroxypyrene during consumption of grilled meat. *Carcinogenesis* 15: 2263–2268.
107. Peluso M, Ceppi M, Munnia A, Puntoni R and Parodi S (2001) Analysis of 13 ^{32}P-DNA postlabeling studies on occupational cohorts exposed to air pollution. *Am. J. Epidemiol.* 153: 546–558.
108. IARC (2004) *Monographs on the Evaluation of Carcinogenic Risks to Humans.* Volume 83: Tobacco smoke and involuntary smoking. International Agency for Research on Cancer, Lyon, France.

5 DNA Damage and Mutagenesis Induced by Polycyclic Aromatic Hydrocarbons

Ahmad Besaratinia[†] and Gerd P. Pfeifer[‡]

Division of Biology, Beckman Research Institute of the City of Hope National Medical Center, Duarte, CA, USA
E-mails: [†]ania@coh.org; [‡]gpfeifer@coh.org

5.1 Introduction

Environmental factors play a pivotal role in human cancer. A myriad of pollutants present in the air we breathe, in the water we drink, and in the food we eat are known or suspected cancer-causing agents. Polycyclic aromatic hydrocarbons (PAHs) represent a prototype class for such environmental toxins. PAHs are formed as a result of incomplete combustion of organic matter e.g., fossil fuels and vegetation.[1] Apart from the natural sources of PAHs (e.g., forest fires and volcanoes), other emission sources include power plants, aluminum smelters, coke ovens, petroleum refineries, gas and diesel engines, domestic heating systems, refuse burning, and, more locally, tobacco smoke.[1] PAHs find their way into our food chain either as a contaminant (e.g., in leafy vegetables), or as a natural component of a variety of food stuff (e.g., in cereals), or as a by-product of food processing (e.g., in charbroiled meat).[2] Humans are unavoidably exposed to air-, water- and food-borne PAHs on a daily basis. Not only are many PAHs carcinogenic in various test systems, but also occupational, dietary, medicinal and recreational exposures to PAHs in humans are linked to cancer.[3] The ever occurring PAHs can therefore best exemplify the environmental determinants of human cancer, and be used as a model to unravel the underlying mechanisms of chemical carcinogenesis. In this chapter we survey the current knowledge on the mechanistic involvement of PAHs in carcinogenesis accounted for by their DNA damaging properties.

5.2 Evolution of Research on PAHs

The first implication of the involvement of PAHs in human cancer dates back to 1775, when Percival Pott reported an increased incidence of scrotal cancer among the chimney sweeps, presumably due to their occupational exposure to soot.[4,5] This was followed by a number of reports showing the high rate of skin and lung cancers in workers of coal tar, gas and shale oil industries.[5,6] These observations were experimentally confirmed in the early half of the 19th century when it was demonstrated that dermal application of coal tar in rabbits and mice could induce malignant skin tumors. The chemical identity of the tumor-inducing agents in coal tar was however, not revealed until 1930, when it was shown that painting of the skin of mice with synthetic dibenz[*a*,*h*]anthracene and its 3-methyl derivative led to tumorigenesis[6] (cf. Chapter 1 and Figure 2.1 in Chapter 2).

Initially, it was thought that the parent PAHs were the ultimate carcinogens and the metabolism of PAHs was simply a detoxification process, so attempts were made to correlate the carcinogenic potency of PAHs to their chemical structures (cf. Chapter 2). Accordingly, the highly reactive K-region- and non-reactive L-region-containing PAHs were identified on the basis of angular fusion of their aromatic rings on the benzene rings (Figure 5.1).[7] However, such structural specificity of PAHs could

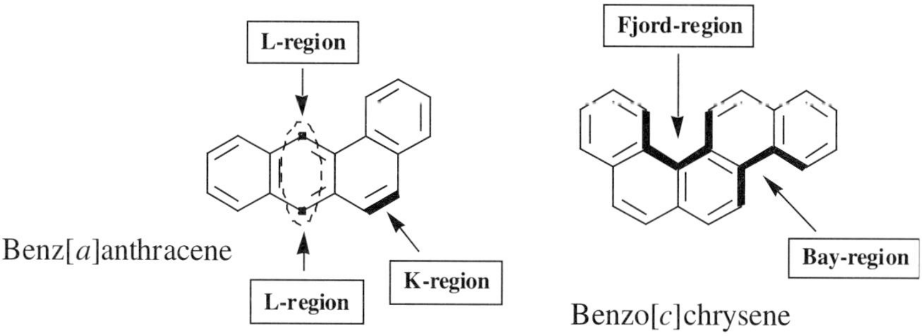

Figure 5.1: Molecular regions of polycyclic aromatic hydrocarbons (cf. Figure 2.1 in Chapter 2).

not consistently predict their carcinogenicity.[8] A major breakthrough came about in late 1960s, when the Millers disclosed the metabolic pathway of aromatic amines to the ultimate carcinogenic esters.[9] Based on this finding, the authors proposed that chemical carcinogens require metabolic activation to form electrophilic reactants, which are capable of covalently binding to macromolecules, e.g., DNA, thereby exerting their biological effects[10] (cf. Chapter 2). The theory gained support from animal studies in which the carcinogenic potencies of several PAHs were correlated to their extent of binding to cellular DNA.[11,12] Constitutive proof came in when the parent PAHs were shown to induce mutations in bacteria in the presence of metabolizing enzymes catalyzing their covalent binding to DNA. The concepts of DNA binding and mutagenicity of PAHs gained special momentum when it was realized that transferring only the DNA of a chemically transformed cell to a normal cell leads to its transformation.[13] A new era of PAH research that focuses on the inter-related aspects of carcinogenesis evolved. Investigations explore the continuum of 'PAH exposure, DNA damage, mutagenesis and neoplasia' in a multi-disciplinary manner so as to elucidate the underlying interplay amongst its components.

5.3 Chemistry and Biological Effects

Earlier quantum mechanical theories followed by the nuclear magnetic resonance (NMR) and mass spectral analyses have mainly defined our current understanding of PAH chemical structures in relation to their specific bioactivities.[14] Initial categorization of the parent PAHs on the basis of possession of the quasi-olefinic bonds, i.e., K-region bonds, could not satisfactorily be predictive of their carcinogenicity.[8] With the recognition of the role of metabolic activation in PAH carcinogenicity,[9,10] the attention was shifted to the metabolites of PAHs and consequently, the metabolic pathways of a number of PAHs were extensively scrutinized. It appeared that PAHs are generally metabolized in the endoplasmic reticulum by two distinct groups of enzymes: the phase-I (activating) and phase-II (detoxifying) enzymes[8,15] (cf. Chapter 2). The phase-I enzymes mediate oxidation of PAHs by introducing polar groups (e.g., hydroxyl groups) into these hydrophobic molecules. The phase-II enzymes catalyze the conjugation of the primary oxidized metabolites to polar groups such as glutathione (GSH), or further oxidize them, thereby making excretable end products. The metabolism of benzo[*a*]pyrene (B[*a*]P) best typifies the phase-I- and phase-II-mediated biotransformation of PAHs (cf. Chapter 2).[8,15]

In vivo, B[*a*]P is mainly oxidized by the family of cytochrome P450-dependent monooxygenases (CYP) at different aromatic bonds, yielding various unstable epoxides.[16] These intermediates undergo (i) conjugation with GSH catalyzed by glutathione *S*-transferases (GST);[17] (ii) non-enzymatic isomerization to phenols;[8] or (iii) hydration catalyzed by microsomal epoxide hydrolase (mEH) to form vicinal *trans*-dihydrodiols.[18] The GSH-conjugated metabolites are preferentially excreted in the bile or, alternatively, transported to the kidney where they are converted to the urinary excretable mercapturic acid conjugates.[15,17] The phenolic isomers may undergo further oxidation to quinones or conjugation reactions with glucuronic acid or sulfate catalyzed by UDP-glucuronosyltransferases (UGT) and sulfotransferases (SULT), respectively. The glucuronide and sulfate conjugates are water soluble and are readily excretable end products.[8,19,20] The dihydrodiols are further oxidized by CYP enzymes yielding diastereomeric *anti*- and *syn*-diol-epoxides. These highly reactive metabolites may subsequently undergo enzymatic conjugation with GSH, or non-enzymatic hydration to tetrahydrotetraols ('tetraols'). Alternatively, diol-epoxides may also covalently bind to cellular nucleophiles, e.g., DNA, thereby forming various DNA adducts.[8,15] The DNA adducts refractory to the cellular repair can cause mispairing during DNA replication and as a result, induce mutations (cf. Chapter 6).[21,22] The very simplified metabolic pathway of B[*a*]P is shown in Figure 5.2 (see also Figure 2.11 in Chapter 2).

In addition to the diol-epoxide-rendering pathway of PAHs, which has offered the initial insights into the mode of carcinogenicity of these chemicals, other metabolic pathways have also been elucidated, including one-electron oxidation of PAHs to radical cations, dehydrogenase-dependent oxidation of PAH dihydrodiols to *ortho*-quinones, and biomethylation ('alkylation') of *meso*- or L-region-containing PAHs[8] (cf. Chapter 2). Nonetheless, the common feature of most carcinogenic PAHs generated via different pathways is proven to be their electrophilicity and DNA adduct formation property.[8,15] Enzymatic digestion of the carcinogenic PAH-bound DNA and subsequent chromatographic analysis have shown that the resultant PAH-nucleosides consist mostly of bay-region-containing diol-epoxides.[8,23–25] The high reactivity of the bay-region molecules has been verified by NMR spectroscopy showing the sterically dense protons in this region with a characteristic downfield shift. The fjord-region, with its more severely hindered protons and drastic NMR downfield shift, has later been found to possess even more reactivity. Evidently, the repulsive interaction between the two opposing hydrogen bonds in the fjord-region distorts the molecule and yields a non-planar chemical structure[26,27] (Figure 5.1). Accordingly, the fjord-region-containing PAHs, e.g., dibenzo[*a*,*l*]pyrene (DB[*a*,*l*]P) and benzo[*c*]phenanthrene (B[*c*]Ph), have the greatest tumorigenicity in all test systems.[28–32]

The fate of PAHs in different metabolic pathways is governed by a battery of enzymes, many of which are stereoselective. As the ratio of these enzymes varies in different species, organs and cell types, and as a multitude of factors induce/inhibit the expression or activity of these enzymes, it is conceivable to find different stereoisomers of PAHs in various test systems.[8] For example, in mammalian cells the dihydrodiol metabolites of PAH are exclusively *trans*-isomers (the hydroxyl groups are on the opposite face of the planar hydrocarbon molecules), whereas in bacteria they are generally

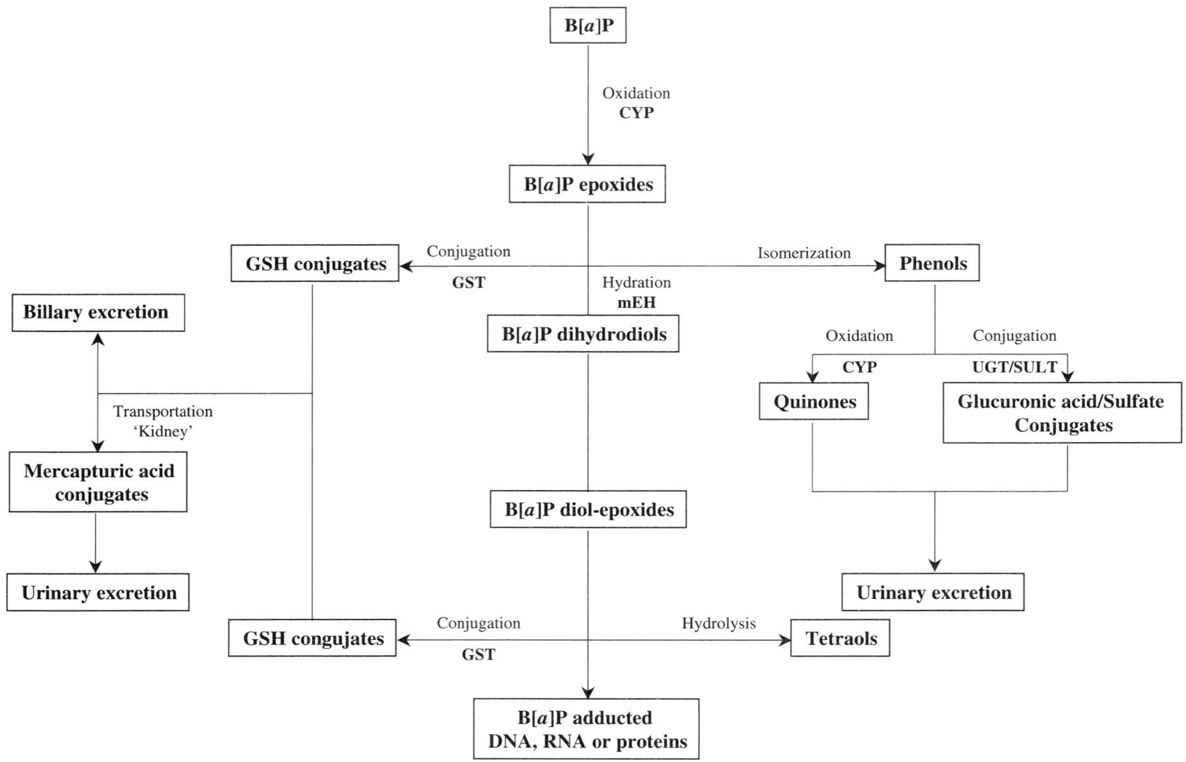

Figure 5.2: Simplified metabolic pathway of benzo[*a*]pyrene (B[*a*]P). Major enzymes involved in the reactions are: CYP, cytochrome P450-dependent monooxygenase(s); mEH, microsomal epoxide hydrolase; GST, glutathione *S*-transferase(s); SULT, sulfotransferase(s); UGT, UDP-glucuronosyltransferase(s) (cf. Figures 2.2 and 2.11 in Chapter 2).

cis-isomers (the hydroxyl groups are on the same face of the planar hydrocarbon molecules).[33,34] This stereoselectivity of PAH metabolism determines the configuration of the formed DNA adducts in each individual case.[8,33] For instance, the diol-epoxide derivatives of B[*a*]P yield differently *anti*-diastereomeric adducts (the epoxide oxygen is on the opposite face of the molecule relative to the benzylic hydroxyl group) and *syn*-diastereomeric adducts (the epoxide oxygen is on the same face of the molecule relative to the benzylic hydroxyl group) in different test systems[33] (cf. Figures 2.5 and 2.6 in Chapter 2). Depending on the absolute configuration of certain groups around a chiral center (R/S), there will theoretically be multiple conformations for any given DNA adduct.[33] However, not all conformations in each class of DNA adducts have the same biological significance. A classical example is B[*a*]P-7,8-diol-9,10-epoxide (B[*a*]PDE), which may get trapped in 16 different conformations in double stranded DNA; however, the (+)-*anti*-B[*a*]PDE stereoisomer (Figure 5.3) that binds covalently to the exocyclic N^2 amino group of guanine via *trans* opening at the C10-position of the hydrocarbon, renders an exceptionally high tumorigenicity.[6,8,35–38]

Currently, the greatest challenge in the study of PAH carcinogenicity is the daunting task of investigating the biochemical and biophysical properties of PAH-induced DNA adducts.

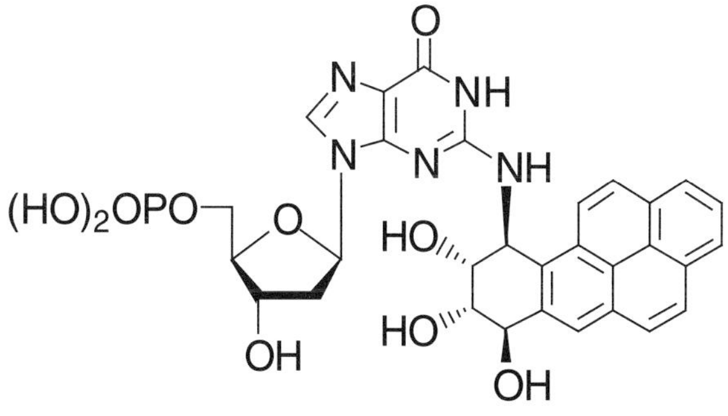

Figure 5.3: Major DNA adduct of benzo[*a*]pyrene, the (+)-*trans*-*anti*-B[*a*]PDE-N^2-deoxyguanosine (cf. Figure 2.7 in Chapter 2).

5.4 Significance of Stable *versus* Unstable PAH-DNA Adducts

In addition to the stable covalently bound PAH-DNA adducts, which are generated mainly via the diol-epoxide-rendering metabolic pathway, unstable PAH-DNA adducts are also formed, for example, via one-electron oxidation of PAHs to radical cations[8] (cf. Chapter 2). The lability of the latter adducts causes destabilization of the glycosidic bond in the DNA backbone resulting in depurination of the modified bases and formation of apurinic (AP) sites.[8] To compare the proportionality of stable and unstable DNA adducts, Chinese hamster ovary B11 cells, extracted DNA from such cells, and human mammary carcinoma MCF-7 cells were treated with DB[*a*,*l*]P and/or DB[*a*,*l*]P-11,12-diol-13,14-epoxides (DB[*a*,*l*]PDE). The formation of stable and unstable DNA adducts was investigated by ^{33}P-postlabeling coupled with high-pressure liquid chromatography (HPLC) and a sensitive AP sites detection assay based on Southern blotting analysis of the integrity of a genomic restriction fragment after alkaline hydrolysis, respectively.[39] The diol-epoxide derivative, DB[*a*,*l*]PDE, induced readily detectable stable DNA adducts in all three systems, whilst there was no increase in the level of unstable DNA adducts in either system. Treatment of the pure B11 cell-derived DNA with the parent compound, DB[*a*,*l*]P, in the presence of metabolizing enzymes generating radical cations yielded detectable levels of both stable and unstable DNA adducts. However, DB[*a*,*l*]P treatment of MCF-7 cells, which possess the required metabolic machinery for producing radical cation intermediates, did not result in any detectable level of unstable DNA adducts. Instead, there was a readily detectable level of stable DNA adducts in the treated MCF-7 cells. Altogether, it was calculated that the stable DNA adducts induced by both DB[*a*,*l*]P and DB[*a*,*l*]PDE in the tested systems contributed to over 99% of the overall DNA adducts formed. The non-significant proportion of unstable DNA adducts was ascribed to the very efficient mammalian repair systems, which can handle over 10,000 AP sites per cell per day.[40]

Confirming results were observed in a related study in which female SENCAR mice received topical administrations of DB[*a*,*l*]P, B[*a*]P and 7,12-dimethylbenz[*a*]anthracene (DMBA).[41] Interestingly, the formation of stable adducts in epidermal DNA induced by all of these compounds was dose-dependent and also the proportion of stable to unstable DNA adducts followed the respective tumorigenicity of these PAHs in mice. In other words, the levels of stable DNA adducts detected followed the order: fjord-region-containing DB[*a*,*l*]P > non-planar sterically hindered DMBA > bay-region-containing B[*a*]P. Conversely, DB[*a*,*l*]P produced the lowest level of unstable DNA adducts from all of these three carcinogens. In contrast to these findings, other studies reported that 99% of all DB[*a*,*l*]P- and DMBA-induced DNA adducts and more than 70% of all DNA adducts induced by B[*a*]P detected in mouse skin were unstable and could be released from the sugar phosphodiester backbone by depurination.[42–44] The respective percentages of such depurinated adducts in mammary glands from rats treated with DB[*a*,*l*]P and DMBA were estimated to be 97% and 52%, respectively.[45,46]

In conclusion, although the available data do not unequivocally undermine the significance of unstable DNA adducts in tumorigenesis, they mostly support the notion that stable PAH-DNA adducts are more likely to be responsible for PAH carcinogenicity.

5.5 Mutagenicity of PAH-DNA Adducts

Early works of Benzer and Freese gave rise to the idea that various mutagens produce different yet, unique mutational spectra.[47,48] Subsequent breakthroughs in DNA technology including inventions of DNA sequencing techniques and polymerase chain reaction (PCR) methods enabled analysis of phenotypically expressed mutant fragment(s) of DNA at the nucleotide level.[49–51] So far, mutational spectrum analysis of PAHs has been performed in a number of systems using various target or reporter genes, and more recently in different transgenic animal models. These include studying of PAH-induced mutations in target genes, e.g., *supF* of bacterial plasmids or mammalian shuttle vectors, as well as in endogenous genes, e.g., *lacI* and hypoxanthine-guanine phosphoribosyltransferase (*hprt*) of bacteria or mammalian cells, respectively (cf. Chapter 8).[52–71] The majority of these studies have established the specificity of individual PAHs to induce mutations (e.g., G $\rightarrow$ T transversions by bay-region PAHs and

A → T transversions by fjord-region PAHs) in a given system dependent upon a variety of factors including exposure dose, metabolic activity and DNA repair capacity of the tested system.[61–63,65,72] Unlike in experimental settings wherein most of these variables can be controlled, exposure to multi-PAH mixtures in humans occurs with an inter- and intra-individual variability for most, if not all, of these influential factors. Therefore, the current focus of interest is on fully characterizing the impact of these factors in validated test systems, which can mimic human exposure to PAHs. To achieve this, 'site-specific mutagenesis' approaches have been developed during the past two decades, which have remarkably helped shape our understanding of PAH-DNA adduct-induced mutagenicity.[73–75]

5.5.1 Site-Specific Mutagenicity of PAH-DNA Adducts

Site-specific mutagenesis studies use synthesized oligonuleotides containing genomic sequences of the target genes (e.g., *ras* and *p53*) wherein most mutations occur *in vivo*. The adduct of interest is placed at a specific site along the sequence and the resulting oligonucleotide is incorporated into a single- or double-stranded vector. The vector is replicated in an appropriate cellular system and the progenies are analyzed for the occurrence as well as the type of mutations.[73–75] Conceptually, the single-stranded vector-based studies are relevant for mutagenesis because *in vivo* mutations occur at a single/double-stranded DNA junction in the presence of DNA polymerases during the translesional synthesis (cf. Chapter 6). Inherent in the double-stranded vector-based studies are the DNA repair and strand bias of the constructed vector. Such drawbacks can however, be rectified by placing the adduct in one strand of the vector and ultra violet (UV)-irradiating the complementary strand (to saturate with photoproducts), thereby preventing progeny formation from the latter strand after the vector is replicated in the respective host cell.[76–79]

To date, *ras* oncogene sequences have frequently served as the target in site-specific studies, although many studies have also been carried out utilizing defined sequences derived from *p53*, *supF*, *lacZ*, etc.[73–75] Unlike *p53*, in which multitude mutations throughout exons 5–9 can lead to inactivation of the gene,[80,81] mutations of the *ras* genes can convert them into active oncogenes only when they occur at a few specific codons.[82,83] The *ras* oncogene family is comprised of three members: N-*ras*, H-*ras* and K-*ras* located on chromosomes 1, 11 and 12, respectively. All three genes encode a 21 kDa protein (p21), which possesses GTPase activity. The p21 protein has four domains (non-essential, essential, nucleotide binding and effector) with the first domain identical in all three *ras* genes. Activated *ras* oncogenes harbor mutations at codons 12, 13, 59 or 61 that cause conformational changes in the other three domains of the protein and as a result, the p21 protein loses its GTPase activity. The activated *ras* oncogenes are thought to be involved in aberrant cell proliferation, altered cell checkpoint control, and cell differentiation[82–85] (cf. Chapters 6 and 7).

5.5.1.1 *Effects of host cell co-factors, DNA sequence context, and stereochemistry of DNA adducts*

Synthetically defined 15-mer oligonucleotides were modified with both (+)- and (−)-*anti*-B[*a*]PDE enantiomers at 1st and 2nd guanines (G_1 and G_2) of their sequences, corresponding to the 1st base of codon 61 and to the 3rd base of codon 60 of the human H-*ras* oncogene, respectively. The modified oligonucleotides were incorporated into a single-stranded vector, pMS2. The constructs were used to either directly transform *E. coli* AB1157 cells, or to transfect COS cells and then to transform *E. coli* DH10B after being amplified in the COS host. The induced mutation frequencies in most parts were significantly different between the two host cells. Except for the (+)-*cis-anti*-B[*a*]PDE-N^2-dG_1 (i.e., (+)-*anti*-B[*a*]PDE-N^2-dG_1 adduct with *cis*-opened epoxide ring; cf. Chapter 6), which showed similar mutation frequency in both hosts, all other lesions were significantly more mutagenic in COS cells than in *E. coli* at the respective sequences (13- to 70-folds). Moreover, in *E. coli* both (+)-*cis-anti*-B[*a*]PDE-N^2-dG_1 and (+)-*cis-anti*-B[*a*]PDE-N^2-dG_2 showed higher mutagenicities compared to all of the (−)-*anti*-B[*a*]PDE-derived adducts. However, there was no appreciable difference in the mutagenicity of different stereoisomeric adducts in COS cells. Additionally, the mutagenicity of adducts varied markedly in different sequence contexts in both hosts. For example, in *E. coli* the (+)-*cis-anti*-B[*a*]PDE-N^2-dG_1 was 10 times (SOS-induced) and 6 times

(SOS-non-induced) more mutagenic than (+)-*cis-anti*-B[*a*]PDE-N^2-dG_2 (sequence contexts: TG_1G_2 and G_1G_2C, respectively). Conversely, in COS cells the (+)-*cis-anti*-B[*a*]PDE-N^2-dG_2 adduct was 3.5-fold more mutagenic compared to the (+)-*cis-anti*-B[*a*]PDE-N^2-dG_1 adduct.[86]

To further investigate the sequence-dependent mutagenicity of DNA adducts, a series of studies have been conducted with specifically defined *trans*-opened (+)-*anti*-B[*a*]PDE adduct-containing oligonucleotides incorporated into the relevant vectors and the appropriate host cells.[79,87–90] It was shown that the (+)-*trans-anti*-B[*a*]PDE-N^2-dG (i.e., (+)-*anti*-B[*a*]PDE-N^2-dG adduct with *trans*-opened epoxide ring; cf. Chapter 6) induced ~97% G to T transversions in a 5′–TGC–3′ sequence context and ~82% G to A transitions in another 5′–CGT–3′ sequence context.[79] The distinct patterns of mutations induced by an individual adduct in different sequence contexts were subsequently studied by using NMR and molecular modeling/computational chemistry approaches.[91–94] It was demonstrated that a given adduct in the duplex DNA can adopt multiple conformations, some of which may retain their base pairing potential whereas others, e.g., in the intercalative or base-displaced mode, may have impaired base pairing potential (cf. Chapter 6). It was hypothesized that the base-displaced conformation of (+)-*trans-anti*-B[*a*]PDE-N^2-dG, in which the B[*a*]P moiety is stacked in the helix DNA and the guanine moiety is displaced in the major *versus* minor grooves, would be responsible for the respective G → T or G → A mutations. Assumingly, the base-displaced conformations that are recognized as non-coding can trigger the DNA polymerases to insert dATP and dTTP opposite the lesions when the bulk is located in the major and minor grooves, respectively.[89,90,93–96]

5.5.1.2 *Effects of leading versus lagging strands of DNA*

An 11-mer oligonucleotide representing the sequence around the codon 61 (5′–CAA) of the N-*ras* gene was adducted with racemic *anti*-B[*a*]PDE at the N^6 position of the first adenine yielding *cis*-opened DNA adducts, i.e., (+)-*cis-anti*-B[*a*]PDE- and (−)-*cis-anti*-B[*a*]PDE-N^6-dA. The modified oligonucleotide was incorporated into a longer double-stranded oligomer and afterwards inserted unidirectionally into the pre-linearized RF

M13mp2SV vector. This vector contains an origin of replication from SV40, located either in the right or in the left proximal region of the insertion site, thereby replicating the adducted strand either in the leading or lagging fashions. It was shown that the (−)-*cis*-*anti*-B[*a*]PDE-N^6-dA adduct was over two times more mutagenic compared with the (+)-*cis*-*anti*-B[*a*]PDE-N^6-dA adduct. The difference in the mutagenic potencies of these two stereoisomeric adducts was even more pronounced when they were placed on the lagging strand. It was suggested that the highest tumorigenic diastereomer of B[*a*]PDE, the (+) *anti*-B[*a*]PDE, mainly binds to the N^2 amino group of guanine by *trans* opening of the 9,10-epoxide, thereby overshadowing the minor (+)- or (−)-*cis*-*anti*-B[*a*]PDE-adenine adducts to manifest their distinct mutagenicity.[97]

5.5.1.3 *Effects of structural bioactivity and DNA reparability: fjord-region versus bay-region-containing PAHs*

The codon 61 of human H-*ras* and N-*ras* oncogenes was used as a target to compare the removal of bay-region-containing B[*a*]PDE and fjord-region-containing B[*c*]Ph-3,4-diol-1,2-epoxides (B[*c*]PhDE) by the human nucleotide excision repair (NER) system (cf. Chapter 6). Eight different oligonucleotides containing the (+)- or (−)-*trans*-*anti*-B[*a*]PDE-N^6-dA and (+)- or (−)-*trans*-*anti*-B[*c*]PhDE-N^6-dA (11-mer to 18-mer) were synthesized (adduction sites* in the N-*ras* and H-*ras* were within the sequences of 5′–ACA*AG–3′ and 5′–CCA*GG–3′, respectively). The modified oligonucleotides were subsequently annealed and ligated with partially overlapping oligonucleotides to generate duplex DNA of 139–146 bp as the substrates for a standard NER-proficient HeLa cell extract. In all cases, the (+/−)-*trans*-*anti*-B[*a*]PDE-N^6-dA was excised efficiently, whereas the (+/−)-*trans*-*anti*-B[*c*]PhDE-N^6-dA was completely refractory to repair (see following chapter). Complementary analyses for the (+/−)-*trans*-*anti*-B[*a*]PDE-N^6-dA adducts located in two other sequence contexts, as well as subjected to a mouse NER system yielded similar results.[98] The observed difference in the reparability of DNA adducts induced by bay- *versus* fjord-region diol-epoxides of PAHs, which is in accordance with their respective magnitudes of tumorigenicity, was further investigated by NMR and molecular modeling analyses. It was shown that the (−)-*trans-anti*-B[*a*]PDE-N^6-dA is inserted into the DNA double helix in a way that distorts the base pairing at the site of covalent modification.[99] On the other hand, both (+)- and (−)-*trans-anti*-B[*c*]PhDE-N^6-dA adducts are intercalated into the helix without disturbing the Watson-Crick base pairing at the site of adduction.[100,101] Accordingly, the (+/−)-*trans-anti*-B[*a*]PDE-N^6-dA adducts can significantly reduce the melting point of the DNA duplex, which is an indicator for the weakening of the Watson-Crick hydrogen bonds.[102] The DNA duplex melting point however, remains unchanged after being modified with (+/−)-*trans-anti*-B[*c*]PhDE-N^6-dA.[103] These structural differences may explain the differences in repair of the two adduct types. Such differences in repair between different types of adducts or between different conformations of the same adduct will have a strong impact on site-specific mutagenesis.[98]

5.5.2 Translesional Synthesis

Replicative DNA polymerases, e.g. Pol α, Pol δ and Pol ε, are devised with a complex proofreading machinery to ensure the faithful replication of the genome.[104–106] The fidelity and processivity of DNA replication can however be compromised, e.g., when a repair resistant PAH adducted DNA is used as a template.[107–109] Consequently, the adducted nucleotide can promote miscoding during a process called 'translesional synthesis'.[109–112] In principle, when replicating past a bulky adduct, the DNA polymerases often pause or terminate replication as a function of enzyme blockage, which is highly dependent on the sequence context of the adduct.[113–116] In some cases however, the stalled or blocked polymerases are transiently replaced by another class of polymerases, collectively designated as Y-family polymerases, which can bypass the lesion. Because these Y-family enzymes do not possess the proofreading machinery (i.e., they all lack the $3' \rightarrow 5'$ exonuclease activity), their lesion-bypassing may result in inaccurate insertion of a mispairing nucleotide, depending on the type of lesion and the recruited enzyme.[108–110,117–119] The replication of such mispaired nucleotides translates into mutation, which is of relevance for carcinogenesis if the mutational events occur in genes encoding proteins crucial for cell-cycle control and growth, e.g., oncogenes or tumor suppressor genes.[21,22,120–122]

Currently, a number of mammalian Y-family enzymes are known, including Pol η, Pol κ, Rev1, Pol ζ and Pol ι.[123–129] All these polymerases can variably bypass the adducted DNA either in an error-free or error-prone fashion.[123–129] Pol η has intrinsically low fidelity and processivity.[129] It efficiently and accurately bypasses UV radiation-induced thymine–thymine dimers by inserting two dAMPs opposite the lesions; however, it is incapable of bypassing (6–4) photoproducts.[127,130] The extremely high susceptibility of *Xeroderma pigmentosum* (XP) patients to sunlight-associated skin cancer has been attributed to their lack of a functional *Pol* η gene.[128,131] The Pol η enzyme has also been shown to bypass various other lesions such as cisplatin G-G intrastrand cross-links, 2-acetylaminofluorene(AAF)-dG adducts and 8-oxo-dG (8-OH-dG) with relatively high fidelity,[127,132,133] whereas the *trans-anti*-B[*a*]PDE-N^2-dG adduct is erroneously bypassed by this enzyme with dAMP, dGMP and dTMP.[134]

The Pol κ is a low-fidelity polymerase with moderate processivity, which is highly inaccurate when replicating undamaged DNA.[135–137] The overexpression of this enzyme has recently been shown in tumorous lung tissues as compared with the adjacent normal lung tissues in cancer patients.[138] Also, the promoter region of the mouse *Pol* κ gene has been found to contain two copies of the arylhydrocarbon receptor (AhR) binding site, hence Pol κ expression can be induced by treatment of the animal with 3-methylcholanthrene[139] (cf. Figure 2.1 in Chapter 2). *In vitro*, Pol κ is capable of efficiently bypassing all stereoisomeric *anti*-B[*a*]PDE-N^2-dG adducts in both error-free and error-prone modes.[137] However, the overall bypass efficiency with dCMP opposite the lesion has been calculated to be 2–6 times higher than that with other mispairs.[140] More recently, it has been shown that mouse embryonic stem cells deficient in *Pol* κ are highly sensitive to the lethal and mutagenic effects of B[*a*]P. Most notably, mutational spectrum analysis of the *hprt* gene revealed a distinct pattern of mutations induced by B[*a*]P in the *Pol* κ-deficient stem cells as compared with the wild-type-containing control cells. The predominant occurrence of base substitutions at dG residues, with an overwhelming majority being G $\rightarrow$ T transversions, and mostly resulting from having the adducts on the non-transcribed strand of the *hprt* gene, characterizes the unique pattern of B[*a*]P-induced mutations in these *Pol* κ-deficient cells.[141] Given the uncertain efficacy of both Pol ζ and Pol ι in bypassing the *anti*-B[*a*]PDE-N^2-dG adduct,[142] and the lack of processivity of Rev1 despite its successful insertion of a dCMP opposite the lesion,[143] it was hypothesized that the induced mutations in the *Pol* κ-deficient stem cells might have arisen from the activity of Pol η alone or in combination with Rev1. The latter is thought to be carried out through a 'two-polymerase two-step' mechanism in which after the insertion of a correct deoxynucleoside triphosphate by one polymerase, the extension and processing procedures are carried out by another polymerase.[143–145]

Molecular dynamics simulation studies have shown that the (+)-*trans*-*anti*-B[*a*]PDE-N^2-dG adduct can adopt a '*syn*' conformation in the active site when bypassed by specific DNA polymerases, allowing the modified guanine base to mispair with an incoming deoxynucleoside triphosphate and the polymerases to adopt a closed conformation necessary for the formation of phosphodiester bonds.[91] The '*syn*' conformation of the (+)-*trans-anti*-B[*a*]PDE-N^2-dG adduct places the B[*a*]P moiety on the major groove of DNA in an open cavity of the polymerases,[91] thereby preventing the steric interference of the residues required for the replicative and fidelity-ensuring function of the enzymes.[104–106,146] Since the '*syn*' (+)-*trans-anti*-B[*a*]PDE-N^2-dG:dATP mismatch closely resembles the size and shape of a normal Watson-Crick base pair, it can easily trick the geometric selection machinery of the polymerase active site into accepting the mismatch and proceeding with the formation of the phosphodiester bond and extension of the primer.[147–149] Conversely, an '*anti*' conformation of the (+)-*trans-anti*-B[*a*]PDE-N^2-dG places the B[*a*]P moiety on the minor groove of DNA, which is densely packed with crucial residues for polymerase activities. Such obstruction of the active site of the polymerases inhibits the enzymes closing motion that is necessary for catalysis.[150] Therefore, the '*anti*' conformation can account for the predominant biological consequence of the (+)-*trans-anti*-B[*a*]PDE-N^2-dG adduct, namely replication blockage, whereas the '*syn*' conformation can explain the rare and infrequent bypass of the lesion resulting in a G $\rightarrow$ T transversion after replication.[89,90,93–96,113,114,116]

Despite the wealth of information gained from *in vitro* studies, we still know very little about which specific DNA polymerases are involved in error-prone bypass of PAH-DNA adducts *in vivo*, and how error-free *versus* error-prone bypass is regulated. Future work in this field are deemed necessary to elucidate in particular the interplay between DNA adducts and mutagenesis, one major aspect in chemical carcinogenesis.

5.6 Cancer Epidemiology and PAH-DNA Adducts

Although laboratory animal studies constitute the main forum for directly investigating the carcinogenicity of PAHs, epidemiological studies have also been instrumental. Except for a few early clinical trails on B[*a*]P efficacy to inhibit skin malignancy in humans,[6] and for a single recent prospective lung cancer study,[151] all other epidemiological data are compiled from case-control and retrospective cohort studies.[152] Methodologically, the nature and design of the latter studies however, enable them to elicit information on the association rather than causal relationship between a variable (e.g., PAH-DNA adducts) and an endpoint (e.g., cancer).[153] So far, PAH-DNA adducts have been successfully associated with cancer in patients as well as in model populations, e.g., occupationally, environmentally, medicinally and recreationally exposed individuals to PAHs[3] (cf. Chapter 4). To draw a causative link between PAH-DNA adduct formation and tumorigenesis in humans however, it is imperative to follow-up healthy populations non-experimentally exposed to PAHs for developing cancer (given that the experimental exposure of human subjects to carcinogenic PAHs is no longer considered to be ethical and thus simply out of question). Being informative, this approach however, has inherently major drawbacks, i.e., the lengthy follow-up period needed for cancer, a disease with long latency, leads to high costs and loss of information over the course of observation. For example, in the above-mentioned nested case-control study in which PAH-DNA adduct levels in white blood cells (WBC) predicted the risk for lung cancer, the unavailability of the follow-up data for smoking behaviors could easily bias the results toward the null.[151]

An indirect alternative approach is to use the epidemiological data to identify specific genetic alterations in cancers with known history of

PAH exposure and, subsequently, to investigate the involvement of PAH-DNA adducts in causing such genetic changes. For example, lung cancer is unequivocally associated with tobacco smoking, which imposes a substantial burden of PAHs on smokers.[154] Mutations of the *p53* tumor suppressor gene and *ras* oncogene are among the most frequent events occurring in lung cancer of smokers.[155–159] Mutational spectrum analyses of the *p53* and *ras* from the lung tumors of smokers reveal a distinct pattern of mutations for both genes. Non-random mutation sites predominantly at CpG dinucleotides (codons: 157, 158, 245, 248, 249 and 273; except codon 249 all other codons consist of CpGs) and a preponderance of G to T transversions occurring >90% on the non-transcribed strand are the major characteristics of *p53* mutations in smoking-associated lung cancer.[155,156] For the *ras* oncogenes, the respective mutations are strictly clustered at codons 12, 13, 59 and 61.[157–159] Such unique occurrence and specific pattern of mutations along both the *ras* and *p53* genes provide an invaluable venue for studying the etiological relevance of PAH-DNA adducts in lung carcinogenesis. In fact, the contribution of PAH-DNA adducts to the hypermutability of these two genes can be explored by co-localizing the sites of mutations and PAH-DNA adduct formations along the two genes in smokers' lung cancer. In addition, the involvement of PAH-DNA adducts in mutating these genes can be investigated by examining the similarities between the types of mutation induced by PAH-DNA adducts in experimental systems and those observed in the *p53* and *ras* genes in lung cancer from smokers. The former approach offers suggestive evidence for a correlation between PAH-DNA adducts and mutations in the two genes, whereas the latter approach implies a tentative causal relationship between these two events. To date, the 'adduct mapping' and 'mutagenesis' assays have been commonly used to apply these approaches in different experimental settings.

5.6.1 Mapping of PAH-DNA Adducts

Distribution of PAH-DNA adducts has mostly been mapped in the *p53* tumor suppressor gene[160–164] and most recently also in the *ras* oncogene.[165] The multifunctional *p53* gene plays a key role in cell survival and acts as a safeguard against genetic instability imposed by endogenous and/or exogenous stressors. The *p53* encodes a 53 kDa nuclear protein consisting of 393 amino acids. There are five major domains in the p53 protein characterized by their specific structures and functions. The sequence-specific DNA binding domain (amino acid residues 97–300) is the main target in the majority of human cancers including lung cancer. Such mutations of the *p53* gene are mostly missense and map to this domain of the protein causing its loss of function as a transcription factor.[80,81] The central part of the *p53* gene that contains this DNA-binding domain has been scrutinized by the UvrABC-coupled ligation-mediated polymerase chain reaction (LM-PCR) for the existence of PAH-DNA adducts. In this method, genomic DNA is cleaved at the site of adducted bases with *E. coli* UvrABC nuclease complex, which makes a dual incision six to seven bases 5′ and four bases 3′ to the modification site. The cleavage sites are subsequently subjected to the *p53*-specific oligonucleotide primer extension, ligation and amplification. The amplified products are denatured by gel electrophoresis, transferred to a membrane by electroblotting, and finally hybridized with radioactively labeled *p53* probes for visualization.[166]

Using the UvrABC-coupled LM-PCR, the distribution of DNA adducts was mapped along the exons 5, 7 and 8 of the *p53* gene in normal human bronchial epithelial cells and HeLa cells as well as in isolated DNA from normal human fibroblasts, all pre-treated with racemic *anti*-B[*a*]PDE. In all cases, three major adduct formation sites (codons 157, 248 and 273) were found that perfectly matched the frequently mutated sites ('hotspots') observed in human lung cancer. Moreover, the preferential formation of *anti*-B[*a*]PDE-DNA adducts occurred almost exclusively within the CpG dinucleotides on the non-transcribed strand of the *p53* gene.[162] This is in line with the sequence specificity of mutations and the strand bias of G to T transversions in lung cancers of smokers. Such accordance was pointed back to the transcription-coupled repair (TCR) phenomenon when it was shown that the removal of *anti*-B[*a*]PDE-DNA adducts in the non-transcribed strand of the *p53* gene in normal human fibroblasts is two to four times slower than that in the transcribed strand.[161] The selectivity of *anti*-B[*a*]PDE adduct formation was confirmed in another experiment in which *anti*-B[*a*]PDE treatment of the unmethylated plasmid pAT153P53π containing a genomic sequence of human *p53* (encompassing exons 2–11), resulted in a different adduct formation profile compared to that of the *anti*-B[*a*]PDE-treated genomic DNA. However, when the CpG dinucleotides of the plasmid were methylated by methylase *Sss*I, the profile of *anti*-B[*a*]PDE-DNA adducts was identical to that of the genomic DNA.[160] Additionally, it is known that the promoter region of the X-linked *PGK1* gene is a CpG island whose methylation status differs between the active and inactive X chromosomes, which are completely unmethylated at all 120 CpG sites and highly methylated at 118 CpGs respectively.[167,168] Treatment of hamster-human hybrid cell lines Y162–11C (X-active) and X8–6T2 (X-inactive) with *anti*-B[*a*]PDE showed a strong preference of DNA adduct formation at methylated CpGs over their unmethylated counterparts in this gene.[160] In another study, it was demonstrated that cytosine methylation greatly enhances the binding of B[*a*]PDE and benzo[*g*]chrysene 11,12-diol-13,14-epoxide (B[*g*]CDE) to guanine residues at all CpG sites in the *p53* gene.[164] To date, the exact mechanism(s) by which methylation of cytosine at CpG sites promotes the covalent binding of PAHs to guanine bases is not fully understood; however, it is widely believed that the higher nucleophilicity of guanine in CpG sequences as a result of an electron-donating effect of 5-methyl cytosine, and the increased pre-stacking of some PAHs such as B[*a*]PDE prior to their DNA binding, might be of influence.[169,170]

Subsequently, the distribution of DNA adducts was mapped along exons 5, 7 and 8 of the *p53* gene in normal human bronchial epithelial cells treated with five other diol-epoxides of PAHs including chrysene 1,2-diol-3,4-epoxide (CDE), 5-methylchrysene 1,2-diol-3,4-epoxide (5-MeCDE), 6-methylchrysene 1,2-diol-3,4-epoxide (6-MeCDE), B[*g*]CDE and B[*c*]PhDE. Overall, the DNA adduct profiles of all these five diol-epoxides were in most parts similar to that of B[*a*]PDE. Nonetheless, some additional DNA adduction sites were observed depending on the compound used. Interestingly, the fjord-region diol-epoxides (B[*g*]CDE and B[*c*]PhDE) and the bay-region diol-epoxides (CDE, 5-MeCDE and 6-MeCDE) had more comparable adduct profiles relative to each other. With the exception of codon 249, all major *p53* mutational hotspots in lung cancer were preferential binding sites for all or at least some of the PAH diol-epoxides.[163] It seems that mutations of the *p53* gene in lung cancer might be determined by the burden of multi-PAH exposure with additive or multiplicative interactions. This is in accordance with the fact that tobacco smoking, as the single most prominent risk factor for lung cancer, imposes a mixture of PAHs; some of which can act as promoters for the other tumor initiators.[154] Selection may also play a delicate role in shaping the overall spectrum of mutations in the *p53* gene in lung cancer because not all PAH-DNA adduction sites are necessarily major mutational hotspots.[156]

Recently, DNA adduct distribution was mapped in the N-, H- and K-*ras* oncogenes in normal human bronchial epithelial cells, the extracted DNA from such cells and normal human fibroblasts, all pre-treated with racemic *anti*-B[*a*]PDE. In all cases, preferential formation of DNA adducts was observed at guanine residues of codons 12 and 14 in the non-transcribed strand of the K-*ras* oncogene; however, removal of the adducts at codon 14 was twice as fast as that at codon 12. There was no substantial selectivity for DNA adducts formation along the N- or H-*ras* oncogenes. To further explore DNA adducts targeting specific sites of the K-*ras* gene, normal human bronchial epithelial cells were treated with three other bulky adduct-inducing agents, including B[*g*]CDE, *N*-acetoxy-AAF and aflatoxin B_1 8,9-epoxide (AFB_1-E). Similar results as seen with B[*a*]PDE were also observed for both B[*g*]CDE and *N*-acetoxy-AAF, whilst for the AFB_1-E, the intensity of DNA binding at codon 12 was higher than that at codon 14.[165] Altogether, these findings are in line with the epidemiological data documenting an association greater than 30% between smoking-related lung cancer and mutations at codon 12 of the K-*ras* oncogene.[157–159]

5.6.2 Additional Evidence for the Etiological Relevance of PAHs in Human Carcinogenesis: The Exemplary Case of *p53* Mutations in Lung Cancer

(I) DNA adduct formation and induction of mutations by B[*a*]PDE were studied in a *supF*-based vector as well as in the *lacI* and *cII* transgenes in embryonic mouse fibroblasts. Although there was no appreciable difference between the mutant frequencies of unmethylated and methylated non-treated vectors, treatment of both vectors with B[*a*]PDE significantly increased their respective mutant frequencies over the background (27-fold and 32-fold). Mutational spectrum analysis showed that of all induced mutations in unmethylated and methylated *supF* vectors, 25% and 42%, respectively, were at CpG sites. Although the predominant type of mutations in both unmethylated and methylated vectors were G → T transversions (65% and 68%, respectively), the percentage of these mutations at CpGs was significantly higher in methylated vectors than unmethylated vectors (42% *versus* 25%). Likewise, treatment of embryonic mouse fibroblasts with B[*a*]PDE increased the mutant frequencies of both the *lacI* and *cII* genes over the background (∼17- and 7-fold, respectively). Also, G → T transversions comprised 68% and 56% of all induced mutations in the *lacI* and *cII* transgenes, respectively. Of these, 77% and 58%, respectively, were at CpG sites, which were shown to be highly methylated in both genes *in vivo*. Mapping of B[*a*]PDE-DNA adducts along the *cII* gene showed a co-localization of several preferential adduct formation sites and frequently mutated sites in this gene, particularly at CpG-containing sequences, which harbored G to T transversions. The overall data in these three reporter genes mirrored the situation of *p53* mutations targeted by PAH-DNA adducts in smoking-associated lung cancer.[171]

(II) Chinese women using smoky coal for residential heating and cooking are highly susceptible to lung cancer[172,173] (cf. Chapter 4). The composition of the smoky coal combustion emissions is mainly of organic matter, of which over 40% are PAHs. It is estimated that the daily dose of B[*a*]P inhaled by a non-smoker exposed to household smoky coal is 30 times more than that of a regular smoker.[174] Accordingly, the urinary excretion of the depurinated B[*a*]P-DNA adducts was shown to be 600-fold higher in non-smoking Chinese women exposed to smoky coal as compared with a control group of smokers.[174,175] Determination of mutations in the lung tumors of these women showed that 71% of all samples had base substitutions in the *p53* gene, of which 76% were G → T transversions.[176] All these G to T transversions occurred in the non-transcribed strand of the *p53* gene, and the C → T transitions, which are commonly thought to have an endogenous origin[177,178] comprised <10% of the total mutations.[176] In addition, 33% of mutations in the *p53* gene clustered within a GC rich region along codons 153–158, with codon 154 being a hotspot. There were two other hotspots at codons 249 and 273 of the *p53* gene. Altogether, the data well reflected the impact of exogenously inhaled PAHs on forming the overall mutational spectrum of the *p53* gene in this study population.[176]

(III) Some critics have raised the concern that in mapping of DNA adducts by UvrABC-coupled LM-PCR, it is the enzymes-nicking site rather than the adduct location *per se* that is being quantified. As the cleavage of the modified DNA by UvrABC complex can be sequence-dependent, this might bias the quantification of DNA adducts by this technique. Moreover, the method is unable to reveal the structural identity and stereochemistry of the quantified DNA adducts.[166] In an attempt to address these issues, a highly specific HPLC electrospray ionization tandem mass spectrometry (ESI-MS/MS) technique was developed, where the extent of B[*a*]PDE-N^2-dG formation at ^{15}N-labeled guanine in a defined sequence from exon five of the *p53* gene could be directly determined. A clustering of the B[*a*]PDE-N^2-dG(^{15}N) adducts at codons 154, 156, 157 and 158, together with an enhancing effect of cytosine methylation on the formation of adducts at these sites were observed. Although the higher extent of adduct formation at codon 156 relative to codons 157 and 158 did not perfectly correspond with their known respective mutability in lung cancers (most likely due to limited selection of the codon 156 mutation), the overall results were in good agreement with those previously obtained by the UvrABC-coupled LM-PCR.[179]

(IV) Mutational load of the *p53* gene at codons 157, 248, 249 and 250 was determined in non-tumorous lung tissues from smoking lung cancer patients. A high frequency of G $\rightarrow$ T transversions at both codons 157 and 248 together with G $\rightarrow$ T transversions and G $\rightarrow$ A transitions at codon 249 were observed. Additionally, B[*a*]PDE treatment of BEAS-2B bronchial epithelial cells resulted in a dose-dependent induction of G $\rightarrow$ T transversions at both codons 157 and 248, as well as C $\rightarrow$ T transitions and G $\rightarrow$ T transversions at codon 249 of the *p53* gene.[180]

5.7 Concluding Remarks

Throughout the years, ‘PAH-DNA adduct-induced mutagenesis’ has evolved as a major theme for research in chemical carcinogenesis. Initial theories on the involvement of parent PAHs *per se* in carcinogenesis faded away[7] after the recognition that reactive metabolites of PAHs are capable of forming adduct complexes with DNA;[9] the moiety required for neoplastic transformation of the cell.[13] The triggering of mutagenesis by DNA adducts is now believed to be a major mechanism by which PAHs exert their carcinogenic effects. Currently, the focus of attention is on the identity of DNA adducts, which specifically induce mutations in cancer related genes, e.g., oncogenes and tumor suppressor genes in the human genome.[21,156] The advent of technology has enabled sensitive and specific detection of DNA adducts and their three-dimensional chemical structures, as well as prediction of the processes by which formation of any given DNA adduct can lead to mutagenesis. The goal is to validate certain types of DNA adducts that are of most relevance for carcinogenesis as the biomarkers of early effects and susceptibility for cancer (cf. Chapter 4). The ultimate hope is to use such validated biomarkers in evaluating the effectiveness of preventive strategies for cancer.

Acknowledgment

Work of the authors was supported by the National Institute of Health (NIH) grant CA84469.

References

1. Harvey RG (1991) Environmental occurrence and cancer. In: *Polycyclic Aromatic Hydrocarbons: Chemistry and Carcinogenicity* [Harvey RG (ed.)] pp 11–25, Cambridge University Press, New York.
2. Phillips DH (1999) Polycyclic aromatic hydrocarbons in the diet. *Mutat. Res.* 443: 139–147.
3. Phillips DH (2002) Smoking-related DNA and protein adducts in human tissues. *Carcinogenesis* 23: 1979–2004.
4. Kennaway E (1955) The identification of a carcinogenic compound in coal-tar. *Br. Med. J.* 2: 749–752.
5. Alloway BJ and Ayres DC (1994) *Chemical Principles of Environmental Pollution*. Blackie Academic & Professional, Glasgow, UK.
6. Phillips DH (1983) Fifty years of benzo[*a*]pyrene. *Nature* 303: 468–472.

7. Pullman A and Pullman B (1955) Electronic structure and carcinogenic activity of aromatic molecules. In: *Advances in Cancer Research* [Greenstein A and Haddow PP (eds.)] pp 117–169, Wiley-Interscience, New York.
8. Harvey RG (1991) Metabolic activation, DNA binding, and mechanisms of carcinogenesis. In: *Polycyclic Aromatic Hydrocarbons: Chemistry and Carcinogenicity* [Harvey RG (ed.)] pp 50–95, Cambridge University Press, New York.
9. Miller JA and Miller EC (1969) The metabolic activation of carcinogenic aromatic amines and amides. *Prog. Exp. Tumor Res.* 11: 273–301.
10. Miller JA (1970) Carcinogenesis by chemicals: an overview — G.H.A. Clowes Memorial Lecture. *Cancer Res.* 30: 559–576.
11. Brookes P (1966) Quantitative aspects of the reaction of some carcinogens with nucleic acids and the possible significance of such reactions in the process of carcinogenesis. *Cancer Res.* 26: 1994–2003.
12. Lawley PD and Brookes P (1964) Methylation of adenine in deoxyadenylic acid or deoxyribonucleic acid at N-7. *Biochem. J.* 92: 19c–20c.
13. Shih C, Shilo BZ, Goldfarb MP, Dannenberg A and Weinberg RA (1979) Passage of phenotypes of chemically transformed cells via transfection of DNA and chromatin. *Proc. Natl. Acad. Sci. USA* 76: 5714–5718.
14. Harvey RG (1991) Molecular properties of polyarenes. In: *Polycyclic Aromatic Hydrocarbons: Chemistry and Carcinogenicity* [Harvey RG (ed.)] pp 96–116, Cambridge University Press, New York.
15. Hall M and Grover PL (1990) Polycyclic aromatic hydrocarbons: metabolism, activation and tumour initiation. In: *Chemical Carcinogenesis and Mutagenesis* [Cooper CS and Grover PL (eds.)] Volume I, pp 327–372, Springer-Verlag, Berlin.
16. Nebert DW and Gonzalez FJ (1987) *P450* genes: structure, evolution, and regulation. *Annu. Rev. Biochem.* 56: 945–993.
17. Armsrong RN (1997) Glutathione-*S*-transferases. In: *Comprehensive Toxicology: Biotransformation* [Guengerich FP (ed.)] Volume 3, pp 307–327, Elsevier Science, Oxford, UK.
18. Guenther TM and Oesch F (1981) Microsomal epoxide hydrolase and its role in polycyclic aromatic hydrocarbon transformation. In: *Polycyclic Aromatic Hydrocarbons and Cancer* [Gelboin HV and Ts'o POP (eds.)] Volume 3, pp 183–212, Academic Press, New York.
19. Burchell B, McGurk K, Brierley CH and Clarke DJ (1997) UDP-glucuronosyltransferases. In: *Comprehensive Toxicology: Biotransformation* [Guengerich FP (ed.)] Volume 3, pp 401–436, Elsevier Science, Oxford, UK.

20. Duffel MW (1997) Sulfotransferases. In: *Comprehensive Toxicology: Biotransformation* [Guengerich FP (ed.)] Volume 3, pp 365–384, Elsevier Science, Oxford, UK.
21. Pfeifer GP and Denissenko MF (1998) Formation and repair of DNA lesions in the *p53* gene: relation to cancer mutations? *Environ. Mol. Mutagen.* 31: 197–205.
22. Seo KY, Jelinsky SA and Loechler EL (2000) Factors that influence the mutagenic patterns of DNA adducts from chemical carcinogens. *Mutat. Res.* 463: 215–246.
23. Jeffrey AM, Weinstein IB, Jennette KW, Grzeskowiak K, Nakanishi K, Harvey RG, Autrup H and Harris C (1977) Structures of benzo[*a*]pyrene-nucleic acid adducts formed in human and bovine bronchial explants. *Nature* 269: 348–350.
24. Nakanishi K, Kasai H, Cho H, Harvey RG, Jeffrey AM, Jennette KW and Weinstein I (1977) Absolute configuration of a ribonucleic acid adduct formed *in vivo* by metabolism of benzo[*a*]pyrene. *J. Am. Chem. Soc.* 99: 258–260.
25. Sims P and Grover PL (1981) Involvement of dihydrodiols and diol epoxides in the metabolic activation of polycyclic aromatic hydrocarbons other than benzo[*a*]pyrene. In: *Polycyclic Hydrocarbons and Cancer* [Gelboin HV and Ts'o POP (eds.)] pp 117–181, Academic Press, New York.
26. Memory JD and Wilson NK (1982) *NMR of Aromatic Compounds.* Wiley, New York.
27. Karcher W, Fordham RJ, Dubois JJ, Glaude PG and Lighthart JA (1985) *Spectral Atlas of Polycyclic Aromatic Compounds.* Reidel, Dordrecht, The Netherlands.
28. Levin W, Chang RL, Wood AW, Thakker DR, Yagi H, Jerina DM and Conney AH (1986) Tumorigenicity of optical isomers of the diastereomeric bay-region 3,4-diol-1,2-epoxides of benzo[*c*]phenanthrene in murine tumor models. *Cancer Res.* 46: 2257–2261.
29. Hecht SS, El-Bayoumy K, Rivenson A and Amin S (1994) Potent mammary carcinogenicity in female CD rats of a fjord region diol-epoxide of benzo[*c*]phenanthrene compared to a bay region diol-epoxide of benzo[*a*]pyrene. *Cancer Res.* 54: 21–24.
30. Cavalieri EL, Rogan EG, Higginbotham S, Cremonesi P and Salmasi S (1989) Tumor-initiating activity in mouse skin and carcinogenicity in rat mammary gland of dibenzo[*a*]pyrenes: the very potent environmental carcinogen dibenzo[*a*,*l*]pyrene. *J. Cancer Res. Clin. Oncol.* 115: 67–72.

31. Cavalieri EL, Higginbotham S, RamaKrishna NV, Devanesan PD, Todorovic R, Rogan EG and Salmasi S (1991) Comparative dose-response tumorigenicity studies of dibenzo[*a*,*l*]pyrene versus 7,12-dimethylbenz-[*a*]anthracene, benzo[*a*]pyrene and two dibenzo[*a*,*l*]pyrene dihydrodiols in mouse skin and rat mammary gland. *Carcinogenesis* 12: 1939–1944.
32. Higginbotham S, RamaKrishna NV, Johansson SL, Rogan EG and Cavalieri EL (1993) Tumor-initiating activity and carcinogenicity of dibenzo[*a*,*l*]pyrene versus 7,12-dimethylbenz[*a*]anthracene and benzo[*a*]-pyrene at low doses in mouse skin. *Carcinogenesis* 14: 875–888.
33. Harvey RG (1991) Nomenclature. In: *Polycyclic Aromatic Hydrocarbons: Chemistry and Carcinogenicity* [Harvey RG (ed.)] pp 1–10, Cambridge University Press, New York.
34. Harvey RG (1991) Dihydrodiols. In: *Polycyclic Aromatic Hydrocarbons: Chemistry and Carcinogenicity* [Harvey RG (ed.)] pp 306–329, Cambridge University Press, New York.
35. Weinstein IB, Jeffrey AM, Jennette KW, Blobstein SH, Harvey RG, Harris C, Autrup H, Kasai H and Nakanishi K (1976) Benzo[*a*]pyrene diol epoxides as intermediates in nucleic acid binding *in vitro* and *in vivo*. *Science* 193: 592–595.
36. Conney AH (1982) Induction of microsomal enzymes by foreign chemicals and carcinogenesis by polycyclic aromatic hydrocarbons: G.H.A. Clowes Memorial Lecture. *Cancer Res.* 42: 4875–4917.
37. Huberman E, Sachs L, Yang SK and Gelboin V (1976) Identification of mutagenic metabolites of benzo[*a*]pyrene in mammalian cells. *Proc. Natl. Acad. Sci. USA* 73: 607–611.
38. Newbold RF and Brookes P (1976) Exceptional mutagenicity of a benzo-[*a*]pyrene diol epoxide in cultured mammalian cells. *Nature* 261: 52–54.
39. Melendez-Colon VJ, Smith CA, Seidel A, Luch A, Platt KL and Baird WM (1997) Formation of stable adducts and absence of depurinating DNA adducts in cells and DNA treated with the potent carcinogen dibenzo[*a*,*l*]pyrene or its diol epoxides. *Proc. Natl. Acad. Sci. USA* 94: 13542–13547.
40. Loeb LA (1989) Endogenous carcinogenesis: molecular oncology into the twenty-first century — Presidential address. *Cancer Res.* 49: 5489–5496.
41. Melendez-Colon VJ, Luch A, Seidel A and Baird WM (1999) Cancer initiation by polycyclic aromatic hydrocarbons results from formation of stable DNA adducts rather than apurinic sites. *Carcinogenesis* 20: 1885–1891.
42. Rogan EG, Devanesan PD, RamaKrishna NV, Higginbotham S, Padmavathi NS, Chapman K, Cavalieri EL, Jeong H, Jankowiak R and Small GJ (1993)

Identification and quantitation of benzo[*a*]pyrene-DNA adducts formed in mouse skin. *Chem. Res. Toxicol.* 6: 356–363.

43. Devanesan PD, RamaKrishna NV, Padmavathi NS, Higginbotham S, Rogan EG, Cavalieri EL, Marsch GA, Jankowiak R and Small GJ (1993) Identification and quantitation of 7,12-dimethylbenz[*a*]anthracene-DNA adducts formed in mouse skin. *Chem. Res. Toxicol.* 6: 364–371.
44. Li KM, Todorovic R, Cavalieri EL, Rogan EG, Jankowiak R and Small G (1997) Depurinating DNA adducts detected in mouse skin treated with dibenzo[*a*,*l*]pyrene (DB[*a*,*l*]P), DB[*a*,*l*]P-11,12-dihydrodiol, (±)-*anti*-DB[*a*,*l*]P diol epoxide (DE) or (±)-*syn*-DB[*a*,*l*]PDE. *Proceedings of the American Association for Cancer Research.* American Association for Cancer Research, Volume 38, p 127.
45. Todorovic R, Ariese F, Devanesan P, Jankowiak R, Small GJ, Rogan E and Cavalieri E (1997) Determination of benzo[*a*]pyrene- and 7,12-dimethylbenz[*a*]anthracene-DNA adducts formed in rat mammary glands. *Chem. Res. Toxicol.* 10: 941–947.
46. Todorovic R, Li KM, Rogan EG and Cavalieri EL (1996) Identification and quantitation of DNA adducts of benzo[*a*]pyrene (BP), 7,12-dimethylbenz[*a*]anthracene (DMBA) and dibenzo[*a*,*l*]pyrene (DB[*a*,*l*]P) formed in rat mammary glands. *Proceedings of the American Association for Cancer Research.* American Association for Cancer Research, Volume 37, p 119.
47. Benzer S and Freese E (1958) Induction of specific mutations with 5-bromouracil. *Proc. Natl. Acad. Sci. USA* 44: 112–119.
48. Benzer S (1961) On the topography of genetic fine structure. *Proc. Natl. Acad. Sci. USA* 47: 403–415.
49. Maxam AM and Gilbert W (1977) A new method for sequencing DNA. *Proc. Natl. Acad. Sci. USA* 74: 560–564.
50. Sanger F, Nicklen S and Coulson AR (1977) DNA sequencing with chain-terminating inhibitors. *Proc. Natl. Acad. Sci. USA* 74: 5463–5467.
51. Saiki RK, Scharf S, Faloona F, Mullis KB, Horn GT, Erlich HA and Arnheim N (1985) Enzymatic amplification of beta-globin genomic sequences and restriction site analysis for diagnosis of sickle cell anemia. *Science* 230: 1350–1354.
52. Mizusawa H, Lee CH, Kakefuda T, McKenney K, Shimatake H and Rosenberg M (1981) Base insertion and deletion mutations induced in an *Escherichia coli* plasmid by benzo[*a*]pyrene-7,8-dihydrodiol-9,10-epoxide. *Proc. Natl. Acad. Sci. USA* 78: 6817–6820.

53. Chakrabarti S, Mizusawa H and Seidman M (1984) Survival and mutagenesis of bacterial plasmids with localized carcinogen adducts. *Mutat. Res.* 126: 127–137.
54. Rodriguez H and Loechler EL (1993) Mutational specificity of the (+)-*anti*-diol epoxide of benzo[*a*]pyrene in a *supF* gene of an *Escherichia coli* plasmid: DNA sequence context influences hotspots, mutagenic specificity and the extent of SOS enhancement of mutagenesis. *Carcinogenesis* 14: 373–383.
55. Roilides E, Gielen JE, Tuteja N, Levine AS and Dixon K (1988) Mutational specificity of benzo[*a*]pyrene diol epoxide in monkey cells. *Mutat. Res.* 198: 199–206.
56. Yang JL, Chen RH, Maher VM and McCormick JJ (1991) Kinds and location of mutations induced by (±)-7β,8α-dihydroxy-9α,10α-epoxy-7,8,9,10-tetrahydrobenzo[*a*]pyrene in the coding region of the *hypoxanthine (guanine) phosphoribosyltransferase* gene in diploid human fibroblasts. *Carcinogenesis* 12: 71–75.
57. Yang JL, Maher VM and McCormick JJ (1987) Kinds of mutations formed when a shuttle vector containing adducts of (±)-7β,8α-dihydroxy-9α,10α-epoxy-7,8,9,10-tetrahydrobenzo[*a*]pyrene replicates in human cells. *Proc. Natl. Acad. Sci. USA* 84: 3787–3791.
58. Yang JL, Maher VM and McCormick JJ (1987) Kinds of mutations formed when a shuttle vector containing adducts of benzo[*a*]pyrene-7,8-diol-9,10-epoxide replicates in COS7 cells. *Mol. Cell. Biol.* 7: 1267–1270.
59. Eisenstadt E, Warren AJ, Porter J, Atkins D and Miller JH (1982) Carcinogenic epoxides of benzo[*a*]pyrene and cyclopenta[*c,d*]pyrene induce base substitutions via specific transversions. *Proc. Natl. Acad. Sci. USA* 79: 1945–1949.
60. Bernelot-Moens C, Glickman BW and Gordon AJ (1990) Induction of specific frameshift and base substitution events by benzo[*a*]pyrene diol epoxide in excision-repair-deficient *Escherichia coli*. *Carcinogenesis* 11: 781–785.
61. Wei SJ, Chang RL, Wong CQ, Bhachech N, Cui XX, Hennig E, Yagi H, Sayer JM, Jerina DM, Preston BD and Conney AH (1991) Dose-dependent differences in the profile of mutations induced by an ultimate carcinogen from benzo[*a*]pyrene. *Proc. Natl. Acad. Sci. USA* 88: 11227–11230.
62. Wei SJ, Chang RL, Bhachech N, Cui XX, Merkler KA, Wong CQ, Hennig E, Yagi H, Jerina DM and Conney AH (1993) Dose-dependent differences in the profile of mutations induced by (+)-7*R*,8*S*-dihydroxy-9*S*,10*R*-epoxy-7,8,9,10-tetrahydrobenzo[*a*]pyrene in the coding region of the *hypoxanthine*

(guanine) phosphoribosyltransferase gene in Chinese hamster V-79 cells. *Cancer Res.* 53: 3294–3301.

63. Chen RH, Maher VM, Brouwer J, van de Putte P and McCormick JJ (1992) Preferential repair and strand-specific repair of benzo[*a*]pyrene diol epoxide adducts in the *HPRT* gene of diploid human fibroblasts. *Proc. Natl. Acad. Sci. USA* 89: 5413–5417.
64. Chen RH, Maher VM and McCormick JJ (1991) Lack of a cell cycle-dependent strand bias for mutations induced in the *HPRT* gene by (±)-7β,8α-dihydroxy-9α,10α-epoxy-7,8,9,10-tetrahydrobenzo[*a*]pyrene in excision repair-deficient human cells. *Cancer Res.* 51: 2587–2592.
65. Chen RH, Maher VM and McCormick JJ (1990) Effect of excision repair by diploid human fibroblasts on the kinds and locations of mutations induced by (±)-7β,8α-dihydroxy-9α,10α-epoxy-7,8,9,10-tetrahydrobenzo[*a*]pyrene in the coding region of the *HPRT* gene. *Proc. Natl. Acad. Sci. USA* 87: 8680–8684.
66. Wei SJ, Chang RL, Hennig E, Cui XX, Merkler KA, Wong CQ, Yagi H, Jerina DM and Conney AH (1994) Mutagenic selectivity at the *HPRT* locus in V-79 cells: comparison of mutations caused by bay-region benzo[*a*]pyrene 7,8-diol-9,10-epoxide enantiomers with high and low carcinogenic activity. *Carcinogenesis* 15: 1729–1735.
67. Keohavong P and Thilly WG (1992) Mutational spectrometry: a general approach for hot-spot point mutations in selectable genes. *Proc. Natl. Acad. Sci. USA* 89: 4623–4627.
68. Andersson B, Falt S and Lambert B (1992) Strand specificity for mutations induced by (+)-*anti*-BPDE in the *hprt* gene in human T-lymphocytes. *Mutat. Res.* 269: 129–140.
69. Jakubczak JL, Merlino G, French JE, Muller WJ, Paul B, Adhya S and Garges S (1996) Analysis of genetic instability during mammary tumor progression using a novel selection-based assay for *in vivo* mutations in a bacteriophage lambda transgene target. *Proc. Natl. Acad. Sci. USA* 93: 9073–9078.
70. Skopek TR, Kort KL, Marino DR, Mittal LV, Umbenhauer DR, Laws GM and Adams SP (1996) Mutagenic response of the endogenous *hprt* gene and *lacI* transgene in benzo[*a*]pyrene-treated Big Blue B6C3F1 mice. *Environ. Mol. Mutagen.* 28: 376–384.
71. Bigger CA, Ponten I, Page JE and Dipple A (2000) Mutational spectra for polycyclic aromatic hydrocarbons in the *supF* target gene. *Mutat. Res.* 450: 75–93.
72. Chen J and Thilly WG (1996) Mutational spectra vary with exposure conditions: benzo[*a*]pyrene in human cells. *Mutat. Res.* 357: 209–217.

73. Basu AK and Essigmann JM (1988) Site-specifically modified oligodeoxynucleotides as probes for the structural and biological effects of DNA-damaging agents. *Chem. Res. Toxicol.* 1: 1–18.
74. Loechler EL (1996) The role of adduct site-specific mutagenesis in understanding how carcinogen-DNA adducts cause mutations: perspective, prospects and problems. *Carcinogenesis* 17: 895–902.
75. Singer B and Essigmann JM (1991) Site-specific mutagenesis: retrospective and prospective. *Carcinogenesis* 12: 949–955.
76. Mackay W, Benasutti M, Drouin E and Loechler EL (1992) Mutagenesis by (+)-*anti*-B[*a*]P-N^2-Gua, the major adduct of activated benzo[*a*]pyrene, when studied in an *Escherichia coli* plasmid using site-directed methods. *Carcinogenesis* 13: 1415–1425.
77. Jelinsky SA, Liu T, Geacintov NE and Loechler EL (1995) The major N^2-Gua adduct of the (+)-*anti*-benzo[*a*]pyrene diol epoxide is capable of inducing G → A and G → C, in addition to G → T, mutations. *Biochemistry* 34: 13545–13553.
78. Shukla R, Jelinsky S, Liu T, Geacintov NE and Loechler EL (1997) How stereochemistry affects mutagenesis by N^2-deoxyguanosine adducts of 7,8-dihydroxy-9,10-epoxy-7,8,9,10-tetrahydrobenzo[*a*]pyrene: configuration of the adduct bond is more important than those of the hydroxyl groups. *Biochemistry* 36: 13263–13269.
79. Shukla R, Liu T, Geacintov NE and Loechler EL (1997) The major N^2-dG adduct of (+)-*anti*-B[*a*]PDE shows a dramatically different mutagenic specificity (predominantly, G → A) in a 5′-C<u>G</u>T-3′ sequence context. *Biochemistry* 36: 10256–10261.
80. Guimaraes DP and Hainaut P (2002) *TP53*: a key gene in human cancer. *Biochimie* 84: 83–93.
81. Robles AI, Linke SP and Harris CC (2002) The *p53* network in lung carcinogenesis. *Oncogene* 21: 6898–6907.
82. Sills RC, Boorman GA, Neal JE, Hong HL and Devereux TR (1999) Mutations in *ras* genes in experimental tumours of rodents. *IARC Sci. Publ.* 55–86.
83. Maronpot RR, Fox T, Malarkey DE and Goldsworthy TL (1995) Mutations in the *ras* proto-oncogene: clues to etiology and molecular pathogenesis of mouse liver tumors. *Toxicology* 101: 125–156.
84. Bos JL (1989) *Ras* oncogenes in human cancer: a review. *Cancer Res.* 49: 4682–4689.
85. Lowy DR and Willumsen BM (1993) Function and regulation of *ras*. *Annu. Rev. Biochem.* 62: 851–891.

86. Moriya M, Spiegel S, Fernandes A, Amin S, Liu T, Geacintov N and Grollman AP (1996) Fidelity of translesional synthesis past benzo[*a*]pyrene diol epoxide-2′-deoxyguanosine DNA adducts: marked effects of host cell, sequence context, and chirality. *Biochemistry* 35: 16646–16651.
87. Page JE, Zajc B, Oh-hara T, Lakshman MK, Sayer JM, Jerina DM and Dipple A (1998) Sequence context profoundly influences the mutagenic potency of *trans*-opened benzo[*a*]pyrene 7,8-diol 9,10-epoxide-purine nucleoside adducts in site-specific mutation studies. *Biochemistry* 37: 9127–9137.
88. Tsao H, Mao B, Zhuang P, Xu R, Amin S and Geacintov NE (1998) Sequence dependence and characteristics of bends induced by site-specific polynuclear aromatic carcinogen-deoxyguanosine lesions in oligonucleotides. *Biochemistry* 37: 4993–5000.
89. Kozack RE and Loechler EL (1997) Molecular modeling of the conformational complexity of (+)-*anti*-B[*a*]PDE-adducted DNA using simulated annealing. *Carcinogenesis* 18: 1585–1593.
90. Shukla R, Geacintov NE and Loechler EL (1999) The major N^2-dG adduct of (+)-*anti*-B[*a*]PDE induces G → A mutations in a 5′-AGA-3′ sequence context. *Carcinogenesis* 20: 261–268.
91. Perlow RA and Broyde S (2001) Evading the proofreading machinery of a replicative DNA polymerase: induction of a mutation by an environmental carcinogen. *J. Mol. Biol.* 309: 519–536.
92. Lin CH, Huang X, Kolbanovskii A, Hingerty BE, Amin S, Broyde S, Geacintov NE and Patel DJ (2001) Molecular topology of polycyclic aromatic carcinogens determines DNA adduct conformation: a link to tumorigenic activity. *J. Mol. Biol.* 306: 1059–1080.
93. Kozack RE and Loechler EL (1999) Molecular modeling of the major adduct of (+)-*anti*-B[*a*]PDE (N^2-dG) in the eight conformations and the five DNA sequences most relevant to base substitution mutagenesis. *Carcinogenesis* 20: 85–94.
94. Kozack RE, Shukla R and Loechler EL (1999) A hypothesis for what conformation of the major adduct of (+)-*anti*-B[*a*]PDE (N^2-dG) causes G → T *versus* G → A mutations based upon a correlation between mutagenesis and molecular modeling results. *Carcinogenesis* 20: 95–102.
95. Seo KY, Jelinsky SA and Loechler EL (2000) Factors that influence the mutagenic patterns of DNA adducts from chemical carcinogens. *Mutat. Res.* 463: 215–246.
96. Kozack R, Seo KY, Jelinsky SA and Loechler EL (2000) Toward an understanding of the role of DNA adduct conformation in defining mutagenic mechanism based on studies of the major adduct (formed at N^2-dG) of

the potent environmental carcinogen, benzo[*a*]pyrene. *Mutat. Res.* 450: 41–59.

97. Lavrukhin OV and Lloyd RS (1998) Mutagenic replication in a human cell extract of DNAs containing site-specific and stereospecific benzo[*a*]pyrene-7,8-diol-9,10-epoxide DNA adducts placed on the leading and lagging strands. *Cancer Res.* 58: 887–891.
98. Buterin T, Hess MT, Luneva N, Geacintov NE, Amin S, Kroth H, Seidel A and Naegeli H (2000) Unrepaired fjord region polycyclic aromatic hydrocarbon-DNA adducts in *ras* codon 61 mutational hot spots. *Cancer Res.* 60: 1849–1856.
99. Zegar IS, Kim SJ, Johansen TN, Horton PJ, Harris CM, Harris TM and Stone MP (1996) Adduction of the human N-*ras* codon 61 sequence with (−)-(7*S*,8*R*,9*R*,10*S*)-7,8-dihydroxy-9,10-epoxy-7,8,9,10-tetrahydro-benzo[*a*]pyrene: structural refinement of the intercalated SRSR(61,2) (−)-(7*S*,8*R*,9*S*,10*R*)-N^6-[10-(7,8,9,10-tetrahydrobenzo[*a*]pyrenyl)]-2′-deoxyadenosyl adduct from ^{1}H NMR. *Biochemistry* 35: 6212–6224.
100. Cosman M, Laryea A, Fiala R, Hingerty BE, Amin S, Geacintov NE, Broyde S and Patel DJ (1995) Solution conformation of the (−)-*trans-anti*-benzo[*c*]phenanthrene-dA ([BPh]dA) adduct opposite dT in a DNA duplex: intercalation of the covalently attached benzo[*c*]phenanthrenyl ring to the 3′-side of the adduct site and comparison with the (+)-*trans-anti*-[BPh]dA opposite dT stereoisomer. *Biochemistry* 34: 1295–1307.
101. Cosman M, Fiala R, Hingerty BE, Laryea A, Lee H, Harvey RG, Amin S, Geacintov NE, Broyde S and Patel DJ (1993) Solution conformation of the (+)-*trans-anti*-[BPh]dA adduct opposite dT in a DNA duplex: intercalation of the covalently attached benzo[*c*]phenanthrene to the 5′-side of the adduct site without disruption of the modified base pair. *Biochemistry* 32: 12488–12497.
102. Schurter EJ, Yeh HJ, Sayer JM, Lakshman MK, Yagi H, Jerina DM and Gorenstein DG (1995) NMR solution structure of a nonanucleotide duplex with a dG mismatch opposite a 10*R* adduct derived from *trans* addition of a deoxyadenosine N^6-amino group to (−)-(7*S*,8*R*,9*R*,10*S*)-7,8-dihydroxy-9,10-epoxy-7,8,9,10-tetrahydrobenzo[*a*]pyrene. *Biochemistry* 34: 1364–1375.
103. Laryea A, Cosman M, Lin JM, Liu T, Agarwal R, Smirnov S, Amin S, Harvey RG, Dipple A and Geacintov NE (1995) Direct synthesis and characterization of site-specific adenosyl adducts derived from the binding of a 3,4-dihydroxy-1,2-epoxybenzo[*c*]phenanthrene stereoisomer to an 11-mer oligodeoxyribonucleotide. *Chem. Res. Toxicol.* 8: 444–454.

104. Doublie S and Ellenberger T (1998) The mechanism of action of T7 DNA polymerase. *Curr. Opin. Struct. Biol.* 8: 704–712.
105. Beard WA and Wilson SH (1998) Structural insights into DNA polymerase beta fidelity: hold tight if you want it right. *Chem. Biol.* 5: R7–R13.
106. Steitz TA (1999) DNA polymerases: structural diversity and common mechanisms. *J. Biol. Chem.* 274: 17395–17398.
107. Goodman MF, Creighton S, Bloom LB and Petruska J (1993) Biochemical basis of DNA replication fidelity. *Crit. Rev. Biochem. Mol. Biol.* 28: 83–126.
108. Friedberg EC, Walker GC and Siede W (1995) *DNA Repair and Mutagenesis*. ASM Press, Washington, D.C.
109. Woodgate R (1999) A plethora of lesion-replicating DNA polymerases. *Genes Dev.* 13: 2191–2195.
110. Friedberg EC, Wagner R and Radman M (2002) Specialized DNA polymerases, cellular survival, and the genesis of mutations. *Science* 296: 1627–1630.
111. Goodman MF (2002) Error-prone repair DNA polymerases in prokaryotes and eukaryotes. *Annu. Rev. Biochem.* 71: 17–50.
112. Hübscher U, Maga G and Spadari S (2002) Eukaryotic DNA polymerases. *Annu. Rev. Biochem.* 71: 133–163.
113. Lipinski LJ, Ross HL, Zajc B, Sayer JM, Jerina DM and Dipple A (1998) Effect of single benzo[*a*]pyrene diol epoxide-deoxyguanosine adducts on the action of DNA polymerases *in vitro*. *Int. J. Oncol.* 13: 269–273.
114. Keohavong P, Shukla R, Melacrinos A, Day BW and Reha-Krantz L (1998) Effects of bulky polycyclic aromatic hydrocarbon adducts on DNA replication by exonuclease-deficient T7 and T4 DNA polymerases. *DNA Cell Biol.* 17: 541–549.
115. Thrall BD, Mann DB, Smerdon MJ and Springer DL (1992) DNA polymerase, RNA polymerase and exonuclease activities on a DNA sequence modified by benzo[*a*]pyrene diol epoxide. *Carcinogenesis* 13: 1529–1534.
116. Hruszkewycz AM, Canella KA, Peltonen K, Kotrappa L and Dipple A (1992) DNA polymerase action on benzo[*a*]pyrene-DNA adducts. *Carcinogenesis* 13: 2347–2352.
117. Friedberg EC and Gerlach VL (1999) Novel DNA polymerases offer clues to the molecular basis of mutagenesis. *Cell* 98: 413–416.
118. Burgers PM, Koonin EV, Bruford E, Blanco L, Burtis KC, Christman MF, Copeland WC, Friedberg EC, Hanaoka F, Hinkle DC, Lawrence CW, Nakanishi M, Ohmori H, Prakash L, Prakash S, Reynaud CA, Sugino A,

Todo T, Wang Z, Weill JC and Woodgate R (2001) Eukaryotic DNA polymerases: proposal for a revised nomenclature. *J. Biol. Chem.* 276: 43487–43490.

119. Johnson RE, Prakash S and Prakash L (2000) The human *DINB1* gene encodes the DNA polymerase Pol θ. *Proc. Natl. Acad. Sci. USA* 97: 3838–3843.
120. Ross JA and Nesnow S (1999) Polycyclic aromatic hydrocarbons: correlations between DNA adducts and *ras* oncogene mutations. *Mutat. Res.* 424: 155–166.
121. Nesnow S, Ross JA, Stoner GD and Mass MJ (1995) Mechanistic linkage between DNA adducts, mutations in oncogenes and tumorigenesis of carcinogenic environmental polycyclic aromatic hydrocarbons in strain A/J mice. *Toxicology* 105: 403–413.
122. Ross JA, Nelson GB, Wilson KH, Rabinowitz JR, Galati A, Stoner GD, Nesnow S and Mass MJ (1995) Adenomas induced by polycyclic aromatic hydrocarbons in strain A/J mouse lung correlate with time-integrated DNA adduct levels. *Cancer Res.* 55: 1039–1044.
123. McDonald JP, Rapic-Otrin V, Epstein JA, Broughton BC, Wang X, Lehmann AR, Wolgemuth DJ and Woodgate R (1999) Novel human and mouse homologs of *Saccharomyces cerevisiae* DNA polymerase eta. *Genomics* 60: 20–30.
124. Gibbs PE, McGregor WG, Maher VM, Nisson P and Lawrence CW (1998) A human homolog of the *Saccharomyces cerevisiae REV3* gene, which encodes the catalytic subunit of DNA polymerase zeta. *Proc. Natl. Acad. Sci. USA* 95: 6876–6880.
125. Ogi T, Kato T Jr, Kato T and Ohmori H (1999) Mutation enhancement by DINB1, a mammalian homologue of the *Escherichia coli* mutagenesis protein dinB. *Genes Cells* 4: 607–618.
126. Gerlach VL, Aravind L, Gotway G, Schultz RA, Koonin EV and Friedberg EC (1999) Human and mouse homologs of *Escherichia coli* DinB (DNA polymerase IV), members of the UmuC/DinB superfamily. *Proc. Natl. Acad. Sci. USA* 96: 11922–11927.
127. Masutani C, Araki M, Yamada A, Kusumoto R, Nogimori T, Maekawa T, Iwai S and Hanaoka F (1999) Xeroderma pigmentosum variant (XP–V) correcting protein from HeLa cells has a thymine dimer bypass DNA polymerase activity. *EMBO J.* 18: 3491–3501.
128. Masutani C, Kusumoto R, Yamada A, Dohmae N, Yokoi M, Yuasa M, Araki M, Iwai S, Takio K and Hanaoka F (1999) The *XPV* (xeroderma

pigmentosum variant) gene encodes human DNA polymerase eta. *Nature* 399: 700–704.
129. Matsuda T, Bebenek K, Masutani C, Hanaoka F and Kunkel TA (2000) Low fidelity DNA synthesis by human DNA polymerase-eta. *Nature* 404: 1011–1013.
130. Johnson RE, Washington MT, Prakash S and Prakash L (2000) Fidelity of human DNA polymerase eta. *J. Biol. Chem.* 275: 7447–7450.
131. Johnson RE, Kondratick CM, Prakash S and Prakash L (1999) *hRAD30* mutations in the variant form of xeroderma pigmentosum. *Science* 285: 263–265.
132. Haracska L, Yu SL, Johnson RE, Prakash L and Prakash S (2000) Efficient and accurate replication in the presence of 7,8-dihydro-8-oxoguanine by DNA polymerase eta. *Nat. Genet.* 25: 458–461.
133. Masutani C, Kusumoto R, Iwai S and Hanaoka F (2000) Mechanisms of accurate translesion synthesis by human DNA polymerase eta. *EMBO J.* 19: 3100–3109.
134. Chiapperino D, Kroth H, Kramarczuk IH, Sayer JM, Masutani C, Hanaoka F, Jerina DM and Cheh AM (2002) Preferential misincorporation of purine nucleotides by human DNA polymerase eta opposite benzo[*a*]pyrene 7,8-diol 9,10-epoxide deoxyguanosine adducts. *J. Biol. Chem.* 277: 11765–11771.
135. Ohashi E, Ogi T, Kusumoto R, Iwai S, Masutani C, Hanaoka F and Ohmori H (2000) Error-prone bypass of certain DNA lesions by the human DNA polymerase kappa. *Genes Dev.* 14: 1589–1594.
136. Ohashi E, Bebenek K, Matsuda T, Feaver WJ, Gerlach VL, Friedberg EC, Ohmori H and Kunkel TA (2000) Fidelity and processivity of DNA synthesis by DNA polymerase kappa, the product of the human *DINB1* gene. *J. Biol. Chem.* 275: 39678–39684.
137. Zhang Y, Yuan F, Xin H, Wu X, Rajpal DK, Yang D and Wang Z (2000) Human DNA polymerase kappa synthesizes DNA with extraordinarily low fidelity. *Nucleic Acids Res.* 28: 4147–4156.
138. O-Wang J, Kawamura K, Tada Y, Ohmori H, Kimura H, Sakiyama S and Tagawa M (2001) DNA polymerase kappa, implicated in spontaneous and DNA damage-induced mutagenesis, is overexpressed in lung cancer. *Cancer Res.* 61: 5366–5369.
139. Ogi T, Mimura J, Hikida M, Fujimoto H, Fujii-Kuriyama Y and Ohmori H (2001) Expression of human and mouse genes encoding pol κ: testis-specific

developmental regulation and AhR-dependent inducible transcription. *Genes Cells* 6: 943–953.

140. Suzuki N, Ohashi E, Kolbanovskiy A, Geacintov NE, Grollman AP, Ohmori H and Shibutani S (2002) Translesion synthesis by human DNA polymerase kappa on a DNA template containing a single stereoisomer of dG-(+)- or dG-(−)-*anti*-N^2-BPDE (7,8-dihydroxy-anti-9,10-epoxy-7,8,9,10-tetrahydrobenzo[*a*]pyrene). *Biochemistry* 41: 6100–6106.
141. Ogi T, Shinkai Y, Tanaka K and Ohmori H (2002) Pol kappa protects mammalian cells against the lethal and mutagenic effects of benzo[*a*]pyrene. *Proc. Natl. Acad. Sci. USA* 99: 15548–15553.
142. Rechkoblit O, Zhang Y, Guo D, Wang Z, Amin S, Krzeminsky J, Louneva N and Geacintov NE (2002) *Trans*-lesion synthesis past bulky benzo[*a*]pyrene diol epoxide N^2-dG and N^6-dA lesions catalyzed by DNA bypass polymerases. *J. Biol. Chem.* 277: 30488–30494.
143. Zhang Y, Wu X, Rechkoblit O, Geacintov NE, Taylor JS and Wang Z (2002) Response of human REV1 to different DNA damage: preferential dCMP insertion opposite the lesion. *Nucleic Acids Res.* 30: 1630–1638.
144. Zhang Y, Wu X, Guo D, Rechkoblit O and Wang Z (2002) Activities of human DNA polymerase kappa in response to the major benzo[*a*]pyrene DNA adduct: error-free lesion bypass and extension synthesis from opposite the lesion. *DNA Repair* 1: 559–569.
145. Zhang Y, Wu X, Guo D, Rechkoblit O, Geacintov NE and Wang Z (2002) Two-step error-prone bypass of the (+)- and (−)-*trans-anti*-BPDE-N^2-dG adducts by human DNA polymerases eta and kappa. *Mutat. Res.* 510: 23–35.
146. Doublie S, Tabor S, Long AM, Richardson CC and Ellenberger T (1998) Crystal structure of a bacteriophage T7 DNA replication complex at 2.2 Å resolution. *Nature* 391: 251–258.
147. Kool ET (1998) Replication of non-hydrogen bonded bases by DNA polymerases: a mechanism for steric matching. *Biopolymers* 48: 3–17.
148. Morales JC and Kool ET (1998) Efficient replication between non-hydrogen-bonded nucleoside shape analogs. *Nat. Struct. Biol.* 5: 950–954.
149. Kool ET, Morales JC and Guckian KM (2000) Mimicking the structure and function of DNA: insights into DNA stability and replication. *Angew. Chem. Int. Ed. Engl.* 39: 990–1009.
150. Alekseyev YO, Dzantiev L and Romano LJ (2001) Effects of benzo[*a*]pyrene DNA adducts on *Escherichia coli* DNA polymerase I (Klenow fragment)

primer-template interactions: evidence for inhibition of the catalytically active ternary complex formation. *Biochemistry* 40: 2282–2290.

151. Tang D, Phillips DH, Stampfer M, Mooney LA, Hsu Y, Cho S, Tsai WY, Ma J, Cole KJ, She MN and Perera FP (2001) Association between carcinogen-DNA adducts in white blood cells and lung cancer risk in the physicians health study. *Cancer Res.* 61: 6708–6712.
152. IARC (1994) *Monographs on the Evaluation of the Carcinogenic Risk of Chemicals to Humans.* Volume 60: Some Industrial Chemicals. International Agency for Research on Cancer, Lyon, France.
153. Altman D (1991) *Practical Statistics for Medical Research.* Chapman and Hall, London, UK.
154. Besaratinia A, Kleinjans JC and van Schooten FJ (2002) Biomonitoring of tobacco smoke carcinogenicity by dosimetry of DNA adducts and genotyping and phenotyping of biotransformational enzymes: a review on polycyclic aromatic hydrocarbons. *Biomarkers* 7: 209–229.
155. Hainaut P and Pfeifer GP (2001) Patterns of p53 G → T transversions in lung cancers reflect the primary mutagenic signature of DNA-damage by tobacco smoke. *Carcinogenesis* 22: 367–374.
156. Pfeifer GP, Denissenko MF, Olivier M, Tretyakova N, Hecht SS and Hainaut P (2002) Tobacco smoke carcinogens, DNA damage and *p53* mutations in smoking-associated cancers. *Oncogene* 21: 7435–7451.
157. Wistuba II, Gazdar AF and Minna JD (2001) Molecular genetics of small cell lung carcinoma. *Semin. Oncol.* 28: 3–13.
158. Rodenhuis S and Slebos RJ (1992) Clinical significance of *ras* oncogene activation in human lung cancer. *Cancer Res.* 52, Suppl: 2665s–2669s.
159. Slebos RJ, Hruban RH, Dalesio O, Mooi WJ, Offerhaus GJ and Rodenhuis S (1991) Relationship between K-*ras* oncogene activation and smoking in adenocarcinoma of the human lung. *J. Natl. Cancer Inst.* 83: 1024–1027.
160. Denissenko MF, Chen JX, Tang MS and Pfeifer GP (1997) Cytosine methylation determines hot spots of DNA damage in the human *p53* gene. *Proc. Natl. Acad. Sci. USA* 94: 3893–3898.
161. Denissenko MF, Pao A, Pfeifer GP and Tang M (1998) Slow repair of bulky DNA adducts along the nontranscribed strand of the human *p53* gene may explain the strand bias of transversion mutations in cancers. *Oncogene* 16: 1241–1247.
162. Denissenko MF, Pao A, Tang M and Pfeifer GP (1996) Preferential formation of benzo[*a*]pyrene adducts at lung cancer mutational hotspots in *p53*. *Science* 274: 430–432.

163. Smith LE, Denissenko MF, Bennett WP, Li H, Amin S, Tang M and Pfeifer GP (2000) Targeting of lung cancer mutational hotspots by polycyclic aromatic hydrocarbons. *J. Natl. Cancer Inst.* 92: 803–811.
164. Chen JX, Zheng Y, West M and Tang MS (1998) Carcinogens preferentially bind at methylated CpG in the *p53* mutational hot spots. *Cancer Res.* 58: 2070–2075.
165. Feng Z, Hu W, Chen JX, Pao A, Li H, Rom W, Hung MC and Tang MS (2002) Preferential DNA damage and poor repair determine *ras* gene mutational hotspot in human cancer. *J. Natl. Cancer Inst.* 94: 1527–1536.
166. Pfeifer GP, Chen HH, Komura J and Riggs AD (1999) Chromatin structure analysis by ligation-mediated and terminal transferase-mediated polymerase chain reaction. *Methods Enzymol.* 304: 548–571.
167. Pfeifer GP, Tanguay RL, Steigerwald SD and Riggs AD (1990) *In vivo* footprint and methylation analysis by PCR-aided genomic sequencing: comparison of active and inactive X chromosomal DNA at the CpG island and promoter of human *PGK-1*. *Genes Dev.* 4: 1277–1287.
168. Pfeifer GP, Steigerwald SD, Hansen RS, Gartler SM and Riggs AD (1990) Polymerase chain reaction-aided genomic sequencing of an X chromosome-linked CpG island: methylation patterns suggest clonal inheritance, CpG site autonomy, and an explanation of activity state stability. *Proc. Natl. Acad. Sci. USA* 87: 8252–8256.
169. Geacintov NE, Shahbaz M, Ibanez V, Moussaoui K and Harvey RG (1988) Base-sequence dependence of noncovalent complex formation and reactivity of benzo[*a*]pyrene diol epoxide with polynucleotides. *Biochemistry* 27: 8380–8387.
170. Johnson WS, He QY and Tomasz M (1995) Selective recognition of the m5CpG dinucleotide sequence in DNA by mitomycin C for alkylation and cross-linking. *Bioorg. Med. Chem.* 3: 851–860.
171. Yoon JH, Smith LE, Feng Z, Tang M, Lee CS and Pfeifer GP (2001) Methylated CpG dinucleotides are the preferential targets for G-to-T transversion mutations induced by benzo[*a*]pyrene diol epoxide in mammalian cells: similarities with the *p53* mutation spectrum in smoking-associated lung cancers. *Cancer Res.* 61: 7110–7117.
172. Mumford JL, He XZ, Chapman RS, Cao SR, Harris DB, Li XM, Xian YL, Jiang WZ, Xu CW, Chuang JC, Wilson WE and Cooke M (1987) Lung cancer and indoor air pollution in Xuan Wei, China. *Science* 235: 217–220.
173. Lan Q, He X, Costa DJ, Tian L, Rothman N, Hu G and Mumford JL (2000) Indoor coal combustion emissions, *GSTM1* and *GSTT1* genotypes, and lung

cancer risk: a case-control study in Xuan Wei, China. *Cancer Epidemiol. Biomarkers Prev.* 9: 605–608.

174. Mumford JL, Li X, Hu F, Lu XB and Chuang JC (1995) Human exposure and dosimetry of polycyclic aromatic hydrocarbons in urine from Xuan Wei, China with high lung cancer mortality associated with exposure to unvented coal smoke. *Carcinogenesis* 16: 3031–3036.
175. Casale GP, Singhal M, Bhattacharya S, RamaNathan R, Roberts KP, Barbacci DC, Zhao J, Jankowiak R, Gross ML, Cavalieri EL, Small GJ, Rennard SI, Mumford JL and Shen M (2001) Detection and quantification of depurinated benzo[*a*]pyrene-adducted DNA bases in the urine of cigarette smokers and women exposed to household coal smoke. *Chem. Res. Toxicol.* 14: 192–201.
176. DeMarini DM, Landi S, Tian D, Hanley NM, Li X, Hu F, Roop BC, Mass MJ, Keohavong P, Gao W, Olivier M, Hainaut P and Mumford JL (2001) Lung tumor *KRAS* and *TP53* mutations in nonsmokers reflect exposure to PAH-rich coal combustion emissions. *Cancer Res.* 61: 6679–6681.
177. Rideout WM 3rd, Coetzee GA, Olumi AF and Jones PA (1990) 5-Methylcytosine as an endogenous mutagen in the human LDL receptor and *p53* genes. *Science* 249: 1288–1290.
178. Shen JC, Rideout WM 3rd and Jones PA (1992) High frequency mutagenesis by a DNA methyltransferase. *Cell* 71: 1073–1080.
179. Tretyakova N, Matter B, Jones R and Shallop A (2002) Formation of benzo[*a*]pyrene diol epoxide-DNA adducts at specific guanines within *K-ras* and *p53* gene sequences: stable isotope-labeling mass spectrometry approach. *Biochemistry* 41: 9535–9544.
180. Hussain SP, Amstad P, Raja K, Sawyer M, Hofseth L, Shields PG, Hewer A, Phillips DH, Ryberg D, Haugen A and Harris CC (2001) Mutability of *p53* hotspot codons to benzo[*a*]pyrene diol epoxide (BPDE) and the frequency of *p53* mutations in nontumorous human lung. *Cancer Res.* 61: 6350–6355.

6 Mechanisms of Repair of Polycyclic Aromatic Hydrocarbon-Induced DNA Damage

Hanspeter Naegeli[1] and Nicholas E. Geacintov[2]

[1] *Institute of Pharmacology and Toxicology, University of Zürich-Tierspital, Zürich, Switzerland*
E-mail: naegelih@vetpharm.unizh.ch
[2] *Chemistry Department, New York University, New York, NY, USA*
E-mail: nicholas.geacintov@nyu.edu

6.1 Introduction

As part of physiological excretory processes, hydrophobic xenobiotics are metabolized to hydrophilic species, which can be more easily eliminated from the organism. The oxidative metabolism of polycyclic aromatic hydrocarbons (PAHs) yields highly reactive diol-epoxide derivatives[1–3] that react chemically with DNA, mostly with the purine bases, to form covalently linked bulky adducts[4] (cf. Chapter 2). This genotoxic reaction is mediated through either *trans* or, less frequently, *cis* opening of the epoxide ring (Figure 6.1), generating base adducts mainly at the exocyclic positions N^2 of

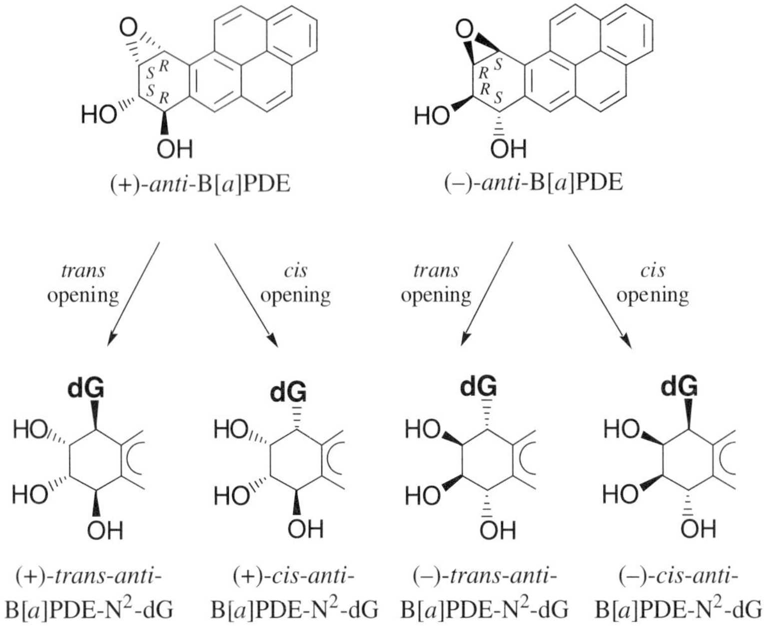

Figure 6.1: Stereochemistry of the reaction of enantiomeric *anti*-B[*a*]P-7,8-diol-9,10-epoxides (B[*a*]PDE) with guanine. Each enantiomer produces two diasteromeric products by either *cis* or *trans* opening of the epoxide ring. The (+)-*trans-anti*-B[*a*]PDE-N^2-dG isomer is the major product in double-stranded DNA (cf. Figure 11.1 in Chapter 11).

guanine and N^6 of adenine.[5] Such adducts perturb the normal Watson-Crick double helix, thus compromising essential nuclear functions, including transcription and replication. If not repaired, DNA adducts suppress the transcriptional activity in the damaged genes and reduce cell survival with the possible consequence of accelerated aging processes in the host organism[6]. Also, PAH-damaged templates can result in nucleotide misincorporation during the next round of DNA replication and, therefore, have the potential of inducing irreversible changes in the sequence of key genetic targets.[7–9] Mutations that activate proto-oncogenes or inactivate tumor suppressor genes constitute the molecular basis for pathological alterations of regulatory processes, ultimately leading to neoplastic cell transformation.[10]

6.2 Nucleotide Excision Repair

Stable adducts, predominantly involving covalent linkages of the PAH residues with either N^2 of guanine or N^6 of adenine are targeted by a universal cut and patch repair apparatus, referred to as the nucleotide excision repair (NER) system, that operates through a double endonucleolytic cleavage of damaged DNA strands. In humans, like in other mammalian species, NER represents the only repair pathway that is able to remove PAH adducts or other large chemical modifications of DNA in an accurate manner, i.e., without any loss of genetic information. However, PAH diol-epoxides also produce labile adducts with linkages at the N^7 position of guanine and the N^3 position of adenine. The presence of these bulky substituents destabilizes the N-glycosidic bond between the deoxyribose and the bases and hence promotes depurination.[11,12] It has been proposed that the formation of apurinic sites is mostly responsible for the mutagenic burden associated with the reaction of PAH compounds with DNA.[13] However, subsequent studies led to the conclusion that this pathway contributes only marginally to the mutagenicity of PAH diol-epoxides because apurinic sites are rapidly repaired by apurinic/apyrimidinic DNA endonucleases, DNA polymerase β, DNA ligase III and other components of the base excision repair pathway.[14,15] Since the risk of mutations due to PAH adducts increases progressively with their persistence in the genome, the present chapter is exclusively devoted to the slow mechanism of repair of the major, stable and mutagenic PAH adducts that are bound covalently to the DNA bases. The protection from such persistent PAH adducts is critically dependent on the NER proficiency of the damaged cells or organisms.

6.2.1 Mammalian DNA Nucleotide Excision Repair

A versatile NER system has evolved in all organisms, including bacteria, yeast and mammals, to cope with bulky DNA adducts induced by environmental carcinogens such as UV light and electrophilic chemicals (reviewed by Sancar[16] and Wood[17], cf. Figure 2.7 in Chapter 2). A general scheme illustrating the principal reaction intermediates of the NER pathway is shown in Figure 6.2.

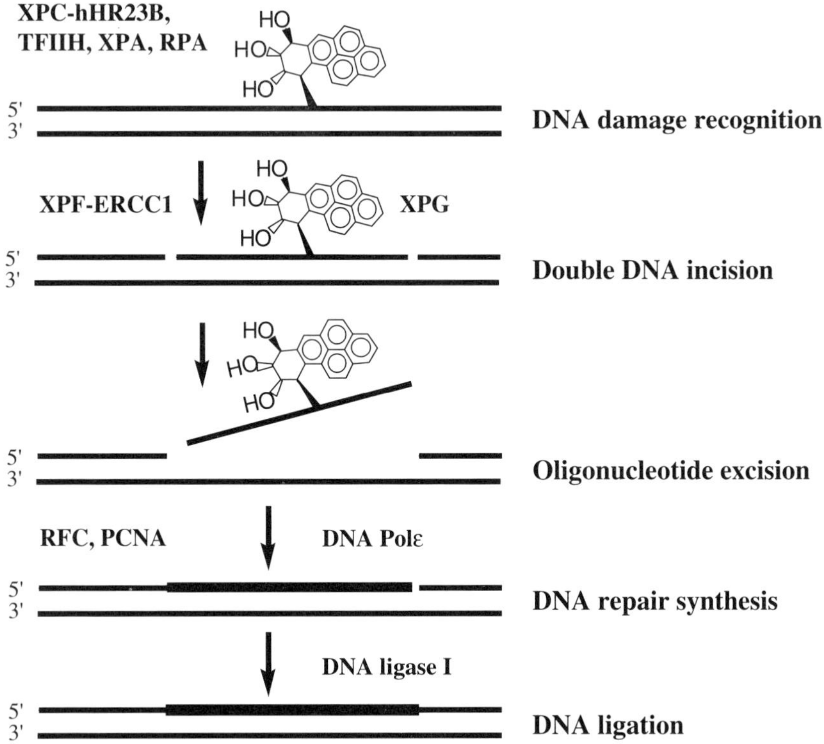

Figure 6.2: Reaction mechanism and core components of the human NER machinery. This cut and patch DNA repair process is executed through the stepwise assembly of two multiprotein complexes, i.e., an incision complex composed of XPC-hHR23B, TFIIH (containing XPB and XPD), XPA, RPA, XPG and XPF-ERCC1, and a repair synthesis complex involving RFC, PCNA, DNA Polδ or ε, and DNA ligase I.

A most notable feature of the NER mechanism is its exquisite flexibility in processing a multitude of damaged DNA targets. Despite a limited repertoire of molecular recognition subunits, the NER machinery is able to detect an extremely wide range of base adducts and other base lesions. Characteristic to the NER mechanism is also its pattern of damage removal by DNA incision at two distinct sites flanking the lesion. Such dual incisions lead to the release of damaged residues as part of oligonucleotide segments that, in human cells, are 24–32 nucleotides in length.[18] The single-stranded gap in the double helix produced by this excision reaction is processed by

DNA synthesis enzymes that restore the correct nucleotide sequence using the complementary, undamaged strand as a template. Finally, the NER pathway is completed by ligation of the newly synthesized repair patches to the pre-existing strands (Figure 6.2). Defects in this particular excision repair system result in failure to remove bulky DNA adducts and cause *Xeroderma pigmentosum* (XP) in humans, an inherited syndrome characterized by about a 10,000-fold increased risk of sunlight-induced cancer mainly of the skin. Individuals affected by XP are classified into 7 repair-deficient complementation groups designated XP-A through XP-G.[19,20]

6.2.2 Subunits of the Human NER Machinery

All core components that carry out the NER reaction in both prokaryotic and eukaryotic organisms have been identified. Only 6 polypeptides (UvrA, UvrB, UvrC and UvrD, DNA polymerase I, and DNA ligase) are required for the NER process in prokaryotes.[21] However, as many as 10 different factors and a minimal set of 30 distinct subunits are necessary for human NER activity.[20,22,23] These NER components can be divided into two groups: the factors necessary for damage recognition and double DNA incision, and others required for DNA repair synthesis. In particular, the core factors involved in the initial recognition and incision reactions include XPA, which forms a complex with replication protein A (RPA); XPC together with a human homolog of RAD23 (hHR23B); transcription factor IIH (TFIIH); the 3′ endonuclease XPG; and the 5′ endonuclease, which is a heterodimer composed of XPF and the excision repair cross complementing 1 (ERCC1) protein.

It is becoming increasingly clear that in mammalian cells, NER is executed by the sequential assembly of repair proteins at the site of the DNA lesion, rather than by the action of a pre-assembled 'repairosome' complex. XPC-hHR23B is generally thought to represent the key molecular recognition component that initiates the NER pathway through binding to adducted sites,[24–27] but the structural determinants underlying its interaction with damaged DNA have not been elucidated. Another factor with an affinity for UV-irradiated DNA, the UV-damaged DNA-binding protein (UV-DDB), stimulates excision of bulky UV lesions in humans and, therefore, also has been implicated in the damage recognition process.[28] XPA and RPA are additional subunits that have in common a weak preference for damaged DNA, although recent studies argue against a direct participation of XPA-RPA as a sensor of defective nucleotides. In fact, XPA displays a strong affinity for rigidly kinked DNA reaction intermediates, suggesting that this key NER factor adopts a specialized function as an architectural protein; that coordinates and controls the correct assembly of incision complexes before activation of the two nuclease components.[29] The single-stranded DNA binding properties of its interaction partner, RPA, are compatible with an additional role in protecting the single strand segments that arise transiently after oligonucleotide excision until completion of DNA repair synthesis occurs.[30] However, a different function of RPA in forming close contacts with the DNA lesion has been proposed recently.[31]

After recruitment to damaged sites, XPC-hHR23B associates with TFIIH, thereby attracting this large transcription factor to the target lesion.[27] The subunits of TFIIH are organized in a ring-like structure with a 2.5–3.0 nm wide central hole.[32] The enzymatic activities of this TFIIH complex lead to unwinding of the damaged site by 20–25 nucleotides, generating an open intermediate that precedes DNA incision.[33–35] At this stage, XPC and hHR23B are displaced from the pre-incision intermediate[36] while XPA and RPA become involved in the nucleoprotein complex, presumably to trigger the DNA incision reaction and to protect the undamaged strand from unspecific degradation. Central to the local unwinding process at the site of the lesion are the two DNA helicases localized in the ring-like domain of TFIIH, XPB and XPD, which promote partial DNA denaturation during both transcription and NER. In transcription, DNA unwinding by TFIIH allows the nascent mRNA molecules to progress from initiation to the elongation phase.[37] In NER, local DNA unwinding by TFIIH produces double-stranded to single-stranded transitions at the edges of a central bubble, thus providing a substrate for incision by the two structure-specific endonucleases XPG and XPF-ERCC1 (Figure 6.3).

The molecular complexity of the mammalian NER system is enhanced by the subunit composition of several factors. UV-DDB, for example, consists of two polypeptides (p48 and p125), TFIIH is made up of nine distinct polypeptides (XPB, XPD, p62, p52, p44, cdk7, cyclin H, p34 and MAT1),

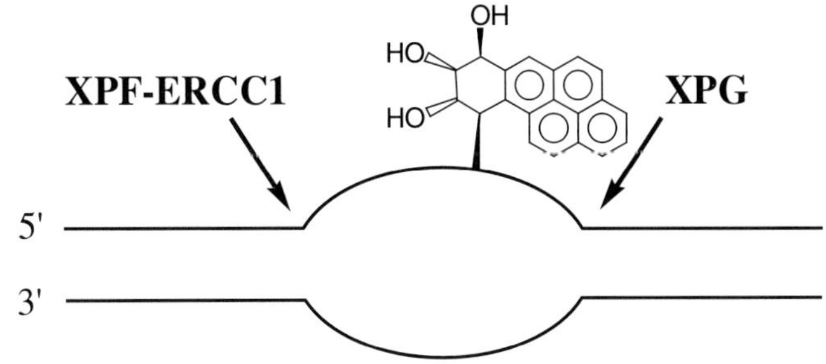

Figure 6.3: Open DNA intermediate generated prior to dual incision of damaged strands. Local denaturation around the lesion by the two DNA helicases XPB and XPD generates a bubble substrate that is targeted by structure-specific endonucleases. XPG cuts the damaged DNA strand 3′ to the site of base damage, whereas XPF-ERCC1 cuts the damaged strand 5′ to such sites. The damaged base is always located closer to the 3′ incision than to the 5′ incision.

and RPA consists of three polypeptides (p70, p34 and p14), such that at least 20 different subunits are involved in the stepwise assembly of a partially open DNA complex that culminates in dual DNA incision. The subsequent release of an oligonucleotide segment containing the target lesion generates a gapped substrate that is further processed by the second group of NER factors, i.e., DNA replication enzymes assisted by their accessory factors (Figure 6.2). In particular, replication factor C (RFC) is a molecular matchmaker that mediates the loading of proliferating nuclear antigen (PCNA) in the proximity of the 3′ end of the DNA gap, which in turn acts as a sliding clamp to promote the processive and error-free synthesis of repair patches by DNA polymerase δ and ε.[22,38] Finally, the remaining nick on the other side of the repair patch is ligated by the action of DNA ligase I.

6.2.3 Transcription-Coupled DNA Repair

DNA repair is highly non-uniform in the context of mammalian chromosomes, and excision activity that operates on transcriptionally active DNA strands differs considerably from that in transcriptionally silent DNA (cf. Chapter 5). This observation has led to the recognition of two distinct excision modes referred to as 'transcription-coupled' and 'global genome' repair pathways, respectively. Transcription-coupled repair is dependent on RNA polymerase II activity but requires no XPC-hHR23B, whereas global NER is independent of transcription but needs the XPC complex.[20]

The chromosomal distribution of mammalian NER activity is best understood for the repair of cyclobutane pyrimidine dimers induced by UV radiation, as very efficient techniques have been developed to monitor the formation and removal of these lesions. In most studies, a prokaryotic DNA glycosylase was exploited as a molecular tool to generate nicks at cyclobutane dimers in the DNA of UV-exposed cells. The relative frequency of cuts — indicating the presence of UV lesions — at particular genomic sites was determined by electrophoretic separation of cleavage products and hybridization to specific DNA probes or, in more recent studies, by ligation-mediated PCR reactions.[39–41] The decrease in the number of DNA glycosylase-induced incisions following different recovery times after UV irradiation was used to estimate site-specific kinetics of repair. These studies demonstrated that, in both rodent and human cells, cyclobutane dimers are removed more rapidly in actively transcribed genes than from the genome as a whole,[39] and in particular are removed from the template strand of RNA polymerase II-transcribed genes more rapidly than from the non-transcribed coding strand.[42] When (A)BC excinuclease, the prokaryotic NER system, was used as a tool to introduce DNA cuts at lesion sites, the same technique demonstrated that also the removal of PAH-DNA adducts can occur with vastly different efficiencies in mammalian chromosomes. For example, DNA damage produced by the prototypical representative of PAH metabolites, benzo[*a*]pyrene (B[*a*]P) 7,8-diol-9,10-epoxides (B[*a*]PDE), is again excised more readily from transcribed genes in human cells[43,44] but in hamster cells, the same lesion is removed from active genes at the same rate as from the overall genome.[45] In contrast, the damage produced by diol-epoxides of a related PAH compound, benzo[*c*]phenanthrene (B[*c*]Ph), is preferentially repaired in the transcribed strand of active genes in hamster cells.[46]

The molecular mechanism of preferential transcription-coupled NER has not been studied in the detail just described in the previous section for the global genome pathway. Humans and mice that are genetically defective in XPC retain the capacity for transcription-coupled NER, indicating that this particular factor is not essential for the preferential repair of template strands in transcribed genes.[47] A widely accepted model assumes that XPC is not needed in transcription-coupled NER because the damaged base physically blocks the transcription machinery. Thus, progression of RNA polymerase II is arrested by damaged DNA bases, eventually leading to the recruitment of a large protein complex that includes CSA and CSB, several NER proteins (XPB, XPD and XPG), and one or more proteins involved in mismatch repair.[20,48–50] Inactivation of *CSA* or *CSB* results in the genetic disease Cockayne syndrome, characterized by the early onset of an aging phenotype,[51] while inactivation of mismatch repair is associated with hereditary colon cancer. Although the composition and function of each individual component remains to be established, the multi-protein complex assembled during transcription-repair coupling is thought to dislocate the stalled transcription machinery from the site of damage in the transcribed strand, allowing access to this site to proteins required for NER or other DNA repair pathways. The facilitated recruitment of repair factors to such damaged sites accounts for the accelerated removal of DNA lesions from transcriptionally active sequences.

6.2.4 Global NER Deficiency and Cancer

The potential for DNA mutations following genotoxic stress derives from error-prone replicative synthesis across base lesions such as PAH adducts.[43,52–55] In fact, as already indicated, the pathogenesis of cancer is initiated by mutations arising in critical proto-oncogenes and tumor suppressor genes that regulate cell-cycle progression, apoptosis or senescence. Due to the fast excision promoted by transcription-coupled repair, DNA lesions in transcribed strands are unlikely to cause deleterious effects, and mutations are more likely to be induced by the longer persisting lesions located in non-transcribed sequences.[46,56] For example, the *p53* mutational spectrum in lung tumors associated with smoking is characterized by G to T transversions in the non-transcribed coding strand (cf. Chapter 5). The slower repair of B[*a*]PDE- or other PAH-DNA adducts in this non-transcribed strand, as compared to the more efficient removal of the same lesions from the complementary sequence,[44] accounts for the observed strand bias in the induction of mutations. An important consequence of these strand-specific DNA repair mechanisms is that the magnitude of the mutational risk arising from genotoxic insults is dictated by the rate and efficiency of damage removal by the global NER pathway, which processes the lesions induced in non-transcribed sequences or transcriptionally silent regions. This view is confirmed by the susceptibility of XP patients, who are deficient in global NER, to cancer. In contrast, the absence of transcription-coupled repair in patients with the genetic disease Cockayne syndrome is not associated with an increased tumor incidence.[51,57]

Mouse DNA repair knock-out experiments lend support to the notion that global NER activity is an important part of the cellular defense system that protects the genome against somatic mutations leading to cancer. In fact, several *XP* genes have been inactivated in mice with the advent of targeted gene replacement in embryonic stem cells. In general, these mouse XP models recapitulate the phenotype of extreme predisposition to UV-induced cancer of the skin, eyes and tongue. Additionally, these animal experiments also confirmed the expectation that a deficiency in NER may result in increased cancer risk in organs such as the lung, stomach, liver and colon; all of which are known targets of chemical carcinogens. This conclusion was not deduced from the analysis of human XP patients, where cancer susceptibility at sites other than the body surface remains controversial.[58] For example, the homozygous $XPA^{-/-}$ mutant mouse, completely deficient in NER, is prone to tumors not only of the skin but also in the lymphoid system and the liver after treatment with B[*a*]P or 7,12-dimethylbenz[*a*]anthracene.[59–61]

Recent studies showed that expression of XPC, the initiator of global genome NER, is induced by the p53 tumor suppressor protein after exposure to DNA damaging agents. Analysis of the human *XPC* gene sequence revealed a nucleotide sequence element in the promoter region that mediates DNA binding by p53 protein, indicating that p53 acts as a crucial transcription factor not only in regulating cell-cycle and triggering apoptosis in response to DNA damage, but also by increasing the efficiency of bulky lesion repair.[62] Not surprisingly, it has been found for benzo[*g*]chrysene (B[g]C) 11,12-diol-13,14-epoxide (B[g]CDE) adducts that the repair of PAH-induced lesions is dependent on the *p53* status of human cells.[63] Functional p53 tumor suppressor protein facilitates the removal of PAH adducts by at least two mechanisms. First, p53 stabilization in response to checkpoint activation delays cell-cycle progression, thereby providing more time for the slow global NER pathway — between induction of the DNA damage and the initiation of replicative DNA synthesis — to restore template integrity by removing potentially mutagenic lesions. Second, the p53 protein is able to stimulate global NER activity directly by up-regulating the expression of the *XPC* gene. The increased production of XPC protein accelerates one of the early rate-limiting steps during the assembly of incision complexes. Upregulation by p53 has also been reported for UV-DDB, another factor implicated in damage recognition in the human NER pathway (see above).[64,65] These observations enforce the notion of a direct molecular link between global NER activity and cancer predisposition.

6.3 Repair of PAH-DNA Adducts

6.3.1 The *In Vitro* Oligonucleotide Excision Reaction

In human cells, DNA repair by nucleotide excision is accomplished by cleavage of the damaged strands on either side of the lesion, thus releasing oligomeric products 24–32 residues in length.[18] NER activity in response to PAH adducts was determined by an *in vitro* excision assay performed in cell extracts prepared from human cell lines.[66] This cell-free system exploits the unique dual DNA incision pattern of human NER and, as a consequence, provides high levels of specificity and sensitivity. Also, this cell-free approach rules out any participation of transcription factors or other chromatin components, thereby revealing the intrinsic capacity of global NER to recognize and process a particular type of bulky lesion.

Appropriate DNA substrates of 130–150 base pairs with a site-directed PAH adduct on a single deoxyguanosine or deoxyadenosine residue were constructed from 6 different oligomers as illustrated in Figure 6.4A. Before substrate assembly, the central oligonucleotide was labeled with [γ-^{32}P]ATP at its 5′ end so that the resulting duplex contained an internal radiolabel in the vicinity of the adduct (Figure 6.4B). After electrophoretic purification, the double-stranded and internally labeled fragments were

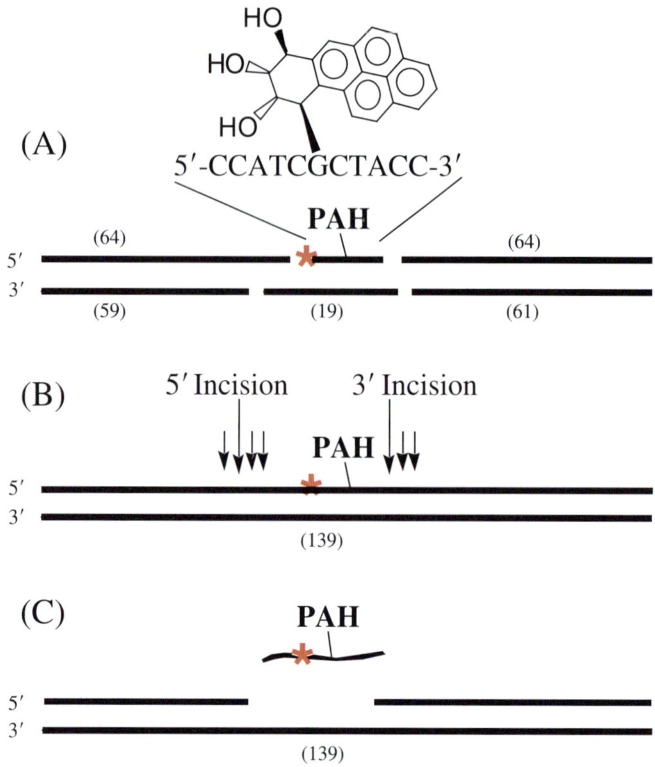

Figure 6.4: Construction of site-specific substrates for *in vitro* oligonucleotide excision assays. (A) Assembly of internally radiolabeled duplexes. (B) Sites of incision in human cell extracts. (C) Excision of oligonucleotides containing the unique adduct. The asterisk denotes a ^{32}P-labeled deoxyribonucleotide residue.

incubated in a HeLa cell extract that is proficient in NER activity when supplemented with ATP and all 4 deoxyribonucleoside triphosphates. Damage-specific dual DNA incision by human NER enzymes generates radioactive products 24 to 32 nucleotides in length (Figure 6.4C), which were separated from substrate DNA by denaturing gel electrophoresis and visualized by autoradiography. This standard assay yields linear excision kinetics for incubation times of up to 40 min at 30°C (Figure 6.5). No excision products are released from undamaged control DNA (Figure 6.5, lane 1), although intact full-length substrate as well as radioactive bands

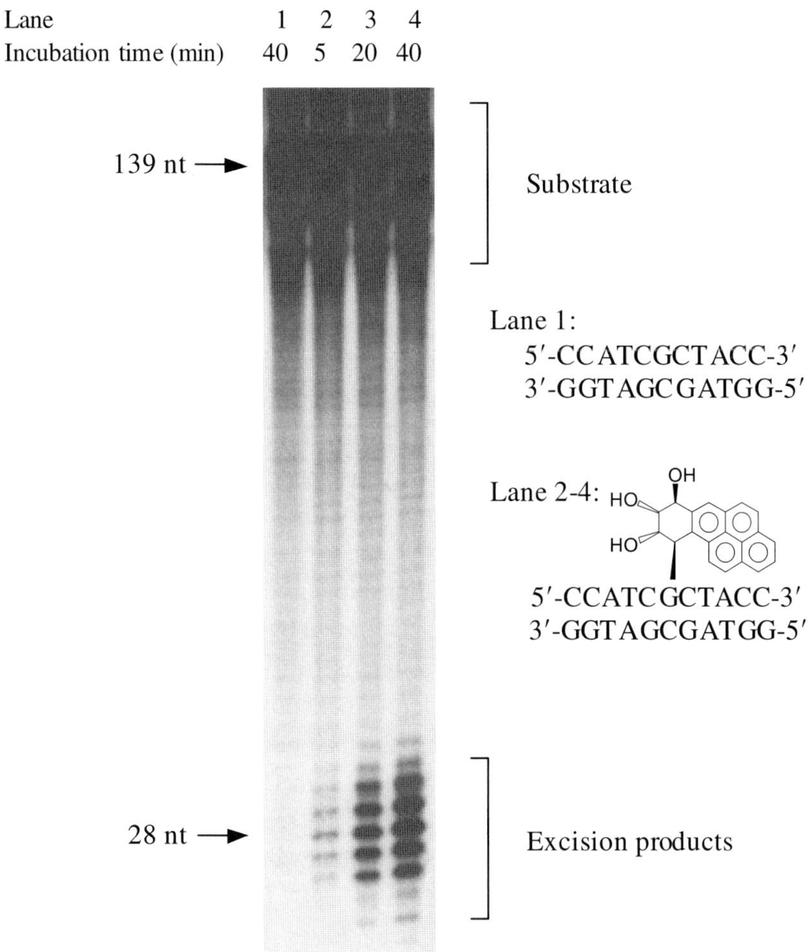

Figure 6.5: Time course of oligonucleotide excision in HeLa cell extract. Lane 1 shows a control reaction with undamaged DNA fragments.

generated by non-specific nuclease activity can be observed at similar levels on top of the gel.

Various *in vitro* assays have been used in the literature to assess mammalian NER activity. For example, repair reactions elicited by PAH adducts have also been probed by measuring the incorporation of radiolabeled nucleotides during DNA repair synthesis.[14,67] Unlike the approach based on oligonucleotide excision, which is highly specific, measurements of repair patch synthesis may reflect alternative excision repair mechanisms and, therefore, should not be considered as diagnostic indicators of NER reactions.

6.3.2 Base Pair Conformation-Dependent Excision of B[*a*]PDE-dG Adducts

A comparison between structurally highly defined PAH-modified substrates demonstrated that human NER efficiency is largely determined by the type and degree of local DNA distortion. In particular, fast excision of PAH adducts has been found to correlate with displacement of the modified base and its partner residue in the complementary strand from their normal intra-helical position.[68,69]

An initial structure-activity relationship was established by exploiting the stereochemistry of guanine adducts formed by the enantiomeric pair of B[*a*]PDE metabolites. The combined stereoselectivity of cytochrome P450-dependent monooxygenases and microsomal epoxide hydrolase results in the formation of (+)-*anti*-B[*a*]PDE as the major diol-epoxide stereoisomer in the liver[1–3,62,70] (cf. Figures 2.5 and 2.6 in Chapter 2). Interestingly, this same stereoisomer is the only one of the set of four optically active diol-epoxides that has high tumorigenic potency in mammals.[71,72] The primary targets for covalent modification of DNA by the diol-epoxides are the exocyclic amino groups of the purine bases. In the case of B[*a*]PDE, binding to N^2-dG is favored over N^6-dA.[73,74] As outlined in Figure 6.1, racemic *anti*-B[*a*]PDE generates mainly (+)-*trans-anti*-B[*a*]PDE-N^2-dG adducts with a minor proportion of (+)-*cis*-, (−)-*trans*- and (−)-*cis-anti*-B[*a*]PDE-N^2-dG isomers (the prefix '*cis*' and '*trans*' refers to the *cis* and *trans* opening of the epoxide ring, respectively).[73] These four stereochemically different lesions impose distinct different structural alterations on the host DNA (summarized in Table 6.1). In the 5′–TCGCT–3′ sequence context, nuclear magnetic resonance (NMR) analysis showed that (+)-*trans*- and (−)-*trans-anti*-B[*a*]PDE-N^2-dG adopt external adduct conformations that accommodate the benzo[*a*]pyrenyl moiety in the minor DNA groove

Table 6.1: Structure-activity relationship established with PAH-modified DNA substrates incubated in NER-proficient HeLa cell extracts. The NMR solution structures of the PAH-DNA adducts are reviewed by Geacintov *et al.*[82]

PAH adduct	*Sequence context*[a]	*Structural motif*[b]	*Local base pairing*[c]	*NER rate*[d]
(+)-*trans-anti*-B[*a*]PDE-N^2-dG·dC	5′-ATC**G**CTA-3′	minor groove adduct	destabilized	moderate
(−)-*trans-anti*-B[*a*]PDE-N^2-dG·dC	5′-ATC**G**CTA-3′	minor groove adduct	destabilized	moderate
(+)-*cis-anti*-B[*a*]PDE-N^2-dG·dC	5′-ATC**G**CTA-3′	base-displaced intercalation	completely disrupted	high
(−)-*cis-anti*-B[*a*]PDE-N^2-dG·dC	5′-ATC**G**CTA-3′	base-displaced intercalation	completely disrupted	high
(−)-*trans-anti*-B[*a*]PDE-N^6-dA·dT	5′-CTC**A**CTT-3′	intercalative insertion	destabilized	moderate
(+)-*trans-anti*-B[*c*]PhDE-N^6-dA·dT	5′-CTC**A**CTT-3′	intercalative insertion	intact base pairing	low
(−)-*trans-anti*-B[*c*]PhDE-N^6-dA·dT	5′-CTC**A**CTT-3′	intercalative insertion	intact base pairing	low
(+)-*trans-anti*-B[*a*]PDE-N^2-dG·deletion	5′-ATC**G**CTA-3′	wedge-shaped insertion	partner dC missing	low; decoy adduct[e]
(+)-*cis-anti*-B[*a*]PDE-N^2-dG·deletion	5′-ATC**G**CTA-3′	wedge-shaped insertion	partner dC missing	low, decoy adduct[e]

[a]The covalently modified purine is indicated with large bold letters; [b]Based on NMR models; [c]Hydrogen bonding interaction between the covalently modified purine and its partner base; [d]Fraction of excised substrate after 40-min incubations. High $\approx$10%; Moderate $\approx$1%; Low $<$0.1%; [e]Decoy adducts inhibit the concomitant repair of NER substrates in co-incubation reactions.

leaving a Watson-Crick alignment at all base pairs, including the modified one.[75,76] In contrast, (+)-*cis*- and (−)-*cis-anti*-B[*a*]PDE-N^2-dG in the same 5′–TCGCT–3′ sequence adopt intercalative, internal adduct conformations characterized by benzo[*a*]pyrenyl insertion into the double helix and a concomitant disruption of the Watson-Crick hydrogen bonding at the modified base pair. As a consequence, these helix-inserted (+)-*cis* and (−)-*cis* configurations cause displacement of the covalently modified guanine and its cytosine partner into the major or minor grooves.[77,78] On the other hand, the extent of bending associated with these *cis* configurations is lower than in the case of the stereoisomeric *trans-anti*-B[*a*]PDE-N^2-dG adducts.[79]

DNA repair reactions in HeLa cell extracts performed with linear 139-mer substrates containing one of four stereoselective *anti*-B[*a*]PDE-N^2-dG adducts in the 5′–TCGCT–3′ sequence context yielded excision products that migrated on polyacrylamide gels as oligomeric ladders with dominant lengths around 28 nucleotides (Figure 6.6A). Excision repair activity was quantified by laser scanning densitometry of the autoradiograms, and confirmed by measuring the *Cerenkov* radiation emitted from the gel slices containing the excision products and corresponding full-length substrates (Figure 6.6B). Unexpectedly, these experiments revealed that *anti*-B[*a*]PDE-N^2-dG adducts with *trans* configuration at the linkage sites were excised one order of magnitude more slowly than the corresponding stereoisomeric lesions with *cis* configuration. Also surprisingly, the direct comparison between (+)-*trans*- and (−)-*trans-anti*-B[*a*]PDE-N^2-dG yielded identical excision levels, although these stereoisomeric adducts differ in their extent of DNA bending and unwinding,[79–81] as well as in the orientation of the benzo[*a*]pyrenyl moiety within the minor groove.[76,82] Similarly, in the two *cis-anti*-B[*a*]PDE-N^2-dG adducts, the benzo[*a*]pyrenyl ring systems are oppositely oriented at their respective intercalation sites and the modified guanines are displaced into different grooves.[78] However, these structural differences between the two *cis-anti*-B[*a*]PDE-N^2-dG isomers exert only minor effects on the overall excision efficiency.

Combined with the available detailed structural information (Table 6.1), the 10-fold higher excision repair rate induced by (+)-*cis*- and (−)-*cis-anti*-B[*a*]PDE-N^2-dG adducts relative to the chemically identical (+)-*trans* and

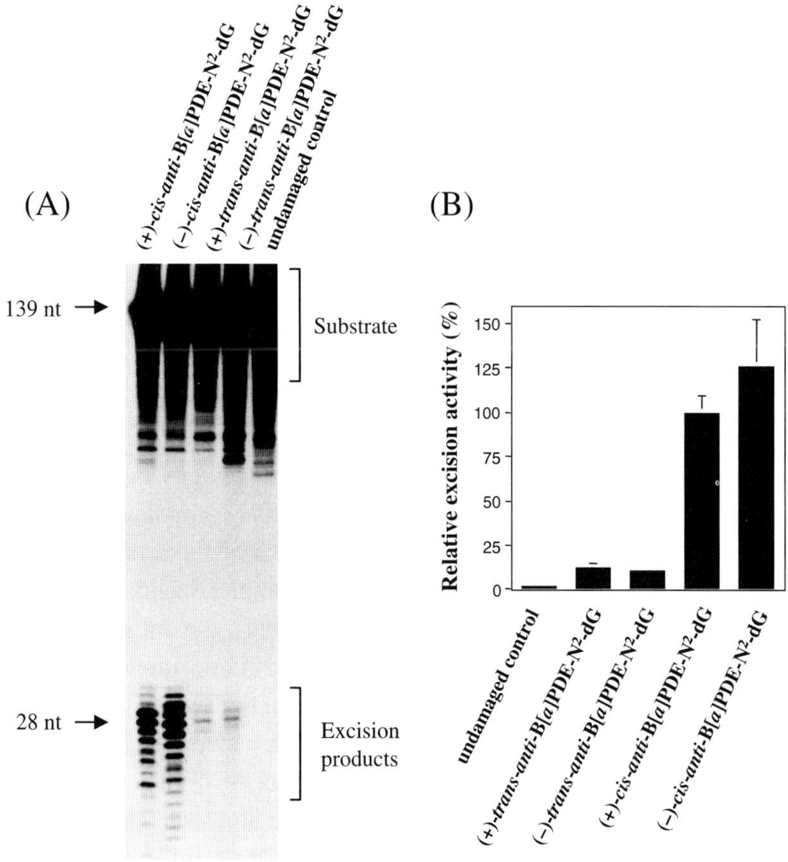

Figure 6.6: Excision of stereochemically distinct *anti*-B[*a*]PDE-N^2-dG adducts in human cell extracts. (A) Representative polyacrylamide gel showing the varying levels of excision products after 40-minutes reactions. (B) Quantitative evaluation obtained by laser scanning densitometry. The structure of each adduct is illustrated in Figure 6.1.

(−)-*trans* isomeric lesions indicates that, in the human system, efficient NER of PAH-modified substrates is dependent on the degree of local base displacement at the lesion site. Interestingly, a distinctly different hierarchy of excision, with (+)-*cis*- and (+)-*trans-anti*-B[*a*]PDE-N^2-dG being processed more efficiently than the corresponding (−)-isomers, was observed when (A)BC excinuclease (the prokaryotic NER system) was tested with the same conformationally defined adducts in the identical sequence context.[83]

These differences in the preference of damage excision by prokaryotic and eukaryotic NER systems could possibly reflect diverging mechanisms of substrate recognition. The principle of a preferential targeting of human NER activity to displaced bases is supported in the next section by the comparison between bay-region B[*a*]PDE-N^6-dA adducts and fjord-region analogs in *ras* codon 61 mutational hotspots.

6.3.3 Unrepaired Fjord-Region PAH-DNA Adducts in *Ras* Codon 61 Mutational Hotspots

The observed modulation of human NER activity due to the local base pairing characteristics prompted us to test the hypothesis that inefficient NER activity may contribute to the exceptional carcinogenicity of fjord-region PAH diol-epoxides.[68] Two different classes of PAH compounds have been recognized that differ from one another in terms of the arrangements of the aromatic hexameric ring systems (see for example, Szeliga and Dipple[5] and Chapter 2). In the bay-region PAH compounds, the epoxide group in the metabolically most active PAH diol-epoxides is positioned in a sterically unhindered region called the bay-region. In the second class of PAH diol-epoxides, the epoxide group is positioned in a sterically hindered site called the fjord-region (Figure 6.7). The epoxide in the fjord-region causes a distortion from planarity of the adjacent aromatic rings, whereas in the case of the bay-region compounds, steric hindrance effects are less pronounced and the aromatic ring systems remain planar. Both bay- and fjord-region diol-epoxides can exist in a number of different stereoisomeric configurations that display variable biological activities (cf. Figure 2.5 in Chapter 2). Among the bay-region B[*a*]P derivatives, the (+)-*anti*-B[*a*]PDE shown in Figure 6.7 is the most genotoxic isomer in mammalian systems (as indicated previously in this chapter). The biological activity of fjord-region B[*c*]Ph-3,4-diol-1,2-epoxides (B[*c*]PhDE) also depends on their stereochemical configuration, with the (−)-*anti*-B[*c*]PhDE illustrated in Figure 6.7 being the most potent isomer in animal tumorigenesis assays.[84,85] It is interesting to note that the absolute configurations of these potent substituents in the saturated ring are identical, i.e., (+)-*anti*-7*R*,8*S*,9*S*,10*R* in the B[*a*]PDE and (−)-*anti*-1*R*,2*S*,3*S*,4*R* in the B[*c*]PhDE.[5,82]

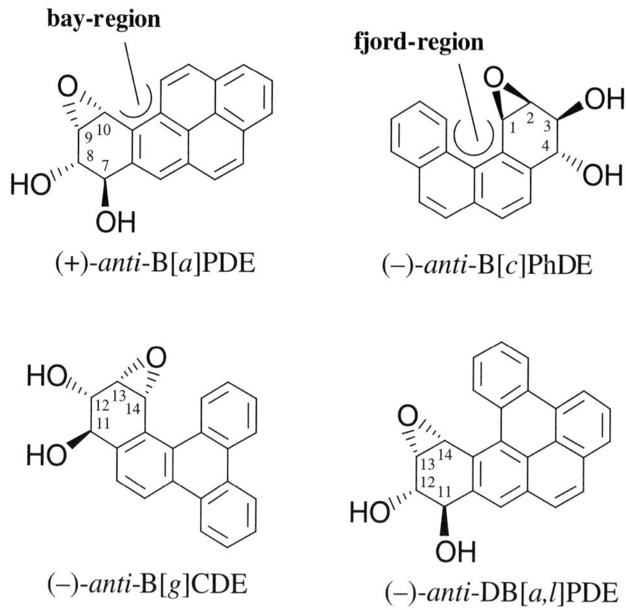

Figure 6.7: Stereochemical configuration of the major genotoxic diol-epoxides of B[*a*]P, B[*c*]Ph, B[*g*]C, and DB[*a,l*]P. Note that each of these diol-epoxides possesses '*R,S,S,R*' configuration: (+)-*anti*-B[*a*]P-7*R*,8*S*-diol-9*S*,10*R*-epoxide, (−)-*anti*-B[*c*]Ph-4*R*,3*S*-diol-2*S*,1*R*-epoxide, (−)-*anti*-B[*g*]C-11*R*,12*S*-diol-13*S*,14*R*-epoxide, and (−)-*anti*-DB[*a,l*]P-11*R*,12*S*-diol-13*S*,14*R*-epoxide (cf. Figures 2.5 and 2.6 in Chapter 2). Diol-epoxide metabolites of fjord-region PAHs display exceptional tumorigenic activities relative to diol-epoxides of PAHs with bay-regions.

Previous studies have demonstrated that (−)-*anti*-B[*c*]PhDE (in addition to stereochemically analogous metabolites of other fjord-region compounds) is up to 10 times more active in inducing tumors in animals than any of the known bay-region diol-epoxide derivatives.[84–88] Of course, this exceptional tumorigenic activity exerted by diol-epoxide metabolites with sterically hindered fjord-regions, relatively to structurally similar compounds with unhindered bay-regions, is of critical importance for understanding the mechanisms of tumor initiation by PAHs. These findings have therefore stimulated a collaborative effort between the two authors of this article to study the susceptibilities to DNA repair of stereochemically analogous DNA adducts derived from bay- and fjord-region PAH compounds. A DNA target of particular interest is the mutational hotspot sequence of a proto-oncogene, such as the *ras* mutation hotspot. The *ras* proto-oncogene is a key molecular switch in signal transduction pathways that control cell proliferation in response to external signals[89–91] (cf. Chapter 7). Activating mutations in H-*ras*, K-*ras* or N-*ras* are among the most frequent genetic changes associated with rodent or human cancers,[92–94] and various PAH-DNA adducts have been implicated in cell transformation processes mediated by *ras* oncogenes. For example, exposure of plasmids carrying a human *ras* proto-oncogene to diol-epoxide derivatives of PAHs generated a transforming oncogene when the damaged DNA was introduced into cultured fibroblasts.[95] Subsequent analysis of DNA isolated from transformed cells showed that the resulting *ras* mutations are mainly confined to codon 61.[96] Exactly the same mutations have been found in rodent tumors induced by PAH treatment, supporting the view that *ras* codon 61 is an important molecular target of this class of ubiquitous carcinogens (see for example Balmain and Pragnell;[97] cf. Chapters 5 and 7).

In contrast to B[*a*]PDE stereoisomers, which react preferentially with guanine, all B[*c*]PhDE stereoisomers display a bias for the N^6 exocyclic amino group of adenine.[98] Because induction of carcinomas in rodents by PAHs has been shown to involve A to T transversions in the second position of *ras* codon 61,[97,99] we constructed (+)-*trans*- and (−)-*trans*-PAH diol-epoxide adducts at the exocyclic N^6 residue of adenine at this particular hotspot site in two different *ras* codon 61 sequences. A first series of substrates contained the sequence for codons 60–62 of the human N-*ras* gene, and a total synthesis method was used to generate bay-region (+)-*trans*- or (−)-*trans*-B[*a*]PDE-N^6-dA as well as fjord-region (+)-*trans*- or (−)-*trans*-B[*c*]PhDE-N^6-dA adducts at the second position of codon 61. In these constructs, the site-specific adducts were located in the context 5′–ACA*AG–3′, where the modified adenosine in N-*ras* codon 61 (CA*A) is indicated by the asterisk. Distinct conformational properties have been associated with such bay- and fjord-region diol-epoxide-DNA adducts (Table 6.1). For example, NMR analysis showed that the bay-region (−)-*trans-anti*-B[*a*]PDE-N^6-dA adduct is inserted into the Watson-Crick double helix, thereby inducing a modest but detectable distortion of base pairing interactions at the site of covalent modification.[100] The structure of the (+)-*trans-anti*-B[*a*]PDE-N^6-dA isomer has not been amenable to determination by NMR, although molecular dynamics simulations suggest a similar motif of adduct insertion but with more destabilization of Watson-Crick base pairing.[101,102] In contrast, the corresponding fjord-region (+)-*trans*- or (−)-*trans-anti*-B[*c*]PhDE-N^6-dA adducts are incorporated into the double helix by an intercalative mode that retains normal Watson-Crick base pairing throughout the modified duplexes.[103–105] These molecular models are supported by differences in the thermal stabilities of short double-stranded fragments containing single bay- or fjord-region PAH-N^6-dA adducts (Table 6.2). In fact, the DNA duplex melting point is significantly lowered by the presence of a (+)-*trans-anti*-B[*a*]PDE-N^6-dA lesion and, to a lesser extent, also by the (−)-*trans-anti*-B[*a*]PDE-N^6-dA isomeric adduct,[101,106] most likely indicating a weakening of Watson-Crick hydrogen bonding. In contrast, the melting point is unchanged in the presence of the corresponding fjord-region B[*c*]PhDE-N^6-dA adducts.[107,108]

Incubations in HeLa cell extracts of PAH-modified DNA duplexes demonstrated that both the (+)-*trans*- or (−)-*trans-anti*-B[*a*]PDE-N^6-dA adducts in the second position of N-*ras* codon 61 were able to elicit NER activity (Figure 6.8A, lanes 2 and 3) with excision products in the expected

Table 6.2: Differences in melting temperature associated with the presence of a single bay- or fjord-region PAH diol-epoxide-dA adduct in the 11-mer duplex 5′–CTCTCA*CTTCC–3′/5′–GGAAGTGAGAG-3′

PAH adduct	*Reactive region*	*Changes in melting temperature*[a]
(+)-*trans-anti*-B[*a*]PDE-N^6-dA	bay	−17.0
(−)-*trans-anti*-B[*a*]PDE-N^6-dA	bay	−9.6
(+)-*trans-anti*-B[*c*]PhDE-N^6-dA	fjord	−0.9
(−)-*trans-anti*-B[*c*]PhDE-N^6-dA	fjord	−0.9
(+)-*trans-anti*-B[*g*]CDE-N^6-dA	fjord	+8.7
(−)-*trans-anti*-B[*g*]CDE-N^6-dA	fjord	−3.9
(+)-*trans-anti*-DB[*a,l*]PDE-N^6-dA	fjord	+7.2
(−)-*trans-anti*-DB[*a,l*]PDE-N^6-dA	fjord	−5.9

[a]The melting temperature of unmodified duplexes is 42.5°C. The melting points are defined as the temperature at which 50% of the duplexes are dissociated.

size range of 24–32 nucleotides. In contrast to these DNA lesions derived from B[*a*]PDE, we found that fjord-region (+)-*trans*- or (−)-*trans-anti*-B[*c*]PhDE-N^6-dA adducts located at the same position of N-*ras* codon 61 failed to stimulate detectable levels of oligonucleotide excision (Figure 6.8A, lanes 5 and 6). To rule out the possibility that the DNA preparations with B[*c*]PhDE-N^6-dA adducts contained a chemical contaminant that may inhibit DNA repair in a non-specific manner, we performed co-incubation reactions in which the actively repaired (+)-*trans-anti*-B[*a*]PDE-N^6-dA adduct was mixed with equal amounts of substrate containing the unrepaired (+)-*trans-anti*-B[*c*]PhDE-N^6-dA adduct. These co-incubation experiments yielded the expected amount of B[*a*]PDE adduct excision (Figure 6.8B), thereby excluding the presence of a repair inhibitor in the fractions of B[*c*]PhDE-damaged DNA.

A second series of substrates contained the sequence for codons 60–65 of the human H-*ras* gene, and a total synthesis method was again used to generate bay-region B[*a*]PDE-N^6-dA and fjord-region B[*c*]PhDE-N^6-dA adducts at the second position of codon 61. In these constructs, the site-specific adducts were located in the context 5′–CCA*GG–3′, where the modified adenosine in H-*ras* codon 61 (CA*G) is indicated by the asterisk. Consistent with the previous experiments performed in the N-*ras* environment, we found that both (+)-*trans*- or (−)-*trans-anti*-B[*a*]PDE-N^6-dA adducts are removed from the H-*ras* codon 61 sequence upon incubation in HeLa cell extract (Figure 6.8A, lanes 7 and 8). We also noted similar levels of oligonucleotide excision when the configuration of these B[*a*]PDE-N^6-dA adducts was changed from *anti* to *syn* (Figure 6.8A, lanes 9 and 10). However, excision repair of adenine lesions in the identical sequence context was abolished when the bay-region B[*a*]PDE adduct was replaced by the fjord-region B[*c*]PhDE residue. In fact, (+)-*trans*- and (−)-*trans-anti*-B[*c*]PhDE-N^6-dA adducts located at the same second position of H-*ras* codon 61 were unable to induce any detectable excision of the modified sequence (Figures 6.8A, lanes 11 and 12, and 6.8C).

The resistance of fjord-region adducts to excision by human DNA repair enzymes in HeLa cell extracts is not restricted to the (+)-*trans*- and (−)-*trans-anti*-B[*c*]PhDE-N^6-dA lesions. A survey of duplex DNA melting temperatures[107,108] showed that fjord-region PAH adducts cause

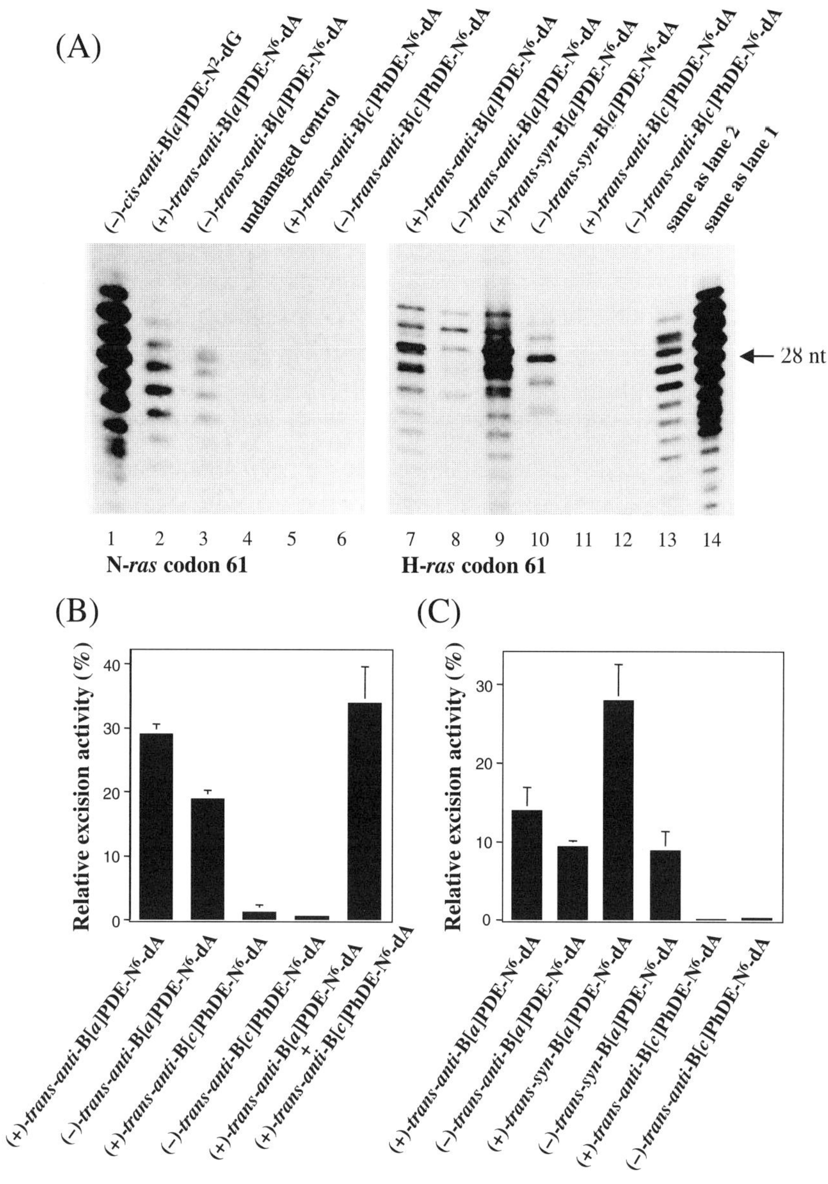

Figure 6.8: Differential repair of B[*a*]PDE and B[*c*]PhDE adducts in *ras* codon 61 hotspot sequences. (A) Representative polyacrylamide gel showing the specific excision products obtained after 40-minutes incubations in HeLa cell extracts. The prefix '*syn*' refers to diol-epoxides with the epoxide ring on the same side of the aromatic plane as the benzylic hydroxy group (cf. Figure 2.6 in Chapter 2). (B) Quantification of excision rates in the N-*ras* codon 61 context. (C) Quantification of excision rates in the H-*ras* codon 61 context.

less destabilization than the bay-region B[*a*]PDE prototype (Table 6.2). Furthermore, an unusual thermal stabilization is observed for DNA duplexes containing the *trans*-opened isomers of (+)-*anti*-B[*g*]CDE-N^6-dA and (+)-*anti*-dibenzo[*a,l*]pyrene (DB[*a,l*]P) 11,12-diol-13,14-epoxide (DB[*a,l*]PDE)-N^6-dA. B[*g*]C consists of the same number of aromatic ring systems as B[*a*]P, whereas DB[*a,l*]P is composed of an additional aromatic ring as compared to B[*a*]P (Figure 6.7). Oligonucleotide excision assays in cell extracts demonstrated that poor distortion of the native Watson-Crick double helix, with little or no destabilization of local base pairing, correlates with low repair efficiency. In fact, four representative DNA adducts with the fjord-region structural motif, i.e., (+)-*trans*- and (−)-*trans-anti*-B[*g*]CDE-N^6-dA, as well as (+)-*trans*- and (−)-*trans-anti*-DB[*a,l*]PDE-N^6-dA, do not elicit detectable excision repair reactions in the sequences examined (Figure 6.9). These findings suggest that the different

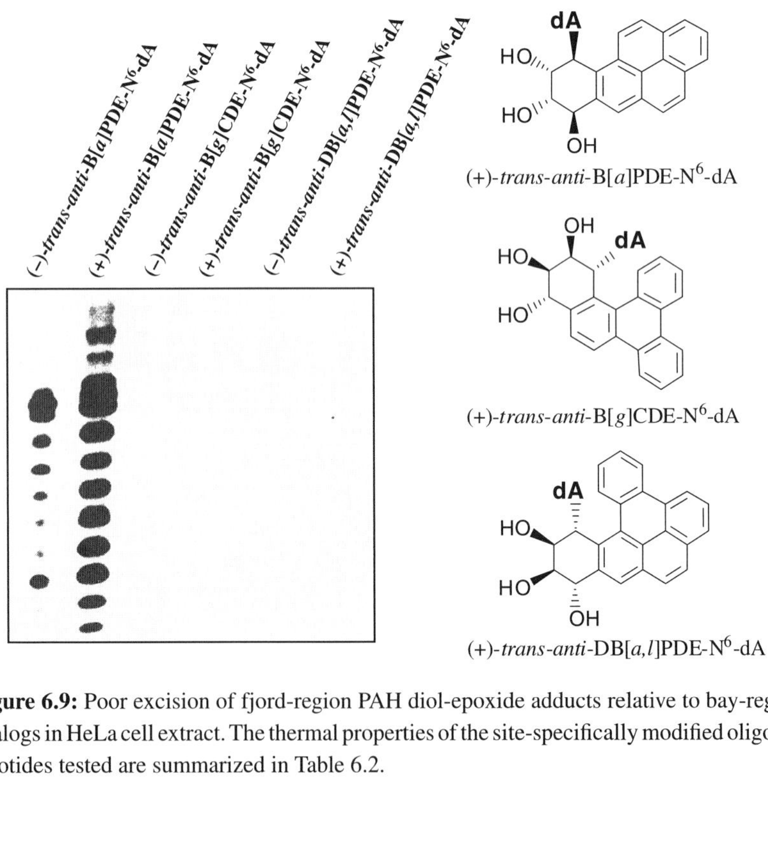

Figure 6.9: Poor excision of fjord-region PAH diol-epoxide adducts relative to bay-region analogs in HeLa cell extract. The thermal properties of the site-specifically modified oligonucleotides tested are summarized in Table 6.2.

thermodynamic properties of bay- and fjord-region PAH adducts are responsible for the observed variability in NER rates, and raise the question of whether fjord-region PAH adducts might be generally more refractory than B[*a*]P adducts to removal by the human NER system.[68]

Taken together, the exceptional tumorigenic activity of B[*c*]PhDE or similar fjord-region metabolites of other PAHs may well be associated with the formation of non-distorting DNA adducts that, at least in some critical target sequences such as H-*ras* or N-*ras* codon 61, are resistant to NER enzymes. These findings are consistent with the hypothesis that inefficient excision constitutes an important determinant of mutagenic and tumorigenic potency.[56] As a consequence, even minor PAH lesions that are formed with only low probabilities may exert disproportionately large genotoxic effects because of extremely slow repair by the NER system. This view is supported by a recent study describing the poor excision of guanine adducts of PAHs from codon 12, relative to other sites of the K-*ras* gene, in human bronchial epithelial and fibroblast cells.[109]

6.3.4 Bipartite Recognition of PAH-DNA Adducts

The evasion of human NER, observed in the case of non-distorting fjord-region PAH adducts, indicates that local disruption of Watson-Crick base pairing, which is a major manifestation of damage-induced structural perturbations, may be one of the critical determinants for the recognition of bulky lesions. The more efficient excision of *cis-anti*-B[*a*]PDE-N^2-dG adducts, characterized by base displacement and disruption of local base pairing, as compared to excision of the stereoisomeric *trans-anti*-B[*a*]PDE-N^2-dG adducts, is in accord with this hypothesis since normal Watson-Crick geometry of base pairs is retained in the latter (see section 6.3.2.).

A model supported by experiments performed with artificially manipulated DNA substrates indicates that the molecular signal leading to recognition of base adducts during NER requires two fundamental elements, i.e., disruption of normal Watson-Crick base pairing and altered chemistry in the damaged strand.[110,111] Neither defective base pairing in duplex DNA nor defective chemistry of its deoxyribonucleotide components induces NER activity, but the combination of these two substrate alterations generates the molecular trigger for assembly of active incision complexes. The term of 'bipartite recognition' has been proposed to indicate that human NER factors utilize two principal levels of discrimination by recognizing both disruption of Watson-Crick base pairing (as the manifestation of local DNA distortion) and alterations in the normal chemistry of the DNA nucleotides.[111]

Further evidence supporting the bipartite model of substrate discrimination was sought by disrupting base pairing in the proximity of a fjord-region (+)-*trans-anti*-B[*c*]PhDE-N^6-dA adduct, which is resistant to repair.

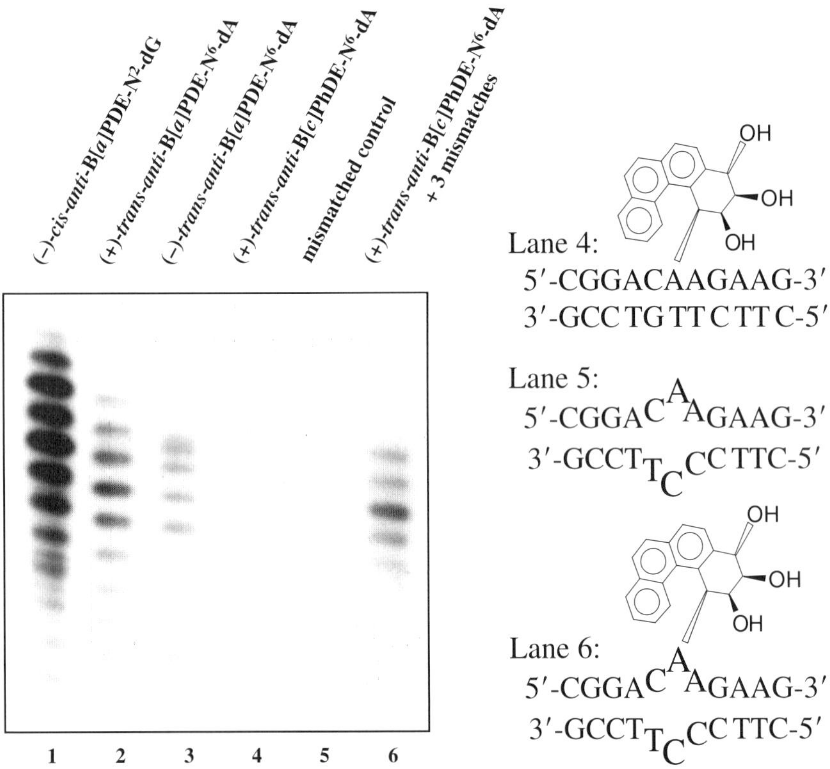

Figure 6.10: Bipartite substrate discrimination by the human NER system. The non-distorting B[*c*]PhDE-dA adduct is not excised when all the base pairs in the duplex are complementary. However, if some mismatched base pairs are introduced at the lesion site, this same adduct is efficiently excised.

As observed previously, the (+)-*trans-anti*-B[*c*]PhDE-N^6-dA lesion was not excised during incubation in HeLa cell extracts when situated in a fully complementary duplex (Figure 6.10, lane 4). Similarly, an artificial distortion consisting of three consecutive mismatches in an unmodified duplex was not processed by the human NER system (Figure 6.10, lane 5). However, the same (+)-*trans-anti*-B[*c*]PhDE-N^6-dA adduct in combination with three mismatched bases led to the formation of oligonucleotide excision products that are characteristic of NER activity (Figure 6.10, lane 6). This experiment with a composite substrate supports the hypothesis that the NER susceptibility of a particular adduct is determined by changes in the thermodynamic properties associated with local disruption of base pairing.

6.3.5 Modulation of Human NER Activity by 5-Methylcytosines

Mammalian DNA methylation patterns are generated by 5-methylcytosine modifications in approximately 70–80% of CpG dinucleotide sequences (reviewed by Razin[112]). Previous reports indicated that PAH adducts are preferentially formed at CpG sequences and that this predisposition for damage formation is attributable to the presence of 5-methylcytosines instead of cytosines in the CpG targets. For example, the distribution of B[*a*]P lesions in human genes is influenced by DNA methylation because the presence of 5-methylcytosine in CpG dinucleotides increases the reactivity of B[*a*]PDE with the neighboring guanine, thus inducing hotspots for covalent B[*a*]PDE binding[113,114] (cf. Chapter 5). These findings prompted us to examine possible effects of cytosine methylation on B[*a*]PDE-DNA adduct repair in detail using the human *p53* codon 273 mutational hotspot as a model sequence for *in vitro* excision assays.[115] All mutational hotspots of the *p53* gene are normally methylated, but early stages of tumor progression are associated with widespread DNA hypomethylation.[10] Thus, we were interested in the question as to whether CpG methylation may modulate the efficiency of NER reactions in an important mutational target.

A synthetic oligonucleotide with the sequence corresponding to codons 272–274 of the non-transcribed strand in the *p53* gene was exposed to racemic *anti*-B[*a*]PDE. The resulting products were separated from one another and purified by reverse phase HPLC. The positions of the modified guanines were determined by Maxam-Gilbert sequencing, and the nature of the modified bases were verified by MS and circular dichroism methods. Oligonucleotides with the (+)-*trans-anti*-B[*a*]PDE-N^2-dG adduct situated within codon 273 (5′–CGT–3′) were selected and incorporated in the center of 139-mer DNA duplexes that served as a substrate for NER reactions in human HeLa cell extracts. Unexpectedly, the B[*a*]PDE adduct was removed from the methylated sequences more efficiently than from the corresponding unmethylated substrate (Figure 6.11A, compare lanes 3, 4 and 5). Even more surprisingly, about the same stimulatory effect was observed regardless of whether a single 5-methylcytosine substitution was introduced in the CpG dinucleotide of the damaged strand (lane 4), or in the opposite CpG of the undamaged strand (lane 6). Excision of the (+)-*trans-anti*-B[*a*]PDE-N^2-dG was further enhanced when both cytosines present across the lesion carried a methyl group (lane 8). Finally, stimulation of NER activity was maintained when either strand carried a 5-methylcytosine substitution (Figure 6.11B).

The addition of a methyl group to a cytidine flanking the lesion is potentially able to induce a striking conformational change of B[*a*]PDE-N^2-dG adducts, such that a minor groove conformation of these adducts may be changed to the more distorting intercalated structure.[116] Thus, increased repair of (+)-*trans-anti*-B[*a*]PDE-N^2-dG located in the methylated codon 273 hotspot sequence is likely to result from alterations in the damage-induced DNA distortion, generating a lesion that is more susceptible to recognition by the NER complex. It appears therefore, that CpG methylation not only modulates the pattern of adduct formation but it also regulates adduct removal by stimulating NER activity, at least at some sites.

6.3.6 Antagonistic Interaction of NER Factors between Substrate and Decoy DNA Adducts

PAHs are generally found in mixtures often containing several different representative compounds that generate structurally distinct DNA adducts. However, the NER system of all living organisms lacks specificity for

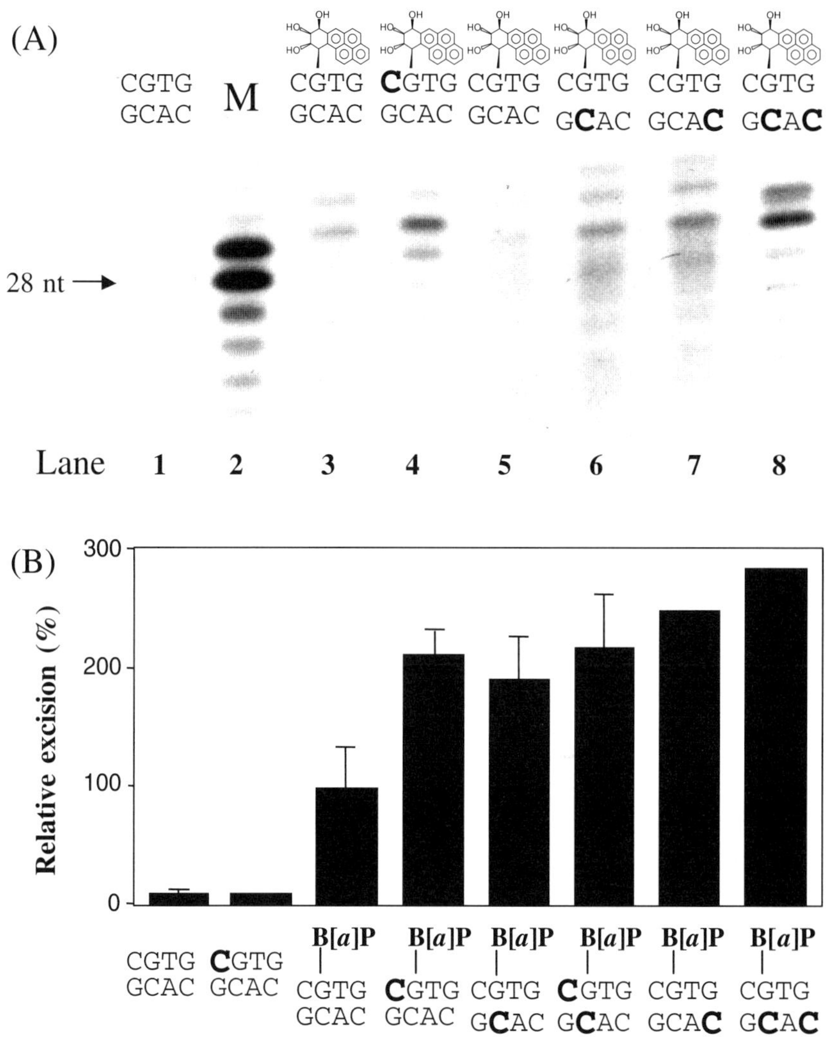

Figure 6.11: Modulation of NER activity by cytosine methylation. (A) Representative polyacrylamide gel electrophoretic analysis of excision fragments after incubation of the DNA substrates with human NER enzymes for 40 minutes. A portion of the 139-mer duplex substrates is indicated at the top, and the modified guanine residues are defined in each case. M, excision reaction with substrate containing a (+)-*cis*-*anti*-B[*a*]PDE-N^2-dG adduct. (B) Quantitative evaluation of 2–3 independent experiments, demonstrating stimulation of NER activity associated with the B[*a*]PDE-modified guanine positioned at the *p53* codon 273 mutation hotspot. The bold 'C' indicates the methylated cytosine residue.

particular types of DNA damage and, hence, recognizes a wide range of base abnormalities. In view of this broad repair activity, we analyzed the distribution of human NER factors between structurally different DNA lesions.[117] Surprisingly, challenging the activity of repair proteins with mixtures of different DNA adducts instead of single types of lesions revealed a novel mechanism of repair inhibition involving the sequestration (or 'hijacking') of critical NER subunits by DNA adducts that are either poorly repaired or not repaired at all. Such 'decoy DNA adducts' are partially or completely refractory to excision, but nevertheless immobilize NER factors.[117]

We were able to construct decoy PAH-DNA adducts by manipulating the nucleotide sequence environment near the lesion site. In HeLa cell extracts, excision of (+)-*trans*- or (+)-*cis-anti*-B[*a*]PDE-N^2-dG adducts (as monitored by the appearance of the characteristic oligonucleotide fragments 24–32 nucleotides long) was abolished when the whole dCMP residue, which is normally located across the lesion in the fully complementary duplex, was removed (Figure 6.12A, lanes 6 and 9). As indicated in Table 6.1, the conformation of such dCMP deletion duplexes containing (+)-*trans*- or (+)-*cis-anti*-B[*a*]PDE-N^2-dG adducts have been determined by NMR studies, thereby revealing the formation of these wedge-shaped intercalation complexes with an extra-helical displacement of the modified guanine residue.[118,119]

The complete lack of excision from such deletion duplexes provided the basis for the design of competition experiments, in which the site-specifically modified deletion duplexes were incubated in HeLa cell extracts with a PAH-damaged DNA substrate that is normally excised when present on its own. In the experiment shown in Figure 6.12B, DNA substrates (0.2 nM) containing either a (+)-*trans*- or (+)-*cis-anti*-B[*a*]PDE-N^2-dG

Figure 6.12: Construction of decoy DNA adducts. (A) Comparison of DNA repair of stereochemically different lesions (defined at the top) in duplexes with different sequence contexts (defined at the right of the figure) when incubated with NER enzymes individually; the lesions in the deletion duplexes (lanes 6 and 9) are not excised. (B) Competition experiment revealing that damaged deletion duplexes sequester NER activity (lanes 3 and 6). (C) Competitor dose response obtained from two independent 40-minutes experiments in HeLa cell extract.

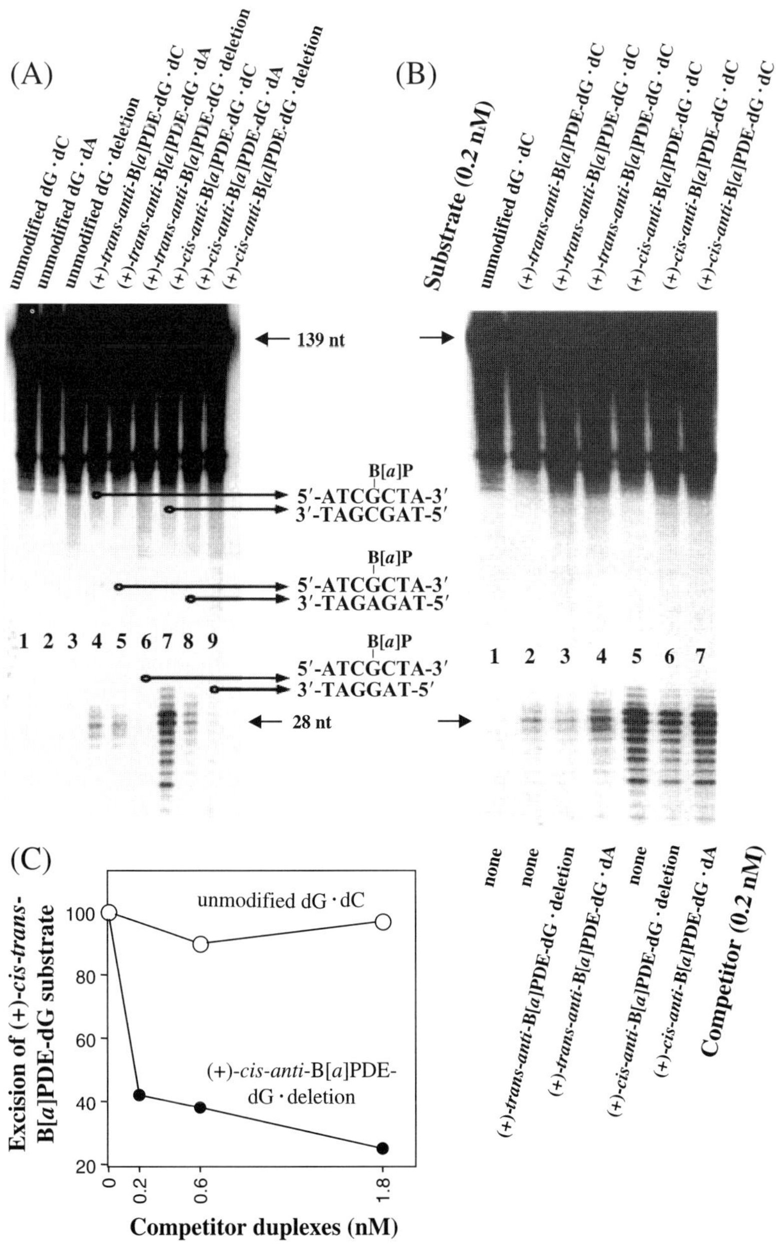

(A)
unmodified dG · dC
unmodified dG · dA
unmodified dG · deletion
(+)-trans-anti-B[a]PDE-dG · dC
(+)-trans-anti-B[a]PDE-dG · dA
(+)-trans-anti-B[a]PDE-dG · deletion
(+)-cis-anti-B[a]PDE-dG · dC
(+)-cis-anti-B[a]PDE-dG · dA
(+)-cis-anti-B[a]PDE-dG · deletion
139 nt
B[a]P
5′-ATCGCTA-3′
3′-TAGCGAT-5′
B[a]P
5′-ATCGCTA-3′
3′-TAGAGAT-5′
1 2 3 4 5 6 7 8 9
B[a]P
5′-ATCGCTA-3′
3′-TAGGAT-5′
28 nt
(B)
Substrate (0.2 nM)
unmodified dG · dC
(+)-trans-anti-B[a]PDE-dG · dC
(+)-trans-anti-B[a]PDE-dG · dC
(+)-trans-anti-B[a]PDE-dG · dC
(+)-cis-anti-B[a]PDE-dG · dC
(+)-cis-anti-B[a]PDE-dG · dC
(+)-cis-anti-B[a]PDE-dG · dC
1 2 3 4 5 6 7
none
none
(+)-trans-anti-B[a]PDE-dG · deletion
(+)-trans-anti-B[a]PDE-dG · dA
none
(+)-cis-anti-B[a]PDE-dG · deletion
(+)-cis-anti-B[a]PDE-dG · dA
Competitor (0.2 nM)
(C)
Excision of (+)-cis-trans-B[a]PDE-dG substrate
100
80
60
40
20
0
0.2
0.6
1.8
unmodified dG · dC
(+)-cis-anti-B[a]PDE-dG · deletion
Competitor duplexes (nM)

adduct were mixed with equimolar amounts of the respective competitor DNA fragments of identical lengths. There was inhibition of repair of the (+)-*trans-anti*-B[*a*]PDE-N^2-dG substrate when the competitor DNA carried the same B[*a*]PDE adduct at a deletion site lacking the cytosine residue opposite the lesion (compare lanes 2 and 3). This inhibition was not observed when the substrate was co-incubated with a competitor that contained a dA residue in the complementary strand opposite the lesion (dG-dA duplex, lane 4). Inhibition of repair was also detected when the (+)-*cis-anti*-B[*a*]PDE-N^2-dG substrate was mixed with equimolar amounts of competitor fragments carrying the same adduct opposite a deletion site (compare lanes 5 and 6). Again, this inhibition was not detected when the substrate was co-incubated with the same lesion in the dG-dA context as a competitor (lane 7). A dose response was established by mixing internally labeled substrates containing the *cis*-opened B[*a*]PDE adduct: the addition of increasing amounts of deletion duplexes containing the same adduct but with a missing dCMP residue opposite the lesion in the complementary strand, resulted in progressive inhibition of substrate repair. However, excision was only marginally reduced when the same amounts of unmodified duplexes were added to the assay mixtures (Figure 6.12C). These results demonstrate that human NER factors are prone to hijacking by decoy DNA adducts. A similar highjacking concept was previously inferred in the case of transcription factor binding to B[*a*]PDE-modified DNA by McLeod and co-workers.[120]

The prototype decoy adduct consisting of a bulky guanine adduct opposite dCMP deletions is not an artificial construct. Exactly this kind of lesion can be generated as a mutagenic intermediate after the replicative bypass of PAH-induced damage in template DNA strands. In fact, translesion synthesis across PAH adducts is facilitated by a mechanism in which one strand can slip with respect to the other strand, generating misalignment frameshifts in the daughter strand. Depending on the sequence context, single nucleotide deletions across the adduct constitute the predominant source of carcinogen-induced mutations in replication systems based on purified DNA polymerases[121–123] as well as in site-directed mutagenesis experiments in mammalian cells.[53,124] Such single nucleotide deletions have also been found in the chromosomes of B[*a*]PDE-exposed human cells.[43,125]

Thus, the B[*a*]PDE-modified deletion duplex identified by Buterin *et al.* as a decoy substrate[117] is normally generated in many cases of environmental genotoxic stress due to B[*a*]P.

Trapping of NER factors was described before only in the context of the analogous system in prokaryotes exposed to reactive cross-linking agents.[126] However, it has already been observed that chronic, low dose application of B[*a*]P, at conditions to which humans are normally exposed, leads to the accumulation of DNA adducts that are poorly repaired.[127] The discovery of a prototype decoy adduct suggests that particular components of genotoxic mixtures may induce bulky DNA lesions that not only persist in the genome, but also exert negative effects on DNA repair. The resulting antagonism between substrate and decoy adducts could be responsible for hyper-additive interactions between multiple genotoxic agents, which may further aggravate the mutagenic risk of environmental genotoxic insults.

6.3.7 Mechanism of PAH Adduct Recognition by the Human NER Machinery

How the human NER machinery recognizes many kinds of bulky DNA base adducts, and discriminates between these lesions and undamaged DNA (including the undamaged strand directly opposite the site of base lesion), is still poorly understood and poses an experimental challenge that has not yet been fully resolved. The results summarized in this chapter converge on the concept that the efficiency of bulky lesion removal by NER enzymes is determined by the base pairing properties in the immediate vicinity of the damaged nucleotide. This striking dependence on the local base pair geometry has been demonstrated by the manipulation of PAH-modified substrates, for example, by changing the predominant minor groove *trans*-conformation of (+)-*anti*-B[*a*]PDE-N^2-dG adducts to a base-displaced intercalative conformation that is characteristic of the *cis*-opened stereoisomeric adducts (section 6.3.2.); by replacing the lesions derived from B[*a*]PDE with adducts derived from B[*c*]PhDE (section 6.3.3.); by altering the bases in the complementary strand opposite the adduct (section 6.3.4.); by changing cytosines to 5-methylcytosines near an adducted base (section 6.3.5.); or by entirely removing the nucleotide opposite the lesion (section 6.3.6.). All of these substrate manipulations alter the degree by which Watson-Crick hydrogen bonding is perturbed, and these manipulations are associated with large changes in the damage recognition and excision capacity.

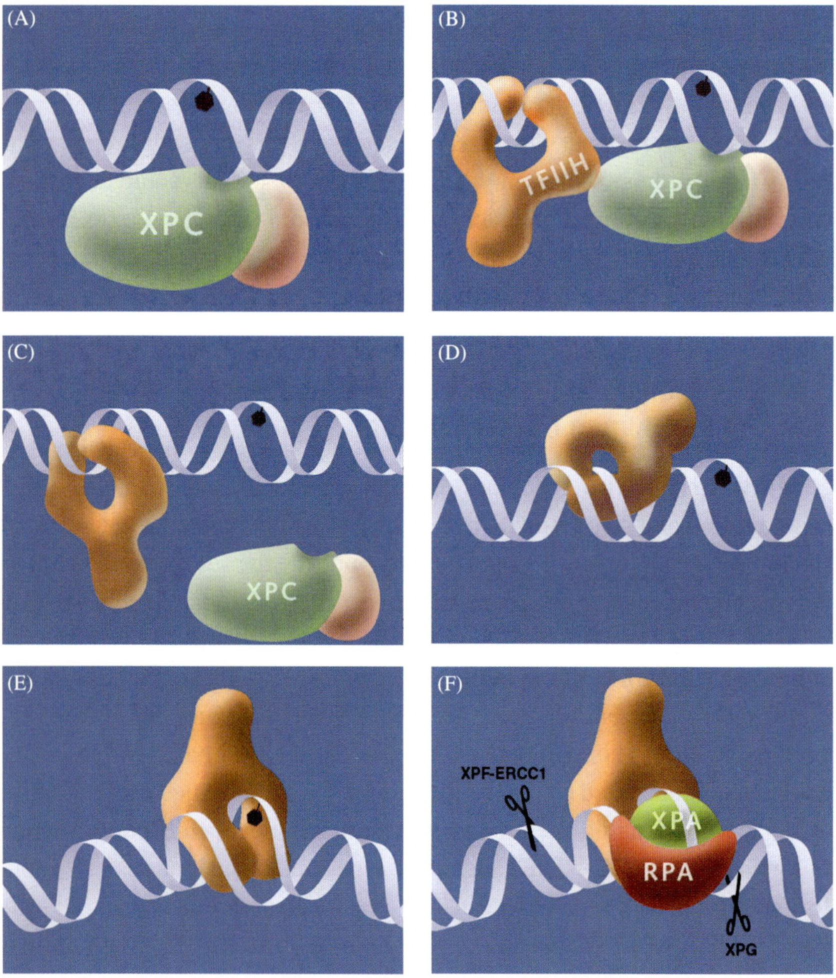

Figure 6.13: Model of bipartite substrate discrimination by the human NER machinery. Recognition of defective base pairing (A) is followed by recruitment of TFIIH (B), localization of the damaged nucleotide residue (C–E), and activation of DNA endonucleases (F).

The heterogeneity of NER reactions upon formation of PAH-DNA adducts can be accommodated within the framework of the following model of damage recognition and damage-specific incision (Figure 6.13):

1. *Recognition of defective base pairing.* XPC-hHR23B, the initiator of global NER activity, probes the stability of duplex DNA and senses thermodynamically abnormal sites characterized by a weakening of Watson-Crick base pairing. Recent studies suggest that XPC uses the undamaged strand opposite to the adduct as a docking site (Figure 6.13A) to initiate assembly of the NER complex (T. Buterin and H. Naegeli, unpublished results).
2. *Recruitment of TFIIH.* XPC attracts TFIIH to the distorted site and loads the ring-like helicase domain of this large multifunctional factor onto the damaged target strand (Figure 6.13B). Improper loading of TFIIH on the complementary undamaged strand could result in non-productive complexes that, in some cases, may lead to hijacking of NER factors.
3. *Recognition of defective chemistry.* The two DNA helicases (XPB and XPD) promote translocation of TFIIH, which serves to probe the chemical composition of the target strand. In the presence of a DNA adduct, TFIIH translocation is blocked, the DNA helicase and ATPase activities are suppressed, and the entire factor is frozen in a stable nucleoprotein complex near the damaged site (Figure 6.13C, D; see Naegeli *et al.*,[128] and Villani and Le Gac[129] for the role of DNA helicases in damage recognition). Conversely, spurious loading of TFIIH onto undamaged sites in the absence of any adducts would result in continued translocation and, eventually, dissociation of TFIIH and its helicases from DNA.
4. *Formation of a distorted intermediate.* The stalled TFIIH complex at sites of DNA damage causes a further distortion of the DNA duplex, generating the kinked and partially unwound intermediate that is recognized by XPA and RPA (Figure 6.13E).
5. *Licensing by XPA-RPA.* Originally, the 'licensing' concept has been introduced to describe the finding that a particular factor, which is required for DNA replication, is used in higher eukaryotes to prevent the semiconservative DNA synthesis at inappropriate sites or at improper stages of the cell-cycle.[130] The same concept is used here for the role of the XPA-RPA

complex during assembly of the incision machinery. As outlined briefly in the introductory sections, the biochemical properties of XPA and RPA argue against a direct role of these subunits in recognizing or verifying the damage. Instead, our results[29] indicate that XPA acquires an essential function in monitoring the degree of DNA bending, while RPA monitors local DNA unwinding, thereby double-checking the three-dimensional architecture of the pre-incision complex. If this nucleoprotein intermediate is correctly assembled and the DNA strands are properly positioned, the XPA-RPA complex enables the two structure-specific endonucleases to incise DNA, leading to damage excision (Figure 6.13F). Thus, the molecular 'licensing' function of XPA-RPA excludes the risk that DNA may be processed by endonucleases at an inappropriate (undamaged) site or at an improper (premature) step during assembly of the incision complex.

6.4 Conclusion

There is growing appreciation that DNA repair modulates cellular mutagenesis, since only those DNA lesions that survive repair have the chance to induce mutations upon error-prone DNA replication. In particular, several lines of evidence demonstrate that the global genome NER pathway provides an essential line of defense against the mutagenic and carcinogenic activity of PAH diol-epoxide metabolites. However, the rates of excision of individual adducts vary widely in non-transcribed sequences, supporting the notion that inefficient DNA repair plays an important role in the formation of PAH-induced mutational hotspots. The results summarized in this present chapter lead to the conclusion that improved risk assessment procedures, particularly in the problematic low-dose range, require a more complete knowledge of the molecular mechanisms by which relevant PAH-DNA adducts are recognized and repaired in critical mutational hotspots.

Acknowledgment

Research in the authors' laboratories is supported by grants from the Swiss National Science Foundation (31-61494.00), and the National Institute of Health (NIH) grant CA76660.

References

1. Conney AH (1982) Induction of microsomal enzymes by foreign chemicals and carcinogenesis by polycyclic aromatic hydrocarbons. *Cancer Res.* 42: 4875–4917.
2. Phillips DH and Grover PL (1994) Polycyclic aromatic hydrocarbon activation: bay regions and beyond. *Drug Metab. Rev.* 26: 443–467.
3. Sims P, Grover PL, Swaisland A, Pal K and Hewer A (1974) Metabolic activation of benzo[*a*]pyrene proceeds by a diol-epoxide. *Nature* 252: 326–328.
4. Peltonen K and Dipple A (1995) Polycyclic aromatic hydrocarbons: chemistry of DNA adduct formation. *J. Occup. Environ. Med.* 37: 52–58.
5. Szeliga J and Dipple A (1998) DNA adduct formation by polycyclic aromatic hydrocarbon dihydrodiol epoxides. *Chem. Res. Toxicol.* 11: 1–11.
6. De Boer J, Andressoo JO, de Wit J, Huijmans J, Beems RB, van Steeg H, Weeda G, van der Horst GTJ, van Leeuwen W, Themmen APN, Meradji M and Hoeijmakers JHJ (2002) Premature aging in mice deficient in DNA repair and transcription. *Science* 296: 1276–1279.
7. Bigger CA, Flickinger DJ, St John J, Harvey RG and Dipple A (1991) Preferential mutagenesis at G.C base pairs by the *anti* 3,4-dihydrodiol 1,2-epoxide of 7-methylbenz[*a*]anthracene. *Mol. Carcinog.* 4: 176–179.
8. Glatt H, Harvey RG, Phillips DH, Hewer A and Grover PL (1989) Influence of the alkyl substituent on mutagenicity and covalent DNA binding of bay region diolepoxides of 7-methyl- and 7-ethylbenz[*a*]anthracene in Salmonella and V79 Chinese hamster cells. *Cancer Res.* 49: 1778–1782.
9. Yan S, Shapiro R, Geacintov NE and Broyde S (2001) Stereochemical, structural, and thermodynamic origins of stability differences between stereoisomeric benzo[*a*]pyrene diol epoxide deoxyadenosine adducts in a DNA mutational hot spot sequence. *J. Am. Chem. Soc.* 123: 7054–7066.
10. Vogelstein B and Kinzler KW (1993) The multistep nature of cancer. *Trends Genet.* 9: 138–141.
11. Drinkwater NR, Miller EC and Miller JA (1980) Estimation of apurinic/apyrimidinic sites and phosphotriesters in deoxyribonucleic acid treated with electrophilic carcinogens and mutagens. *Biochemistry* 19: 5087–5092.
12. Lawley PD and Brookes P (1963) Further studies on the alkylation of nucleic acids and their constituent nucleotides. *Biochem J.* 89: 127–138.
13. Chakravarti D, Pelling JC, Cavalieri EL and Rogan EG (1995) Relating aromatic hydrocarbon-induced DNA adducts and c-H-*ras* mutations in mouse

skin papillomas: the role of apurinic sites. *Proc. Natl. Acad. Sci. USA* 92: 10422–10426.

14. Braithwaite E, Wu X and Wang Z (1998) Repair of DNA lesions induced by polycyclic aromatic hydrocarbons in human cell-free extracts: involvement of two excision repair mechanisms *in vitro*. *Carcinogenesis* 19: 1239–1246.
15. Melendez-Colon VJ, Luch A, Seidel A and Baird WM (1999) Cancer initiation by polycyclic aromatic hydrocarbons results from formation of stable DNA adducts rather than apurinic sites. *Carcinogenesis* 20: 1885–1891.
16. Sancar A (1996) DNA excision repair. *Annu. Rev. Biochem.* 65: 43–81.
17. Wood RD (1997) Nucleotide excision repair in mammalian cells. *J. Biol. Chem.* 272: 23465–23468.
18. Huang JC, Svoboda D, Reardon JT and Sancar (1992) Human nucleotide excision nuclease removes thymine dimers from DNA by incising the 22nd phosphodiester bond 5′ and the 6th phosphodiester bond 3′ to the photodimer. *Proc. Natl. Acad. Sci. USA* 89: 3664–3668.
19. De Boer J and Hoeijmakers JH (2000) Nucleotide excision repair and human syndromes. *Carcinogenesis* 21: 453–460.
20. Friedberg EC (2001) How nucleotide excision repair protects against cancer. *Nat. Rev. Cancer* 1: 22–33.
21. Lin JJ and Sancar A (1992) (A)BC excinuclease: the *Escherichia coli* nucleotide excision repair enzyme. *Mol. Microbiol.* 6: 2219–2224.
22. Araujo SJ, Tirode F, Coin F, Pospiech H, Syväoja JE, Stucki M, Hübscher U, Egly JM and Wood RD (2000) Nucleotide excision repair of DNA with recombinant human proteins: definition of the minimal set of factors, active forms of TFIIH, and modulation by CAK. *Genes Dev.* 14: 349–359.
23. Mu D, Park CH, Matsunaga T, Hsu DS, Reardon JT and Sancar A (1995) Reconstitution of human DNA repair excision nuclease in a highly defined system. *J. Biol. Chem.* 270: 2415–2418.
24. Batty D, Rapic'-Otrin V, Levine AS and Wood RD (2000) Stable binding of human XPC complex to irradiated DNA confers strong discrimination for damaged sites. *J. Mol. Biol.* 300: 275–290.
25. Sugasawa K, Ng JM, Masutani C, Iwai S, van der Spek PJ, Eker AP, Hanaoka F, Bootsma D and Hoeijmakers JH (1998) Xeroderma pigmentosum group C protein complex is the initiator of global genome nucleotide excision repair. *Mol. Cell* 2: 223–232.
26. Volker M, Moné MJ, Karmakar P, van Hoffen A, Schul W, Vermeulen W, Hoeijmakers JH, van Driel R, van Zeeland AA and Mullenders LH (2001)

Sequential assembly of the nucleotide excision repair factors *in vivo*. *Mol. Cell* 8: 213–224.

27. Yokoi M, Masutani C, Maekawa T, Sugasawa K, Ohkuma Y and Hanaoka F (2000) The xeroderma pigmentosum group C protein complex XPC-HR23B plays an important role in the recruitment of transcription factor IIH to damaged DNA. *J. Biol. Chem.* 275: 9870–9875.
28. Tang JY, Hwang BJ, Ford JM, Hanawalt PC and Chu J (2000) Xeroderma pigmentosum p48 gene enhances genomic repair and suppresses UV-induced mutagenesis. *Mol. Cell* 5: 737–744.
29. Missura M, Buterin T, Hindges R, Hübscher U, Kaspárková J, Brabec V and Naegeli H (2001) Double-check probing of DNA bending and unwinding by XPA-RPA: an architectural function in DNA repair. *EMBO J.* 20: 3554–3564.
30. De Laat WL, Appeldoorn E, Sugasawa K, Weterings E, Jaspers NG and Hoeijmakers JH (1998) DNA binding polarity of human replication protein A positions nucleases in nucleotide excision repair. *Genes Dev.* 12: 2598–2609.
31. Reardon JT and Sancar A (2002) Molecular anatomy of the human excision nuclease assembled at sites of DNA damage. *Mol. Cell. Biol.* 22: 5938–5945.
32. Schultz P, Fribourg S, Poterszman A, Mallouh V, Moras D and Egly JM (2000) Molecular structure of human TFIIH. *Cell* 102: 599–607.
33. Evans E, Moggs JG, Hwang JR, Egly JM and Wood RD (1997) Mechanism of open complex and dual incision formation by human nucleotide excision repair factors. *EMBO J.* 16: 6559–6573.
34. Mu D, Wakasugi M, Hsu DS and Sancar A (1997) Characterization of reaction intermediates of human excision repair nuclease. *J. Biol. Chem.* 272: 121971–121979.
35. Jawhari A, Laine JP, Dubaele S, Lamour V, Poterszman A, Coin F, Moras D and Egly JM (2002) P52 mediates XPB function within the transcription/repair factor TFIIH. *J. Biol. Chem.* 277: 31761–31767.
36. You JS, Wang M and Lee SH (2003) Biochemical analysis of the damage recognition process in nucleotide excision repair. *J. Biol. Chem.* 278: 7476–7485.
37. Drapkin R and Reinberg D (1994) The multifunctional TFIIH complex and transcriptional control. *Trends Biochem. Sci.* 19: 504–508.
38. Shivji MK, Podust VN, Hübscher U and Wood RD (1995) Nucleotide excision repair DNA synthesis by DNA polymerase epsilon in the presence of PCNA, RFC, and RPA. *Biochemistry* 34: 5011–5017.

39. Bohr VA, Smith CA, Okumoto DS and Hanawalt PC (1985) DNA repair in an active gene: removal of pyrimidine dimers from the DHFR gene of CHO cells is much more efficient than in the genome overall. *Cell* 40: 359–369.
40. Ford JM, Lommel L and Hanawalt PC (1994) Preferential repair of ultraviolet light-induced DNA damage in the transcribed strand of the human *p53* gene. *Mol. Carcinog.* 10: 105–109.
41. Tornaletti S and Pfeifer GP (1994) Slow repair of pyrimidine dimers at p53 mutation hotspots in skin cancer. *Science* 263: 1436–1438.
42. Mellon I, Spivak G and Hanawalt PC (1987) Selective removal of transcription-blocking DNA damage from the transcribed strand of the mammalian DHFR gene. *Cell* 51: 241–249.
43. Chen RH, Maher VM and McCormick JJ (1990) Effect of excision repair by diploid human fibroblasts on the kinds and locations of mutations induced by (±)-7β,8α-dihydroxy-9α,10α-epoxy-7,8,9,10-tetrahydrobenzo[*a*]pyrene in the coding region of the *HPRT* gene. *Proc. Natl. Acad. Sci. USA* 87: 8680–8684.
44. Denissenko MF, Pao A, Pfeifer GP and Tang MS (1998) Slow repair of bulky DNA adducts along the nontranscribed strand of the human *p53* gene may explain the strand bias of transversion mutations in cancers. *Oncogene* 16: 1241–1247.
45. Tang MS, Pao A and Zhang XS (1994) Repair of benzo[*a*]pyrene diol epoxide- and UV-induced DNA damage in dihydrofolate reductase and adenine phosphoribosyltransferase genes of CHO cells. *J. Biol. Chem.* 269: 12749–12754.
46. Carothers AM, Zhen W, Mucha J, Zhang YJ, Santella RM, Grunberger D and Bohr VA (1992) DNA strand-specific repair of (±)-3α,4β-dihydroxy-1α,2α-epoxy-1,2,3,4-tetrahydrobenzo[*c*]phenanthrene adducts in the hamster dihydrofolate reductase gene. *Proc. Natl. Acad. Sci. USA* 89: 11925–11929.
47. Venema J, van Hoffen A, Karcagi V, Natarajan AT, van Zeeland AA and Mullenders LH (1991) Xeroderma pigmentosum complementation group C cells remove pyrimidine dimers selectively from the transcribed strand of active genes. *Mol. Cell. Biol.* 11: 4128–4134.
48. Gowen LC, Avrutskaya AV, Latour AM, Koller BH and Leadon SA (1998) BRCA1 required for transcription-coupled repair of oxidative DNA damage. *Science* 281: 1009–1012. (retracted)
49. Le Page F, Kwoh EE, Avrutskaya A, Gentil A, Leadon SA, Sarasin A and Cooper PK (2000) Transcription-coupled repair of 8-oxoguanine:

requirement for XPG, TFIIH, and CSB and implications for Cockayne syndrome. *Cell* 101: 159–171.

50. Mellon I, Fajpal DK and Koi M (1996) Transcription-coupled repair deficiency and mutations in human mismatch repair genes. *Science* 272: 557–560.
51. Hasty P, Campisi J, Hoeijmakers JH, van Steeg H and Vijg J (2002) Aging and genome maintenance: lessons from the mouse? *Science* 299: 1355–1359.
52. Chary P, Latham GJ, Robberson DL, Kim SJ, Han S, Harris CM, Harris TM and Lloyd RS (1995) *In vivo* and *in vitro* replication consequences of stereoisomeric benzo[*a*]pyrene-7,8-dihydrodiol 9,10-epoxide adducts on adenine N^6 at the second position of N-*ras* codon 61. *J. Biol. Chem.* 10: 4990–5000.
53. Moriya M, Spiegel S, Fernandes A, Amin S, Liu T, Geacintov NE and Grollman AP (1996) Fidelity of translesional synthesis past benzo[*a*]pyrene diol epoxide-2′-deoxyguanosine DNA adducts: marked effects of host cell, sequence context, and chirality. *Biochemistry* 35: 16646–16651.
54. Seo KY, Jelinsky SA and Loechler EL (2000) Factors that influence the mutagenic patterns of DNA adducts from chemical carcinogens. *Mutat. Res.* 463: 215–246.
55. Wei SJ, Chang RL, Hennig E, Cui XX, Merkler KA, Wong CQ, Yagi H, Jerina DM and Conney AH (1994) Mutagenic selectivity at the *HPRT* locus in V-79 cells: comparison of mutations caused by bay-region benzo[*a*]pyrene 7,8-diol-9,10-epoxide enantiomers with high and low carcinogenic activity. *Carcinogenesis* 15: 1729–1735.
56. Wei D, Maher VM and McCormick JJ (1995) Site-specific rates of excision repair of benzo[*a*]pyrene diol epoxide adducts in the hypoxanthine phosphoribosyltransferase gene of human fibroblasts: correlation with mutation spectra. *Proc. Natl. Acad. Sci. USA* 92: 2204–2208.
57. Friedberg EC, Walker GC and Siede W (1995) *DNA Repair and Mutagenesis.* ASM Press, Washington D.C.
58. Kraemer KH, Lee MM, Andrews AD and Lambert WC (1994) The role of sunlight and DNA repair in melanoma and nonmelanoma skin cancer. *Arch. Dermatol.* 130: 1018–1021.
59. De Vries A, van Oostrom CT, Hofhuis FM, Dortant PM, Berg RJ, de Gruijl FR, Wester PW, van Kreijl CF, Capel PJ, van Steeg H and Verbeek SJ (1995) Increased susceptibility to ultraviolet-B and carcinogens of mice lacking the DNA excision repair gene *XPA*. *Nature* 377: 169–173.

60. De Vries A, van Oostrom CT, Dortant PM, Beems RB, van Kreijl CF, Capel PJ and van Steeg H (1997) Spontaneous liver tumors and benzo[*a*]pyrene-induced lymphomas in *XPA*-deficient mice. *Mol. Carcinog.* 19: 46–53.
61. Nakane H, Takeuchi S, Yuba S, Saijo M, Nakatsu Y, Murai H, Nakatsuru Y, Ishikawa T, Hirota S, Kitamura Y, Kato Y, Tsunodo Y, Miyauchi H, Horio T, Tokunaja T, Matsunaga T, Nikaido O, Nishimune Y, Okada Y and Tanaka K (1995) High incidence of ultraviolet-B- or chemical-carcinogen-induced skin tumors in mice lacking the xeroderma pigmentosum group A gene. *Nature* 377: 165–168.
62. Adimoolam S and Ford JM (2002) P53 and DNA damage-inducible expression of the xeroderma pigmentosum group C gene. *Proc. Natl. Acad. Sci. USA* 99: 12985–12990.
63. Lloyd DR and Hanawalt PC (2002) P53 controls global nucleotide excision repair of low levels of structurally diverse benzo[*g*]chrysene-DNA adducts in human fibroblasts. *Cancer Res.* 62: 5288–5294.
64. Ford JM and Hanawalt PC (1995) Li-Fraumeni syndrome fibroblasts homozygous for p53 mutations are deficient in global DNA repair but exhibit normal transcription-coupled repair and enhanced UV resistance. *Proc. Natl. Acad. Sci. USA* 92: 8876–8880.
65. Ford JM and Hanawalt PC (1999) Expression of p48 xeroderma pigmentosum gene is p53-dependent and is involved in global genomic repair. *Proc. Natl. Acad. Sci. USA* 96: 424–428.
66. Hess MT, Schwitter U, Petretta M, Giese B and Naegeli H (1996) Site-specific DNA substrates for human excision repair: comparison between deoxyribose and base adducts. *Chem. Biol.* 3: 121–128.
67. Custer L, Zajc B, Sayer JM, Cullinane C, Phillips DR, Cheh AM, Jerina DM, Bohr VA and Mazur SJ (1999) Stereospecific differences in repair by human cell extracts of synthesized oligonucleotides containing *trans*-opened 7,8,9,10-tetrahydrobenzo[*a*]pyrene 7,8-diol 9,10-epoxide N^2-dG adduct stereoisomers located within the human K-*ras* codon 12 sequence. *Biochemistry* 38: 569–581.
68. Buterin T, Hess MT, Luneva N, Geacintov NE, Amin S, Kroth H, Seidel A and Naegeli H (2000) Unrepaired fjord region polycyclic aromatic hydrocarbon-DNA adducts in *ras* codon 61 mutational hot spots. *Cancer Res.* 60: 1849–1856.
69. Hess MT, Gunz D, Luneva N, Geacintov NE and Naegeli H (1997) Base pair conformation-dependent excision of benzo[*a*]pyrene diol epoxide-guanine adducts by human nucleotide excision repair enzymes. *Mol. Cell. Biol.* 17: 7069–7076.

70. Thakker DR, Yagi H, Akagi H, Koreeda M, Lu AY, Levin W, Wood AW, Conney AH and Jerina DM (1977) Metabolism of benzo[*a*]pyrene. VI. Stereoselective metabolism of benzo[*a*]pyrene and benzo[*a*]pyrene 7,8-dihydrodiol to diol epoxides. *Chem.-Biol. Interact.* 16: 281–300.
71. Buening MK, Wislocki PG, Levin W, Yagi H, Thakker DR, Akagi H, Koreeda M, Jerina DM and Conney AH (1978) Tumorigenicity of the optical enantiomers of the diastereomeric benzo[*a*]pyrene 7,8-diol-9,10-epoxides in newborn mice: exceptional activity of (+)-7β,8α-dihydroxy-9α,10α-epoxy-7,8,9,10-tetrahydrobenzo[*a*]pyrene. *Proc. Natl. Acad. Sci. USA* 75: 5358–5361.
72. Slaga TJ, Bracken WJ, Gleason G, Levin W, Yagi H, Jerina DM and Conney AH (1979) Marked differences in the skin tumor-initiating activities of the optical enantiomers of the diastereomeric benzo[*a*]pyrene 7,8-diol-9,10-epoxides. *Cancer Res.* 39:67–71.
73. Cheng SC, Hilton BD, Roman JM and Dipple A (1989) DNA adducts from carcinogenic and noncarcinogenic enantiomers of benzo[*a*]pyrene dihydrodiol epoxide. *Chem. Res. Toxicol.* 2, 334–340.
74. Jerina DM, Chadha A, Cheh AM, Schurdak ME, Wood AW and Sayer JM (1991) Covalent bonding of bay-region diol epoxides to nucleic acids. *Adv. Exp. Med. Biol.* 283: 533–553.
75. Cosman M, de los Santos C, Fiala R, Hingerty BE, Singh SB, Ibanez V, Margulis LA, Live D, Geacintov NE, Broyde S and Patel DJ (1992) Solution conformation of the major adduct between the carcinogen (+)-*anti*-benzo-[*a*]pyrene diol epoxide and DNA. *Proc. Natl. Acad. Sci. USA* 89: 1914–1918.
76. De los Santos C, Cosman M, Hingerty BE, Ibanez V, Margulis LA, Geacintov NE, Broyde S and Patel DJ (1992) Influence of benzo[*a*]pyrene diol epoxide chirality on solution conformations of DNA covalent adducts: the (−)-*trans-anti*-[BP]G·C adduct structure and comparison with the (+)-*trans-anti*-[BP]G·C enantiomer. *Biochemistry* 31: 9850–9863.
77. Cosman M, de los Santos C, Fiala R, Hingerty BE, Ibanez V, Luna E, Harvey RG, Geacintov NE, Broyde S and Patel DJ (1993) Solution conformation of the (+)-*cis-anti*-[BP]dG adduct in a DNA duplex: intercalation of the covalently attached benzo[*a*]pyrenyl ring into the helix and displacement of the modified deoxyguanosine. *Biochemistry* 32: 4145–4155.
78. Cosman M, Hingerty BE, Luneva N, Amin S, Geacintov NE, Broyde S and Patel DJ (1996) Solution conformation of the (−)-*cis-anti*-benzo[*a*]pyrenyl-dG adduct opposite dC in a DNA duplex: intercalation of the covalently attached BP ring into the helix with base displacement of the modified deoxyguanosine into the major groove. *Biochemistry* 35: 9850–9863.

79. Wu M, Yan SX, Patel DJ, Geacintov NE and Broyde S (2002) Relating repair susceptibility of carcinogen-damaged DNA with structural distortion and thermodynamic stability. *Nucleic Acids Res.* 30: 3422–3432.
80. Liu T, Xu J, Tsao H, Li B, Xu R, Yang C, Amin S, Moriya M and Geacintov NE (1996) Base sequence-dependent bends in site-specific benzo[*a*]pyrene diol epoxide-modified oligonucleotide duplexes. *Chem. Res. Toxicol.* 9: 255–261.
81. Tsao H, Mao B, Zhuang P, Xu R, Amin S and Geacintov NE (1998) Sequence dependence and characteristics of bends induced by site-specific polynuclear aromatic carcinogen-deoxyguanosine lesions in oligonucleotides. *Biochemistry* 37: 4993–5000.
82. Geacintov NE, Cosman M, Hingerty BE, Amin S, Broyde S and Patel DJ (1997) NMR solution structures of stereoisomeric covalent polycyclic aromatic carcinogen-DNA adducts: principles, patterns, and diversity. *Chem. Res. Toxicol.* 10: 111–146.
83. Zou Y, Liu T, Geacintov NE and van Houten B (1995) Interaction of the UvrABC nuclease with a DNA duplex containing a single stereoisomer of dG-(+)- or dG-(−)-*anti*-BPDE. *Biochemistry* 34: 13582–13593.
84. Hecht SS, El-Bayoumy K, Rivenson A and Amin S (1994) Potent mammary carcinogenicity in female CD rats of a fjord region diol-epoxide of benzo[*c*]phenanthrene compared to a bay region diol-epoxide of benzo[*a*]pyrene. *Cancer Res.* 54: 21–24.
85. Levin W, Chang RL, Wood AW, Thakker DR, Yagi H, Jerina DM and Conney AH (1986) Tumorigenicity of optical isomers of the diastereomeric bay-region 3,4-diol-1,2-epoxides of benzo[*c*]phenanthrene in murine tumor models. *Cancer Res.* 46: 2257–2261.
86. Amin S, Krzeminski J, Rivenson A, Kurtzke C, Hecht SS and El-Bayoumy K (1995) Mammary carcinogenicity in female CD rats of fjord region diol epoxides of benzo[*c*]phenanthrene, benzo[*g*]chrysene, benzo[*a*]pyrene and dibenzo[*a,l*]pyrene. *Carcinogenesis* 16: 1971–1974.
87. Cavalieri EL, Higginbotham S, RamaKrishna NV, Devanesan PD, Todorovic R, Rogan EG and Salmasi S (1991) Comparative dose-tumorigenicity studies of dibenzo[*a,l*]pyrene versus 7,12-dimethylbenz[*a*]anthracene, benzo[*a*]pyrene and two dibenzo[*a,l*]pyrene dihydrodiols in mouse skin and rat mammary gland. *Carcinogenesis* 12: 1939–1944.
88. Higginbotham S, RamaKrishna NV, Johansson SL, Rogan EG and Cavalieri EL (1993) Tumor initiating activity and carcinogenicity of dibenzo[*a,l*]pyrene versus 7,12-dimethylbenz[*a*]anthracene and benzo[*a*]-pyrene at low doses in mouse skin. *Carcinogenesis* 14: 875–878.

89. Barbacid M (1987) *Ras* genes. *Annu. Rev. Biochem.* 56: 779–827.
90. Bos JL (1995) P21ras: an oncoprotein functioning in growth factor-induced signal transduction. *Eur. J. Cancer* 31A: 1051–1054.
91. Lacal JC (1997) Regulation of proliferation and apoptosis by Ras and Rho GTPases through specific phospholipid-dependent signaling. *FEBS Lett.* 410: 73–77.
92. Bos JL (1989) *Ras* oncogenes in human cancer: a review. *Cancer Res.* 49: 4682–4689.
93. Gao HG, Chen JK, Stewart J, Song B, Rayappa C, Whong WZ and Ong T (1997) Distribution of *p53* and K-*ras* mutations in human lung cancer tissues. *Carcinogenesis* 18: 473–478.
94. Zhu D, Keohavong P, Finkelstein SD, Swalsky P, Bakker A, Weissfeld J, Srivastava S and Whiteside TL (1997) K-*ras* gene mutations in normal colorectal tissues from K-*ras* mutation-positive colorectal cancer patients. *Cancer Res.* 57: 2485–2492.
95. Marshall CJ, Vousden KH and Phillips DH (1984) Activation of c-Ha-*ras*-1 proto-oncogene by *in vitro* modification with a chemical carcinogen, benzo[*a*]pyrene diol-epoxide. *Nature* 310: 586–589.
96. Vousden KH, Bos JL, Marshall CJ and Phillips DH (1986) Mutations activating human c-Ha-*ras1* protooncogene (*HRAS1*) induced by chemical carcinogens and depurination. *Proc. Natl. Acad. Sci. USA* 83: 1222–1226.
97. Balmain A and Pragnell IB (1983) Mouse skin carcinomas induced *in vivo* by chemical carcinogens have a transforming Harvey-*ras* oncogene. *Nature* 303: 72–74.
98. Dipple A, Pigott MA, Agarwal SK, Yagi H, Sayer JM and Jerina DM (1987) Optically active benzo[*c*]phenanthrene diol epoxides bind extensively to adenine in DNA. *Nature* 327: 535–536.
99. Quintanilla M, Brown K, Ramsden M and Balmain A (1986) Carcinogen-specific mutation and amplification of Ha-*ras* during mouse skin carcinogenesis. *Nature* 322: 78–80.
100. Zegar IS, Kim SJ, Johansen TN, Horton PJ, Harris CM, Harris T and Stone MP (1996) Adduction of the human N-*ras* codon 61 sequence with (−)-(7*S*,8*R*,9*R*,10*S*)-7.8-dihydroxy-9,10-epoxy-7,8,9,10-tetrahydrobenzo[*a*]pyrene: structural refinement of the intercalated SRSR(61,2) (−)-(7*S*,8*R*,9*R*,10*S*)-N^6-[10-(7,8,9,10-tetrahydrobenzo[*a*]pyrenyl)]-2′-deoxyadenosyl adduct from ^{1}H NMR. *Biochemistry* 35: 6212–6224.
101. Geacintov NE, Broyde S, Buterin T, Naegeli H, Wu M, Yan S and Patel DJ (2002) Thermodynamics and structural factors in the removal of bulky

DNA adducts by the nucleotide excision repair machinery. *Biopolymers* 65: 202–210.

102. Xu R, Mao B, Xu J, Li B, Birke S, Swenberg CE and Geacintov NE (1995) Stereochemistry-dependent bending in oligonucleotide duplexes induced by site-specific covalent benzo[*a*]pyrene diol epoxide-guanine lesions. *Nucleic Acids Res.* 23: 2314–2319.
103. Cosman M, Fiala R, Hingerty BE, Laryea A, Lee H, Harvey RG, Amin S, Geacintov NE, Broyde S and Patel DJ (1993) Solution conformation of the (+)-*trans-anti*-[BPh]dA adduct opposite dT in a DNA duplex: intercalation of the covalently attached benzo[*c*]phenanthrene to the 5′-side of the adduct site without disruption of the modified base pair. *Biochemistry* 32: 12488–12497.
104. Cosman M, Laryea A, Fiala R, Hingerty BE, Amin S, Geacintov NE, Broyde S and Patel DJ (1995) Solution conformation of the (−)-*trans-anti*-benzo[*c*]phenanthrene-dA ([BPh]dA) adduct opposite dT in a DNA duplex: intercalation of the covalently attached benzo[*c*]phenanthrenyl ring to the 3′-side of the adduct site and comparison with the (+)-*trans-anti*-[BPh]dA opposite dT stereoisomer. *Biochemistry* 34: 1295–1307.
105. Krzeminski J, Ni JS, Zhuang P, Luneva N, Amin S and Geacintov NE (1999) Total synthesis, mass spectrometric sequencing, and stabilities of oligonucleotide duplexes with single *trans-anti*-BPDE-N6-dA lesions in the N-*ras* codon 61 and other sequence contexts. *Polycycl. Aromat. Compds.* 17: 1–10.
106. Shurter EJ, Yeh HJ, Sayer JM, Lakshman MK, Yagi H, Jerina DM and Gorenstein DG (1995) NMR solution structure of an nonanucleotide duplex with a dG mismatch opposite a 10*R* adduct derived from *trans* addition of a deoxyadenosine N^6-amino group to (−)-(7*S*,8*R*,9*R*,10*S*)-7,8-dihydroxy-9,10-epoxy-7,8,9,10-tetrahydrobenzo[*a*]pyrene. *Biochemistry* 34: 1364–1375.
107. Laryea A, Cosman M, Lin JM, Liu T, Agarwal R, Smirnov S, Amin S, Harvey RG, Dipple A and Geacintov NE (1995) Direct synthesis and characterization of site-specific adenosyl adducts derived from the binding of a 3,4-dihydroxy-1,2-epoxybenzo[*c*]phenanthrene stereoisomer to an 11-mer oligodeoxyribonucleotide. *Chem. Res. Toxicol.* 8: 444–454.
108. Ruan Q, Kolbanovskiy A, Zhuang P, Chen J, Krzeminski J, Amin S and Geacintov NE (2002) Synthesis and characterization of site-specific and stereoisomeric fjord dibenzo[*a,l*]pyrene diol epoxide-N^6-adenine adducts: unusual thermal stabilization of modified DNA duplexes. *Chem. Res. Toxicol.* 15: 249–261.

109. Feng Z, Hu W, Chen JX, Pao A, Li H, Rom W, Hung MC and Tang MS (2002) Preferential DNA damage and poor repair determine *ras* gene mutational hotspot in human cancer. *J. Natl. Cancer Inst.* 94: 1527–1536.
110. Buschta-Hedayat N, Buterin T, Hess MT, Missura M and Naegeli H (1999) Recognition of nonhybridizing base pairs during nucleotide excision repair of DNA. *Proc. Natl. Acad. Sci. USA* 96: 6090–6095.
111. Hess MT, Schwitter U, Petretta M, Giese B and Naegeli H (1997) Bipartite substrate discrimination by human nucleotide excision repair enzymes. *Proc. Natl. Acad. Sci. USA* 94: 6664–6669.
112. Razin A (1998) CpG methylation, chromatin structure and gene silencing — a three-way connection. *EMBO J.* 17: 4905–4908.
113. Denissenko MF, Chen JX, Tang MS and Pfeifer GP (1997) Cytosine methylation determines hot spots of DNA damage in the human *p53* gene. *Proc. Natl. Acad. Sci. USA* 94: 3893–3898.
114. Weisenberger DJ and Romano LJ (1999) Cytosine methylation in a CpG sequence leads to enhanced reactivity with benzo[*a*]pyrene diol epoxide that correlates with a conformational change. *J. Biol. Chem.* 274: 23948–23955.
115. Muheim R, Buterin T, Colgate KC, Kolbanovskiy A, Geacintov NE and Naegeli H (2003) Modulation of human nucleotide excision repair by 5-methylcytosines. *Biochemistry* 42: 3247–3254.
116. Huang X, Colgate KC, Kolbanovskiy A, Amin S and Geacintov NE (2002) Conformational changes of a benzo[*a*]pyrene diol epoxide-N(2)-dG adduct induced by a 5′-flanking 5-methyl-substituted cytosine in a (Me)CG double-stranded oligonucleotide sequence context. *Chem. Res. Toxicol.* 15: 438–444.
117. Buterin T, Hess MT, Gunz D, Geacintov NE, Mullenders LH and Naegeli H (2002) Trapping of DNA nucleotide excision repair factors by nonrepairable carcinogen adducts. *Cancer Res.* 62: 4229–4235.
118. Cosman M, Fiala R, Hingerty BE, Amin S, Geacintov NE, Broyde S and Patel DJ (1994) Solution conformation of the (+)-*trans-anti*-[BP]dG adduct opposite a deletion site in a DNA duplex: intercalation of the covalently attached benzo[*a*]pyrene into the helix with base displacement of the modified deoxyguanosine into the major groove. *Biochemistry* 33: 11507–11517.
119. Cosman M, Fiala R, Hingerty BE, Amin S, Geacintov NE, Broyde S and Patel DJ (1994) Solution conformation of the (+)-*cis-anti*-[BP]dG adduct opposite a deletion site in a DNA duplex: intercalation of the covalently attached benzo[*a*]pyrene into the helix with base displacement of the modified deoxyguanosine into the minor groove. *Biochemistry* 33: 11518–11527.

120. McLeod MC, Powell KL and Tran N (1995) Binding of the transcription factor Sp1 to non-target sites in DNA modified by benzo[*a*]pyrene diol epoxide. *Carcinogenesis* 16: 975–983.
121. Suzuki N, Ohashi E, Kolbanovskiy A, Geacintov NE, Grollman AP, Ohmori H and Shibutani S (2002) Translesion synthesis by human DNA polymerase κ on a DNA template containing a single stereoisomer of dG-(+)- or dG-(−)-*anti*-N^2-BPDE (7,8-dihydroxy-*anti*-9,10-epoxy-7,8,9,10-tetrahydrobenzo[*a*]pyrene). *Biochemistry* 41: 6100–6106.
122. Huang X, Kolbanovskiy A, Wu X, Zhang Y, Wang Z, Zhuang P, Amin S and Geacintov NE (2003) Effects of base sequence context on translesion synthesis past a bulky (+)-*trans-anti*-B[*a*]P-N^2-dG lesion catalyzed by the Y-family polymerase pol kappa. *Biochemistry* 42: 2456–2466.
123. Fernandes A, Liu T, Amin S, Geacintov NE, Grollman AP and Moriya M (1998) Mutagenic potential of stereoisomeric bay region (+)- and (−)-*cis-anti*-benzo[*a*]pyrene diol epoxide-N^2-2′-deoxyguanosine adducts in *Escherichia coli* and simian kidney cells. *Biochemistry* 37: 10164–10172.
124. Yang JL, Maher VM and McCormick JJ (1987) Kinds of mutation formed when a shuttle vector containing adducts of (+)-7β,8α-dihydroxy-9α,10α-epoxy-7,8,9,10-tetrahydrobenzo[*a*]pyrene replicates in human cells. *Proc. Natl. Acad. Sci. USA* 84: 3787–3791.
125. Zhu Y, Bye S, Stambrook PJ and Tischfield JA (1994) Single-base deletion induced by benzo[*a*]pyrene diol epoxide at the adenine phosphoribosyltransferase locus in human fibrosarcoma cell lines. *Mutat. Res.* 321: 73–79.
126. Van Houten B, Illenye S, Yen Q and Farrell N (1993) Homodinuclear (Pt-Pt) and heterodinuclear (Ru-Pt) complexes as protein-DNA cross-linking agents: potential suicide DNA lesions. *Biochemistry* 32: 11794–11801.
127. Talaska G, Jaeger M, Reilman R, Collins T and Warshawsky D (1996) Chronic, topical exposure to benzo[*a*]pyrene induces relatively high steady-state levels of DNA adducts in target tissues and alters kinetics of adduct loss. *Proc. Natl. Acad. Sci. USA* 93: 7789–7793.
128. Naegeli H, Bardwell L and Friedberg EC (1992) The DNA helicase and adenosine triphosphatase activities of yeast Rad3 protein are inhibited by DNA damage. *J. Biol. Chem.* 267: 392–398.
129. Villani G and Le Gac NT (2000) Interactions of DNA helicases with damaged DNA: possible biological consequences. *J. Biol. Chem.* 275: 33185–33188.
130. Blow JJ (1993) Preventing re-replication of DNA in a single cell cycle: evidence for a replication licensing factor. *J. Cell Biol.* 122: 993–1002.

7 Aberrant Gene Expression and Cell Signalling/Epigenetic Effects Induced by Polycyclic Aromatic Hydrocarbons

Pablo Steinberg

Institute of Nutritional Science, University of Potsdam,
Bergholz-Rehbrücke, Germany
E-mail: steinber@rz.uni-potsdam.de

7.1 Introduction

During the last twenty years it has become evident that polycyclic aromatic hydrocarbons (PAHs) can lead to malignant transformation of cells in mammary gland, skin and/or lung, and that this transformation process is accompanied by a variety of genetic alterations including protooncogene activation and tumor suppressor gene inactivation. However, evidence is also accumulating that PAHs might contribute to tumor formation by epigenetic mechanisms (i.e., events not involving the induction of mutations leading to gene activation or inactivation). Furthermore, PAHs have been recently implicated in the development of atherosclerosis and in the induction of apoptosis (programmed cell death) in certain

conclusion that in general PAHs forming DNA adducts with deoxyguanosine (B[*a*]P, B[*b*]F, 5-MeC, CP[*c,d*]P and benz[*j*]aceanthrylene) lead to mutations in codon 12 of Ki-*ras*, while those reacting with deoxyguanosine as well as deoxyadenosine (aceanthrylene and DB[*a,l*]P) induce mutations in codons 12 and 61 of Ki-*ras* in mouse lung tissue. Dibenz[*a,h*]anthracene was the only PAH tested in strain A/J mice that formed adducts with deoxyguanosine but did not induce mutations in Ki-*ras*.[22] Wessner *et al.*[23] reported that in mouse lung tumors arising after transplacental exposure to 3-methylcholanthrene, which is known to bind to deoxyguanosine,[24] 13 out of 16 tumors exhibited point mutations in Ki-*ras*, including seven tumors with a GGT → GTT and four tumors with a GGT → TGT transversion at codon 12 as well as two tumors with a GGC → CGC transversion at codon 13.

The fact that PAHs preferentially react with certain bases in DNA, thereby inducing specific mutations, also applies to the H-*ras* gene. Bay-region diol-epoxides of B[*a*]P preferentially bind to guanine residues in DNA[25,26] and activate the mouse H-*ras* gene mainly by G → T transversions in codons 12 (GGA) and 13 (GGC).[27] Conversely, fjord-region diol-epoxides of DMBA, 7,14-dimethyldibenz[*a,j*]anthracene and B[*c*]Ph consistently induce A → T transversions in codon 61 (CAA) of the mouse H-*ras* gene.[8,10,28,29] Similarly, benzo[*c*]chrysene 9,10-diol-11,12-epoxides bind to guanine and adenine residues in DNA and lead to G → T transversions at codons 12 and 13 as well as A → T transversions at codon 61 of the H-*ras* gene.[30]

A further factor to be considered is the species in which the tumor develops. Racemic *anti*-B[*c*]PhDE is a potent rat mammary carcinogen.[10] However, Ronai *et al.*[10] detected a CAA → CTA mutation in codon 61 of the H-*ras* gene in only one out of 42 rat mammary tumors analyzed; the analysis of 35 other mammary tumors did not reveal any mutations in codons 12 or 13 of the H-*ras* gene or in codons 12, 13 or 61 of the Ki-*ras* gene. In accordance with this observation it has consistently been reported that mutations in codon 61 of the H-*ras* gene are relatively rare events in DMBA-induced rat mammary tumors.[31–33] In contrast, H-*ras* activation through codon 61 mutations is common in mouse mammary tumors.[8,34]

7.2.2 Effects on *p53*

The tumor suppressor protein p53 is a nuclear phosphoprotein that plays a central role in at least three processes: cell-cycle arrest, DNA repair and apoptosis,[35–37] thereby being able to suppress the growth of cancer cells.[38,39] Consistent with its pivotal role in tumor suppression is the observation that mutations within the *p53* gene, which lead to an inactive p53 protein, are the most common alterations detected in human cancer cells.[40]

A number of studies[40–48] have shown that in tobacco-related human neoplasias, including small-cell and non-small-cell lung cancers, esophageal carcinomas and squamous cell carcinomas of the head and neck, a much higher frequency of G → T transversions among *p53* mutations are observed in lung tumors as compared to non-lung tumors, with the exception of hepatocellular carcinomas. It has been suggested that the frequent *p53* G → T transversions in tobacco-related human tumors are related to PAHs such as B[*a*]P present in tobacco smoke (cf. Chapter 5). In accordance with this hypothesis, Puisieux *et al.*[49] used DNA polymerase fingerprint analysis to show that B[*a*]P causes a high frequency of G → T transversions in the wild-type *p53* gene. Denissenko *et al.*[50] incubated normal human bronchial epithelial cells with *anti*-B[*a*]PDE and analyzed the distribution of DNA adducts along the exons 5, 7 and 8 of the *p53* gene. The authors showed that a strong and selective adduct formation occurred at guanine positions in codons 157, 248 and 273 of the *p53* gene. Since these positions are the major mutational hotspots in human lung cancer, Denissenko *et al.*[50] concluded that targeted adduct formation rather than the selection of pre-existing endogenous mutations determines the *p53* mutational spectrum in lung cancer. It should be noted that two recent reports[51,52] questioned the differences in *p53* mutation spectra between smokers and non-smokers. Using the latest update of the *p53* mutation database of the *International Agency for Research on Cancer* and taking into account recent data on non-smokers, Hainaut and Pfeifer[53] reinforced their view that *p53* mutation spectra are indeed different in lung tumors from smokers and non-smokers, and that *p53* mutations in lung cancer can be attributed to direct DNA damage from carcinogenic cigarette smoke constituents (cf. Chapter 5).

Ruggeri *et al.*[54] detected *p53* mutations in 7 out of 20 skin tumors induced in mice by topical application of B[*a*]P, whereby 5 of the 7 missense mutations were G $\rightarrow$ T transversions. In the same study, 8 out of 36 skin tumors induced by DMBA carried *p53* mutations, but a G $\rightarrow$ T transversion was only observed in 1 out of 8 tumors. Furthermore, alterations in the *p53* gene seem to be a late event in skin tumorigenesis induced by PAHs; *p53* mutations predominate in carcinomas, but are not detected in papillomas and hyperplastic lesions.[54–57]

Very little information is available on the type and/or role of *p53* mutations induced by PAHs in tissues other than lung and skin. Cherpillod and Amstad[58] showed that in the human hepatocellular carcinoma cell line HepG2 B[*a*]P preferentially induces a G $\rightarrow$ T transversion in the second and third position of codon 248 and a C $\rightarrow$ A transversion in the first position of codon 248 within the *p53* gene. In rat esophageal epithelial cells transformed *in vitro* by B[*a*]P-7,8-dihydrodiol a frameshift mutation in codon 176 of the *p53* gene was detected.[59] Only one out of 40 DMBA-induced rat mammary tumors, however, were found to carry a *p53* mutation (AGC $\rightarrow$ GGC, at codon 307).[31] Thus, the induction of *p53* mutations does not seem to be a prerequisite to induce mammary tumors in rats by PAHs.

Several research groups have shown that treatment of cultured cells[60–62] or mouse skin *in vivo*[63] with PAHs or the corresponding diol-epoxides results in the accumulation of DNA adducts and a concomitant short-term up-regulation of p53 protein expression. The enhanced p53 levels lead to a cell-cycle arrest at the G_1/S boundary and it has been hypothesized that the G_1 arrest is necessary to repair the damaged DNA before DNA replication occurs. At the molecular level it is p53 that transcriptionally activates the synthesis of the cyclin-dependent kinase inhibitor $p21^{WAF1/CIP1}$, which in turn prevents the phosphorylation of the retinoblastoma protein required for cells to enter the DNA synthesis phase of the cell-cycle. While Baird and co-workers[60,61] have shown that DB[*a,l*]P-11,12-diol-13,14-epoxide enantiomers lead to DNA adduct formation and G_1 arrest in human diploid fibroblasts, Dipple and colleagues[64–67] reported that MCF-7 cells treated with benzo[*g*]chrysene 11,12-diol-13,14-epoxides are delayed in the S phase and the induced p53 protein is not able to transcriptionally activate the $p21^{WAF1/CIP1}$ required for a G_1 arrest. This discrepancy could possibly result from using different cell types that may respond in a different manner to PAH-induced DNA damage.

7.2.3 Effects on Mitogen-Activated Protein Kinases

Eukaryotic cells contain various functionally equivalent proteins collectively referred to as mitogen-activated protein kinases, including the extracellular signal-regulated kinases 1 and 2, c-*jun* *N*-terminal kinases/stress-activated protein kinases and p38 kinases. All these proteins are serine/threonine kinases that are activated in the cytosol in response to specific extracellular signals and can be translocated to the nucleus, where they trigger specific cellular responses by regulating gene transcription (e.g., by phosphorylation of specific transcription factors).

If primary cultures of small airway epithelial cells, an *in vitro* model of normal human airway epithelium, were treated with 1 nM *anti*-B[*a*]PDE, the differentiation of the small airway epithelial cells into Clara cell-like, non-ciliated epithelial cells was inhibited without affecting their proliferation rate. This effect was in part mediated by the activation of a signal transduction pathway involving phosphatidylinositol-3-kinase as well as the extracellular signal-regulated kinases 1 and 2.[68] Based on these results it has been suggested that the inhibition of airway epithelial cell differentiation may represent an important early step in lung tumorigenesis induced by *anti*-B[*a*]PDE.[68] Treatment of HT29 (human colon carcinoma) cells with 10 nM B[*a*]P also led to an activation of extracellular signal-regulated kinases 1 and 2 as well as p38 kinases, whereas c-*jun* *N*-terminal kinases were not affected.[69] Furthermore, the authors could show that [^{3}H]-thymidine incorporation into DNA was increased under these conditions, an effect specifically due to the activation of the extracellular signal-regulated kinases 1 and 2.

7.2.4 Disruption of BRCA1 Expression

Familial breast cancer is characterized by the clustering of a high number of breast cancer cases in a single family and accounts for 5–10% of all breast tumors. A substantial number of familial breast cancer cases is linked to mutations in the tumor susceptibility genes *BRCA1* and *BRCA2*.[70,71] Both genes encode large proteins encompassing 1,863 and 3,359 amino acids, respectively. Neither gene bears significant homology with other known genes. The majority of *BRCA1* and *BRCA2* mutations identified so far lead to protein truncation and it is believed that cancer develops when the second copy of the corresponding gene is lost. Thus, *BRCA1* and *BRCA2* behave like classic tumor suppressor genes. Although the function of *BRCA1* is still under investigation, several lines of evidence indicate that the BRCA1 protein is localized in the nucleus and is involved in multiple cellular responses to DNA damage, such as transcription-coupled repair of oxidative DNA damage and recombinational repair of double strand breaks (cf. Chapter 6).[72–77]

About 90–95% of breast cancers are sporadic and occur in the absence of *BRCA1* and *BRCA2* mutations.[78,79] In this context it has been postulated that environmental contaminants such as PAHs could contribute to the induction of sporadic breast cancer by disrupting the expression of BRCA1 (cf. Chapter 9).[80] Exposure of MCF-7 cells to B[*a*]P led to a time- and concentration-dependent reduction of BRCA1 mRNA and protein levels.[81] Moreover, in B[*a*]P-treated MCF-7 cells accumulation of p53 was shown to occur before BRCA1 expression was downregulated.[82] Transient expression of mutant p53 in MCF-7 cells neutralized the B[*a*]P-mediated repression of BRCA1 expression.[83] Furthermore, the known ultimate carcinogenic metabolite of B[*a*]P, *anti*-B[*a*]PDE, proved to be much more potent than the parent compound in reducing BRCA1 levels in MCF-7 cells.[81,83] Taken together, these studies show that PAHs may disrupt BRCA1 expression by a p53-dependent pathway. BRCA1 downregulation could in turn lead to a reduced DNA repair capacity, thereby facilitating the selection (outgrowth) of cells (i.e., 'initiated cells') that carry cancer-prone mutations induced by PAHs.

7.2.5 Inhibition of Intercellular Communication

Cell-cell communication via gap junctions is a way by which neighboring cells exchange a number of molecules in order to control homeostasis, cell growth and differentiation.[84,85] It has been postulated that downregulation of gap junctional intercellular communication results in uncontrolled cell growth leading to the development of tumors.[86,87] This hypothesis is supported by various lines of evidence:

(i) most, if not all, tumor cells show dysfunctional gap junction intercellular communication;
(ii) tumor promoting agents reversibly inhibit gap junction intercellular communication;
(iii) oncogenes downregulate gap junction intercellular communication;
(iv) tumor suppressor genes upregulate gap junction intercellular communication; and
(v) transfection of cDNAs encoding gap junction proteins into gap junction intercellular communication-deficient and tumorigenic cells restores gap junction intercellular communication and normal growth regulation.[87–89]

Trosko and colleagues[90–92] analyzed the effect of twelve PAHs on gap junctional intercellular communication by using the scrape loading/ dye transfer technique. The PAHs used were naphthalene, 1-methylnaphthalene, 2-methylnaphthalene, anthracene, 1-methylanthracene, 2-methylanthracene, 9-methylanthracene, 9,10-dimethylanthracene, phenanthrene, fluorene, 1-methylfluorene and fluoranthene. PAHs containing bay- (phenanthrene, fluoranthene) or baylike regions (1-methylnaphthalene, 1-methylanthracene, 9-methylanthracene, 9,10-dimethylanthracene, 1-methylfluorene) inhibited gap junction intercellular communication stronger than did the linear compounds (naphthalene, 2-methylnaphthalene, anthracene, 2-methylanthracene, fluorene) (see Figure 7.1; for definition of PAHs with bay- or baylike regions, see Weis *et al.*[92] and Figure 2.1 in Chapter 2). The non-naphthalene compounds were not cytotoxic as determined by a vital dye uptake assay, but the naphthalene derivatives were cytotoxic at the higher doses, indicating that the downregulation of gap junction intercellular communication by these naphthalene derivatives could be a consequence of general membrane damage. Thus, PAHs possessing structural determinants such as bay- or baylike regions contribute to the tumor promoting activity of these compounds via an epigenetic effect (i.e., the inhibition of gap junction intercellular communication).

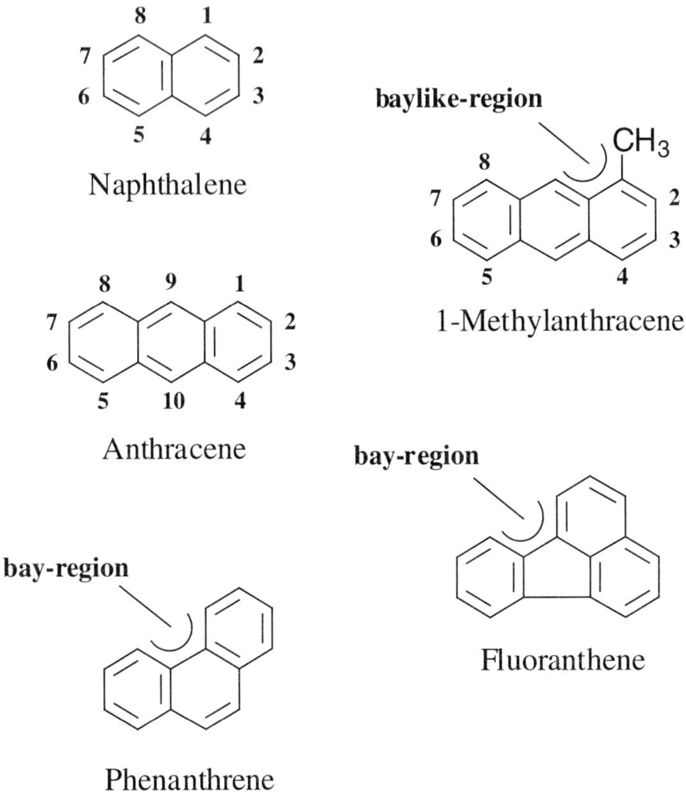

Figure 7.1: Structures of naphthalene, anthracene, 1-methylanthracene, fluoranthene, and phenanthrene. The baylike-region is formed by methyl substitution of linear condensed compounds such as naphthalene, anthracene, or fluoranthene. Examples are 1-methylnaphthalene or 1-methylanthracene. For definition of bay- and baylike regions and further examples, see also Weis *et al.*[92] and Figure 2.1 in Chapter 2.

7.3 Atherosclerosis-Related Alterations Induced by PAHs

A number of studies suggest that B[*a*]P accelerates smooth muscle proliferation and promotes atherosclerosis in animals ranging from chicken to rats.[93–97] In addition, B[*a*]P-DNA adducts have been detected in atherosclerotic arteries.[96,98,99] Regarding the mechanisms involved in the deregulation of vascular smooth muscle cell growth and differentiation during the course of atherogenesis and the role of PAHs in this process, Ou *et al.*[100] reported that exposure of quail smooth muscle cells to B[*a*]P inhibited inositol phospholipid turnover and protein kinase C activity and concluded that the ability of B[*a*]P to modulate growth and differentiation in quail aortic smooth muscle cells involved early interference with protein kinase C-related signal transduction.

Thereafter, Sadhu *et al.*[101] demonstrated that B[*a*]P enhances c-H-*ras* expression in rat aortic smooth muscle cells, an effect that was inhibited by the arylhydrocarbon receptor (AhR) antagonist α-naphthoflavone (cf. Chapter 2). A subsequent study showed that B[*a*]P upregulates c-H-*ras* expression via a transcriptional mechanism that involves the transactivation of regulatory sequences within the c-H-*ras* promoter region.[102]

Yan *et al.*[103] reported that B[*a*]P induces cyclooxygenase-2 (COX-2) mRNA and protein levels as well as prostaglandin synthesis in smooth muscle cells from aortas of normal rats or from atherosclerotic human arteries. Furthermore, the authors demonstrated that B[*a*]P induced COX-2 by activating the extracellular signal-regulated kinases 1 and 2 signalling pathway and that NF-κB mediates in part the induction of COX-2 by B[*a*]P. In contrast, p38 did not appear to be involved in this effect. COX-2 is able to catalyze the conversion of B[*a*]P-7,8-dihydrodiol to highly reactive B[*a*]PDE, which in turn can bind to DNA (cf. Chapter 2). Whether B[*a*]PDE-DNA adducts in atherosclerotic lesions[96,98,99] produce mutations that contribute to plaque development or represent tissue markers for oxidative vascular injury, remains to be clarified. Since COX-2 contributes to the formation of toxic B[*a*]P metabolites, B[*a*]P-mediated induction of this enzyme in vascular tissues suggests that a feedback mechanism capable of amplifying the toxic effects of B[*a*]P does in fact exist, thereby resulting in vascular injury and potentiating atherogenesis.

In an attempt to identify further genetic alterations leading to atherosclerotic lesions, Lu and Ramos[104] applied differential display polymerase chain reaction to murine vascular smooth muscle cells treated with B[*a*]P *in vitro* for 8 hours. It could be demonstrated that the expression of unique retrotransposon cDNAs was enhanced in vascular smooth cell cultures shortly after PAH treatment. A recent report by Lu *et al.*[105] revealed that mRNAs for ribosomal protein L31 and Zis (zinc-finger, splicing) were suppressed, while gas (growth arrest-specific)-5 and mitochondrial mRNAs were overexpressed in vascular smooth muscle cells isolated from mice and treated for eight weeks with B[*a*]P.

7.4 Apoptosis-Related Alterations Induced by PAHs

The capacity of PAHs to suppress immune cell function has been extensively documented. In this context, it has been shown that B[*a*]P and DMBA suppress B cell lymphopoiesis and it has been suggested that this toxic effect is mediated at least in part by the induction of apoptosis (i.e., programmed cell death).[106,107] Furthermore, several PAHs have been shown to induce apoptosis in Hepa1c1c7 hepatoma cells, Daudi human B cells, human ectocervical cells, A20.1 murine B cells and murine T cell hybridomas.[108–112]

In a limited number of recent reports information on the signalling pathways involved in PAH-mediated apoptosis has been gained. Chin *et al.*[113] showed that B[*a*]P adsorbed on carbon black induced the release of tumor necrosis factor α in RAW 264.7 macrophages, thereby causing apoptotic cell death via the activation of extracellular signal-regulated kinases 1 and 2. In the murine hepatoma cell line Hepa1c1c7, B[*a*]P induced apoptosis via a caspase 3-dependent pathway,[109] while c-*jun* *N*-terminal kinase 1 appeared not to be involved in this process. In contrast to these findings, Yoshii *et al.*[114] very recently showed that the α-p21 activated kinase 1-interacting exchange factor (α-PIX) plays a crucial role in the induction of apoptosis by B[*a*]P in 293T (human embryonal kidney) and in HeLa (human cervix carcinoma) cells, and that c-*jun* *N*-terminal kinase 1 activation is indeed involved in triggering apoptosis in these cell lines. B[*a*]P enhanced the mRNA and protein levels of α-PIX, thereby leading to the activation of c-*jun* *N*-terminal kinase 1 through Cdc42/Rac1, p21-activated kinase 1 (PAK1) and stress-activated protein kinase 1 (SEK1) (for activation cascade see Figure 7.2).

In conclusion, these studies show that PAHs are able to induce programmed cell death in a variety of cell types by activating several different signalling pathways. However, the signalling pathway being activated varies among the different cell types.

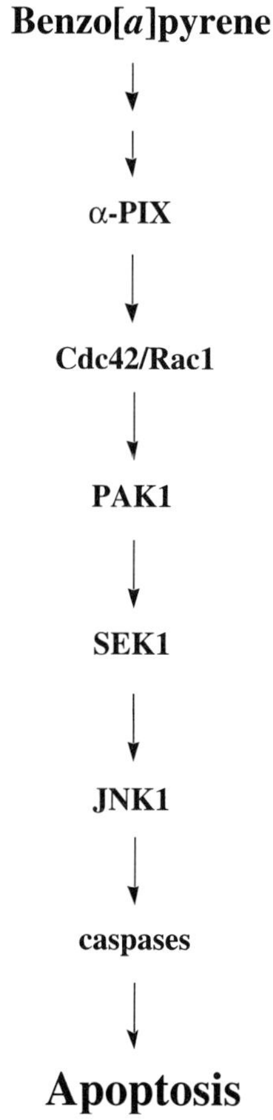

Figure 7.2: B[*a*]P-mediated apoptotic pathway (adapted from Yoshii *et al.*[114]). α-PIX, α-p21 activated kinase 1 interacting exchange factor; PAK1, p21-activated kinase 1; SEK1, stress-activated protein kinase 1; JNK1, c-*jun* *N*-terminal kinase 1.

7.5 Summary

PAHs are known to be carcinogenic in various tissues and more recently have been implicated in the development of atherosclerosis and in the induction of apoptosis. Point mutations in the *ras* genes play an important role in the carcinogenic process induced by PAHs, whereby the *ras* gene to be activated and the mutation to be induced may differ, depending on the tissue in which the tumor develops. Mutations in the *p53* tumor suppressor gene are detected in tobacco-related lung tumors as well as in skin tumors from PAH-treated mice. Recent studies show that PAHs may disrupt the expression of the breast cancer susceptibility gene *BRCA1*. BRCA1 protein downregulation could in turn lead to a reduced DNA repair capacity and to the outgrowth of cells initiated by PAHs. PAHs may also contribute to the carcinogenic process via inhibition of gap junction intercellular communication.

On the other hand, PAHs have been shown to promote atherosclerosis in various animal species through interference with protein kinase C-related signal transduction and upregulation of H-*ras* expression as well as induction of COX-2. Recent studies show that PAHs are also able to induce programmed cell death in a variety of cell types by activating several different signalling pathways. However, the signalling pathway being activated may vary depending on the cell type.

Acknowledgment

The studies performed in the laboratory of the author were generously supported by the Deutsche Forschungsgemeinschaft (Heisenberg-Program grant Ste 493/3-2).

References

1. Barbacid M (1987) *Ras* genes. *Annu. Rev. Biochem.* 56: 779–827.
2. Bollag G and McCormick F (1991) Regulators and effectors of ras proteins. *Annu. Rev. Cell. Biol.* 7: 601–632.
3. Downward J (1992) Regulatory mechanisms for ras proteins. *Bioessays* 14: 177–184.
4. Bos JL (1988) The *ras* gene family and human carcinogenesis. *Mutat. Res.* 195: 255–271.
5. Marshall CJ, Vousden KH and Phillips DH (1983) Activation of c-H-*ras*-1 proto-oncogene by *in vitro* modification with a chemical carcinogen, benzo[*a*]pyrene diol-epoxide. *Nature* 310: 586–589.

6. Balmain A and Pragnell IB (1983) Mouse skin carcinomas induced *in vivo* by chemical carcinogens have a transforming Harvey-*ras* oncogene. *Nature* 303: 72–74.
7. Bizub D, Wood AW and Skalka AM (1986) Mutagenesis of the H-*ras* oncogene in mouse skin tumors induced by PAHs. *Proc. Natl. Acad. Sci. USA* 83: 6048–6052.
8. Dandekar S, Sukumar S, Zarbl H, Young LJ and Cardiff RD (1986) Specific activation of the cellular Harvey-*ras* oncogene in dimethylbenzanthracene-induced mouse mammary tumors. *Mol. Cell. Biol.* 6: 4104–4108.
9. Balmain A, Ramsden M, Bowden GT and Smith J (1984) Activation of the mouse cellular Harvey-*ras* gene in chemically induced benign skin papillomas. *Nature* 307: 658–660.
10. Ronai ZA, Gradia S, El-Bayoumy K, Amin S and Hecht SS (1994) Contrasting incidence of *ras* mutations in rat mammary and mouse skin tumors induced by *anti*-benzo[*c*]phenanthrene-3,4-diol-1,2-epoxide. *Carcinogenesis* 15: 2113–2116.
11. Loktionov A, Hollstein M, Martel N, Galendo D, Cabral JR, Tomatis L and Yamasaki H (1990) Tissue-specific activating mutations of Ha- and Ki-*ras* oncogenes in skin, lung, and liver tumors induced in mice following transplacental exposure to DMBA. *Mol. Carcinog.* 3: 134–140.
12. Manam S, Storer RD, Prahalada S, Leander KR, Kraynak AR, Ledwith BJ, van Zwieten MJ, Bradley MO and Nichols WW (1992) Activation of the Ha-, Ki-, and N-*ras* genes in chemically induced liver tumors from CD-1 mice. *Cancer Res.* 52: 3347–3352.
13. Manam S, Shinder GA, Joslyn DJ, Kraynak AR, Hammermeister CL, Leander KR, Ledwith BJ, Prahalada S, van Zwieten MJ and Nichols WW (1995) Dose-related changes in the profile of *ras* mutations in chemically induced CD-1 mouse liver tumors. *Carcinogenesis* 16: 1113–1119.
14. Osaka M, Matsuo S, Koh T, Liang P, Kinoshita H, Maeda S and Sugiyama T (1995) N-*ras* mutation in 7,12-dimethylbenz[*a*]anthracene (DMBA)-induced erythroleukemia in Long-Evans rats. *Cancer Lett.* 91: 25–31.
15. You M, Candrian U, Maronpot RR, Stoner GD and Anderson MW (1989) Activation of the Ki-*ras* protooncogene in spontaneously occurring and chemically induced lung tumors of the strain A mouse. *Proc. Natl. Acad. Sci. USA* 86: 3070–3074.
16. Mass MJ, Jeffers AJ, Ross JA, Nelson G, Galati AJ, Stoner GD and Nesnow S (1993) Ki-*ras* oncogene mutations in tumors and DNA adducts formed by

benz[*j*]aceanthrylene and benzo[*a*]pyrene in the lungs of strain A/J mice. *Mol. Carcinog.* 8: 186–192.

17. Mass MJ, Abu-Shakra A, Roop BC, Nelson G, Galati AJ, Stoner GD, Nesnow S and Ross JA (1996) Benzo[*b*]fluoranthene: tumorigenicity in strain A/J mouse lungs, DNA adducts and mutations in the Ki-*ras* oncogene. *Carcinogenesis* 17: 1701–1704.
18. Nesnow S, Ross JA, Nelson G, Wilson K, Roop BC, Jeffers AJ, Galati AJ, Stoner GD, Sangaiah R, Gold A and Mass MJ (1994) Cyclopenta[*c,d*]pyrene-induced tumorigenicity, Ki-*ras* codon 12 mutations and DNA adducts in strain A/J mouse lung. *Carcinogenesis* 15: 601–606.
19. Prahalad AK, Ross JA, Nelson GB, Roop BC, King LC, Nesnow S and Mass MJ (1997) Dibenzo[*a,l*]pyrene-induced DNA adduction, tumorigenicity, and Ki-*ras* oncogene mutations in strain A/J mouse lung. *Carcinogenesis* 18: 1955–1963.
20. Ross JA and Nesnow S (1999) PAHs: correlation between DNA adducts and *ras* oncogene mutations. *Mutat. Res.* 424: 155–166.
21. You L, Wang D, Galati AJ, Ross JA, Mass MJ, Nelson GB, Wilson KH, Amin S, Stoner JC, Nesnow S and Stoner GD (1994) Tumor multiplicity, DNA adducts and K-*ras* mutation pattern of 5-methylchrysene in strain A/J mouse lung. *Carcinogenesis* 15: 2613–2618.
22. Nesnow S, Ross JA, Stoner GD and Mass MJ (1995) Mechanistic linkage between DNA adducts, mutations in oncogenes and tumorigenesis of carcinogenic environmental PAHs in strain A/J mice. *Toxicology* 105: 403–413.
23. Wessner LL, Fan M, Schaeffer DO, McEntee MF and Miller MS (1996) Mouse lung tumors exhibit specific Ki-*ras* mutations following transplacental exposure to 3-methylcholanthrene. *Carcinogenesis* 17: 1519–1526.
24. Eastman A, Sweetenham J and Bresnick E (1978) Comparison of *in vivo* and *in vitro* binding of polycyclic hydrocarbons to DNA. *Chem.-Biol. Interact.* 23: 345–353.
25. Koreeda M, Moore PD, Wislocki PG, Levin W, Yagi H and Jerina DM (1978) Binding of benzo[*a*]pyrene 7,8-diol-9,10-epoxides to DNA, RNA, and protein of mouse skin occurs with high stereoselectivity. *Science* 199: 778–780.
26. Pelling JC, Slaga TJ and DiGiovanni J (1984) Formation and persistence of DNA, RNA, and protein adducts in mouse skin exposed to pure optical enantiomers of 7β,8α-dihydroxy-9α,10α-epoxy-7,8,9,10-tetrahydrobenzo-[*a*]pyrene *in vivo*. *Cancer Res.* 44: 1081–1086.
27. Colapietro AM, Goodell AL and Smart RC (1993) Characterization of benzo[*a*]pyrene-initiated mouse skin papillomas for H-*ras* mutations and protein kinase C levels. *Carcinogenesis* 14: 2289–2295.

28. Gill RD, Beltran L, Nettikumara AN, Harvey RG, Kootstra A and DiGiovanni J (1992) Analysis of point mutations in murine c-H-*ras* of skin tumors initiated with dibenz[*a,j*]anthracene and derivatives. *Mol. Carcinog.* 6: 53–59.
29. Quintanilla M, Brown K, Ramsden M and Balmain A (1986) Carcinogen-specific mutation and amplification of H-*ras* during mouse skin carcinogenesis. *Nature* 322: 78–80.
30. Du MQ, Seidel A and Phillips DH (1995) Activating mutations in human c-H-*ras*-1 protooncogene induced by stereoisomeric fjord-region benzo[*c*]chrysene diol-epoxides. *Mol. Carcinog.* 14: 160–169.
31. Kito K, Kihana T, Sugita A, Murao S, Akehi S, Sato M, Tachibana M, Kimura S and Ueda N (1996) Incidence of *p53* and H-*ras* gene mutations in chemically induced rat mammary carcinomas. *Mol. Carcinog.* 17: 78–83.
32. Vuorio T, Warri A, Sandberg M, Alitalo K and Vuorio E (1988) Expression of the c-H-*ras* and *neu* oncogenes in DMBA-induced, anti-estrogen-treated rat mammary tumors. *Int. J. Cancer* 42: 774–779.
33. Waldmann V, Suchy B and Rabes HM (1993) Cell proliferation and prevalence of *ras* gene mutations in 7,12-dimethylbenz[*a*]anthracene (DMBA)-induced rat mammary tumors. *Res. Exp. Med.* 193: 143–151.
34. Kumar R, Medina D and Sukumar S (1990) Activation of H-*ras* oncogenes in preneoplastic mouse mammary tissues. *Oncogene* 5: 1271–1277.
35. Kastan MB, Onyekwere O, Sidransky D, Vogelstein B and Craig RW (1991) Participation of p53 protein in the cellular response to DNA damage. *Cancer Res.* 51: 6304–6311.
36. Marx J (1994) New link found between p53 and DNA repair. *Science* 266: 1321–1322.
37. Shaw P, Bovey R, Tardy S, Sahli R, Sordat B and Costa J (1992) Induction of apoptosis by wild-type p53 in a human colon tumor-derived cell line. *Proc. Natl. Acad. Sci. USA* 89: 4495–4499.
38. Baker SJ, Markowitz S, Fearon ER, Willson JK and Vogelstein B (1990) Suppression of human colorectal carcinoma cell growth by wild-type p53. *Science* 249: 912–915.
39. Finlay CA, Hinds PW and Levine AJ (1989) The p53 proto-oncogene can act as a suppressor of transformation. *Cell* 57: 1083–1093.
40. Hollstein M, Sidransky D, Vogelstein B and Harris CC (1991) *P53* mutations in human cancers. *Science* 253: 49–53.
41. Casson AG, Mukhopadhyay T, Cleary KR, Ro JY, Levin B and Roth JA (1991) *P53* gene mutations in Barrett's epithelium and esophageal cancer. *Cancer Res.* 51: 4495–4499.

42. Field JK, Spandidos DA, Malliri A, Gosney JR, Yiagnisis M and Stell PM (1991) Elevated p53 expression correlates with a history of heavy smoking in squamous cell carcinoma of the head and neck. *Br. J. Cancer* 64: 573–577.
43. Hollstein MC, Metcalf RA, Welsh JA, Montesano R and Harris CC (1990) Frequent mutation of the *p53* gene in human esophageal cancer. *Proc. Natl. Acad. Sci. USA* 87: 9958–9961.
44. Hollstein MC, Peri L, Mandard AM, Welsh JA, Montesano R, Metcalf RA, Bak M and Harris CC (1991) Genetic analysis of human esophageal tumors from two high incidence geographic areas: frequent *p53* base substitutions and absence of *ras* mutations. *Cancer Res.* 51: 4102–4106.
45. Mitsudomi T, Steinberg SM, Nau MM, Carbone D, D'Amico D, Bodner S, Oie HK, Linnoila RI, Mulshine JL, Minna JD and Gazdar AF (1992) *P53* gene mutations in non-small-cell lung cancer cell lines and their correlation with the presence of *ras* mutations and clinical features. *Oncogene* 7: 171–180.
46. Chiba I, Takahashi T, Nau MM, D'Amico D, Curiel DT, Mitsudomi T, Buchhagen DL, Carbone D, Piantadosi S, Koga H, Reissman PT, Slamon DJ, Holmes EC and Minna JD (1990) Mutations in the *p53* gene are frequent in primary, resected non-small cell lung cancer. Lung Cancer Study Group. *Oncogene* 5: 1603–1610.
47. Suzuki H, Takahashi T, Kuroishi T, Suyama M, Ariyoshi Y, Takahashi T and Ueda R (1992) *P53* mutations in non-small cell lung cancer in Japan: association between mutations and smoking. *Cancer Res.* 52: 734–736.
48. Takahashi T, Takahashi T, Suzuki H, Hida T, Sekido Y, Ariyoshi Y and Ueda R (1991) The *p53* gene is very frequently mutated in small-cell lung cancer with a distinct nucleotide substitution pattern. *Oncogene* 6: 1775–1778.
49. Puisieux A, Lim S, Groopman J and Ozturk M (1991) Selective targeting of *p53* gene mutational hotspots in human cancers by etiologically defined carcinogens. *Cancer Res.* 51: 6185–6189.
50. Denissenko MF, Pao A, Tang M and Pfeifer GP (1996) Preferential formation of benzo[*a*]pyrene adducts at lung cancer mutational hotspots in *p53*. *Science* 274: 430–432.
51. Paschke T (2000) Analysis of different versions of the IARC *p53* database with respect to G → T transversion mutation frequencies and mutation hotspots in lung cancer of smokers and non-smokers. *Mutagenesis* 15: 457–458.

52. Rodin SN and Rodin AS (2000) Human lung cancer and *p53*: the interplay between mutagenesis and selection. *Proc. Natl. Acad. Sci. USA* 97: 12244–12249.
53. Hainaut P and Pfeifer GP (2001) Patterns of p53 G → T transversions in lung cancers reflect the primary mutagenic signature of DNA-damage by tobacco smoke. *Carcinogenesis* 22: 367–374.
54. Ruggeri B, DiRado M, Zhang SY, Bauer B, Goodrow T and Klein-Szanto AJ (1993) Benzo[*a*]pyrene-induced murine skin tumors exhibit frequent and characteristic G to T mutations in the *p53* gene. *Proc. Natl. Acad. Sci. USA* 90: 1013–1017.
55. Burns PA, Kemp CJ, Gannon JV, Lane DP, Bremner R and Balmain A (1991) Loss of heterozygosity and mutational alterations of the *p53* gene in skin tumours of interspecific hybrid mice. *Oncogene* 6: 2363–2369.
56. Mitsunaga SI, Zhang SY, Ruggeri BA, Gimenez-Conti I, Robles AI, Conti CJ and Klein-Szanto AJ (1995) Positive immunohistochemical staining of p53 and cyclin D in advanced mouse skin tumors, but not in precancerous lesions produced by benzo[*a*]pyrene. *Carcinogenesis* 16: 1629–1635.
57. Ruggeri B, Caamano J, Goodrow T, DiRado M, Bianchi A, Trono D, Conti CJ and Klein-Szanto AJ (1991) Alterations of the *p53* tumor suppressor gene during mouse skin tumor progression. *Cancer Res.* 51: 6615–6621.
58. Cherpillod P and Amstad PA (1995) Benzo[*a*]pyrene-induced mutagenesis of *p53* hot-spot codons 248 and 249 in human hepatocytes. *Mol. Carcinog.* 13: 15–20.
59. Wang D, You L, Sneddon J, Cheng SJ, Jamasbi R and Stoner GD (1995) Frameshift mutation in codon 176 of the *p53* gene in rat esophageal epithelial cells transformed by benzo[*a*]pyrene dihydrodiol. *Mol. Carcinog.* 14: 84–93.
60. Luch A, Kudla K, Seidel A, Doehmer J, Greim H and Baird WM (1999) The level of DNA modification by (+)-*syn*-(11*S*,12*R*,13*S*,14*R*)- and (−)-*anti*-(11*R*,12*S*,13*S*,14*R*)-dihydrodiol epoxides of dibenzo[*a,l*]pyrene determined the effect on the proteins p53 and $p21^{WAF1}$ in the human mammary carcinoma cell line MCF-7. *Carcinogenesis* 20: 859–865.
61. Mahadevan B, Luch A, Seidel A, Pelling JC and Baird WM (2001) Effects of the (−)-*anti*-11*R*,12*S*-dihydrodiol 13*S*,14*R*-epoxide of dibenzo[*a,l*]pyrene on DNA adduct formation and cell cycle arrest in human diploid fibroblasts. *Carcinogenesis* 22: 161–169.
62. Rämet M, Castrén K, Järvinen K, Pekkala K, Turpeenniemi-Hujanen T, Soini Y, Pääkkö P and Vähäkangas K (1995) P53 protein expression is

correlated with benzo[*a*]pyrene-DNA adducts in carcinoma cell lines. *Carcinogenesis* 16: 2117–2124.

63. Bjelogrlic NM, Mäkinen M, Stenbäck F and Vähäkangas K (1994) Benzo[*a*]pyrene-7,8-diol-9,10-epoxide-DNA adducts and increased p53 protein in mouse skin. *Carcinogenesis* 15: 771–774.
64. Dipple A, Khan QA, Page JE, Pontén I and Szeliga J (1999) DNA reactions, mutagenic action and stealth properties of PAH carcinogens (review). *Int. J. Oncol.* 14: 103–111.
65. Khan QA, Vousden KH and Dipple A (1997) Cellular response to DNA damage from a potent carcinogen involves stabilization of p53 without induction of $p21^{waf1/cip1}$. *Carcinogenesis* 18: 2313–2318.
66. Khan QA, Agarwal R, Seidel A, Frank H, Vousden KH and Dipple A (1998) DNA adduct levels associated with p53 induction and delay of MCF-7 cells in S phase after exposure to benzo[*g*]chrysene dihydrodiol epoxide enantiomers. *Mol. Carcinog.* 23: 115–120.
67. Khan QA, Vousden KH and Dipple A (1999) Lack of p53-mediated G_1 arrest in response to an environmental carcinogen. *Oncology* 57: 258–264.
68. Jyonouchi H, Sun S, Iijima K, Wang M and Hecht SS (1999) Effects of *anti*-7,8-dihydroxy-9,10-epoxy-7,8,9,10-tetrahydrobenzo[*a*]pyrene on human small airway epithelial cells and the protective effects of *myo*-inositol. *Carcinogenesis* 20: 139–145.
69. Patten Hitt E, DeLong MJ and Merrill AH Jr (2002) Benzo[*a*]pyrene activates extracellular signal-related and p38 mitogen-activated protein kinases in HT29 colon adenocarcinoma cells: involvement in NAD(P)H: quinone reductase activity and cell proliferation. *Toxicol. Appl. Pharmacol.* 183: 160–167.
70. Couch FJ, DeShano ML, Blackwood MA, Calzone K, Stopfer J, Campeau L, Ganguly A, Rebbeck T and Weber BL (1997) *BRCA1* mutations in women attending clinics that evaluate the risk of breast cancer. *N. Engl. J. Med.* 336: 1409–1415.
71. Ford D, Easton DF, Stratton M, Narod S, Goldgar D, Devilee P, Bishop DT, Weber B, Lenoir G, Chang-Claude J, Sobol H, Teare MD, Struewing J, Arason A, Scherneck S, Peto J, Rebbeck TR, Tonin P, Neuhausen S, Barkardottir R, Eyfjord J, Lynch H, Ponder BA, Gayther SA, Zelada-Hedman M and the Breast Cancer Linkage Consortium (1998) Genetic heterogeneity and penetrance analysis of the *BRCA1* and *BRCA2* genes in breast cancer families. The Breast Cancer Linkage Consortium. *Am. J. Hum. Genet.* 62: 676–689.

72. Abbott DW, Thompson ME, Robinson-Benion C, Tomlinson G, Jensen RA and Holt JT (1999) *BRCA1* expression restores radiation resistance in *BRCA1*-defective cancer cells through enhancement of transcription-coupled DNA repair. *J. Biol. Chem.* 274: 18808–18812.
73. Bhattacharyya A, Ear US, Koller BH, Weichselbaum RR and Bishop DK (2000) The breast cancer susceptibility gene *BRCA1* is required for subnuclear assembly of Rad51 and survival following treatment with the DNA cross-linking agent cisplatin. *J. Biol. Chem.* 275: 23899–23903.
74. Gowen LC, Avrutskaya AV, Latour AM, Koller BH and Leadon SA (1998) BRCA1 required for transcription-coupled repair of oxidative DNA damage. *Science* 281: 1009–1012. (retracted)
75. Moynahan ME, Chiu JW, Koller BH and Jasin M (1999) BRCA1 controls homology-directed DNA repair. *Mol. Cell* 4: 511–518.
76. Scully R, Chen J, Ochs RL, Keegan K, Hoekstra M, Feunteun J and Livingston DM (1997) Dynamic changes of BRCA1 subnuclear location and phosphorylation state are initiated by DNA damage. *Cell* 90: 425–435.
77. Snouwaert JN, Gowen LC, Latour AM, Mohn AR, Xiao A, DiBiase L and Koller BH (1999) BRCA1 deficient embryonic stem cells display a decreased homologous recombination frequency and an increased frequency of non-homologous recombination that is corrected by expression of a *brca1* transgene. *Oncogene* 18: 7900–7907.
78. Peto J, Collins N, Barfoot R, Seal S, Warren W, Rahman N, Easton DF, Evans C, Deacon J and Stratton MR (1999) Prevalence of *BRCA1* and *BRCA2* gene mutations in patients with early-onset breast cancer. *J. Natl. Cancer Inst.* 91: 943–949.
79. Shih HA, Nathanson KL, Seal S, Collins N, Stratton MR, Rebbeck TR and Weber BL (2000) *BRCA1* and *BRCA2* mutations in breast cancer families with multiple primary cancers. *Clin. Cancer Res.* 6: 4259–4264.
80. Jeffy BD, Chen EJ, Gudas JM and Romagnolo DF (2000) Disruption of cell cycle kinetics by benzo[*a*]pyrene: inverse expression patterns of BRCA-1 and p53 in MCF-7 cells arrested in S and G2. *Neoplasia* 2: 460–470.
81. Jeffy BD, Chirnomas RB and Romagnolo DF (2002) Epigenetics of breast cancer: PAHs as risk factors. *Environ. Mol. Mutagen.* 39: 235–244.
82. Jeffy BD, Chirnomas RB, Chen EJ, Gudas JM and Romagnolo DF (2002) Activation of the aromatic hydrocarbon receptor pathway is not sufficient for transcriptional repression of BRCA-1: requirements for metabolism of benzo[*a*]pyrene to 7*r*,8*t*-dihydroxy-9*t*,10-epoxy-7,8,9,10-tetrahydrobenzo[*a*]pyrene. *Cancer Res.* 62: 113–121.

83. Jeffy BD, Schultz EU, Selmin O, Gudas JM, Bowden GT and Romagnolo D (1999) Inhibition of BRCA-1 expression by benzo[*a*]pyrene and its diol epoxide. *Mol. Carcinog.* 26: 100–118.
84. Caveney S (1985) The role of gap junctions in development. *Annu. Rev. Physiol.* 47: 319–335.
85. Loewenstein WR (1981) Junctional intercellular communication: the cell-to-cell membrane channel. *Physiol. Rev.* 61: 829–913.
86. Trosko JE, Chang CC, Madhukar BV and Oh SY (1990) Modulators of gap junction function: the scientific basis of epigenetic toxicology. *In Vitro Toxicol.* 3: 9–26.
87. Yamasaki H and Naus CC (1996) Role of connexin genes in growth control. *Carcinogenesis* 17: 1199–1213.
88. Trosko JE, Chang CC, Madhukar BV and Klaunig JE (1990) Chemical, oncogene and growth factor inhibition gap junctional intercellular communication: an integrative hypothesis of carcinogenesis. *Pathobiology* 58: 265–278.
89. Trosko JE, Madhukar BV and Chang CC (1993) Endogenous and exogenous modulation of gap junctional intercellular communication: toxicological and pharmacological implications. *Life Sci.* 53: 1–19.
90. Upham BL, Masten SJ, Lockwood BR and Trosko JE (1994) Nongenotoxic effects of PAHs and their oxygenation by-products on the intercellular communication of rat liver epithelial cells. *Fundam. Appl. Toxicol.* 23: 470–475.
91. Upham BL, Weis LM, Rummel AM, Masten SJ and Trosko JE (1996) The effects of anthracene and methylated anthracenes on gap junctional intercellular communication in rat liver epithelial cells. *Fundam. Appl. Toxicol.* 34: 260–264.
92. Weis LM, Rummel AM, Masten SJ, Trosko JE and Upham BL (1998) Bay or baylike regions of PAHs were potent inhibitors of gap junctional intercellular communication. *Environ. Health Perspect.* 106: 17–22.
93. Albert RE, Vanderlaan M, Burns FJ and Nishizumi M (1977) Effect of carcinogens on chicken atherosclerosis. *Cancer Res.* 37: 2232–2235.
94. Bond JA, Kocan RM, Benditt EP and Juchau MR (1979) Metabolism of benzo[*a*]pyrene and 7,12-dimethylbenz[*a*]anthracene in cultured human fetal aortic smooth muscle cells. *Life Sci.* 25: 425–430.
95. Hough JL, Baird MB, Sfeir GT, Pacini CS, Darrow D and Wheelock C (1993) Benzo[*a*]pyrene enhances atherosclerosis in White Carneau and Show Racer pigeons. *Arterioscler. Thromb.* 13: 1721–1727.

96. Izzotti A, De Flora S, Petrilli GL, Gallagher J, Rojas M, Alexandrov K, Bartsch H and Lewtas J (1995) Cancer biomarkers in human atherosclerotic lesions: detection of DNA adducts. *Cancer Epidemiol. Biomarkers Prev.* 4: 105–110.
97. Ou X and Ramos KS (1992) Proliferative responses of quail aortic smooth muscle cells to benzo[*a*]pyrene: implications in PAH-induced atherogenesis. *Toxicology* 74: 243–258.
98. Izzotti A, D'Agostini F, Bagnasco M, Scatolini L, Rovida A, Balansky RM, Cesarone CF and De Flora S (1994) Chemoprevention of carcinogen-DNA adducts and chronic degenerative diseases. *Cancer Res.* 54: 1994s–1998s.
99. Zhang YJ, Weksler BB, Wang L, Schwartz J and Santella RM (1998) Immunohistochemical detection of PAH-DNA damage in human blood vessels of smokers and non-smokers. *Atherosclerosis* 140: 325–331.
100. Ou X, Weber TJ, Chapkin RS and Ramos KS (1995) Interference with protein kinase C-related signal transduction in vascular smooth muscle cells by benzo[*a*]pyrene. *Arch. Biochem. Biophys.* 318: 122–130.
101. Sadhu DN, Merchant M, Safe SH and Ramos KS (1993) Modulation of protooncogene expression in rat aortic smooth muscle cells by benzo[*a*]pyrene. *Arch. Biochem. Biophys.* 300: 124–131.
102. Bral CM and Ramos KS (1997) Identification of benzo[*a*]pyrene-inducible *cis*-acting elements within c-H-*ras* transcriptional regulatory sequences. *Mol. Pharmacol.* 52: 974–982.
103. Yan Z, Subbaramaiah K, Camilli T, Zhang F, Tanabe T, McCaffrey TA, Dannenberg AJ and Weksler B (2000) Benzo[*a*]pyrene induces the transcription of cyclooxygenase-2 in vascular smooth muscle cells. Evidence for the involvement of extracellular signal-regulated kinase and NF-κB. *J. Biol. Chem.* 275: 4949–4955.
104. Lu KP and Ramos KS (1998) Identification of genes differentially expressed in vascular smooth muscle cells following benzo[*a*]pyrene challenge: implications for chemical atherogenesis. *Biochem. Biophys. Res. Commun.* 253: 828–833.
105. Lu KP, Alejandro NF, Taylor KM, Joyce MM, Spencer TE and Ramos KS (2002) Differential expression of ribosomal L31, Zis, gas-5 and mitochondrial mRNAs following oxidant induction of proliferative vascular smooth muscle cell phenotypes. *Atherosclerosis* 160: 273–280.
106. Hardin JA, Hinoshita F and Sherr DH (1992) Mechanisms by which benzo[*a*]pyrene, an environmental carcinogen, suppresses B cell lymphopoiesis. *Toxicol. Appl. Pharmacol.* 117: 155–164.

107. Yamaguchi K, Matulka RA, Shneider AM, Toselli P, Trombino AF, Yang S, Hafer LJ, Mann KK, Tao XJ, Tilly JL, Near RI and Sherr DH (1997) Induction of preB cell apoptosis by 7,12-dimethylbenz[*a*]anthracene in long-term primary murine bone marrow cultures. *Toxicol. Appl. Pharmacol.* 147: 190–203.
108. Burchiel SW, Davis DA, Ray SD and Barton SL (1993) DMBA induces programmed cell death (apoptosis) in the A20.1 murine B cell lymphoma. *Fundam. Appl. Toxicol.* 21: 120–124.
109. Lei W, Yu R, Mandlekar S and Kong AN (1998) Induction of apoptosis and activation of interleukin 1 beta-converting enzyme/Ced-3 protease (caspase-3) and c-*jun* NH_2-terminal kinase 1 by benzo[*a*]pyrene. *Cancer Res.* 58: 2102–2106.
110. Rorke EA, Sizemore N, Mukhtar H, Couch LH and Howard PC (1998) PAHs enhance terminal cell death of human ectocervical cells. *Int. J. Oncol.* 13: 557–563.
111. Salas VM and Burchiel SW (1998) Apoptosis in Daudi human B cells in response to benzo[*a*]pyrene and benzo[*a*]pyrene-7,8-dihydrodiol. *Toxicol. Appl. Pharmacol.* 151: 367–376.
112. Yamaguchi K, Near R, Shneider A, Cui H, Ju ST and Sherr DH (1996) Fluoranthene-induced apoptosis in murine T cell hybridomas is independent of the aromatic hydrocarbon receptor. *Toxicol. Appl. Pharmacol.* 139: 144–152.
113. Chin BY, Choi ME, Burdick MD, Strieter RM, Risby TH and Choi AM (1998) Induction of apoptosis by particulate matter: role of TNF-α and MAPK. *Am. J. Physiol.* 275: L942–949.
114. Yoshii S, Tanaka M, Otsuki Y, Fujiyama T, Kataoka H, Arai H, Hanai H and Sugimura H (2001) Involvement of alpha-PAK-interacting exchange factor in the PAK1-c-jun NH_2-terminal kinase 1 activation and apoptosis induced by benzo[*a*]pyrene. *Mol. Cell. Biol.* 21: 6796–6807.

8 Indicator Assays for Polycyclic Aromatic Hydrocarbon-Induced Genotoxicity

Hansruedi Glatt

German Institute of Human Nutrition, Department of Toxicology, Potsdam-Rehbrücke, Germany
E-mail: glatt@mail.dife.de

8.1 Introduction

Dibenz[*a*,*h*]anthracene and benzo[*a*]pyrene (B[*a*]P) were the first pure chemicals to show carcinogenic activity in an experimental animal model.[1,2] 7,12-Dimethylbenz[*a*]anthracene (DMBA) and 3-methylcholanthrene (3-MC) are still widely used to induce tumors in certain tissues, such as skin, lung, mammary gland, large intestine and the hemopoietic system,[3–8] in order to study the individual steps of carcinogenesis or to find preventive factors.

As early as 1914, Theodor Boveri[9] postulated that somatic mutations may be critically involved in carcinogenesis. This concept has been verified in the meantime, with mutations occurring in oncogenes and tumor suppressor genes being pivotal.[10] Boveri's 'somatic mutation theory' promoted the idea of recognizing chemical carcinogens via their genotoxic activities. The first reports on the induction of germ-cell mutations by polycyclic aromatic hydrocarbons (PAHs) already appeared in the 1940s, but they were not sufficiently documented or not conclusive for other reasons (e.g., Strong[11]). Burdette, who reviewed the possible role of mutations in carcinogenesis in 1955,[12] noticed a poor correlation between the compounds inducing tumors and those producing mutations (in the systems used at that time). A fundamental change in paradigms occurred in the mid 1960s, when it was detected that PAHs form DNA adducts in mouse skin, a target tissue of their carcinogenicity,[13] and when it was recognized that numerous carcinogens require metabolic activation to electrophilic species.[14] Shortly afterwards, a number of new mutagenicity test systems that considered the requirements of metabolic activation were developed. PAHs have been used in the development and the validation of nearly all different types of genotoxicity test systems. It is not the aim of this article to present all these test systems. Rather, we should focus on systems that have been widely used or gave particularly interesting results with PAHs and their metabolites, and also with other structurally related polycyclic aromatic compounds (PACs) that may contain heteroatoms in their ring systems. The endpoint DNA adduct as such is not addressed in this article, as it is treated in detail in other chapters of this book (see Chapters 2, 4 and 5).

8.2 Bacterial Systems

8.2.1 *In Vitro* Mutagenicity Tests Using Bacterial Target Cells

The first use of an external metabolizing system in a bacterial mutagenicity test dates to 1971, when Malling[15] demonstrated mutagenicity of dimethylnitrosamine to an *Escherichia coli* strain in the presence of a murine liver microsomal system. In the following year, Ames *et al.*[16] showed that various epoxides of carcinogenic PAHs are mutagenic to *his*$^-$ *Salmonella typhimurium* strains; shortly afterwards, he found that structurally diverse carcinogens, including representative PAHs, are mutagenic in these bacterial strains when an NADPH-fortified postmitochondrial supernatant fraction (S9) from rat liver was added.[17] The *his*$^-$ *Salmonella* reversion system ('Ames test') developed to the most popular mutagenicity test system in subsequent years. For the investigation of PAHs and other PACs, it has some advantages over similar test systems using *Escherichia coli* or other bacterial species. Ames introduced the deep-rough mutation (*rfa*$^-$) in his strains, facilitating the penetration of large lipophilic molecules, such as PAH derivatives, into the bacteria. He also included frameshift-mutated strains into his test battery, which often are reverted with particular sensitivity by these compounds. Substitution-mutated strains are usually reverted by PAH derivatives only with low efficiency unless the DNA repair systems are perturbed, e.g., by introduction of an error-prone repair factor. Such a factor is encoded, for example, by the *mucAB* genes located in the pKM101 plasmid of *S. typhimurium* strain TA100.[18] Nevertheless, it is possible to detect mutagenicity of PAHs using *E. coli* strains with some limitations in sensitivity. 7-Bromomethylbenz[*a*]anthracene (7-bromomethyl-B[*a*]A),[19] *trans*-B[*a*]P-7,8-dihydrodiol (in the presence of S9),[20] 1-sulfooxymethylpyrene[21] and 7-methyl-B[*a*]A-5,6-oxide[21] are examples of PAH derivatives and metabolites that were shown to be mutagenic in *E. coli* (reversion of substitution-mutated *trp*$^-$ strains).

More than 150 PAHs have been tested for mutagenicity to *S. typhimurium*.[17,22–42] The large majority of these compounds was mutagenic in the presence of an external activating system, but inactive in its absence. Only a small number of PAHs, which belong to the bicyclics or tricyclics (naphthalene, fluorene, and some methylated congeners of phenanthrene, fluorene and anthracene),[22,30,32] or which show extremely poor solubility (2,6-dimethylpyrene[39]) were negative in all studies and under all conditions used. Likewise, direct mutagenic effects to *S. typhimurium* strains were only reported for a few PAHs: 1,2,3,4-tetrahydro-DMBA,[26] 1,2-dihydro-DMBA,[26] 6,7,8,9,10,12b-hexahydro-3-MC,[24] 2-methylanthracene,[43] 9,10-dimethylanthracene,[34] tetracene[43] and benzo[4,5]cyclohept-[1,2,3-*bc*]acenaphthylene.[27] It is difficult to rule out that impurities or photoactivation were involved in some of these effects. Likewise, a large number of PAH derivatives containing heteroatoms has demonstrated mutagenic activity to *S. typhimurium*, again usually only in the presence of an activating system.[44–48] In general, established carcinogens among homo- and heterocyclic PACs were mutagenic, but many congeners that had been inactive in carcinogenicity studies exhibited mutagenic potencies similar to those observed with established carcinogens. Within a given series of structurally related compounds — such as B[*a*]P and the 12 possible monomethylated derivatives;[37] B[*a*]A, DMBA and isomeric monomethyl-B[*a*]As; or 43 heterocyclic PACs[44] — no quantitative association was observed between the strength of the liver S9-mediated mutagenicity and the carcinogenic activity. However, the correlation was substantially improved when intact hepatocytes were used for activation,[37,49] suggesting that the latter system better reflects the activation and inactivation processes occurring *in vivo*.

Light is an alternative to enzyme systems for the activation of some PAHs.[50–52] Exposure of DMBA to light in the presence of oxygen yields the direct mutagen, DMBA-7,12-epidioxide.[50] This compound appears to be involved in the photomutagenicity of DMBA to *S. typhimurium*. Likewise, 9,10-dimethylanthracene-9,10-epidioxide, another ultimate mutagen,[50] may be the actual cause for the weak mutagenic effects observed with 9,10-dimethylanthracene to *S. typhimurium* in the absence of mammalian enzymes.[34] γ-Irradiation of various PAHs in the presence of air resulted in the formation of products that were strongly mutagenic to *S. typhimurium*.[43]

PAHs and their congeners can be metabolized to a large number of products. In a previous review article,[53] we compiled mutagenicity data for 59 metabolites and further similar derivatives of B[*a*]P to bacterial and mammalian V79 cells. A total of 41 of these compounds were mutagenic to *S. typhimurium*, directly and/or in the presence of an activating system, with varying potency. In the meantime, a positive result has been reported for an additional metabolite of B[*a*]P, B[*a*]P-1-sulfate; the effect was strongest in the presence of pulmonary S9 or cytosol fraction, but also occurred in the direct test and in the presence of hepatic enzymes.[54] Strong direct mutagenicity was also observed with 7,8,9-trihydroxy-7,8,9,10-tetrahydro-B[*a*]P-10-sulfonate, a reaction product of the bay-region B[*a*]P-7,8-diol-9,10-epoxide (B[*a*]PDE) with sulfite.[55] Using new tester strains, moderate direct mutagenicity was observed with the B[*a*]P-7,8-quinone (as well as some *o*-quinones of other PAHs).[56] The observation that various types of

PAH metabolites have a mutagenic potential may explain the high percentage of mutagens among PAHs in the liver enzyme/*Salmonella* test system.

Strong direct mutagenicity has been demonstrated for numerous vicinal diol-epoxide metabolites of PAHs,[57–59] well established ultimate carcinogens. However, some K-region oxides, such as B[*a*]P-4,5-oxide and 7-methyl-B[*a*]A-5,6-oxide, are also extremely potent bacterial mutagens.[60] The low carcinogenic activity of these simple arene oxides appears to be due to their efficient detoxification by mammalian epoxide hydrolases and glutathione *S*-transferases (cf. Chapter 2).[61,62]

The addition of metabolites to test systems, or their generation by external metabolizing systems, is only appropriate if these metabolites are capable of permeating into the target cells. The polar polysaccharide layer in the outer cell membrane of bacteria (which is strongly reduced in the *rfa*$^-$ strains of *S. typhimurium*[17]) may act as an efficient barrier for large lipophilic molecules. In addition, there is circumstantial evidence for active export of such molecules by *S. typhimurium* and *E. coli* in a medium-dependent manner.[21,60] The lipid bilayers of the bacterial plasma membranes can form barriers for ionized reactive species. This effect has been demonstrated in particular for sulfuric acid esters derived from PACs.[63–65] For example, 1-thiosulfooxymethylpyrene is not mutagenic when added to *S. typhimurium* (even at a dose of 1000 nmol), but shows clear mutagenic effects in case of being generated from 1-thiomethylpyrene by a cDNA-expressed sulfotransferase (SULT) within the bacteria (a dose of 0.01 nmol of this metabolic precursor being sufficient) (Figure 8.1). With a half-life of hours in aqueous media, 1-thiosulfooxymethylpyrene is substantially more stable than 1-sulfooxymethylpyrene ($t_{1/2} \sim 2.8$ min) and other sulfuric esters. The latter compounds may decompose to the corresponding benzylic cation within the incubation period and react with medium components to form secondary, membrane-penetrating reactive species (such as chloromethylarenes); thus their mutagenicity is strongly influenced by the composition of the exposure medium.[63–65] In any cases, it is more appropriate to use internal metabolizing systems to study this class of PAC derivatives.[66]

The group of Cavalieri has demonstrated that PAHs can be metabolically activated in a one-step reaction to radical cations, which can then form unstable (depurinating) DNA adducts (cf. Chapter 2).[67,68] However, it is not

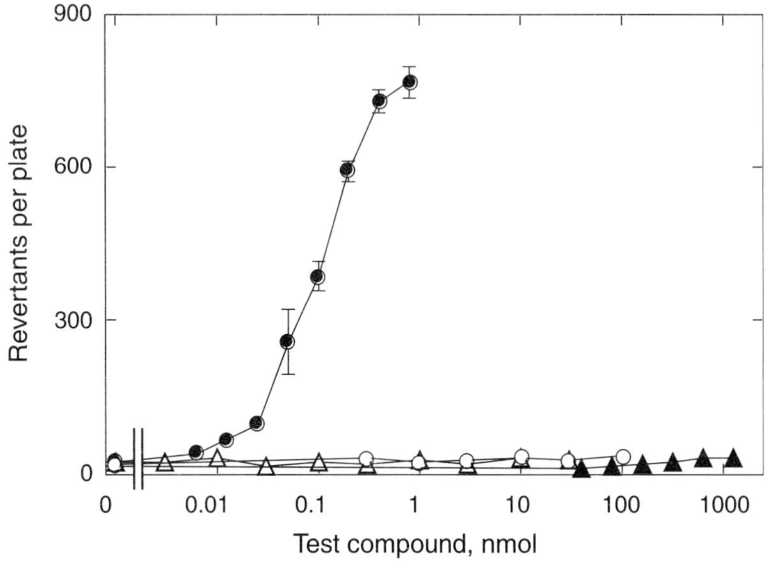

Figure 8.1: Mutagenicity of exogenously added and intracellularly generated 1-thiosulfo-oxymethylpyrene to *S. typhimurium*. Triangles: 1-thiosulfooxymethylpyrene tested in strain TA1538 (two experiments covering different dose ranges); solid circles: 1-thiomethylpyrene tested in a TA1538-derived strain expressing human SULT1A1 (an enzyme converting 1-thiomethylpyrene into its thiosulfate); open circles: 1-thiomethylpyrene tested in strain TA1538. The construction of SULT-expressing strains and the mutagenicity assay used have been described by Meinl *et al.*[155] Values are means and SE of 3 plates.

known whether these adducts are important in PAH-induced mutagenesis and carcinogenesis. It is unlikely that this activation pathway would be detected in a conventional bacterial mutagenicity assay using an external activating system, as the radicals certainly would react with lipids during membrane passage. It would therefore be useful to generate a test system in which this type of reactive species is formed within the target cell, ideally in the absence of any other activation pathways.

Several polynuclear PAH quinones are mutagenic to mammalian cells (cf. Chapter 2).[69] They are usually inactive in bacterial test systems, although some moderate mutagenic effects have been observed in *S. typhimurium* TA104, a strain responsive towards 'reactive oxygen species'.[70] The low sensitivity of bacteria to many quinone mutagens may be due to their high reductive capacity, converting quinones and semiquinones into inactive hydroquinones.

Recently, various human cytochrome P450-dependent monooxygenases (CYP) have been individually expressed together with cytochrome P450 reductase in *E. coli* MX100 (reversion to arginine prototrophy),[71] *E. coli lacZ* (reversion of the *lacZ* gene),[72] and Ames's *Salmonella* strains.[73] B[*a*]P and DMBA were mutagenic in the CYP1A1-expressing MX100 strain, without requiring any exogenous activating system.[71] Their mutagenic activities exceeded those observed in the corresponding control strain in the presence of liver S9 enzymes by factors of approximately 10 and 100, respectively (but were still low compared to those seen with some PAH metabolites in *S. typhimurium*). Human CYP1A1, 1A2 and 1B1 expressed in *E. coli lacZ* activated various heterocyclic aromatic amines, whereas no data have been published with regard to the activation of PAHs.[72] According to a statement in a review article,[73] B[*a*]P was mutagenic to TA1538-derived strains expressing CYP enzymes. The active metabolites have not been identified in these models. Since epoxide hydrolase is not present, diol-epoxides presumably cannot be formed.

In the bacterial test systems described above, reversion mutations are studied. Bacterial test systems detecting forward mutations are also available, but are more laborious and therefore used much less frequently compared to reversion tests. B[*a*]P in the presence of liver enzymes induced forward mutations leading to resistance to arabinose,[74] 8-azaguanine,[75–77] 5-fluorouracil[77] and azetidine carboxylic acid[77] in *S. typhimurium*. Only a small number of PAHs have been studied in these systems. Interestingly, perylene, a PAH without any bay- and K-regions, was more potent than B[*a*]P in inducing resistance to 8-azaguanine.

8.2.2 Other Endpoints of Genotoxicity in Bacterial Target Cells

DNA damage may not only lead to mutations but can also induce a number of defense systems, such as DNA repair enzymes (cf. Chapter 6). These effects can often be recorded much earlier than mutations. Two test systems are frequently used: the SOS Chromotest using *E. coli* PQ37 (*lacZ* reporter gene fused to regulating sequences of *sulA*) and the Umu test using *S. typhimurium* TA1535/pSK1002 (*lacZ* fused to regulating sequences of *umuC*). Induction of SOS repair in *E. coli* PQ37 has been observed with numerous parent PAHs tested in the presence of S9 preparations[42,48,78] as well as with many bay- and fjord-region diol-epoxide derivatives.[58] In general, the results correlated with those obtained in the *Salmonella* mutagenicity assay, which, however, required lower substrate concentrations and detected additional weak genotoxicants. The Umu test has been widely used in combination with membrane fractions of recombinant CYP-expressing *E. coli* strains to study the ability of individual human CYP enzymes to activate various promutagens. The PAHs and PAH dihydrodiols investigated were activated with particular efficiency by human CYP1A1 and CYP1B1.[79] Moderate activation of some of these compounds was observed with CYP1A2, 2C9, 2C19 and 3A4, whereas other forms (2A6, 2B6, 2C8, 2C18, 2D6, 2E1, 3A5, 3A7 and 4A11) were almost inactive towards the PAHs investigated. Furthermore, while human CYP enzymes have been directly expressed in detector strains of the Umu test, they have not yet been used for studying the genotoxicity of PAHs.[80]

8.2.3 Host-Mediated Assays Using Microbial Target Cells

In host-mediated assays, the indicator cells (e.g., appropriate strains of *E. coli*, *S. typhimurium*, *Saccharomyces cerevisiae* or *Neurospora crassa*) are administered intraperitoneally and intrasanguineously to host animals (usually mice), which then are treated with test compounds.[81] Later, the indicator cells are recovered from the peritoneal cavity and the liver (and/or other tissues), respectively, and analyzed for their mutant frequency. It is difficult to use this assay for studying the mutagenicity of PAHs and related compounds, since indicator cells modified in their cell membrane in order to facilitate the uptake of PAH metabolites (such as *rfa*$^-$ in standard *S. typhimurium strains*) are also very susceptible to the defense mechanisms of the host; thus only short treatment periods are possible with these strains. Tester strains with intact cell membranes and walls can survive much longer in the host, but they take up PAH derivatives with rather low efficiency. Only a small number of PAHs and their metabolites were investigated in host-mediated microbial mutagenicity test systems, generally with a negative or weakly positive result. Simmon *et al.*[82] tested 10 PAHs in two *S. typhimurium* strains (TA1530 and TA1538) and in *S. cerevisiae* in the intraperitoneal host-mediated assay in the mouse. B[*a*]A was weakly mutagenic to strain TA1538 at the highest dose used (1600 mg/kg body mass); anthracene was weakly positive in one experiment and inactive in repeat experiments. All other results of that study, including those with B[*a*]P and DMBA, were negative. In another study, reproducible but weak, positive results were obtained with B[*a*]P after intraperitoneal injection of the indicator strains *S. typhimurium* TA98 and TA100 into mice; pretreatment of the mice with 2(3)-*tert*-butyl-4-hydroxyanisole, an inducer of conjugating enzymes, abolished the mutagenicity of B[*a*]P.[83] Using the same strains in an intrasanguineous host-mediated assay, B[*a*]P and its K-region 4,5-oxide gave negative results. In contrast, the proximate mutagen *trans*-B[*a*]P-7,8-dihydrodiol elicited a positive response, which was substantially enhanced when the experiment was conducted in animals pretreated with the CYP inducer 5,6-benzoflavone (β-naphthoflavone).[84]

8.3 Mammalian Systems

8.3.1 Mutations in Mammalian Cells in Culture

Three test systems are widely used for studying the induction of gene mutations in mammalian cells in culture: the acquisition of resistance to 6-thioguanine (or 8-azaguanine) (involving inactivation of the X-chromosomal *hprt* locus) in CHO or V79 Chinese hamster cells, and acquisition of resistance to trifluorothymidine (involving inactivation of the hemizygous *tk* locus) in L5178Y $TK^{+/-}$ cells.[85] CHO and V79 cells are fibroblastoid cell lines isolated from ovary and male lung, respectively, and are similar in many respects. However, the male V79 cell line has advantages when the mutated *hprt* gene is to be analyzed. PAHs, in the presence of an appropriate activating system, are mutagenic in all three test systems as well as in additional mammalian gene mutation assays that are less established in routine genotoxicity testing. A large number of investigations with PAHs, their metabolites and related compounds were conducted in the

V79/HPRT assay. The first study was performed in 1971 by Huberman *et al.*,[86] who found that electrophilic metabolites (K-region oxides and bromomethyl derivatives of PAHs) are mutagenic in this system, whereas parent hydrocarbons are not. Later, feeder layers of lethally irradiated, metabolically competent cells[87] and liver S9 preparations[88] were introduced for the activation of PAHs in the V79/HPRT assay. Eventually, V79 cells were engineered for the expression of heterologous PAH-activating enzymes.[89–93] Relatively high concentrations of PAHs are required for the induction of mutations when external metabolizing systems are used. The concentrations required may be lowered by a factor of 1000 to 10,000 when cells are used that express an appropriate CYP enzyme, such as human CYP1B1 in case of B[*a*]P and dibenzo[*a*,*l*]pyrene (DB[*a*,*l*]P) (Figure 8.2). Relatively strong mutagenic effects were also observed with some benzylic alcohols derived from PAHs in V79-derived cell lines expressing rat or human SULTs.[92,93]

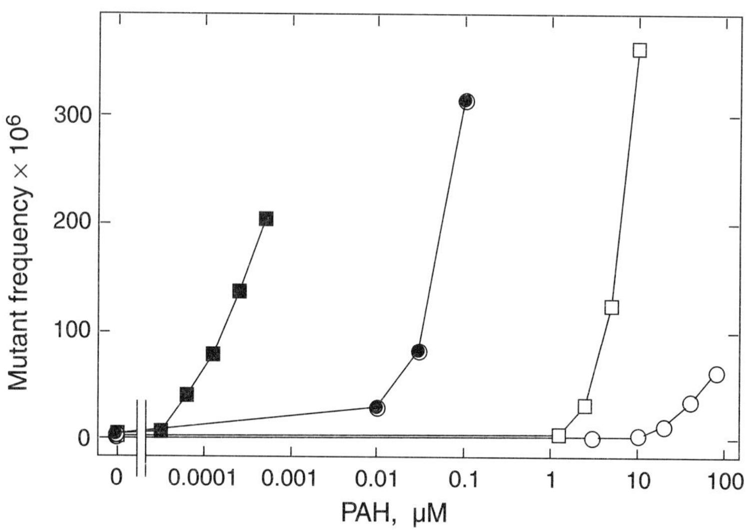

Figure 8.2: Mutagenicity of benzo[*a*]pyrene (circles) and dibenzo[*a*,*l*]pyrene (squares) at the *hprt* locus of V79 cells using exogenously added rat liver S9 (open symbols) or human CYP1B1 expressed in the target cells (solid symbols) for bioactivation. Data taken from Glatt *et al.*,[92] where additional results with other recombinant V79 cell lines are also shown.

Among the PAH metabolites studied in the absence of external or cDNA-expressed enzymes, bay- and fjord-region diol-epoxides were particularly efficient inducers of gene mutations in V79 cells.[53,58,59,94–96] Several simple epoxides were also mutagenic to V79 cells[86,94,97,98] but were substantially less active than the corresponding diol-epoxides. This is opposed by bacterial systems, in which arene oxides sometimes reached similar, or even higher activities. This is due to endogenous epoxide hydrolases and glutathione *S*-transferases that are highly expressed in V79 cells, and that efficiently detoxify simple epoxides (cf. Chapter 2).[99] Unlike the simple epoxides, their aziridine analogues, which are poor substrates of these enzymes, are potent mutagens in V79 cells.[100] On the other hand, diol-epoxides may be detoxified by glutathione *S*-transferases, but they require high enzyme levels. Accordingly, heterologous expression of some human glutathione *S*-transferases led to a decrease in the mutagenicity and DNA binding of PAH diol-epoxides in V79 cells.[101,102]

Although various PAH epoxides, studied in our laboratory, were mutagenic to V79 cells, their potency (mutagenic effect per concentration unit in the linear range) varied over five orders of magnitude (Figure 8.3). We wondered whether these differences were due to differences in the level or the mutagenic potential of the DNA adducts formed. The adduct-forming activity varied over the same five orders of magnitude and correlated with the mutagenic activity (Figure 8.3). Thus, the level of adducts (determined at the end of the exposure time) was the factor dominating the mutagenic activity. Approximately 6×10^7 adducts had to be formed in a cell population in order to induce one 6-thioguanine-resistant mutant. Both endpoints are detected with similar sensitivity, as indicated by the number of events required for a positive test result under standard conditions: 1 mutant per 10^5 cells and 1 adduct per 10^8 nucleotides (approximately 1.2×10^7 adducts per 10^5 cells).

Some pyrene quinones were mutagenic to V79 cells[69] with potencies similar to those of the less active bay-region diol-epoxides, such as *anti*-chrysene-1,2-diol-3,4-oxide (number 6 in Figure 8.3). Other types of PAH metabolites, not mentioned in the preceding sections, have shown negligible mutagenic effects to V79 cells, although they may have been active in other systems, such as the 'Ames assay'.[53]

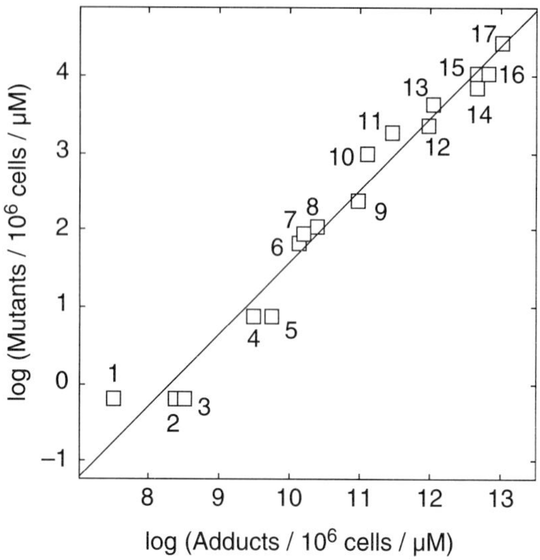

Figure 8.3: Correlation between mutagenic activity (at the *hprt* locus) and DNA adduct-forming activity (determined by the ^{32}P-postlabeling assay; cf. Chapter 4) of various PAH epoxides in V79 cells. 1,3: *syn-* and *anti-*chrysene-3,4-diol-1,2-epoxide (assays close to their limits of detection); 2: chrysene-5,6-oxide; 4,6: *syn-* and *anti-*chrysene-1,2-diol-3,4-epoxide; 5,8: *syn-* and *anti-*9-hydroxychrysene-1,2-diol-3,4-epoxide; 7,10: *anti-*7-ethyl- and *anti-*7-methyl-B[*a*]A-3,4-diol-1,2-epoxide; 9,12: *syn-* and *anti-*B[*a*]P-7,8-diol-9,10-epoxide; 11,13: *syn-* and *anti-*benzo[*c*]phenanthrene-3,4-diol-1,2-epoxide; 14,15: *syn-* and *anti-*benzo[*g*]chrysene-11,12-diol-13,14-epoxide; 16,17: *syn-* and *anti-*benzo[*c*]chrysene-9,10-diol-11,12-epoxide. The diagonal line is the regression curve, indicating that 6×10^7 adducts have to be formed in a cell population to induce one 6-thioguanine-resistant mutant. Data from Phillips *et al.*,[156] Glatt *et al.*,[157] Phillips *et al.*[158] and Glatt *et al.*[159]

8.3.2 Other Endpoints of Genotoxicity Using Mammalian Cells in Culture

Sister chromatid exchange (SCE) is an endpoint that can be studied in any propagatable animal cells in culture. It can also be investigated in animal models with some efforts required for incorporation of 5-bromodeoxyuridine (BrdU) into the DNA (see next section). In earlier times, this endpoint was widely used with PAHs and their metabolites. A large number of PAHs and PAH derivatives have shown positive results.[103,104] In systems in which multiple endpoints can be studied, the ability of a compound to induce SCE is normally associated with its ability to induce gene mutations. The sensitivity of the endpoint SCE is similar to, or sometimes somewhat higher than, the endpoint gene mutations. The hydrolysis products of the bay-region diol-epoxide of B[*a*]P, the B[*a*]P-7,8,9,10-tetraols (7,8,9,10-tetrahydroxy-7,8,9,10-tetrahydro isomers of B[*a*]P), induced SCE in V79-derived cell lines engineered for expression of rat SULTs.[105] The induction of SCE in human lymphocytes or fibroblasts by PAHs has been used to explore inter-individual differences in susceptibility.[106,107] SCE has the disadvantages that an unnatural (BrdU-substituted) DNA is the target of damage and that the toxicological significance of this endpoint as such is unknown.

Representing nucleotide excision repair (NER; cf. Chapter 6), the unscheduled DNA synthesis (UDS) in hepatocytes or other cells is frequently used as an endpoint in genetic toxicology. This assay has also been used to investigate the effects of some PAHs and their derivatives. For example, 1-methylpyrene and 1,6-dimethylpyrene are extremely potent inducers of UDS in primary cultures of rat hepatocytes.[108] The benzylic alcohol 1-hydroxymethylpyrene induced UDS in V79-derived cells expressing various rat and human SULTs, but not in corresponding control cells.[109,110]

Treatment of SV40-transformed Chinese hamster cells with B[*a*]P led to selective amplification of the SV40 DNA.[111] The effect was prevented by 7,8-benzoflavone (α-naphthoflavone), an inhibitor of B[*a*]P metabolism. It was strongly enhanced when *anti*-B[*a*]PDE, rather than the parent compound, was used. The effect was also observed with other carcinogenic PAHs (DMBA, 3-MC), but not with non-carcinogenic PAHs such as phenanthrene and dibenz[*a*,*c*]anthracene.

8.3.3 Genotoxicity in Mammalian Cells *In Vivo*

The bone-marrow micronucleus assay in the mouse is the most popular *in vivo* mutagenicity assay used to date. B[*a*]P and DMBA scored positive in this test; 7,8-benzoflavone, an inhibitor of CYP1 enzymes, reduced the number of DMBA- and B[*a*]P-induced micronucleated cells by 90 and 75%, respectively.[112] After oral administration of B[*a*]P to mice and rats,[113] or intraperitoneal administration of DMBA to mice,[114] elevated levels of micronuclei were found in polychromatic peripheral erythrocytes. Fjord-region diol-epoxides of benzo[*c*]chrysene are potent inducers of micronuclei in bone marrow erythrocytes in the mouse.[115] Although micronuclei can be most easily scored in bone marrow and peripheral erythrocytes, they can also be analyzed in other tissues. For example, elevated levels of micronuclei were observed in mouse skin keratinocytes following treatment with B[*a*]P, DMBA or chrysene, whereas treatment with pyrene showed no effect.[116] Likewise, 15,16-dihydro-11-methylcyclopenta[*a*]phenanthren-17-one, a potent skin carcinogen, induced micronuclei in keratinocytes *in vivo*, whereas its non-carcinogenic unmethylated congener was inactive.[117] Moreover, B[*a*]P and DMBA showed transplacental induction of micronuclei in fetal liver of rodents; actually, the effects in fetal liver were stronger than those observed in maternal bone marrow.[118–120]

Mutations in the *hprt* gene can be scored not only in some cell lines, but also in certain primary cells such as lymphocytes (isolated from blood or spleen). B[*a*]P was mutagenic in this assay; its effect was enhanced in mice deficient in the DNA repair gene *XPA* (*Xeroderma pigmentosum* complementation group A; cf. Chapter 6).[121] Knock-out of another DNA repair gene, *CSB* (Cockayne syndrome complementation group B), also potentiated the mutagenicity of B[*a*]P at the *hprt* gene (a constitutively transcribed gene), but did not affect its mutagenicity at the inactive *lacZ* gene (a transgene) nor its carcinogenicity.[122] Furthermore, B[*a*]P-induced *hprt* mutations were detected in granulation tissue initiated by an air pouch in rats.[123]

B[*a*]P also scored positive in the mouse spot test, i.e., it induced mutations in embryonic somatic cells in genes determining hair pigmentation.[124]

Transgenic animals such as the Muta™ mouse, the Big Blue™ mouse and the Big Blue™ rat contain transgenes (*lacZ* or *lacI*) that can readily be analyzed for the occurrence of mutations. In particular, they provide opportunities for comparing the mutagenicity in different tissues. In a study by Hakura *et al.*,[125] the B[*a*]P-induced mutant frequency in the *lacZ* gene of the Muta™ mouse followed the order: colon > ileum > forestomach > bone marrow, spleen > glandular stomach > liver, lung > kidney, heart > brain (no significant effect). Using the same animal model and treatment protocol, B[*a*]P showed the highest carcinogenic activity in forestomach followed by the lymphatic organs. In addition, bronchiolar-alveolar hyperplasia occurred, whereas no tumors were observed in the other organs examined. In another study, B[*a*]P was painted to the skin of Muta™ mice;[126] the treatment led to a strong increase in the mutant frequency of the *lacZ* gene in the treated area of skin, but did not affect the mutant frequency in lung and liver. Likewise, B[*a*]P and *anti*-B[*a*]PDE painted to the skin of Big Blue™ mice led to the induction of *lacI* mutations in the underlying tissue; this effect was enhanced when a phorbol ester was administered at approximately the same time as the hydrocarbon.[127]

Other PAH-induced genotoxic effects that have been observed in somatic cells *in vivo* are increases in the frequency of SCEs in bone marrow cells,[118,128–131] in fetal liver[118] and, after partial hepatectomy, in hepatocytes of adult animals.[130] Sirianni and Huang[132] implanted V79 cell-containing diffusion chambers into the peritoneal cavity of mice, which were then treated with the test compound (and BrdU). Afterwards, the cells were scored for SCEs. B[*a*]P, DMBA and 3-MC were positive in this test. This result indicates that reactive metabolites of these compounds reached the peritoneal cavity and even penetrated into the diffusion chambers and the indicator cells. Thus, it is evident that these metabolites are widely distributed in the organism and do not necessarily have to be generated at their target sites. Galloway *et al.*[133] treated explanted mouse embryos with B[*a*]P. They found high frequencies of SCEs in arylhydrocarbon (Ah)-responsive mice (in which CYP enzymes are induced by PAHs), but not in non-responsive mice.

Germ-cell mutation assays are usually less sensitive than assays for the detection of genetic damage in somatic cells *in vivo*. In addition, they are laborious. Therefore, only few studies have conducted with PAHs. Nevertheless, B[*a*]P and DMBA induced dominant lethal mutations in spermatogenic cells of mice.[134,135] Thus PAHs may not only induce cancer, but also genetic damage to the progeny.

8.4 Characterization of DNA Sequence Changes Induced by PAHs

Gene mutations are usually scored via phenotypic changes that allow selection of the mutants. It is possible to sequence the corresponding genes in order to determine the molecular nature of the mutations induced. This approach is of limited value with reverse mutations (as monitored, for example, in the 'Ames test'), as only a relatively small number of sequence changes can generate the phenotype investigated. A much larger number of different mutations can lead to forward mutations, i.e., to the inactivation of the investigated gene, such as the *hprt* gene of V79 cells,[136] the *lacZ* gene of the Muta™ mouse,[137] or the *lacI* gene of the Big Blue™ mouse.[127] In all systems, B[*a*]P showed a propensity for G → T transversions. Nevertheless, the spectra of mutations induced by B[*a*]P in *lacZ* of the Muta™ mouse substantially varied between different organs.[137] The pattern of *hprt* mutations induced by *anti*-B[*a*]PDE in V79 cells was also affected by the exposure concentration and the DNA repair competence of the cells.[136]

Critical sequences of oncogenes and tumor suppressor genes in PAH-induced tumors,[138,139] transformed cell clones[138,140] or unselected mass cultures of PAH-treated cells[141] have also been analyzed. For example, it was found that B[*a*]P frequently induces G → T transversions in the middle base of codon 248 of the human *p*53 gene (cf. Chapter 5).[141–143]

Furthermore, shuttle vectors containing a small target gene, such as *supF*, have often been used for studying mutation spectra induced by PAH metabolites.[144–148] The vector is treated in a cell-free system with the ultimate mutagen and then transfected into mammalian or bacterial cells, where it may be replicated and mutations may be fixed. Afterwards mutants are identified using bacterial host cells and then analyzed. In other studies, a defined adduct was incorporated at a specific site of a vector (usually in single-stranded vectors or vectors with an inactivated complementary strand); after replication of the vector in a bacterial or mammalian host cell, the sequence at that site was analyzed.[149–153] The results indicate that the mutational spectrum varies between different PAH-DNA adducts, sequence contexts and host cells (cf. Chapters 5 and 6).

8.5 Summary

B[*a*]P, DMBA and — to a lesser extent — other PAHs and heterocyclic congeners have been investigated for genotoxic effects in numerous systems. They produced a large variety of different types of mutations (e.g., base substitutions, frameshifts, selective DNA amplifications as well as structural and numerical chromosomal aberrations) and other genotoxic effects (e.g., UDS, SCE). They were also found to be active in diverse target cells ranging from bacteria and cultured eukaryotic cells to somatic and germinal cells in laboratory animals. However, nearly all of these effects were only observed in the presence of an endogenous or exogenous metabolic activation system or when activated metabolites rather than the parent compounds were used. A high number of different PAH and PAC metabolites demonstrated genotoxic effects in one or another system. Discrepancies in the results for the same compound between different systems often had toxicokinetic reasons involving differences in activation, inactivation or transmembrane transfer. Bay- and fjord-region diol-epoxides are electrophilic PAH metabolites that are highly genotoxic in numerous test systems, as they readily permeate cell membranes and are relatively resistant towards detoxifying enzymes. They appear to play an important role in the *in vivo* genotoxicity of various PAHs, including the prototypic compounds B[*a*]P and DMBA. Some benzylic alcohol derivatives show strong genotoxicity in SULT-expressing cells in culture and can form exceptionally high levels of DNA adducts in animal models under certain conditions.[154] Various other types of metabolites have demonstrated genotoxicity in special *in vitro* systems, but their role during carcinogenicity and genotoxicity in the whole organism would require further research. Special toxicokinetic properties can render such investigations very difficult. *In vitro* and animal models in which the metabolic pathways of PAHs are triggered into specified directions through genetic manipulation of the xenobiotic-processing system will be useful in clarifying the role of different metabolites in the toxicology of PAHs and related compounds.

Acknowledgment

The author's work on PAHs has been financially supported by the Deutsche Forschungsgemeinschaft (SFB 302 and INK 26).

References

1. Kennaway EL and Hieger I (1930) Carcinogenic substances and their fluorescence spectra. *Br. Med. J.* 1: 1044–1046.
2. Cook JW, Hewett CL and Hieger I (1933) The isolation of a cancer-producing hydrocarbon from coal tar: parts I, II, and III. *J. Chem. Soc.* 395–405.
3. Katsanakis KD, Gorgoulis V, Papavassiliou AG and Zoumpourlis VK (2002) The progression in the mouse skin carcinogenesis model correlates with ERK1/2 signaling. *Mol. Med.* 8: 624–637.
4. Sugiyama T, Osaka M, Koami K, Maeda S and Ueda N (2002) 7,12-DMBA-induced rat leukemia: a review with insights into future research. *Leuk. Res.* 26: 1053–1068.
5. Shan L, He M, Yu M, Qiu C, Lee NH, Liu ET and Snyderwine EG (2002) cDNA microarray profiling of rat mammary gland carcinomas induced by 2-amino-1-methyl-6-phenylimidazo[4,5-*b*]pyridine and 7,12-dimethylbenz[*a*]anthracene. *Carcinogenesis* 23: 1561–1568.
6. Wessner LL, Fan M, Schaeffer DO, McEntee MF and Miller MS (1996) Mouse lung tumors exhibit specific Ki-*ras* mutations following transplacental exposure to 3-methylcholanthrene. *Carcinogenesis* 17: 1519–1526.
7. Kupfer R, Dwyer-Nield LD, Malkinson AM and Thompson JA (2002) Lung toxicity and tumor promotion by hydroxylated derivatives of 2,6-di-*tert*-butyl-4-methylphenol (BHT) and 2-*tert*-butyl-4-methyl-6-iso-propylphenol: correlation with quinone methide reactivity. *Chem. Res. Toxicol.* 15: 1106–1112.
8. Baral RN and Maity P (1992) Induction of colorectal cancer in rats by 20-methylcholanthrene. *Cancer Lett.* 61: 177–183.
9. Boveri T (1929) *The Origin of Malignant Tumors*. Williams and Wilkins Co., Baltimore, MD, USA. (Translation of the German original published in 1914: *Zur Frage der Entstehung maligner Tumore.* Gustav Fischer, Jena, Germany.)
10. Loeb KR and Loeb LA (2000) Significance of multiple mutations in cancer. *Carcinogenesis* 21: 379–385.

11. Strong LC (1945) A genetic analysis of the induction of tumours by methylcholanthrene: XI. Germinal mutations and other sudden biological changes following subcutaneous injection of methylcholanthrene. *Proc. Natl. Acad. Sci. USA* 31: 290–293.
12. Burdette WJ (1955) The significance of mutation in relation to the origin of tumours: a review. *Cancer Res.* 15: 201–226.
13. Brookes P and Lawley PD (1964) Evidence for the binding of polynuclear aromatic hydrocarbons to the nucleic acids of mouse skin: relation between carcinogenic power of the hydrocarbons and their binding to deoxyribonucleic acid. *Nature* 202: 781–784.
14. Miller EC and Miller JA (1966) Mechanism of chemical carcinogenesis: nature of proximate carcinogens and interaction with macromolecules. *Pharmacol. Rev.* 18: 805–838.
15. Malling HV (1971) Dimethylnitrosamine: formation of mutagenic compounds by interaction with mouse liver microsomes. *Mutat. Res.* 13: 425–429.
16. Ames BN, Sims P and Grover PL (1972) Epoxides of carcinogenic polycyclic hydrocarbons are frameshift mutagens. *Science* 176: 47–49.
17. Ames BN, Durston WE, Yamasaki E and Lee FD (1973) Carcinogens are mutagens: a simple test system combining liver homogenates for activation and bacteria for detection. *Proc. Natl. Acad. Sci. USA* 70: 2281–2285.
18. McCann J, Springarn NE, Kobori J and Ames BN (1975) Detection of carcinogens as mutagens: bacterial tester strains with R factor plasmids. *Proc. Natl. Acad. Sci. USA* 72: 979–983.
19. Tarmy EM, Venitt S and Brookes P (1973) Mutagenicity of the carcinogen 7-bromomethylbenz[*a*]anthracene: a quantitative study in repair-deficient strains of *Escherichia coli*. *Mutat. Res.* 19: 153–166.
20. Ivanovic V and Weinstein IB (1980) Genetic factors in *Escherichia coli* that affect cell killing and mutagenesis induced by benzo[*a*]pyrene-7,8-dihydrodiol-9,10-oxide. *Cancer Res.* 40: 3508–3511.
21. Glatt HR, Staffa-Piée A, Enders N, Baidossi W and Blum J (1994) The presence of KCl in the exposure medium strongly influences the mutagenicity of metabolites of polycyclic aromatic hydrocarbons in *Escherichia coli*. *Mutat. Res.* 324: 111–114.
22. McCann J, Choi E, Yamasaki E and Ames BN (1975) Detection of carcinogens as mutagens in the *Salmonella*/microsome test: assay of 300 chemicals. *Proc. Natl. Acad. Sci. USA* 72: 5135–5139.

23. Coombs MM, Dixon C and Kissonerghis AM (1976) Evaluation of the mutagenicity of compounds of known carcinogenicity, belonging to the benz[*a*]anthracene, chrysene, and cyclopenta[*a*]phenanthrene series, using Ames's test. *Cancer Res.* 36: 4525–4529.
24. Andrews AW, Thibault LH and Lijinsky W (1978) The relationship between carcinogenicity and mutagenicity of some polynuclear hydrocarbons. *Mutat. Res.* 51: 311–318.
25. Gold A, Nesnow S, Moore M, Garland H, Curtis G, Howard B, Graham D and Eisenstadt E (1980) Mutagenesis and morphological transformation of mammalian cells by a non-bay-region polycyclic cyclopenta[*c,d*]pyrene and its 3,4-oxide. *Cancer Res.* 40: 4482–4484.
26. Inbasekaran M, Witiak DT, Barone K and Loper JC (1980) Synthesis and mutagenicity of A-ring reduced analogues of 7,12-dimethylbenz[*a*]-anthracene. *J. Med. Chem.* 23: 278–281.
27. Takahashi Y, Nagao M, Sugimura T, Todo E and Murata I (1980) Mutagenicities of new non-alternant isomers of benzo[*a*]pyrene. *Mutat. Res.* 78: 295–299.
28. Malaveille C, Hautefeuille A, Bartsch H, MacNicoll AD, Grover PL and Sims P (1980) Liver microsome-mediated mutagenicity of dihydrodiols derived from dibenz[*a,c*]anthracene in *S. typhimurium* TA100. *Carcinogenesis* 1: 287–289.
29. Oesch F, Bücker M and Glatt HR (1981) Activation of phenanthrene to mutagenic metabolites and evidence for at least two different activation pathways. *Mutat. Res.* 81: 1–10.
30. LaVoie EJ, Tulley-Freiler L, Bedenko V and Hoffmann D (1981) Mutagenicity, tumor-initiating activity, and metabolism of methylphenanthrenes. *Cancer Res.* 41: 3441–3447.
31. LaVoie EJ, Tulley L, Bedenko V and Hoffmann D (1981) Mutagenicity of methylated fluorenes and benzofluorenes. *Mutat. Res.* 91: 167–176.
32. LaVoie EJ, Tulley-Freiler L, Bedenko V and Hoffmann D (1983) Mutagenicity of substituted phenanthrenes in *Salmonella typhimurium*. *Mutat. Res.* 116: 91–102.
33. Sangaiah R, Gold A, Toney GE, Toney SH, Claxton L, Easterling R and Nesnow S (1983) Benz[*j*]aceanthrylene: a novel polycyclic aromatic hydrocarbon with bacterial mutagenic activity. *Mutat. Res.* 119: 259–266.
34. LaVoie EJ, Coleman DT, Tonne RL and Hoffmann D (1983) Mutagenicity, tumor-initiating activity, and metabolism of methylated anthracenes. In: *Polynuclear Aromatic Hydrocarbons: Formation, Metabolism and*

Measurement [Cooke M and Dennis AJ (eds.)] pp 785–798, Battelle Press, Columbus, OH, USA.
35. Nesnow S, Leavitt S, Easterling R, Watts R, Toney SH, Claxton L, Sangaiah R, Toney GE, Wiley J, Fraher P and Gold A (1984) Mutagenicity of cyclopenta-fused isomers of benz[*a*]anthracene in bacterial and rodent cells and identification of the major rat liver microsomal metabolites. *Cancer Res.* 44: 4993–5003.
36. Kohan MJ, Sangaiah R, Ball LM and Gold A (1985) Bacterial mutagenicity of aceanthrylene: a novel cyclopenta-fused polycyclic aromatic hydrocarbon of low molecular weight. *Mutat. Res.* 155: 95–98.
37. Utesch D, Glatt HR and Oesch F (1987) Rat hepatocyte-mediated bacterial mutagenicity in relation to the carcinogenicity of benz[*a*]anthracene, benzo[*a*]pyrene and twenty-five methylated derivatives. *Cancer Res.* 47: 1509–1515.
38. Rice JE, Geddie NG, DeFloria MC and LaVoie EJ (1988) Structural requirements favoring mutagenic activity among methylated pyrenes in *S. typhimurium*. In: *Polynuclear Aromatic Hydrocarbons: A Decade of Progress* [Cooke M and Dennis AJ (eds.)] pp 773–785, Battelle Press, Columbus, OH, USA.
39. Seidel A, Glatt HR, Oesch F and Garrigues P (1990) 2,6-Dimethylpicene: synthesis, mutagenic activity, and identification in natural samples. *Polycycl. Aromat. Compds.* 1: 3–14.
40. Devanesan PD, Cremonesi P, Nunnally JE, Rogan EG and Cavalieri EL (1990) Metabolism and mutagenicity of dibenzo[*a,e*]pyrene and the very potent environmental carcinogen dibenzo[*a,l*]pyrene. *Chem. Res. Toxicol.* 3: 580–586.
41. Cheung YL, Gray TJB and Ioannides C (1993) Mutagenicity of chrysene, its methyl and benzo derivatives, and their interactions with cytochromes P450 and the Ah-receptor: relevance to their carcinogenic potency. *Toxicology* 81: 69–86.
42. Mersch-Sundermann V, Schneider U, Klopman G and Rosenkranz HS (1994) SOS induction in *Escherichia coli* and *Salmonella* mutagenicity: a comparison using 330 compounds. *Mutagenesis* 9: 205–224.
43. Gibson TL, Smart VB and Smitz LL (1978) Non-enzymic activation of polycyclic aromatic hydrocarbons as mutagens. *Mutat. Res.* 49: 153–161.
44. Glatt HR, Schwind H, Zajdela F, Croisy A, Jacquignon PC and Oesch F (1979) Mutagenicity of 43 structurally related heterocyclic compounds and its relationship to their carcinogenicity. *Mutat. Res.* 66: 307–328.

45. Okuda H, Kitahara Y, Shudo K and Okamoto T (1979) Identification of an ultimate mutagen of 10-azabenzo[*a*]pyrene. *Chem. Pharm. Bull.* 27: 2547–2549.
46. LaVoie EJ, Briggs G, Bedenko V and Hoffmann D (1982) Mutagenicity of substituted carbazoles in *Salmonella typhimurium*. *Mutat. Res.* 101: 141–150.
47. Adams EA, LaVoie EJ and Hoffmann D (1983) Mutagenicity and metabolism of azaphenanthrenes. In: *Polynuclear Aromatic Hydrocarbons: Formation, Metabolism and Measurement* [Cooke M and Dennis AJ (eds.)] pp 73–87, Battelle Press, Columbus, OH, USA.
48. Quillardet P, Jenek J, Demerseman P, Royer R and Hofnung M (1986) Genotoxic activity of two furan analogues of benzo[*a*]pyrene and their 2-nitro derivatives. *Mutat. Res.* 172: 223–230.
49. Glatt HR, Billings R, Platt KL and Oesch F (1981) Improvement of the correlation of bacterial mutagenicity with carcinogenicity of benzo[*a*]pyrene and four of its major metabolites by activation with intact liver cells instead of cell homogenate. *Cancer Res.* 41: 270–277.
50. Tu MH, Perry D and Chen C (1979) Mutagenic effect of 7,12-dimethylbenz[*a*]anthracene-epidioxide on *Salmonella typhimurium*. *Mol. Pharmacol.* 15: 189–191.
51. Issaq HD, Andrews AW, Janini GM and Barr EW (1979) Isolation of stable mutagenic photodecomposition products of benzo[*a*]pyrene by thin-layer chromatography. *J. Liquid Chromat.* 2: 319–325.
52. Selby CP, Calkins J, Enoch HG, Wright CW and Wilson BW (1987) Chemical basis for photomutagenicity in synthetic fuels and correlation with carcinogenicity. *Mutat. Res.* 188: 287–299.
53. Glatt HR and Oesch F (1986) Structural and metabolic parameters governing the mutagenicity of polycyclic aromatic hydrocarbons. In: *Chemical Mutagens: Principles and Methods for Their Detection* [De Serres FJ (ed.)] Volume 10, pp 73–127, Plenum Press, New York.
54. Irwin SE, Kwei GY, Blackburn GR, Thurman R and Kauffman FC (1992) Mutagenicity of benzo[*a*]pyrenyl-1-sulfate in the Ames test. *Environ. Mol. Mutagen.* 19: 253–258.
55. Green JL and Reed GA (1990) Benzo[*a*]pyrene bay-region sulfonates, a novel class of reactive intermediates. *Chem. Res. Toxicol.* 3: 59–64.
56. Flowers-Geary L, Bleczinski W, Harvey RG and Penning TM (1996) Cytotoxicity and mutagenicity of polycyclic aromatic hydrocarbon *o*-quinones produced by dihydrodiol dehydrogenase. *Chem.-Biol. Interact.* 99: 55–72.

57. Thakker DR, Yagi H, Levin W, Wood AW, Conney AH and Jerina DM (1985) Polycyclic aromatic hydrocarbons: metabolic activation to ultimate carcinogens. In: *Bioactivation of Foreign Compounds* [Anders MW (ed.)] pp 177–242, Academic Press, London.
58. Glatt HR, Piée A, Pauly K, Steinbrecher T, Schrode R, Oesch F and Seidel A (1991) Fjord- and bay-region diol-epoxides investigated on stability, SOS induction in *Escherichia coli*, and mutagenicity in *Salmonella typhimurium* and mammalian cells. *Cancer Res.* 51: 1659–1667.
59. Luch A, Glatt HR, Platt KL, Oesch F and Seidel A (1994) Synthesis and mutagenicity of the diastereomeric fjord-region 11,12-dihydrodiol 13,14-epoxides of dibenzo[*a*,*l*]pyrene. *Carcinogenesis* 15: 2507–2516.
60. Glatt HR, Martiné U, Baidossi W and Blum J (1991) Strong influence of the exposure medium on mutagenicity in the Ames test: 7-methylbenz[*a*]-anthracene-5,6-oxide. *Mutagenesis* 6: 537–540.
61. Glatt HR, Friedberg T, Grover PL, Sims P and Oesch F (1983) Inactivation of a diol-epoxide and a K-region epoxide with high efficiency by glutathione transferase X. *Cancer Res.* 43: 5713–5717.
62. Glatt HR, Cooper CS, Grover PL, Sims P, Bentley P, Merdes M, Waechter F, Vogel K, Guenthner TM and Oesch F (1982) Inactivation of a diol-epoxide by dihydrodiol dehydrogenase, but not by two epoxide hydrolases. *Science* 215: 1507–1509.
63. Enders N, Seidel A, Monnerjahn S and Glatt HR (1993) Synthesis of 11 benzylic sulfate esters, their bacterial mutagenicity and its modulation by chloride, bromide and acetate anions. *Polycycl. Aromat. Compds.* 3: 887s–894s.
64. Glatt HR, Bartsch I, Doehmer J and Werle-Schneider G (1996) External versus internal metabolic activation of polycyclic aromatic compounds in mutagenicity tests: a comparison using heterologous expression systems. *Polycycl. Aromat. Compds.* 10: 43–50.
65. Landsiedel R, Engst W, Seidel A and Glatt HR (1996) Physico-chemical properties and mutagenicity of benzylic compounds. *Exp. Toxicol. Pathol.* 48, Suppl 2: 215–222.
66. Glatt HR (2000) Sulfotransferases in the bioactivation of xenobiotics. *Chem.-Biol. Interact.* 129: 141–170.
67. Casale GP, Singhal M, Bhattacharya SK, RamaNathan R, Roberts KP, Barbacci DC, Zhao J, Jankowiak R, Gross ML, Cavalieri EL, Small GJ, Rennard SI, Mumford JL and Shen M (2001) Detection and quantification of depurinated benzo[*a*]pyrene-adducted DNA bases in the urine of cigarette

smokers and women exposed to household coal smoke. *Chem. Res. Toxicol.* 14: 192–201.

68. Cavalieri EL and Rogan EG (1995) Central role of radical cations in metabolic activation of polycyclic aromatic hydrocarbons. *Xenobiotica* 25: 677–688.
69. Ludewig G, Dogra S, Setiabudi F, Seidel A, Oesch F and Glatt HR (1991) Quinones derived from polycyclic aromatic hydrocarbons: induction of diverse mutagenic and genotoxic effects in mammalian cells. In: *Polynuclear Aromatic Hydrocarbons: Measurements, Means, and Metabolism* [Cooke M, Loening K, and Merritt J (eds.)] pp 545–556, Battelle Press, Columbus, OH, USA.
70. Chesis PL, Levin DE, Smith MT, Ernster L and Ames BN (1984) Mutagenicity of quinones: pathways of metabolic activation and detoxification. *Proc. Natl. Acad. Sci. USA* 81: 1696–1700.
71. Kranendonk M, Carreira F, Theisen P, Laires A, Fisher CW, Rueff J, Estabrook RW and Vermeulen NPE (1999) *Escherichia coli* MTC, a human NADPH P450 reductase competent mutagenicity tester strain for the expression of human cytochrome P450 isoforms 1A1, 1A2, 2A6, 3A4, or 3A5: catalytic activities and mutagenicity studies. *Mutat. Res.* 441: 73–83.
72. Josephy PD, Batty SM and Boverhof DR (2001) Recombinant human P450 forms 1A1, 1A2, and 1B1 catalyze the bioactivation of heterocyclic amine mutagens in *Escherichia coli* lacZ strains. *Environ. Mol. Mutagen.* 38: 12–18.
73. Kamataki T and Fujita K (2002) Mutagenesis testing based on bacterial expression of human P450s. *Methods Enzymol.* 357: 301–310.
74. Xu J, Whong WZ and Ong T (1984) Validation of the *Salmonella* (SV50)/arabinose-resistant forward mutation assay system with 26 compounds. *Mutat. Res.* 130: 79–86.
75. De Carneri I, Castellino S, Lorenzetti R, Ravenna L and Rivolta R (1980) Quantitative forward mutation to 8-azaguanine resistance and reverse mutation to histidine independence induced in *Salmonella typhimurium* by metabolically activated chemicals. *Microbiologica* 3: 1–14.
76. Penman BW, Kaden DA, Liber HL, Skopek TR and Thilly WG (1980) Perylene is a more potent mutagen than benzo[*a*]pyrene for *S. typhimurium*. *Mutat. Res.* 77: 271–277.
77. Skopek TR and Thilly WG (1983) Rate of induced forward mutation at 3 genetic loci in *Salmonella typhimurium*. *Mutat. Res.* 108: 45–56.
78. Mersch-Sundermann V, Rosenkranz HS and Klopman G (1992) Structural basis of the genotoxicity of polycyclic aromatic hydrocarbons. *Mutagenesis* 7: 211–218.

79. Shimada T, Oda Y, Gillam EMJ, Guengerich FP and Inoue K (2001) Metabolic activation of polycyclic aromatic hydrocarbons and other procarcinogens by cytochromes P450 1A1 and P4501B1 allelic variants and other human cytochromes P450 in *Salmonella typhimurium* NM2009. *Drug Metab. Disp.* 29: 1176–1182.
80. Aryal P, Terashita T, Guengerich FP, Shimada T and Oda Y (2000) Use of genetically engineered *Salmonella typhimurium* OY1002/1A2 strain coexpressing human cytochrome P450 1A2 and NADPH-cytochrome P450 reductase and bacterial *O*-acetyltransferase in SOS/umu assay. *Environ. Mol. Mutagen.* 36: 121–126.
81. Legator MS, Bueding E, Batzinger R, Connor TH, Eisenstadt E, Farrow MG, Ficsor G, Hsie A, Seed J and Stafford RS (1982) An evaluation of the host-mediated assay and body fluid analysis: a report of the U.S. Environmental Protection Agency Gene-Tox Program. *Mutat. Res.* 98: 319–374.
82. Simmon VF, Rosenkranz HS, Zeiger E and Poirier LA (1979) Mutagenic activity of chemical carcinogens and related compounds in the intraperitoneal host-mediated assay. *J. Natl. Cancer Inst.* 62: 911–918.
83. Batzinger RP, Ou SYL and Bueding E (1978) Antimutagenic effects of 2(3)-*tert*-butyl-4-hydroxyanisole and of antimicrobial agents. *Cancer Res.* 38: 4478–4485.
84. Glatt HR, Bücker M, Platt KL and Oesch F (1985) Host-mediated mutagenicity experiments with benzo[*a*]pyrene and two of its metabolites. *Mutat. Res.* 156: 163–169.
85. Glatt HR (1994) Comparison of common gene mutation tests in mammalian cells in culture: a position paper of the GUM Commission for the Development of Guidelines for Genotoxicity Testing. *Mutat. Res.* 313: 7–20.
86. Huberman E, Aspiras L, Heidelberger C, Grover PL and Sims P (1971) Mutagenicity to mammalian cells of epoxides and other derivatives of polycyclic hydrocarbons. *Proc. Natl. Acad. Sci. USA* 68: 3195–3199.
87. Huberman E and Sachs L (1976) Mutability of different genetic loci in mammalian cells by metabolically activated carcinogenic polycyclic hydrocarbons. *Proc. Natl. Acad. Sci. USA* 73: 188–192.
88. Krahn DF and Heidelberger C (1977) Liver homogenate-mediated mutagenesis in Chinese hamster V79 cells by polycyclic aromatic hydrocarbons and aflatoxins. *Mutat. Res.* 46: 27–44.
89. Dogra S, Doehmer J, Glatt HR, Mölders R, Siegert P, Friedberg T, Seidel A and Oesch F (1990) Stable expression of rat cytochrome P450IA1 cDNA

in V79 Chinese hamster cells and their use in mutagenicity testing. *Mol. Pharmacol.* 37: 608–613.

90. Glatt HR, Pudil J, Seidel A and Czich A (1994) Stable expression of heterologous sulfotransferase in V79 cells: activation of primary and secondary benzylic alcohols. *Polycycl. Aromat. Compds.* 7: 153–160.
91. Czich A, Bartsch I, Dogra S, Hornhardt S and Glatt HR (1994) Stable heterologous expression of hydroxysteroid sulphotransferase in Chinese hamster V79 cells and their use for toxicological investigations. *Chem.-Biol. Interact.* 92: 119–128.
92. Glatt HR, Pabel U, Muckel E and Meinl W (2002) Activation of polycyclic aromatic compounds by cDNA-expressed phase I and phase II enzymes. *Polycycl. Aromat. Compds.* 22: 955–967.
93. Teubner W, Meinl W and Glatt HR (2002) Stable expression of rat sulfotransferase 1B1 in V79 cells: activation of benzylic alcohols to mutagens. *Carcinogenesis* 23: 1877–1884.
94. Wislocki PG, Wood AW, Chang RL, Levin W, Yagi H, Hernandez O, Dansette PM, Jerina DM and Conney AH (1976) Mutagenicity and cytotoxicity of benzo[*a*]pyrene arene oxides, phenols, quinones, and dihydrodiols in bacterial and mammalian cells. *Cancer Res.* 36: 3350–3357.
95. Huberman E, Sachs L, Yang SK and Gelboin V (1976) Identification of mutagenic metabolites of benzo[*a*]pyrene in mammalian cells. *Proc. Natl. Acad. Sci. USA* 73: 607–611.
96. Newbold RF and Brookes P (1976) Exceptional mutagenicity of a benzo[*a*]pyrene diol epoxide in cultured mammalian cells. *Nature* 261: 52–54.
97. Chang RL, Wood AW, Levin W, Mah HD, Thakker DR, Jerina DM and Conney AH (1979) Differences in mutagenicity and cytotoxicity of (+)- and (−)-benzo[*a*]pyrene 4,5-oxide: a synergic interaction of enantiomers. *Proc. Natl. Acad. Sci. USA* 76: 4280–4284.
98. Glatt HR, Seidel A, Bochnitschek A, Marquardt H, Hodgson RM, Grover PL and Oesch F (1986) Mutagenic and cell-transforming activities of triol-epoxides as compared to other chrysene metabolites. *Cancer Res.* 46: 4556–4565.
99. Glatt HR, Gemperlein I, Setiabudi F, Platt KL and Oesch F (1990) Expression of xenobiotic-metabolising enzymes in propagatable cell cultures and induction of micronuclei by 13 compounds. *Mutagenesis* 5: 241–249.
100. Glatt HR, Ludewig G, Platt KL, Waechter F, Yona I, Ben-Shoshan S, Jerushalmy P, Blum J and Oesch F (1985) Arene imines, a new class of

exceptionally potent mutagens in bacterial and mammalian cells. *Cancer Res.* 45: 2600–2607.

101. Seidel A, Friedberg T, Löllmann B, Schwierzok A, Funk M, Frank H, Holler R, Oesch F and Glatt HR (1998) Detoxification of optically active bay- and fjord-region polycyclic aromatic hydrocarbon dihydrodiol epoxides by human glutathione transferase P1-1 expressed in Chinese hamster V79 cells. *Carcinogenesis* 19: 1975–1981.
102. Sundberg K, Dreij K, Berntsen S, Seidel A and Jernström B (2001) Expression of human glutathione transferases in V79 cells and the effect on DNA adduct-formation of diol epoxides derived from polycyclic aromatic hydrocarbons. *Chem.-Biol. Interact.* 133: 91–94.
103. Pal K, Grover PL and Sims P (1980) The induction of sister-chromatid exchanges in Chinese hamster ovary cells by some epoxides and phenolic derivatives of benzo[*a*]pyrene. *Mutat. Res.* 78: 193–199.
104. Pal K (1981) The induction of sister-chromatid exchanges in Chinese hamster ovary cells by K-region epoxides and some dihydrodiols derived from benz[*a*]anthracene, dibenz[*a,c*]anthracene and dibenz[*a,h*]anthracene. *Mutat. Res.* 84: 389–398.
105. Glatt HR, Bartsch I, Christoph S, Coughtrie MWH, Falany CN, Hagen M, Landsiedel R, Pabel U, Phillips DH, Seidel A and Yamazoe Y (1998) Sulfotransferase-mediated activation of mutagens studied using heterologous expression systems. *Chem.-Biol. Interact.* 109: 195–219.
106. Rüdiger HW, Harder W, Maack P, Kohl FV and Schmidt-Preuss U (1981) Decreased rate of benzo[*a*]pyrene-induced sister chromatid exchange in fibroblast cultures from patients with lung cancer. *J. Cancer Res. Clin. Oncol.* 102: 169–175.
107. Salama SA, Sierra-Torres CH, Oh HY, Hamada FA and Au WW (2001) Variant metabolizing gene alleles determine the genotoxicity of benzo[*a*]pyrene. *Environ. Mol. Mutagen.* 37: 17–26.
108. Rice JE, Rivenson A, Braley J and LaVoie EJ (1987) Methylated derivatives of pyrene and fluorene: evaluation of genotoxicity in the hepatocyte/DNA repair test and tumorigenic activity in newborn mice. *J. Toxicol. Environ. Health* 21: 525–532.
109. Andrae U, Kreis P, Coughtrie MWH, Pabel U, Meinl W, Bartsch I and Glatt HR (1999) Activation of propane 2-nitronate to a genotoxicant in V79-derived cell lines engineered for the expression of rat hepatic sulfotransferases. *Mutat. Res.* 439: 191–197.

110. Kreis P, Brandner S, Coughtrie MWH, Pabel U, Meinl W, Glatt HR and Andrae U (2000) Human phenol sulfotransferases hP-PST and hM-PST activate propane 2-nitronate to a genotoxicant. *Carcinogenesis* 21: 295–299.
111. Lavi S and Etkin S (1981) Carcinogen-mediated induction of SV40 DNA synthesis in SV40 transformed Chinese hamster embryo cells. *Carcinogenesis* 2: 417–423.
112. Raj AS and Katz M (1983) Inhibitory effect of 7,8-benzoflavone on DMBA- and BaP-induced bone marrow micronuclei in mouse. *Mutat. Res.* 110: 337–342.
113. Shimada H, Suzuki H, Itoh S, Hattori C, Matsuura Y, Tada S and Watanabe C (1992) The micronucleus test of benzo[*a*]pyrene with mouse and rat peripheral blood reticulocytes. *Mutat. Res.* 278: 165–168.
114. Suzuki T, Tamai K, Kodama Y, Asita AO, Matsuoka A, Sofuni T, Kurita M, Ohtsuki H, Hiwatashi T and Hayashi M (1992) Micronucleus induction in mouse peripheral reticulocytes by 7,12-dimethylbenz[*a*]anthracene. *Mutat. Res.* 278: 169–173.
115. Glatt HR, Seidel A, Oesch F and Gumbsch A (1994) Fjord-region diolepoxides of benzo[*c*]chrysene are potent inducers of micronuclei in murine bone marrow. *Mutat. Res.* 309: 37–43.
116. He SL and Baker R (1991) Micronuclei in mouse skin cells following *in vivo* exposure to benzo[*a*]pyrene, 7,12-dimethylbenz[*a*]anthracene, chrysene, pyrene and urethane. *Environ. Mol. Mutagen.* 17: 163–168.
117. Baker RSU, Bonin AM, Arlauskas A, He S and Coombs MM (1992) Tumorigenicity of cyclopenta[*a*]phenanthrene derivatives and micronucleus induction in mouse skin. *Carcinogenesis* 13: 329–332.
118. Cole RJ, Cole J, Henderson L, Taylor NA, Arlett CF and Regan T (1983) Short-term tests for transplacentally active carcinogens: a comparison of sister-chromatid exchange and the micronucleus test in mouse foetal liver erythroblasts. *Mutat. Res.* 113: 61–75.
119. Müller L (1988) Micronucleus induction in mouse and rat fetuses treated transplacentally during histogenesis with mitomycin C and 7,12-dimethylbenz[*a*]anthracene. *Teratog. Carcinog. Mutagen.* 8: 303–313.
120. Wang MY and Lu LJ (1990) Differential effect of gestation stage on benzo[*a*]pyrene-induced micronucleus formation and/or covalent DNA modifications in mice. *Cancer Res.* 50: 2146–2151.
121. Bol SA, van Steeg H, Jansen JG, van Oostrom C, de Vries A, de Groot AJ, Tates AD, Vrieling H, van Zeeland AA and Mullenders LH (1998) Elevated frequencies of benzo[*a*]pyrene-induced *Hprt* mutations in internal tissue of XPA-deficient mice. *Cancer Res.* 58: 2850–2856.

122. Wijnhoven SWP, Kool HJM, van Oostrom CTM, Beems RB, Mullenders LHF, van Zeeland AA, van der Horst GTJ, Vrieling H and van Steeg H (2000) The relationship between benzo[*a*]pyrene-induced mutagenesis and carcinogenesis in repair-deficient Cockayne syndrome group B mice. *Cancer Res.* 60: 5681–5687.
123. Maier P, Manser P and Zbinden G (1980) Granuloma pouch assay: induction of 6-thioguanine resistance by MNNG and benzo[*a*]pyrene *in vivo*. *Mutat. Res.* 77: 165–173.
124. Russell LB, Selby PB, von Halle E, Sheridan W and Valcovic L (1981) Use of the mouse spot test in chemical mutagenesis: interpretation of past data and recommendations for future work. *Mutat. Res.* 86: 355–379.
125. Hakura A, Tsutsui Y, Sonoda J, Kai JK, Imade T, Shimada M, Sugihara Y and Mikami T (1998) Comparison between *in vivo* mutagenicity and carcinogenicity in multiple organs by benzo[*a*]pyrene in the *lacZ* transgenic mouse (Muta™Mouse). *Mutat. Res.* 398: 123–130.
126. Dean SW, Coates A, Brooks TM and Burlinson B (1998) Benzo[*a*]pyrene site of contact mutagenicity in skin of Muta™ mouse. *Mutagenesis* 13: 515–518.
127. Miller ML, Vasunia K, Talaska G Andringa A, de Boer J and Dixon K (2000) The tumor promoter TPA enhances benzo[*a*]pyrene and benzo[*a*]pyrene diol-epoxide mutagenesis in Big Blue mouse skin. *Environ. Mol. Mutagen.* 35: 319–327.
128. Bayer U and Bauknecht T (1977) The dose-dependence of sister chromatid exchanges induced by 3 hydrocarbons, in the *in vivo* bone marrow test with Chinese hamsters. *Experientia* 33: 25.
129. Roszinsky-Köcher G, Basler A and Röhrborn G (1979) Mutagenicity of polycyclic hydrocarbons: V. Induction of sister-chromatid exchanges *in vivo*. *Mutat. Res.* 66: 65–67.
130. Schreck RR and Latt SA (1980) Comparison of benzo[*a*]pyrene metabolism and sister chromatid exchange induction in mice. *Nature* 288: 407–408.
131. Tice RR, Ivett JL and McFee AF (1987) The effect of agent treatment time on the induction of sister-chromatid exchanges in mouse bone marrow cells *in vivo*. *Mutat. Res.* 182: 15–29.
132. Sirianni SR and Huang CC (1978) Sister chromatid exchange induced by promutagens/carcinogens in Chinese hamster cells cultured in diffusion chambers in mice. *Proc. Soc. Exp. Biol. Med.* 158: 269–274.
133. Galloway SM, Perry PE, Meneses J, Nebert DW and Pedersen RA (1980) Cultured mouse embryos metabolize benzo[*a*]pyrene during early gestation: genetic differences detectable by sister chromatid exchange. *Proc. Natl. Acad. Sci. USA* 77: 3524–3528.

134. Murota T, Shibuya T and Iwahara S (1981) Dominant lethal testing of 7,12-dimethylbenz[*a*]anthracene with spermatogenic cells of mice. *Mutat. Res.* 91: 273–277.
135. Generoso WM, Cain KT, Hellwig CS and Cacheiro NLA (1982) Lack of association between induction of dominant-lethal mutations and induction of heritable translocations with benzo[*a*]pyrene in postmeiotic germ cells of male mice. *Mutat. Res.* 94: 155–163.
136. Schiltz M, Cui XX, Lu YP, Yagi H, Jerina DM, Zdzienicka MZ, Chang RL, Conney AH and Wei SJC (1999) Characterization of the mutational profile of (+)-7*R*,8*S*-dihydroxy-9*S*,10*R*-epoxy-7,8,9,10-tetrahydrobenzo[*a*]pyrene at the hypoxanthine (guanine) phosphoribosyltransferase gene in repair-deficient Chinese hamster V-H1 cells. *Carcinogenesis* 20: 2279–2285.
137. Hakura A, Tsutsui Y, Sonoda J, Tsukidate K, Mikami T and Sagami F (2000) Comparison of the mutational spectra of the *lacZ* transgene in four organs of the Muta™ mouse treated with benzo[*a*]pyrene: target organ specificity. *Mutat. Res.* 447: 239–247.
138. Bizub D, Wood AW and Skalka AM (1986) Mutagenesis of the Ha-*ras* oncogene in mouse skin tumors induced by polycyclic aromatic hydrocarbons. *Proc. Natl. Acad. Sci. USA* 83: 6048–6052.
139. Ruggeri B, Dirado M, Zhang SY, Bauer B, Goodrow T and Kleinszanto AJP (1993) Benzo[*a*]pyrene-induced murine skin tumors exhibit frequent and characteristic G-to-T mutations in the *p53*-gene. *Proc. Natl. Acad. Sci. USA* 90: 1013–1017.
140. Wang DA, You L, Sneddon J, Cheng SJ, Jamasbi R and Stoner GD (1995) Frameshift mutation in codon 176 of the *p53* gene in rat esophageal epithelial cells transformed by benzo[*a*]pyrene dihydrodiol. *Mol. Carcinog.* 14: 84–93.
141. Cherpillod P and Amstad PA (1995) Benzo[*a*]pyrene-induced mutagenesis of *p53* hot-spot codons 248 and 249 in human hepatocytes. *Mol. Carcinog.* 13: 15–20.
142. Puisieux A, Lim S, Groopman J and Ozturk M (1991) Selective targeting of *p53* gene mutational hotspots in human cancer by etiologically defined carcinogens. *Cancer Res.* 51: 6185–6189.
143. Krawczak M and Cooper DN (1998) *P53* mutations, benzo[*a*]pyrene and lung cancer. *Mutagenesis* 13: 319–320.

144. Yang JL, Maher VM and McCormick JJ (1987) Kinds of mutations formed when a shuttle vector containing adducts of benzo[*a*]pyrene-7,8-diol-9,10-epoxide replicates in COS7 cells. *Mol. Cell Biol.* 7: 1267–170.
145. Roilides E, Gielen JE, Tuteja N, Levine AS and Dixon K (1988) Mutational specificity of benzo[*a*]pyrene diolepoxide in monkey cells. *Mutat. Res.* 198: 199–206.
146. Gill RD, Rodriguez H, Cortez C, Harvey RG, Loechler EL and DiGiovanni J (1993) Mutagenic specificity of the (+)*anti*-diol epoxide of dibenz[*a*, *j*]anthracene in the *supF* gene of an *Escherichia coli* plasmid. *Mol. Carcinog.* 8: 145–154.
147. Jelinsky SA, Liu TM, Geacintov NE and Loechler EL (1995) The major, N^2-Gua adduct of the (+)-*anti*-benzo[*a*]pyrene diol epoxide is capable of inducing G → A and G → C, in addition to G → T, mutations. *Biochemistry* 34: 13545–13553.
148. Bigger CAH, Ponten I, Page JE and Dipple A (2000) Mutational spectra for polycyclic aromatic hydrocarbons in the *supF* target gene. *Mutat. Res.* 450: 75–93.
149. Shukla R, Liu T, Geacintov NE and Loechler EL (1997) The major, N^2-dG adduct of (+)-*anti*-B[*a*]PDE shows a dramatically different mutagenic specificity (predominantly, G → A) in a 5′-C<u>G</u>T-3′ sequence context. *Biochemistry* 36: 10256–10261.
150. Shukla R, Geacintov NE and Loechler EL (1999) The major, N^2-dG adduct of (+)-*anti*-B[*a*]PDE induces G → A mutations in a 5′-AGA-3′ sequence context. *Carcinogenesis* 20: 261–268.
151. Ponten I, Kroth H, Sayer JM, Dipple A and Jerina DM (2001) Differences between the mutational consequences of replication of *cis*- and *trans*-opened benzo[*a*]pyrene 7,8-diol 9,10-epoxide-deoxyguanosine adducts in M13mp7L2 constructs. *Chem. Res. Toxicol.* 14: 720–726.
152. Ramos LA, Ponten I, Dipple A, Kumar S, Yagi H, Sayer JM, Kroth H, Kalena G and Jerina DM (2002) Site-specific mutagenesis in *Escherichia coli* by N^2-deoxyguanosine adducts derived from the highly carcinogenic fjord-region benzo[*c*]phenanthrene 3,4-diol 1,2-epoxides. *Chem. Res. Toxicol.* 15: 1619–1626.
153. Kramata P, Zajc B, Sayer JM, Jerina DM and Wei CS (2003) A single site-specific *trans*-opened 7,8,9,10-tetrahydrobenzo[*a*]pyrene 7,8-diol 9,10-epoxide N^2-deoxyguanosine adduct induces mutations at multiple sites in DNA. *J. Biol. Chem.* 278: 14940–14948.

154. Ma L, Kuhlow A and Glatt HR (2002) Ethanol enhances the activation of 1-hydroxymethylpyrene to DNA adduct-forming species in the rat. *Polycycl. Aromat. Compds.* 22: 933–946.
155. Meinl W, Meerman JH and Glatt HR (2002) Differential activation of promutagens by alloenzymes of human sulfotransferase 1A2 expressed in *Salmonella typhimurium*. *Pharmacogenetics* 12: 677–689.
156. Phillips DH, Glatt HR, Seidel A, Bochnitschek W, Oesch F and Grover PL (1986) Mutagenic potential and DNA adducts formed by diol-epoxides, triol-epoxides and K-region epoxide of chrysene in mammalian cells. *Carcinogenesis* 7: 1739–1743.
157. Glatt HR, Harvey RG, Phillips DH, Hewer A and Grover PL (1989) Influence of the alkyl substituent on mutagenicity and covalent DNA-binding of bay-region diol-epoxides of 7-methyl- and 7-ethylbenz[*a*]anthracene in *Salmonella* and V79 Chinese hamster cells. *Cancer Res.* 49: 1778–1782.
158. Phillips DH, Hewer A, Seidel A, Steinbrecher T, Schrode R, Oesch F and Glatt HR (1991) Relationship between mutagenicity and DNA adduct formation in mammalian cells for fjord-region and bay-region diol-epoxides of polycyclic aromatic hydrocarbons. *Chem.-Biol. Interact.* 80: 177–186.
159. Glatt HR, Wameling C, Elsberg S, Thomas H, Marquardt H, Hewer A, Phillips DH, Oesch F and Seidel A (1993) Genotoxicity characteristics of reverse diol-epoxides of chrysene. *Carcinogenesis* 14: 11–19.

9 Tumorigenicity of Polycyclic Aromatic Hydrocarbons

"On the Possible Contribution of PAHs and their Nitro-Derivatives to the Development of Human Breast Cancer"

Shantu Amin[†] and Karam El-Bayoumy[‡]

American Health Foundation Cancer Center, Institute for Cancer Prevention, Valhalla, NY, USA
E-mails: [†] samin@ifcp.us; [‡] kelbayou@ifcp.us

9.1 Introduction

Breast cancer is second only to lung cancer as the leading cause of cancer-related deaths in American women.[1] Most of the cases are in industrialized countries (an estimated 180,000 in North America and 220,000 in Europe annually). Breast cancer is diagnosed in 910,000 women worldwide and 376,000 women die from it each year.[2] The etiology of breast cancer remains obscure even though a vast body of literature describes risk factors for breast cancer and proposes various hypotheses for its etiology on the basis of epidemiological and experimental studies. Increased exposure to estrogens can account for only less than one-third of all breast cancer cases.[2–4] A limited number of studies have shown that estrogens can exert a genotoxic effect, though mechanisms of estrogen-induced carcinogenesis remain unclear.[5,6] Inherited mutations in cancer susceptibility genes, such as *BRCA1* and *BRCA2*, are associated with many cases of familial breast cancer which account for about 5% of overall breast cancer incidence.[3–7] Recently, Jeffy *et al.* suggested that exposure to environmental polycyclic aromatic hydrocarbons (PAHs) may be a predisposing factor in the etiology of sporadic breast cancer because PAHs may disrupt the expression of *BRCA1*.[8]

It has been estimated that, in addition to genetic disposition, a significant portion of cancer incidence in the U.S. is related to environmental factors and lifestyle — including diet.[9–13] International studies have repeatedly confirmed that migrants adopt the breast cancer pattern of their new countries within a few generations, which is indicative of the presence of carcinogens in the environment and/or changes in lifestyle.[2] Increased risk of the descendants (2nd and 3rd generations) of migrants coming from low-risk to high-risk regions may be associated with exposure to genotoxic agents during the development of the human breast, since undifferentiated ductal elements of the breast are more susceptible to carcinogenic insult early in life.[2,14–17] Consequently, the search for carcinogens that exist in the human environment challenges both scientists and regulatory agencies.[18–20] There is limited evidence supporting the hypothesis that occupational risk factors for breast cancer are associated with certain industrial activities.[21–23] However, there is sufficient evidence on the causal association of ionizing radiation and breast cancer; this is clearly shown among atomic bomb survivors and in cohorts of women repeatedly exposed to therapeutic doses of radiation.[24] Such an association indicates that the mammary gland may be also susceptible to tumor initiation by environmental carcinogens that can damage DNA.

9.2 Environmental Genotoxic Agents

Chronic exposure to traces of chemical carcinogens in the diet, in polluted air, or in tobacco smoke can be important in the etiology of breast cancer in the presence of host factors that favor the multi-step process of carcinogenesis. At the present time, the association between cigarette smoking and breast cancer risk remains unclear;[25] further investigations are needed to define the molecular mechanisms underlying potential carcinogenic effects of cigarette smoking in mammary tissues. Investigations of interactions of cigarette smoke constituents with genes involved in carcinogen metabolism, as well as those involved in DNA repair, cell proliferation, cell-cycle and apoptosis are highly attractive and plausible areas of future research.[26] Literature data document that nicotine and its metabolite, cotinine, are present in the breast fluid of non-lactating women who smoke;[27] clearly, tobacco smoke constituents are reaching the human breast.[28] The characteristic, high lipid content of the mammary gland makes this organ a reservoir for lipid-soluble, toxic, and carcinogenic agents.[29,30] Mutagenic activity has been detected in nipple aspirates[31] and breast cyst fluid.[32–34] Genotoxicity induced by extracts of mammary lipids from women undergoing breast reduction has been reported; and extracts of human breast milk have also been shown to have genotoxic activity.[35–37] Several studies describe the detection of DNA adducts in human mammary tissue;[38–45] however, future studies should focus on determining the nature of these adducts that remain largely unknown.

Bioassays in laboratory animals can provide important information on the role of environmental agents in the induction of particular types of cancer. Biochemical studies can lead to insights into the nature of interactions of these environmental agents with macromolecules, such as DNA that are necessary, but may not always be sufficient for carcinogenesis. Environmental carcinogens responsible for genomic alterations in human breast cancer have not been clearly defined. Characteristic mutations of human cancer genes (e.g., *p53*) have potential as molecular markers of past exposure to specific carcinogens (cf. Chapter 5). However, as exposures often arise from complex mixtures and multiple other factors that may also be important, it is generally difficult to identify 'signature mutations' of a particular human carcinogen. Transgenic animals and their retrievable 'reporter genes' provide a highly useful system for the identification of signature mutations that can be compared to mutational spectra of cancer genes and such information will be helpful in risk assessment.[46,47] Collectively, information based on biochemical and molecular analysis of human breast tissues, healthy and cancerous cells, or breast fluids can provide insight into the role of environmental carcinogens in breast cancer etiology.

Ubiquitous environmental agents that are known inducers of mammary cancer in rodents should be regarded as potential human risk factors and need to be evaluated more closely. About a decade ago, in a survey that was based on literature data pertaining to the occurrence of carcinogens in the human environment and their relative tumorigenic effect, we hypothesized that certain environmental carcinogens are likely to be involved in human breast cancer etiology.[18] Examples of these environmental carcinogens are the class of PAHs, their nitro-substituted derivatives (NO_2-PAHs), as well as aromatic and heterocyclic amines.[18]

Recently, Gammon *et al.* aimed at determining whether breast cancer risk is increased in relation to exposure to organochlorine compounds with known estrogenic characteristics that were extensively used on Long Island and in other areas of the United States.[48] The results of their study, based on blood analysis, do not support the hypothesis that organochlorine compounds increase breast cancer risk among Long Island women. In a separate report, these investigators describe results of an ELISA of blood samples from breast cancer cases and controls that were assayed for PAH-DNA adducts.[49] This study indicated that PAH-DNA adduct formation may influence breast cancer development, although the association does not appear to be dose-dependent and may have a threshold effect. It was suggested that breast tissue, rather than blood, is the appropriate biological matrix to analyze for PAH-DNA adducts and determine their role in the development of breast cancer.[50]

The present review is intended to update our knowledge and to stimulate further research into the role of PAHs and NO_2-PAHs in the development of human breast cancer. Specifically, we compared levels of these agents in environmental sources, their carcinogenic potency in the rat mammary gland, the capacity of human breast tissue to catalyze their metabolic activation, and types of biomarkers in biological fluids and breast tissue that can be employed in future molecular epidemiological and clinical chemoprevention trials in humans. On the basis of this updated information, future research recommendations are proposed.

9.3 PAHs as Representative Examples of Environmental Mammary Carcinogens

9.3.1 Levels and Carcinogenic Potency of PAHs

In an outstanding report published 20 years ago,[51] it was stated that exposure to PAHs is virtually unavoidable and they are strongly suspected to be causatively related to several human cancers of epithelial origin (e.g., skin, lung and colon). This statement remains as valid now as it was in 1983.[52] On the other hand, the precise role of PAHs in the development of human breast cancer is still unclear.

Human exposure to PAHs is not confined to occupational settings (e.g., workers exposed to coal tar products, to emissions in iron foundries, from coke ovens, and in aluminum production plants, as well as the historical soot exposure of chimney sweeps; cf. Chapters 3 and 4). PAHs are present in virtually all products of incomplete combustion of organic matter; they exist in tobacco smoke, broiled foods, gasoline and diesel engine exhaust emissions, and polluted air. Examples of the most prevalent PAHs (see Figure 9.1), including those known to induce mammary cancer in rodents, are provided in Table 9.1.[53–61] Although the presence of certain PAHs has been established in numerous environmental sources, unequivocal quantitative data were not provided. Furthermore, diet contributes substantially to non-occupational exposure to PAHs[52] and epidemiological studies have suggested that a large portion of human cancers is due, in part, to dietary factors (cf. Chapter 4).[10]

Environmental PAHs that have been tested as inducers of mammary carcinogenesis (Table 9.2)[62–66] in the rat were anthracene, anthanthrene, benz[*a*]anthracene (B[*a*]A), benzo[*a*]pyrene (B[*a*]P), cyclopenta[*c,d*]pyrene (CP[*c,d*]P), dibenz[*a,h*]anthracene, dibenzo[*a,e*]pyrene (DB[*a,e*]P), dibenzo[*a,h*]pyrene (DB[*a,h*]P), dibenzo[*a,i*]pyrene (DB[*a,i*]P), dibenzo-[*a,l*]pyrene (DB[*a,l*]P), 5-methylchrysene, and phenanthrene (see Figure 9.1). These agents were administered either by intramammary injection or by gavage. Doses used for intramammary administration did not exceed 32 μmol/rat. However, substantially higher doses were used for oral administration, ranging from 200 to about 1100 μmol/rat. In contrast to animal model assays for cancer of the lung and of the skin, protocols for assays of most of these compounds in the mammary gland in rats do not provide dose response studies. Furthermore, doses used were extremely high by comparison to the levels to which humans are exposed. It should be also emphasized that humans are usually not exposed to

(A) Pyrene and its derivatives:

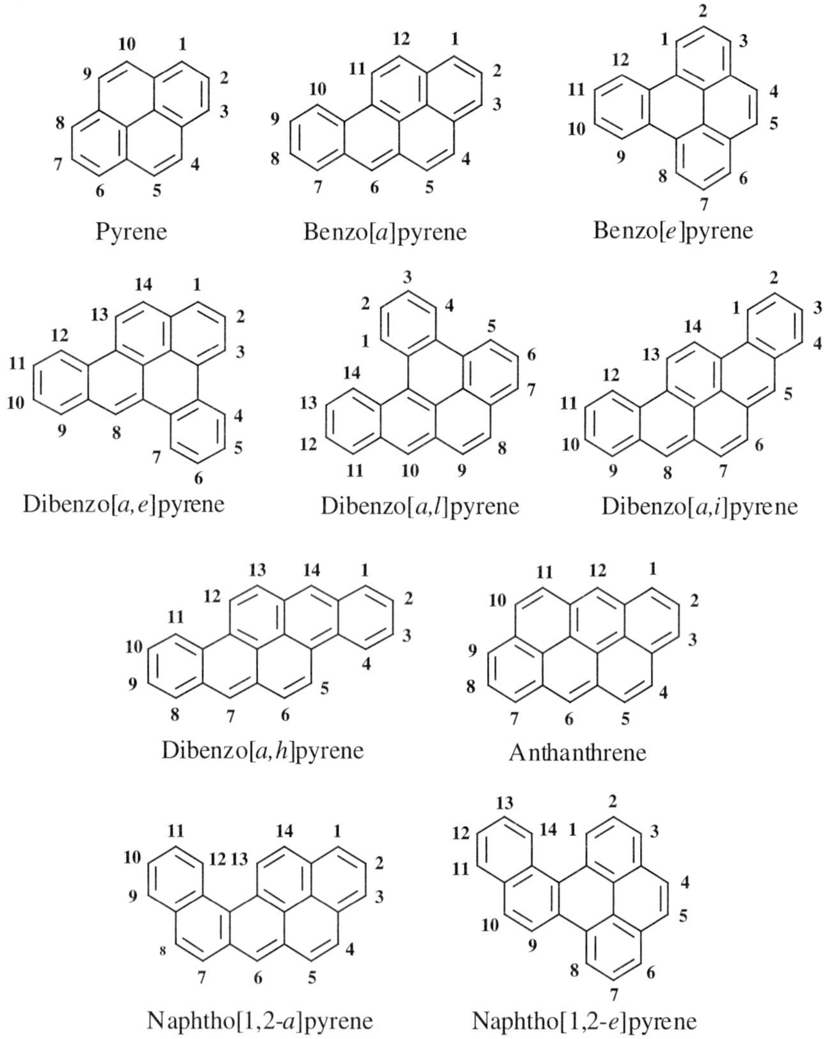

Figure 9.1: Structures of polycyclic aromatic hydrocarbons.

(B) Anthracene and its derivatives:

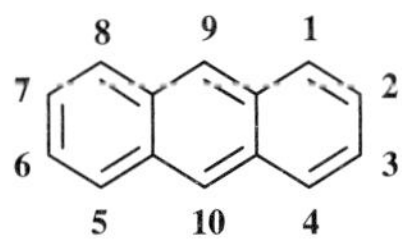

Anthracene

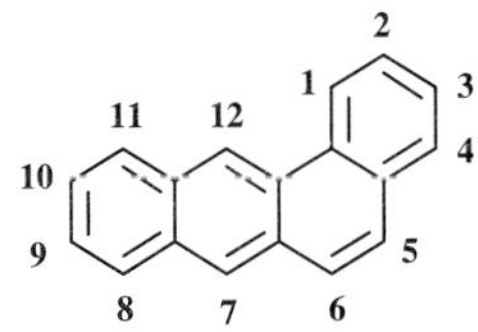

Benz[*a*]anthracene

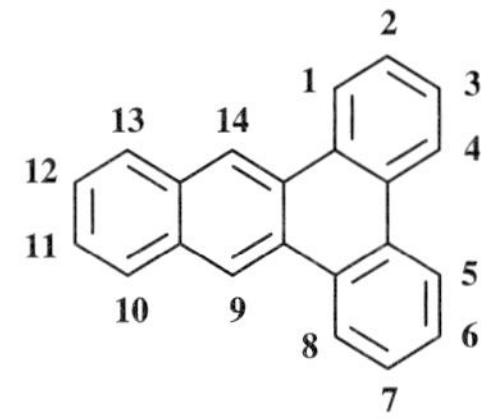

Dibenz[*a,c*]anthracene

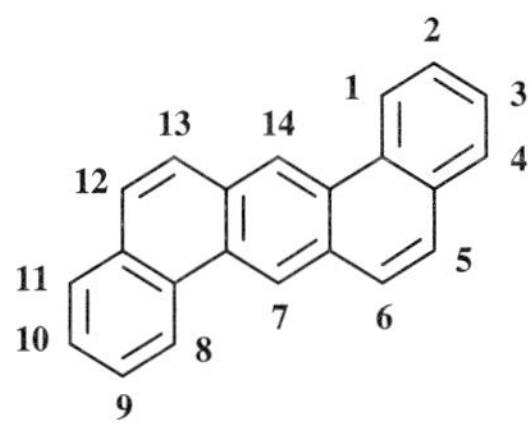

Dibenz[*a,h*]anthracene

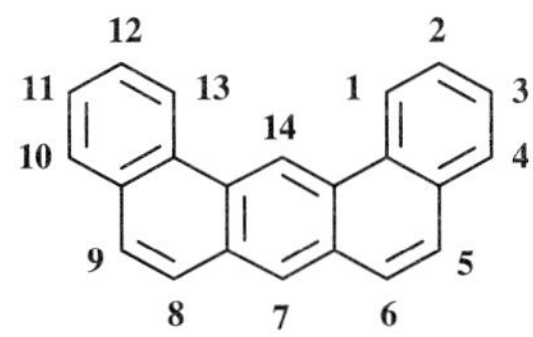

Dibenz[*a,j*]anthracene

(C) Phenanthrene and its derivatives:

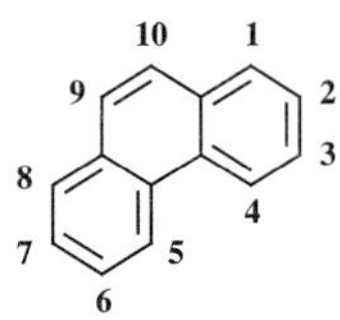

Phenanthrene

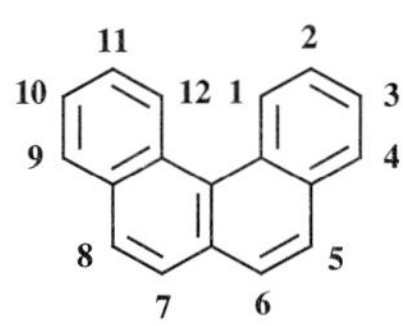

Benzo[*c*]phenanthrene

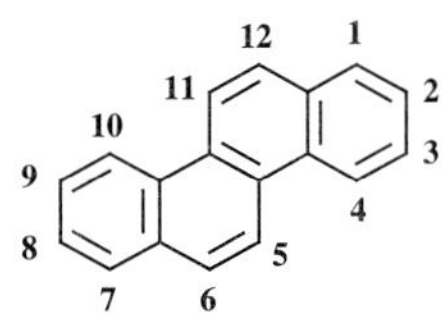

Chrysene

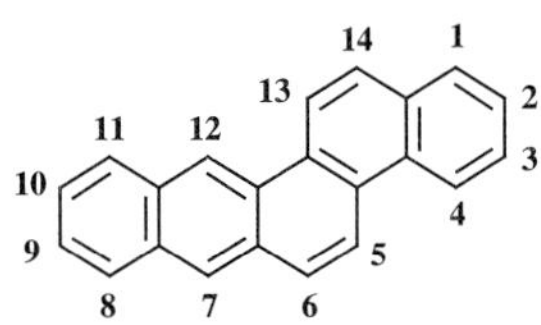

Benzo[*b*]chrysene

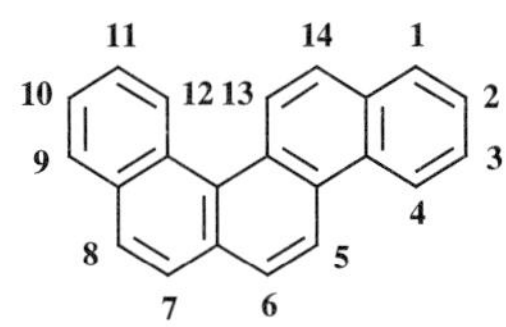

Benzo[*c*]chrysene

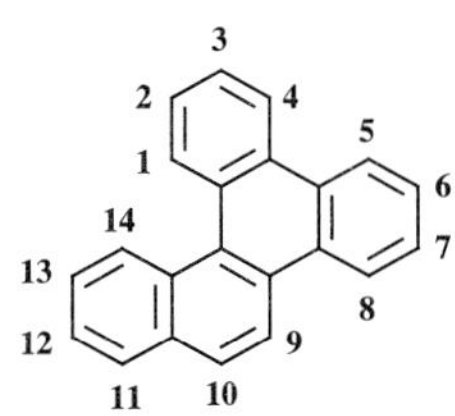

Benzo[*g*]chrysene

Figure 9.1: (continued)

a single PAH, but typically to a mixture of compounds, including NO_2-PAHs.

To realistically assess risks associated with human exposure to environmental mammary carcinogens, it is essential to determine their abundance,

(D) Nonalternant hydrocarbons:

(containing odd-numbered rings) *molecule not fully aromatic

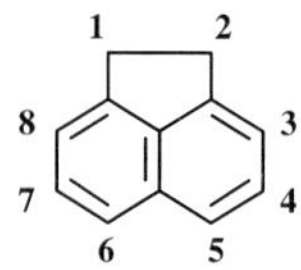

Acenaphthene*

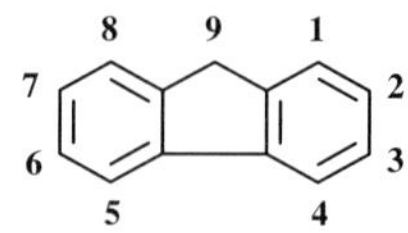

Fluorene*

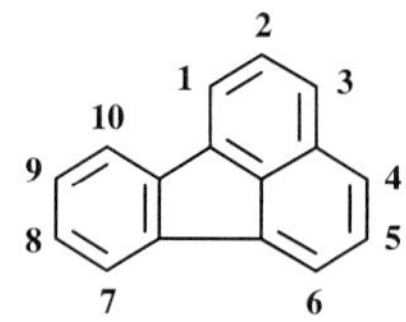

Fluoranthene

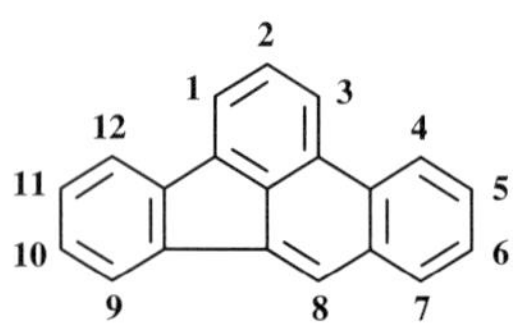

Benzo[*b*]fluoranthene

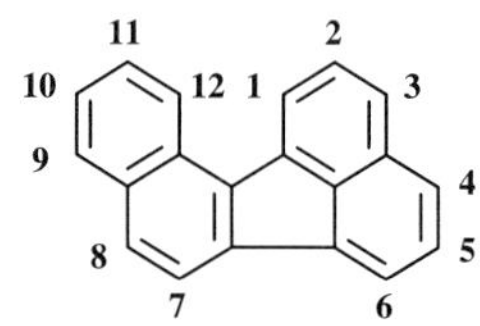

Benzo[*j*]fluoranthene

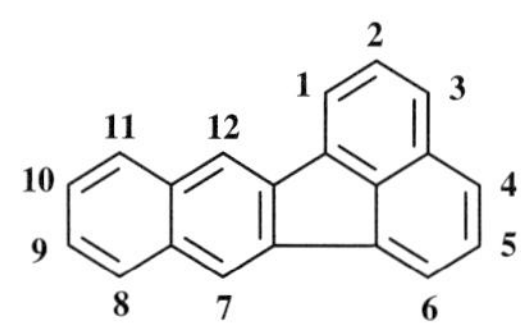

Benzo[*k*]fluoranthene

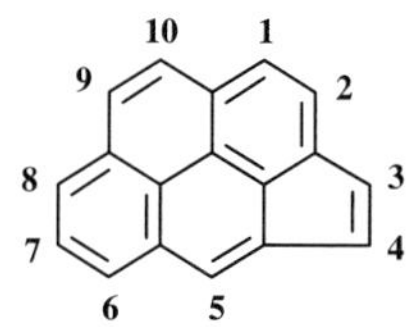

Cyclopenta[*c,d*]pyrene

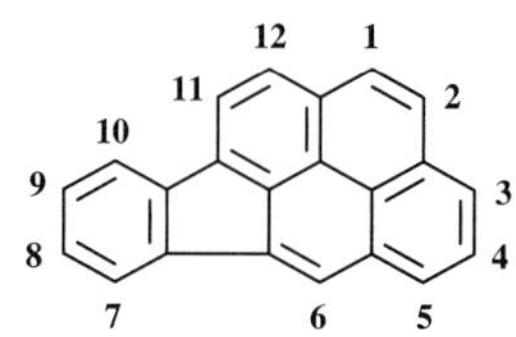

Indeno[1,2,3-*cd*]pyrene

Figure 9.1: (continued)

human uptake and metabolism, and their relative carcinogenic potency in well-defined animal models under identical conditions. Unfortunately, studies reported so far have divergent protocols that do not allow direct comparisons. Furthermore, the differences in dosages, diets and strains of rats in previous bioassays make it impossible to rank each agent according to its true carcinogenic potency. However, making use of existing data, a comparison can be made of tumor incidence obtained with compounds within a given class, and then between different classes, in assays that employed similar routes of administration. At best, a comparison can be made by assuming that the dose response relationship for each compound administered is linear and that the carcinogenic response is independent of the age of the animal used in these studies. With these assumptions in mind,

Table 9.1: Occurrence of polycyclic aromatic hydrocarbons.

	Source							
Compound	*Cigarette smoke μg/100 cigarettes*[55]	*Urban air ng/m³*[53,56]	*Charcoal-broiled steaks μg/kg*[57]	*Gasoline engine exhaust μg/L*[58]	*Sediments μg/kg*[56]	*Burnt coal emissions mg/kg*[56,61]	*Water ng/L*[59]	*Coal tar g/kg*[54,60]
Anthanthrene	0.2–2.2	0.26–6	2	17–26	10–212	0–0.9	0.2–10.9	
Anthracene	2.3–23.5		4.5	534–642	30–650		1.9–1000	2.88–4.35
Benz[*a*]anthracene	0.4–7.6	0.1–21.6	4.5	50–83		0.04–9.4	0–30.6	6.24–6.98
Benzo[*b*]chrysene								0.80–0.93
Benzo[*c*]chrysene								P
Benzo[*g*]chrysene								P
Benzo[*b*]fluoranthene	0.4–2.2	2.3–7.4	0–15.1	29–48		0.05–17.2	0.1–840	0.45–0.63
Benzo[*j*]fluoranthene	0.6–2.1	0.8–4.4		11–27	0.05–17.2		0.6–11.1	0.45–0.63
Benzo[*k*]fluoranthene	0.6–1.2	1.3		7–17		0.05–17.2	0.2–400	1.07–1.08
Benzo[*ghi*]fluoranthene	0.1–0.4			112–244		0.07–8.9	1–11.2	P
Benzo[*a*]fluorene	4.1–18.4			82–136	2–1400		0–0.05	P
Benzo[*b*]fluorene	2	P		65–112			0–0.05	P
Benzo[*c*]fluorene	P	P		65–112				P
Benzo[*ghi*]perylene	0.3–3.9	5.1 ± 1.1	4.5	115–333	1–1930	0.02–2.7	0.02–2.7	1.23–1.89
Benzo[*c*]phenanthrene	P	P		P			1.0–9.1	P
Benzo[*a*]pyrene	0.5–7.8	0.05–74	8	50–81	1–2380	0.39–5.20	1.6–1300	1.76–2.08
Benzo[*e*]pyrene	0.2–2.5	1–37	6	37–59	2–2010	0.02–10.3	0.02–10.3	1.85–1.88
Chrysene	0.6–9.6	1.3–13.3	1.4	85–123	40–240	0.07–21.8	0.07–21.8	2.13–2.86

(continued)

Table 9.2: Mammary tumors induced by polycyclic aromatic hydrocarbons.

Compound	*Source*	*Species & Strain*	*Route*	*Total dose/rat (μmol)*	*Mammary tumor incidence (%)*
Anthanthrene[63]	cigarette smoke, urban air	Sprague-Dawley rats	intramammary	32	5
Benz[*a*]anthracene[62,65]	cigarette smoke, urban air	Sprague-Dawley rats	stomach tube	877	0
			intramammary	16	0
			(powder into incision)	4	0
Benzo[*a*]pyrene[62,64,66]	cigarette smoke, urban air, vehicle exhaust	Sprague-Dawley rats	stomach tube	396.8	89
			intravenous injection	23.8	30
			intramammary	8	0
				2	5
			intramammary	16	50
			(powder into incision)	4	30
		female CD rats	gavage (8 fractions)	400	73
		LEW/MAI rats	gavage (8 fractions)	200	67
Cyclopenta[*c*,*d*]pyrene[62]	cigarette smoke, urban air	Sprague-Dawley rats	intramammary	24	0
			(powder into incision)	8	0
Dibenz[*a*,*h*]anthracene[62]	cigarette smoke, urban air	Sprague-Dawley rats	intramammary	16	0
			(powder into incision)	4	0

Table 9.2: (continued)

Dibenzo[*a,e*]pyrene[63]	cigarette smoke, urban air	Sprague-Dawley rats	intramammary	32	5
Dibenzo[*a,h*]pyrene[63]	cigarette smoke, urban air	Sprague-Dawley rats	intramammary	32	20
Dibenzo[*a,i*]pyrene[63]	cigarette smoke, urban air	Sprague-Dawley rats	intramammary	32	58
Dibenzo[*a,l*]pyrene[63,64]	cigarette smoke, coal tar	Sprague-Dawley rats	intramammary	32	89
			intramammary	8	100
				2	100
5-Methylchrysene[62]	cigarette smoke, vehicle exhaust	Sprague-Dawley rats	intramammary	16	0
			(powder into incision)	4	0
Phenanthrene[65]	cigarette smoke, urban air	Sprague-Dawley rats	stomach tube	1124	0

following intramammary administration, it is possible to rank the carcinogenic potency as follows: DB[*a,l*]P > DB[*a,i*]P ≈ B[*a*]P > DB[*a,h*]P > DB[*a,e*]P = anthracene. However, the carcinogenic potency, if any, in the rat mammary gland of the more abundant PAHs, such as pyrene and fluoranthene, has not been examined. Following intramammary administration, B[*a*]A at 4 and 16 μmol/rat was inactive. In view of the fact that diet is a major source of non-occupational PAH exposure, it is surprising that only a limited number of compounds have been tested by oral administration. B[*a*]P assays at doses ranging between 200 and 400 μmol/rat yielded mammary tumor incidences between 67 and 89%. However, oral administration of phenanthrene at >1100 μmol/rat, or B[*a*]A at 900 μmol/rat showed no effect. The outcome of published studies can be summarized as follows: (1) intramammary administration of several PAHs resulted in mammary cancer induction; however, dose-response studies have not been conducted thus far; (2) only a few studies have employed oral administration; (3) there is a lack of published studies that reflect carcinogenicity assays aimed at mimicking human exposure to low doses of PAH mixtures.

9.3.2 Metabolic Activation of PAHs and Potential Biomarkers

PAHs require metabolic activation to electrophiles that can bind to cellular macromolecules and then exert their deleterious effects (see Chapter 2).[54,67,68] Several metabolic activation pathways have been reported. Catalysis of the parent hydrocarbon by multiple cytochrome P450-dependent monooxygenases (CYP), with CYP1A1 as the dominant form, can lead to the formation of the highly reactive diol-epoxides.[67,69–71] The formation of radical cations catalyzed by peroxidases and that of *o*-quinones catalyzed by dihydrodiol dehydrogenase have also been reported as additional metabolic activation pathways.[72–77] However, a systematic determination of the contributions of diol-epoxides, radical cations, and *o*-quinones, to PAH-induced mammary carcinogenesis, has not been undertaken.

More knowledge about human uptake and metabolism of individual PAHs and admixtures needs to be gathered to enable us to understand how this class of environmental carcinogens may function as human breast carcinogens. Development of biomarkers for these agents is an exciting area of research, but refinement and methods of improvement are still needed. Studies in this area include approaches such as monitoring PAH-protein adducts, PAH-DNA adducts, and urinary metabolites of PAHs[78–89] (cf. Chapters 3 and 4). Although the environmental carcinogen B[*a*]P is ubiquitous, its adducts are frequently undetectable in human tissues. Thus, it cannot be assumed that B[*a*]P adducts are present in human tissues.[90] This conclusion was stated by Boysen and Hecht on the basis of extensive analysis of the literature with regards to studies aimed at developing a reliable biomarker for PAH exposure.[90] Since urinary 1-hydroxypyrene has already been reported to be a useful surrogate of human PAH exposure (see Chapter 3), these authors emphasized that adducts derived from other carcinogenic PAHs and occurring in greater quantities than B[*a*]P, should be considered in future biomonitoring studies; examples are chrysene, anthracene, B[*a*]A, CP[*c,d*]P, and fluoranthene. Phenanthrene is the simplest PAH featuring a bay-region but it is not a carcinogen. Recently, a method was developed for quantitation of phenanthrene tetraols in human urine.[91] The results of this study provided an approach for quantitation of the diol-epoxide pathway of PAH metabolism in humans.[91] DNA adducts are potentially the most accurate biomarkers because they not only reflect metabolism and uptake, but they also could provide knowledge of possible cancer risks. Such adducts could be detected in human blood or in the target tissues. In the case of breast tissue, specimens can be obtained during surgical procedures (cancer or mammoplasty patients) or during autopsies. As described, putative PAH-DNA adducts have been detected in both human blood and breast tissue.[38–45] For a better understanding of the etiology of breast cancer, it is essential to develop and/or improve sensitivity and selectivity of existing methods. Such improved methods are needed to assess the extent of human exposure, determine the ability of breast tissue to metabolically activate PAHs to DNA-damaging intermediates, and define the carcinogenic potential at doses that mimic human exposure to individual PAHs and combinations thereof. The outcome of such studies will unequivocally define the contribution of PAHs to the development of breast cancer.

For more than two decades, several investigators have shown that extrahepatic tissues, such as mammary epithelial cells, contain a number of CYP enzymes that are capable of converting chemical carcinogens, or their proximate metabolites, to reactive electrophilic species.[92–94] Carmichael *et al.* have demonstrated the ability of human breast epithelium (prepared from breast tissues obtained during reduction mammoplasty) to activate a range of carcinogenic compounds that are present in the diet or represent environmental pollutants.[95] For example, using the ^{32}P-postlabelling technique, five DNA adduct spots were detected in breast cells treated with the potent rat mammary carcinogen DB[*a,l*]P.[95] In a separate study, treatment of the human mammary carcinoma MCF-7 cell line with DB[*a,l*]P resulted in the formation of stable DNA adducts. However, the authors could not detect significant amounts of unstable adducts that would lead to apurinic site formation[96] (cf. Chapter 2). Agarwal *et al.* found that benzo[*g*]chrysene is activated in MCF-7 cells to form major DNA adducts through both the fjord-region *syn*- and *anti*-11,12-diol-13,14-epoxide metabolites.[97] Benzo[*c*]phenanthrene led to an induction of CYP1B1 in MCF-7 cells and subsequent enzymatic activation of this carcinogen to DNA-binding 3,4-diol-1,2-epoxides.[98,99] Singletary and MacDonald reported that human mammary epithelial MCF-10F cells can convert B[*a*]P to DNA-damaging species, as measured by ^{32}P-postlabelling.[100] Although DNA adducts were not analyzed, Caruso *et al.* showed that B[*a*]P is capable of inducing a progressively transformed phenotype in an immortalized (MCF-10A) human mammary epithelial cell line.[101] These investigators found that B[*a*]P alters chromosome 8 ('isochromosome 8q abnormality') and increases the expression of c-*myc*; two parameters frequently associated with clinical disease.[101]

9.4 NO_2-PAHs as Representative Examples of Environmental Mammary Carcinogens

9.4.1 Levels and Carcinogenic Potency of NO_2-PAHs

In the late 1970s, Pitts[102] and Jäger[103] were the first to show that PAHs react with nitrogen oxides to form NO_2-PAHs under conditions that might be expected in polluted air and in combustion processes. Following these early investigations, NO_2-PAHs have been detected in numerous environmental sources,[104] including diesel engine emissions, combustion emissions from kerosene heaters, gas fuel, and liquid petroleum, airborne particulates, and coal fly ash. Human exposure to NO_2-PAHs is as unavoidable as that to the ubiquitous PAHs.[105] GC/MS analysis of lung cancer specimens collected from patients in China and Japan showed appreciable contamination with 1-nitropyrene (1-NO_2-P) and B[*a*]P.[106] Hemoglobin adducts derived from 1-NO_2-P, 2-nitrofluorene (2-NO_2-F), 3-nitrofluoranthene (3-NO_2-FA), 9-nitrophenanthrene and 6-nitrochrysene (6-NO_2-C) were analyzed in blood drawn from non-smoking male bus garage workers in Germany, an urban control group and a rural control group.[107] Comparable hemoglobin adduct levels were found in all of these groups with wide inter-individual variations.[107] These results suggest that the NO_2-PAHs are widespread environmental contaminants.[107,108] Levels and sources of known environmental NO_2-PAHs are described in Table 9.3;[53,104,109–116] they occur in the environment as admixtures with PAHs and hundreds of other organic compounds.[104] In general, levels of NO_2-PAHs in the environment are far below those of PAHs. NO_2-PAHs originate primarily as direct or indirect products of incomplete combustion. Only a few NO_2-PAHs (e.g., nitronaphthalenes and 5-nitroacenaphthene) are produced industrially; however, in a recent study, 5-nitroacenaphthene has also been reported in environmental sources.[109] The distribution of NO_2-PAH isomers in samples of ambient air has been found to be significantly different from that in direct emissions from combustion. The most abundant nitro isomers of pyrene, fluorine, and fluoranthene observed in diesel exhaust are 1-NO_2-P, 2-NO_2-F and 3-NO_2-FA; isomers formed from hydroxy radical reactions are 2-NO_2-P, 3-NO_2-F and 2-NO_2-FA. NO_2-PAHs have also been detected in certain foods and beverages.[52,104,111,112] With the exception of spices, smoked and grilled foods, and peanuts, levels of NO_2-PAHs in foods are below 5 μg/kg. In Yakatori grilled chicken with sauce, 1-NO_2-P has been reported to amount to 43 ng/g. In general, dietary exposure to NO_2-PAHs is therefore virtually negligible.

The *National Toxicology Program* listed several NO_2-PAHs as "reasonably anticipated to be human carcinogens";[117] examples are 1-NO_2-P, 4-NO_2-P, and 6-NO_2-C. In addition, 2-NO_2-F was listed as possibly carcinogenic to humans by the *International Agency for Research on Cancer*.[104] The risk associated with human exposure to NO_2-PAHs has not been clearly

Table 9.3: Occurrence of nitropolycyclic aromatic hydrocarbons.

	Source				
Compound	*Air particulate matter ng/gm*[109]	*Diesel emission particulate matter ng/gm*[109,115]	*Urban air ng/m*^{3} [53,110,111,114–116]	*Diesel emission particulate extracts ng/g*[112,113]	*Carbon black ng/kg*[111]
3,7-Dinitrofluoranthene				28	$<3 \times 10^5$
3,9-Dinitrofluoranthene				13	$<3 \times 10^5$
2,5-Dinitrofluorene		P	0.19		
2,7-Dinitrofluorene			1.5	4200	
1,3-Dinitropyrene	<2	44.4 ± 8.5	0.0047	5–600	$<3 \times 10^5$
1,6-Dinitropyrene	<4	84.5 ± 6.2	0–0.0087	33–1200	$<3 \times 10^5$
1,8-Dinitropyrene	<2	<9	0.04–3.8	13–3400	$<3 \times 10^5$
5-Nitroacenaphthene	3.1 ± 0.3	37.0 ± 2.9			
4-Nitroanthracene					
9-Nitroanthracene	35.9 ± 0.6	6080 ± 190	0.063 ± 0.030		
7-Nitrobenz[*a*]anthracene	35.1 ± 3.6	995 ± 68			$<3 \times 10^5$
3-Nitrobenzanthrone			0.0035–0.0172		
1-Nitrobenzo[*a*]pyrene	<5	<5			
3-Nitrobenzo[*a*]pyrene	<5	<5			
6-Nitrobenzo[*a*]pyrene	<40	1442 ± 47	0.04–0.280	40–32,800	$<3 \times 10^5$
1-Nitrobenzo[*e*]pyrene	<7	<10			
3-Nitrobenzo[*e*]pyrene	<5	89 ± 19			
6-Nitrochrysene	4.4 ± 2	44.4 ± 3.4	1		$<3 \times 10^5$
1-Nitrofluoranthene	<2	274 ± 12			

Table 9.3: (continued)

2-Nitrofluoranthene	282 ± 31	201 ± 12	0.091 ± 0.054		
3-Nitrofluoranthene	4.5 ± 1.8	65.2 ± 7.4	0.039 ± 0.020	3500	
7-Nitrofluoranthene	<2	<2			
8-Nitrofluoranthene	8.8 ± 1.4	106 ± 17		1300	
2-Nitrofluorene	<3	46.2 ± 2.6	0–5.2	1200	$<3 \times 10^5$
1-Nitronaphthalene	6.8 ± 0.3	86.4 ± 2.3		950	
2-Nitronaphthalene	10.0 ± 0.5	238 ± 3		350	
3-Nitroperylene	P			P	$<3 \times 10^5$
3-Nitrophenanthrene	22.0 ± 0.6	4350 ± 230			
4-Nitrophenanthrene	0.47 ± 0.03	150 ± 4			
9-Nitrophenanthrene	1.7 ± 0.1	510 ± 9			
1-Nitropyrene	71.5 ± 5.1	18330 ± 340	0–0.000134	75,000	$<3 \times 10^5$
			0.127 ± 0.044		
2-Nitropyrene	24.4 ± 4.0	<4	0.00003–0.00004		$<3 \times 10^5$
			0.020 ± 0.005		
4-Nitropyrene	5.5 ± 0.6	135 ± 8			$<3 \times 10^5$
9-Oxo-2,7-dinitrofluorene		P		3000	
9-Oxo-2-nitrofluorene		P			
9-Oxo-2,4,7-trinitrofluorene		P			
1-Nitrotriphenylene			0.01–0.44		
2-Nitrotriphenylene			0.020–0.36		

P, Detected but not quantified.

defined, even though these agents are widely spread throughout the environment and possibly involved in the etiology of several human cancers.[18,118] Data on carcinogenic effects are available for 28 NO_2-PAHs. Although inhalation is the main exposure route in humans, no long-term inhalation study on any of the NO_2-PAHs is available. Carcinogenicity assays employed oral administration, topical application, intramammary, intraperitoneal and subcutaneous injection, as well as pulmonary implantation, or intratracheal instillation. Table 9.4 summarizes NO_2-PAHs that have been found in environmental sources and the routes employed to induce mammary cancer in the rat.[115,119–130] Oral administration employed significantly higher doses of NO_2-PAHs than those used with other routes of administration. When administered orally, 6-NO_2-C appears to be the most carcinogenic member among all NO_2-PAHs. Intramammary administration showed that the potency as mammary carcinogens of these NO_2-PAHs can be ranked as follows: 6-NO_2-C $>$ 2,7-dinitrofluorene $>$ 9-oxo-2,7-dinitrofluorene $\approx$ 9-oxo-2,4,7-trinitrofluorene $>$ 9-oxo-2-nitrofluorene $>$ dinitropyrene(s) $>$ 4-NO_2-P $>$ 2-NO_2-P $\approx$ 1-NO_2-P. Due to limitations in experimental design, negative data cannot confirm the absence of carcinogenicity of certain NO_2-PAHs.

9.4.2 Metabolic Activation of NO_2-PAHs and Potential Biomarkers

The metabolism of NO_2-PAHs is complex and includes at least five different metabolic activation pathways, as follows: (i) simple nitroreduction (to yield aromatic hydroxyl amines or amines), (ii) nitroreduction followed by acetylation, (iii) ring oxidation, (iv) ring oxidation combined with nitroreduction, and (v) ring oxidation and nitroreduction followed by esterification.[131,132] Various CYP enzymes may be involved in the metabolism of certain NO_2-PAHs and may catalyze the generation of different derivatives or different isomers of the same derivative. Several adducts derived from NO_2-PAHs have been detected both *in vitro* and *in vivo*. These are C8-substituted 2′-deoxyguanosine (dG) and -adenosine (dA) adducts, as well as N^2-substituted dG adducts. In general, DNA adducts derived from nitroreduction of NO_2-PAHs are better characterized than

Table 9.4: Mammary tumors induced by nitropolycyclic aromatic hydrocarbons.

Compound	*Source*	*Species & Strain*	*Route*	*Total dose/rat (μmol)*	*Mammary tumor incidence %*
2,5-Dinitrofluorene[115,129]	urban air, vehicle exhaust	female albino rats	gavage daily for 8 months	3.89×10^3	12
		female CD rats	intramammary	16.3	60
2,7-Dinitrofluorene[115,129]	urban air, vehicle exhaust	female albino rats	gavage daily for 8 months	3.89×10^3	100
		female CD rats	intramammary	12.2	75
		female CD rats	intramammary	16.3	93
1,3-Dinitropyrene[127]	urban air, vehicle exhaust	weaning CD rats	i.p. 3 × weekly for 4 weeks	16	25
			intragastric intubation	16	14
1,6-Dinitropyrene[127]	urban air, vehicle exhaust	weaning CD rats	i.p. 3 × weekly for 4 weeks	16	17
			intragastric intubation	16	31
1,8-Dinitropyrene[127]	urban air, vehicle exhaust	weaning CD rats	i.p. 3 × weekly for 4 weeks	16	42
			intragastric intubation	16	33
5-Nitroacenaphthene[130]	urban air, vehicle exhaust	female Wistar rats	gavage (daily)	1.21×10^5	42
6-Nitrochrysene[119,121,123,124]	urban air, vehicle exhaust, carbon black	female CD rats	gavage (8 fractions)	400	90
				200	83
				100	83
			intramammary	12.2	84
			intramammary	12.3	97
			intramammary	12.3	96
			intramammary	12.2	93

(continued)

Table 9.4: (continued)

Compound	*Source*	*Species & Strain*	*Route*	*Total dose/rat (μmol)*	*Mammary tumor incidence %*
2-Nitrofluorene[115,128]	urban air, vehicle exhaust,	female rats	gavage daily for 8 months	3.89×10^3	44
	carbon black	female CD rats	intramammary	12.2	33
		female CD rats	intramammary	16.3	33
1-Nitropyrene[120,122,125–127]	urban air, vehicle exhaust,	newborn SD rats	gavage weekly for 16 weeks	800	63
	carbon black			320	42
		newborn CD rats	s.c. weekly for 8 weeks	57	31
				29	10
		weanling CD rats	i.p. 3 × weekly for 4 weeks	16	39
			intragastric intubation	16	14
		female CD rats	gavage weekly for 8 weeks	400	6
		female CD rats	i.p. 3 × weekly for 4 weeks	119	7
		female CD rats	i.p. 3 × weekly for 4 weeks	77.3	28
			s.c. weekly for 4 weeks	74.3	21
		female CD rats	intramammary	12.3	7
2-Nitropyrene[126]	urban air, vehicle exhaust,	female CD rats	i.p. 3 × weekly for 4 weeks	119	0
	carbon black		intramammary	12.3	11
4-Nitropyrene[126]	urban air, vehicle exhaust,	female CD rats	i.p. 3 × weekly for 4 weeks	119	45
	carbon black		intramammary	12.3	68
9-Oxo-2,7-dinitrofluorene[115]	vehicle exhaust	female CD rats	intramammary	16.3	87
9-Oxo-2-nitrofluorene[115]	vehicle exhaust	female CD rats	intramammary	16.3	53
9-Oxo-2,4,7-trinitrofluorene[115]	vehicle exhaust	female CD rats	intramammary	16.3	80

those formed via ring oxidation.[108] To evaluate risks, we and others have focused efforts on understanding the mechanistic bases for tumor induction by NO_2-PAHs and on developing sensitive analytical methods for assessment of exposure.[108] Since NO_2-PAHs occur in a complex mixture in the atmosphere and in engine exhausts, including aircraft engine emissions, the exact contribution of NO_2-PAHs to health effects cannot be assessed. Currently, there are no reports on the effects of individual NO_2-PAHs on human health. 1-NO_2-P has been used extensively as a biomarker for occupational exposure to diesel engine exhaust. Urinary metabolites of NO_2-PAHs were determined in workers exposed to diesel engine emissions using the enzyme-linked immunosorbent assay. Other investigators measured metabolites of 1-NO_2-P in the urine of workers in a shipping department.[133] Although limited, studies focused on hemoglobin and plasma adducts of some NO_2-PAHs, including 1-NO_2-P, may provide appropriate biomarkers in future molecular epidemiological studies[107,108] (cf. Chapter 4).

It has been demonstrated that human breast tissues obtained from reduction mammoplasty, as well as cultured immortalized breast epithelial cells (MCF-10A) and breast cancer cells (MCF-7, MDA-MB-435s) are capable of metabolizing certain NO_2-PAHs via ring oxidation, nitroreduction, or both pathways combined.[134] Carmichael *et al.* observed five adduct spots upon ^{32}P-postlabelling when human breast epithelial cells obtained from mammoplasty were exposed to 1-NO_2-P, but fewer adduct spots were formed by other members of this class of compounds, namely, 1,3-, 1,6- and 1,8-dinitropyrenes.[95] The same cells are capable of metabolizing 6-NO_2-C to genotoxic intermediates. MCF-10F cells can convert B[*a*]P to DNA-damaging species.[100] We demonstrated that human breast cancer cell lines can activate 6-NO_2-C to form a major DNA adduct that derives from nitroreduction and ring oxidation.[134] Cultured breast cell lines are known to express constitutively varying levels of CYP1A1 and 1B1; and literature data also reveal a marked difference in the induction of both CYP enzymes.[135] Certain NO_2-PAHs have been shown to induce, although to different extents, CYP1A1, 1A2, and 1B1 mRNAs in various human tissue-derived cells including MCF-7.[136] Inter-individual and cellular specificity of CYP enzyme expression, coupled with selectivity of CYP1A1- and 1B1-mediated bioactivation of NO_2-PAHs, will largely determine carcinogenic potency and inter-individual susceptibility to this class of compounds.

9.5 Summary and Future Recommendations

The etiology of breast cancer remains uncertain. In addition to genetic disposition, a significant portion of cancer incidence in the U.S. is related to environmental factors and lifestyle — including diet. Consequently, the search for carcinogens that exist in the human environment challenges both scientists and regulatory agencies. Chronic exposure to traces of chemical carcinogens in the diet, in polluted air, or in tobacco smoke can be important in the etiology of breast cancer in the presence of host factors that favor the multi-step carcinogenesis process. Presently, the association between cigarette smoking and breast cancer risk remains unclear; further investigations are needed to unravel the molecular mechanisms underlying potential effects of cigarette smoking on this disease. Studying the interaction between cigarette smoke and genes involved in carcinogen metabolism, as well as those involved in DNA repair, cell proliferation, cell-cycle and apoptosis is highly attractive and will yield fruitful areas of future research. Several studies reported that DNA adducts have been detected in human mammary tissue; future studies should focus on determining the nature of these adducts. Environmental carcinogens responsible for genomic alterations in human breast cancer have not been clearly defined; studies in this area will be highly useful for cancer risk assessment. Diet contributes substantially to PAH uptake but dietary exposure to NO_2-PAHs is negligible. It is noted that, in contrast to lung and skin cancer animal models, studies performed in the rat mammary gland for most of these carcinogens lack dose-response protocols. Furthermore, the doses used were extremely high compared to the levels to which humans are exposed. It should be emphasized that humans are not exposed to a single agent, but typically to a mixture of compounds within and outside various classes of chemicals. To realistically assess risks associated with human exposure to individual agents or mixtures of environmental mammary carcinogens, it is essential to determine their abundance, human uptake and metabolism, and their relative carcinogenic potency in well-defined animal models under identical conditions; routes of testing and doses should mimic human exposure. Future studies should also assess the carcinogenic potency of mixtures that consist of environmental mammary carcinogens and hormones. Several metabolic activation pathways have been reported in the literature. However, a systematic study to determine the contribution of individual pathways to the induction of mammary carcinogenesis, has not been undertaken. Although inhalation is the main exposure route in humans, no long-term inhalation study with any NO_2-PAH has been reported. At the present time, there are no reports on the effects of individual NO_2-PAHs on human health. We, as well as other investigators, have demonstrated that human breast tissues obtained from reduction mammoplasty, as well as cultured immortalized breast epithelial cells (MCF-10A) and breast cancer cells are capable of metabolizing PAHs and NO_2-PAHs via several pathways similar to those described for rodents. Inter-individual and cellular specificity of CYP expression, coupled with selectivity of CYP1A1- and 1B1-mediated bioactivation of PAHs and NO_2-PAHs in this target organ, will largely determine carcinogenic potency and inter-individual susceptibility to this class of compounds. The proposed future studies described above are prerequisite to understanding the contribution of PAHs, NO_2-PAHs, and other environmental carcinogens in the development of human breast cancer.

Acknowledgment

The work performed in the laboratories of the authors and presented in this report was supported by the National Cancer Institute grant CA35519, the Cancer Center Support Grant P30 CA17613, and Contract NO_2-CN-95016. The authors thank Christopher D'Souza, Dhimant Desai, and Arun Sharma for helping with literature search. We thank Ms. Patricia Sellazzo for typing the manuscript and Ms. Ilse Hoffmann for editing the manuscript.

References

1. Jemal A, Murray T, Samuels A, Ghafoor A, Ward E and Thun MJ (2003) Cancer Statistics, 2003. *CA Cancer J. Clin.* 53: 5–26.
2. WHO (1997) *The World Health Report.* World Health Organization, Geneva.

3. Willett WC (1995) Diet, nutrition, and avoidable cancer. *Environ. Health Perspect.* 103, Suppl 8: 165–170.
4. Kelsey JL and Bernstein L (1996) Epidemiology and prevention of breast cancer. *Annu. Rev. Public Health* 17: 47–67.
5. Henderson IC (1993) Risk factors for breast cancer development. *Cancer* 71: 2127–2140.
6. Service RF (1998) New role for estrogen in cancer? *Science* 279: 1631–1633.
7. Bennett IC, Gattas M and Teh BT (1999) The genetic basis of breast cancer and its clinical implications. *Aust. N.Z. J. Surg.* 69: 95–105.
8. Jeffy BD, Chirnomas RB and Romagnolo DF (2002) Epigenetics of breast cancer: polycyclic aromatic hydrocarbons as risk factors. *Environ. Mol. Mutagen.* 39: 235–244.
9. Cohen LA, Rose DP and Wynder EL (1993) A rationale for dietary intervention in postmenopausal breast cancer patients: an update. *Nutr. Cancer* 19: 1–10.
10. Doll R (1996) Nature and nurture: possibilities for cancer control (Commentary). *Carcinogenesis* 17: 177–184.
11. Lipworth L (1995) Epidemiology of breast cancer. *Eur. J. Cancer Prev.* 4: 7–30.
12. Willett WC, Coldnitz GA and Mueller N (1996) Strategies for minimizing cancer risk. *Sci. Am.* 88–95.
13. Wynder EL, Cohen LA, Muscat JE, Winters B, Dwyer JT and Blackburn G (1997) Breast cancer: weighing the evidence for a promoting role of dietary fat. *J. Natl. Cancer Inst.* 89: 766–775.
14. Laurence DJ, Monaghan P and Gusterson BA (1991) The development of the normal human breast. *Oxf. Rev. Reprod. Biol.* 13: 149–174.
15. McManus MJ and Welsch CW (1981) Hormone-induced ductal DNA synthesis of human breast tissues maintained in the athymic nude mouse. *Cancer Res.* 41: 3300–3305.
16. Wright WC (1940) Translation of: *De Morbis Artificium Diatriba* ('Diseases of Workers', originally published in 1713 by Bernardino Ramazzini) University of Chicago Press, Chicago.
17. Russo J and Russo IH (1999) Cellular basis of breast cancer susceptibility. *Oncol. Res.* 11: 169–178.
18. El-Bayoumy K (1992) Environmental carcinogens that may be involved in human breast cancer etiology. *Chem. Res. Toxicol.* 5: 585–590.
19. Fung VA, Huff J, Weisburger EK and Hoel DG (1993) Predictive strategies for selecting 379 NCI/NTP chemicals evaluated for carcinogenic

potential: scientific and public health impact. *Fundam. Appl. Toxicol.* 20: 413–436.

20. Morris JJ and Seifter E (1992) The role of aromatic hydrocarbons in the genesis of breast cancer. *Med. Hypoth.* 38: 177–184.
21. Morton WE (1995) Major differences in breast cancer risks among occupations. *J. Occup. Environ. Med.* 37: 328–335.
22. Weiderpass E, Pukkala E, Kauppinen T, Mutanen P, Paakkulainen H, Vasama-Neuvonen K, Boffetta P and Partanen T (1999) Breast cancer and occupational exposures in women in Finland. *Am. J. Ind. Med.* 36: 48–53.
23. Welp EA, Weiderpass E, Boffetta P, Vainio H, Vasama-Neuvonen K, Petralia S and Partanen TJ (1998) Environmental risk factors of breast cancer. *Scand. J. Work Environ. Health* 24: 3–7.
24. Tokunaga M, Land CE, Yamamoto T, Asano M, Tokuoka S, Ezaki H and Nishimori I (1987) Incidence of female breast cancer among atomic bomb survivors, Hiroshima and Nagasaki, 1950–1980. *Radiat. Res.* 112: 243–272.
25. Terry PD and Rohan TE (2002) Cigarette smoking and the risk of breast cancer in women: a review of the literature. *Cancer Epidemiol. Biomarkers Prev.* 11: 953–971.
26. Morabia A (2002) Smoking (active and passive) and breast cancer: epidemiologic evidence up to June 2001. *Environ. Mol. Mutagen.* 39: 89–95.
27. Petrakis NL, Gruen LD, Beelen TC, Castagnoli N Jr and Craig JC (1978) Nicotine in breast fluid of non-lactating women. *Science* 199: 303–305.
28. Hecht SS (2002) Tobacco smoke carcinogens and breast cancer. *Environ. Mol. Mutagen.* 39: 119–126.
29. Beer A and Billingham RE (1978) Adipose tissue, a neglected factor in etiology of breast cancer. *Lancet* 2: 296.
30. Martin RL, Carmichael PL, Crofton-Sleigh C, Venitt S, Phillips DH and Grover PL (1996) Genotoxicity of human mammary lipid. *Cancer Res.* 56: 5342–5346.
31. Petrakis NL, Maack CA, Lee RE and Lyon M (1980) Mutagenic activity in nipple aspirates of human breast fluid. *Cancer Res.* 40: 188–189.
32. Klein PM and Lawrence JA (2002) Lavage and nipple aspiration of breast ductal fluids: a source of biomarkers for environmental mutagenesis. *Environ. Mol. Mutagen.* 39: 127–133.
33. Scott WN and Miller WR (1991) Mutagens in human breast cyst fluid. *J. Cancer Res. Clin. Oncol.* 117: 254–258.

34. Thompson PA, DeMarini DM, Kadlubar FF, McClure GY, Brooks LR, Green BL, Fares MY, Stone A, Josephy PD and Ambrosone CB (2002) Evidence for the presence of mutagenic arylamines in human breast milk and DNA adducts in exfoliated breast ductal epithelial cells. *Environ. Mol. Mutagen.* 39: 134–142.
35. Martin FL, Venitt S, Carmichael PL, Crofton-Sleigh C, Stone EM, Cole KJ, Gusterson BA, Grover PL and Phillips DH (1997) DNA damage in breast epithelial cells: detection by the single-cell gel (comet) assay and induction by human mammary lipid extracts. *Carcinogenesis* 18: 2299–2305.
36. Martin FL, Cole KJ, Weaver G, Williams JA, Millar BC, Grover PL and Phillips DH (1999) Genotoxicity of human milk extracts and detection of DNA damage in exfoliated cells recovered from breast milk. *Biochem. Biophys. Res. Commun.* 259: 319–326.
37. Martin FL, Cole KJ, Harvey DP, Weaver G, Williams JA, Millar BC, Phillips DH and Grover PL (2000) DNA damage in human breast milk cells and its induction by 'early' and 'late' milk extracts. *Carcinogenesis* 21: 799–804.
38. Gorlewska-Roberts K, Green B, Fares M, Ambrosone CB and Kadlubar FF (2002) Carcinogen-DNA adducts in human breast epithelial cells. *Environ. Mol. Mutagen.* 39: 184–192.
39. Li D, Wang M, Dhingra K and Hittleman WN (1996) Aromatic DNA adducts in adjacent tissues of breast cancer patients: clues to breast cancer etiology. *Cancer Res.* 56: 287–293.
40. Li D, Wang M, Firozi PF, Chang P, Zhang W, Baer-Dubowska W, Moorthy B, Vulimiri SV, Goth-Goldstein R, Weyand EH and DiGiovanni J (2002) Characterization of a major aromatic DNA adduct detected in human breast tissues. *Environ. Mol. Mutagen.* 39: 193–200.
41. Perera FP, Estabrook A, Hewer A, Channing K, Rundle A, Mooney LA, Whyatt R and Phillips DH (1995) Carcinogen-DNA adducts in human breast tissue. *Cancer Epidemiol. Biomarkers Prev.* 4: 233–238.
42. Pfau W, Stone EM, Brockstedt U, Carmichael PL, Marquardt H and Phillips DH (1998) DNA adducts in human breast tissue: association with *N*-acetyltransferase-2 (NAT2) and NAT1 genotypes. *Cancer Epidemiol. Biomarkers Prev.* 7: 1019–1025.
43. Rundle A, Tang D, Hibshoosh H, Schnabel F, Kelly A, Levine R, Zhou J, Link B and Perera F (2002) Molecular epidemiologic studies of polycyclic aromatic hydrocarbon-DNA adducts and breast cancer. *Environ. Mol. Mutagen.* 39: 201–207.

44. Seidman LA, Moore CJ and Gould MN (1988) ^{32}P-postlabelling analysis of DNA adducts in human and rat mammary epithelial cells. *Carcinogenesis* 9: 1071–1072.
45. Wang M, Dhingra K, Hittleman WN, Liehr JG, de Andrade M and Li D (1996) Lipid peroxidation-induced putative malondialdehyde-DNA adducts in human breast tissues. *Cancer Epidemiol. Biomarkers Prev.* 5: 705–710.
46. Boyiri, T, Guttenplan J, Khmelnitsky M, Kosinska W, Desai D, Pittman B and El-Bayoumy K (2003) Mammary carcinogenesis and mutations in the *cII* gene of *lacIXCD* transgenic rats induced by the environmental pollutant 6-nitrochrysene. *Proc. Am. Assoc. Cancer Res.* 44: 786, Abstract # 3440.
47. Manjanatha MG, Shelton SD, Cullp SJ, Blankenship LR and Casciano DA (2000) DNA adduct formation and molecular analysis of *in vivo lacI* mutations in the mammary tissue of Big Blue® rats treated with 7,12-dimethylbenz[*a*]anthracene. *Carcinogenesis* 21: 265–273.
48. Gammon MD, Wolff MS, Neugut AI, Eng SM, Teitelbaum SL, Britton JA, Terry MB, Levin B, Stellman SD, Kabat GC, Hatch M, Senie R, Berkowitz G, Bradlow HL, Garbowski G, Maffeo C, Montalvan P, Kemeny M, Citron M, Schnabel F, Schuss A, Hajdu S, Vinceguerra V, Niguidula N, Ireland K and Santella RM (2002) Environmental toxins and breast cancer on Long Island. II. Organochlorine compound levels in blood. *Cancer Epidemiol. Biomarkers Prev.* 11: 686–697.
49. Gammon MD, Santella RM, Neugut AI, Eng SM, Teitelbaum SL, Paykin A, Levin B, Terry MB, Young TL, Wang LW, Wang Q, Britton JA, Wolff MS, Stellman SD, Hatch M, Kabat GC, Senie R, Garbowski G, Maffeo C, Montalvan P, Berkowitz G, Kemeny M, Citron M, Schnabel F, Schuss A, Hajdu S and Vinceguerra V (2002) Environmental toxins and breast cancer on Long Island. I. Polycyclic aromatic hydrocarbon DNA adducts. *Cancer Epidemiol. Biomarkers Prev.* 11: 677–685.
50. Perera F and Rundle A (2003) Correspondence re: Gammon *et al.*, Environmental toxins and breast cancer on Long Island. I. Polycyclic aromatic hydrocarbon DNA adducts. *Cancer Epidemiol. Biomarkers Prev.* 12: 75–76.
51. Phillips DH (1983) Fifty years of benzo[*a*]pyrene. *Nature* 303: 468–472.
52. Phillips DH (1999) Polycyclic aromatic hydrocarbons in the diet. *Mutat. Res.* 443: 139–147.
53. Feilberg A, Poulsen MWB, Nielsen T and Skov H (2000) Occurrence and sources of particulate nitro-polycyclic aromatic hydrocarbons in ambient air in Denmark. *Atmos. Environ.* 35: 353–366.

54. Harvey RG (1991) *Polycyclic Aromatic Hydrocarbons: Chemistry and Carcinogenicity.* Cambridge University Press, Cambridge, UK.
55. IARC (1983) *Monographs on the Evaluation of the Carcinogenic Risk of Chemicals to Humans.* Volume 32: Polynuclear Aromatic Compounds, Part 1, Chemical, Environmental and Experimental Data, pp 37–38. International Agency for Research on Cancer, Lyon, France.
56. IARC (1983) *Monographs on the Evaluation of the Carcinogenic Risk of Chemicals to Humans.* Volume 32: Polynuclear Aromatic Compounds, Part 1, Chemical, Environmental and Experimental Data, pp 95–451. International Agency for Research on Cancer, Lyon, France.
57. IARC (1983) *Monographs on the Evaluation of the Carcinogenic Risk of Chemicals to Humans.* Volume 32: Polynuclear Aromatic Compounds, Part 1, Chemical, Environmental and Experimental Data, p 47. International Agency for Research on Cancer, Lyon, France.
58. IARC (1983) *Monographs on the Evaluation of the Carcinogenic Risk of Chemicals to Humans.* Volume 32: Polynuclear Aromatic Compounds, Part 1, Chemical, Environmental and Experimental Data, pp 41–42. International Agency for Research on Cancer, Lyon, France.
59. IARC (1983) *Monographs on the Evaluation of the Carcinogenic Risk of Chemicals to Humans.* Volume 32: Polynuclear Aromatic Compounds, Part 1, Chemical, Environmental and Experimental Data, pp 44–45. International Agency for Research on Cancer, Lyon, France.
60. Lijinsky W, Domsky I, Mason G, Ramahi HY and Safavi T (1963) The chromatographic determination of trace amounts of polynuclear hydrocarbons in petrolatum, mineral oil, and coal tar. *Anal. Chem.* 35: 952–956.
61. Mumford JL, Harris DB and Williams K (1987) Indoor air sampling and mutagenicity studies of emissions from unvented coal combustion. *Environ. Sci. Technol.* 21: 308–311.
62. Cavalieri EL, Rogan E and Sinha D (1988) Carcinogenicity of aromatic hydrocarbons directly applied to rat mammary gland. *J. Cancer Res. Clin. Oncol.* 114: 3–9.
63. Cavalieri EL, Rogan EG, Higginbotham S and Cremonesi P (1989) Tumor-initiating activity in mouse skin and carcinogenicity in rat mammary gland of dibenzo[*a*]pyrenes: the very potent environmental carcinogen dibenzo[*a*,*l*]-pyrene. *J. Cancer Res. Clin. Oncol.* 115: 67–72.

64. Cavalieri EL, Higginbotham S, RamaKrishna NVS, Devanesan PD, Todorovic R, Rogan EG and Salmasi S (1991) Comparative dose-response tumorigenicity studies of dibenzo[*a,l*]pyrene versus 7,12-dimethylbenz[*a*]-anthracene, benzo[*a*]pyrene and two dibenzo[*a,l*]pyrene dihydrodiols in mouse skin and rat mammary gland. *Carcinogenesis* 12: 1939–1944.
65. Huggins C and Yang NC (1962) Induction and extinction of mammary cancer. *Science* 137: 257–262.
66. Pataki J and Huggins C (1969) Molecular site of substituents of benz[*a*]-anthracene related to carcinogenicity. *Cancer Res.* 29: 506–509.
67. Conney AH (1982) Induction of microsomal enzymes by foreign chemicals and carcinogenesis by polycyclic aromatic hydrocarbons. GHA Clowes Memorial Lecture. *Cancer Res.* 42: 4875–4917.
68. Gelboin HV (1980) Benzo[*a*]pyrene metabolism, activation and carcinogenesis: role and regulation of mixed function oxidases and related enzymes. *Physiol. Rev.* 60: 1107–1166.
69. Jeffrey AM, Jennette KW, Blobstein SH, Weinstein IB, Beland FA, Harvey RG, Kasai H, Miura I and Nakanishi K (1976) Benzo[*a*]pyrene-nucleic acid derivative found *in vivo*: structure of a benzo[*a*]pyrene-tetrahydrodiol epoxide-guanosine adduct. *J. Am. Chem. Soc.* 98: 5714–5715.
70. Thakker DR, Yagi H, Lu AYH, Levin W, Conney AH and Jerina DM (1976) Metabolism of benzo[*a*]pyrene: conversion of (±)-*trans*-7,8-dihydroxy-7,8-dihydrobenzo[*a*]pyrene to highly mutagenic 7,8-diol-9,10-epoxides. *Proc. Natl. Acad. Sci. USA* 73: 3381–3385.
71. Yang SK, McCourt DW, Roller PP and Gelboin HV (1976) Enzymatic conversion of benzo[*a*]pyrene leading predominantly to the diol-epoxide *r*-7,*t*-8-dihydroxy-*t*-9,10-oxy-7,8,9,10-tetrahydrobenzo[*a*]pyrene through a single enantiomer of *r*-7,*t*-8-dihydroxy-7,8-dihydrobenzo[*a*]pyrene. *Proc. Natl. Acad. Sci. USA* 73, 2594–2598.
72. Cavalieri EL and Rogan EG (1995) Central role of radical cations in metabolic activation of polycyclic aromatic hydrocarbons. *Xenobiotica* 25: 677–688.
73. Chen L, Devanesan PD, Higginbotham S, Ariese F, Jankowiak R, Small GJ, Rogan EG and Cavalieri E (1996) Expanded analysis of benzo[*a*]pyrene-DNA adducts formed *in vitro* and in mouse skin: their significance in tumor initiation. *Chem. Res. Toxicol.* 9: 897–903.

74. Penning TM, Oshnishi ST, Ohnishi T and Harvey RG (1996) Generation of reactive oxygen species during the enzymatic oxidation of polycyclic aromatic hydrocarbon *trans*-dihydrodiols catalyzed by dihydrodiol dehydrogenase. *Chem. Res. Toxicol.* 9: 84–92.
75. Penning TM, Burczynski ME, Hung CF, McCoull KD, Palackal NT and Tsuruda LS (1999) Dihydrodiol dehydrogenases and polycyclic aromatic hydrocarbon activation: generation of reactive and redox active *o*-quinones. *Chem. Res. Toxicol.* 12: 1–18.
76. Shou M, Harvey RG and Penning TM (1992) Contribution of dihydrodiol dehydrogenase to the metabolism of (±)-*trans*-7,8-dihydroxy-7,8-dihydrobenzo[*a*]pyrene in fortified rat liver subcellular fractions. *Carcinogenesis* 13: 1575–1582.
77. Smithgall TE, Harvey RG and Penning TM (1988) Spectroscopic identification of *ortho*-quinones as the products of polycyclic aromatic *trans*-dihydrodiol oxidation catalyzed by dihydrodiol dehydrogenase. A potential route of proximate carcinogen metabolism. *J. Biol. Chem.* 263: 1814–1820.
78. Angerer J, Mannschreck C and Gundel J (1997) Biological monitoring and biochemical effect monitoring of exposure to polycyclic aromatic hydrocarbons. *Int. Arch. Occup. Environ. Health* 70: 365–377.
79. Castelao JE, Yuan JM, Skipper PL, Tannenbaum SR, Gago-Dominguez M, Crowder JS, Ross RK and Yu MC (2001) Gender- and smoking-related bladder cancer risk. *J. Natl. Cancer Inst.* 93: 538–545.
80. El-Bayoumy K, Donahue JM, Hecht SS and Hoffmann D (1986) Identification and quantitative determination of aniline and toluidines in human urine. *Cancer Res.* 46: 6064–6067.
81. Gupta RC (1996) ^{32}P-postlabeling for the detection of DNA adducts. In: *Technologies for the Detection of DNA Damage and Mutation* [Pfeifer GP (ed.)] pp 45–61, Plenum Press, New York.
82. Kaderlik KR and Kadlubar FF (1995) Metabolic polymorphisms and carcinogen-DNA adduct formation in human populations. *Pharmacogenetics* 5: S108–S117.
83. Kriek E, Rojas M, Alexandrov K and Bartsch H (1998) Polycyclic aromatic hydrocarbon-DNA adducts in humans: relevance as biomarkers for exposure and cancer risk. *Mutat. Res.* 400: 215–231.
84. Phillips DH (1996) DNA adducts in human tissues: biomarkers of exposure to carcinogens in tobacco smoke. *Environ. Health Perspect.* 104, Suppl 5: 453–458.

85. Phillips DH, Farmer PB, Beland FA, Nath RG, Poirier MC, Reddy MV and Turteltaub KW (2000) Methods of DNA adduct determination and their application to testing compounds for genotoxicity. *Environ. Mol. Mutagen.* 35: 222–233.
86. Poirier MC, Santella RM and Weston A (2000) Carcinogen macromolccular adducts and their measurement. *Carcinogenesis* 21: 353–359.
87. Santella R (1999) Immunological methods for detection of carcinogen-DNA damage in humans. *Cancer Epidemiol. Biomarkers Prev.* 8: 733–739.
88. Skipper PL and Tannenbaum SR (1990) Protein adducts in the molecular dosimetry of chemical carcinogens. *Carcinogenesis* 11: 507–518.
89. Skipper PL, Peng X, SooHoo CK and Tannenbaum SR (1994) Protein adducts as biomarkers of human carcinogen exposure. *Drug Metab. Rev.* 26: 111–124.
90. Boysen G and Hecht SS (2003) Analysis of DNA and protein adducts of benzo[*a*]pyrene in human tissues using structure-specific methods. *Mutat. Res.* 543: 17–30.
91. Chen M, Carmella SG, Yagi H, Jerina DM and Hecht SS (2003) Quantitation of *r*-1,*t*-2,3,*c*-4-tetrahydroxy-1,2,3,4-tetrahydrophenanthrene (PT) in human urine: an approach to phenotying the polycyclic aromatic hydrocarbon (PAH) diol epoxide pathway. *Proc. Am. Assoc. Cancer Res.* 44: 1469, Abstract #6412.
92. King CM, Traub NR, Lortz ZM and Thissen MR (1979) Metabolic activation of arylhydroxamic acids by *N*-O-acyltransferase of rat mammary gland. *Cancer Res.* 39: 3369–3372.
93. Moore CJ, Eldridge SR, Tricomi WA and Gould MN (1987) Quantitation of benzo[*a*]pyrene and 7,12-dimethylbenz[*a*]anthracene binding to nuclear macromolecules in human and rat mammary epithelial cells. *Cancer Res.* 47: 2609–2613.
94. Stampfer MR, Bartholomew JC, Smith HS and Bartley JC (1981) Metabolism of benzo[*a*]pyrene by human mammary epithelial cells: toxicity and DNA adduct formation. *Proc. Natl. Acad. Sci. USA* 78: 6251–6255.
95. Carmichael PL, Stone EM, Grover PL, Gusterson BA and Phillips DH (1996) Metabolic activation and DNA binding of food mutagens and other environmental carcinogens in human mammary epithelial cells. *Carcinogenesis* 17: 1769–1772.
96. Melendez-Colon VJ, Smith CA, Seidel A, Luch A, Platt KL and Baird WM (1997) Formation of stable adducts and absence of depurinating DNA adducts in cells and DNA treated with the potent carcinogen dibenzo[*a,l*]pyrene or its diol epoxides. *Proc. Natl. Acad. Sci. USA* 94: 13542–13547.

97. Agarwal R, Coffing SL, Baird WM, Kiselyov AS, Harvey RG and Dipple A (1997) Metabolic activation of benzo[*g*]chrysene in the human mammary carcinoma cell line MCF-7. *Cancer Res.* 57: 415–419.
98. Einolf HJ, Amin S, Yagi H, Jerina DM and Baird WM (1996) Benzo[*c*]-phenanthrene is activated to DNA-binding diol epoxides in the human mammary carcinoma cell line MCF-7 but only limited activation occurs in mouse skin. *Carcinogenesis* 17: 2237–2244.
99. Einolf HJ, Story WT, Marcus CB, Larsen MC, Jefcoate CR, Greenlee WF, Yagi H, Jerina DM, Amin S, Park SS, Gelboin HV and Baird WM (1997) Role of cytochrome P450 enzyme induction in the metabolic activation of benzo[*c*]phenanthrene in human cell lines and mouse epidermis. *Chem. Res. Toxicol.* 10: 609–617.
100. Singletary K and MacDonald C (2000) Inhibition of benzo[*a*]pyrene- and 1,6-dinitropyrene-DNA adduct formation in human mammary epithelial cells by dibenzoylmethane and sulforaphane. *Cancer Lett.* 155: 47–54.
101. Caruso JA, Reiners JJ Jr, Emond J, Shultz T, Tainsky MA and Alaoui-Jamali M (2001) Genetic alteration of chromosome 8 is a common feature of human mammary epithelial cell lines transformed *in vitro* with benzo[*a*]pyrene. *Mutat. Res.* 473: 85–99.
102. Pitts JN Jr, van Cauwenberghe KA, Frosjean D, Schmidt JP, Fitz DR, Belser WL, Knudson GB and Nynds PM (1978) Atmospheric reaction of polycyclic aromatic hydrocarbons: facile formation of mutagenic nitro derivatives. *Science* 202: 515–519.
103. Jäger J (1978) Detection and characterization of nitro derivatives of some polycyclic aromatic hydrocarbons by fluorescence quenching after thin-layer chromatography: application to air pollution analysis. *J. Chromatogr.* 152: 575–578.
104. Kielhorn J, Wahnschaffe U and Mangelsdorf I (2003) *A Report: Selected Nitro- and Nitro-oxy-polycyclic Aromatic Hydrocarbons.* The World Health Organization 2003, Environmental Health Criteria 229.
105. Fu PP and Herreno-Saenz D (1999) Nitro-polycyclic aromatic hydrocarbons: a class of genotoxic environmental pollutants. *Environ. Carcinog. Ecotox. Rev.* C17: 1–43.
106. Tokiwa H, Sera N, Horikawa K, Nakanishi Y and Shigematu N (1993) The presence of mutagens/carcinogens in the excised lung and analysis of lung cancer induction. *Carcinogenesis* 14: 1933–1938.

107. Zwirner-Baier I and Neumann HG (1999) Genetic toxicology and environmental mutagenesis. *Mutat. Res.* 441: 135–144.
108. El-Bayoumy K, Johnson BE, Roy AK, Upadhyaya P and Partian SJ (1994) *Biomonitoring of Nitropolynuclear Aromatic Hydrocarbons via Protein and DNA Adducts.* Research Report Number 64, April, Health Effects Institute, Boston, MA, USA.
109. Bamford HA, Bezabeh DZ, Schantz MM, Wise SA and Baker JE (2003) Determination and comparison of nitrated-polycyclic aromatic hydrocarbons measured in air and diesel particulate reference materials. *Chemosphere* 50: 575–587.
110. Feilberg A, Ohura T, Nielsen T, Poulsen MWB and Amagai T (2002) Occurrence and photostability of 3-nitrobenzanthrone associated with atmospheric particles. *Atmos. Environ.* 36: 3591–3600.
111. IARC (1989) *Monographs on the Evaluation of the Carcinogenic Risk of Chemicals to Humans.* Volume 46: Diesel and Gasoline Engine Exhausts and Some Nitroarenes, pp 189–373. International Agency for Research on Cancer, Lyon, France.
112. IARC (1989) *Monographs on the Evaluation of the Carcinogenic Risk of Chemicals to Humans.* Volume 46: Diesel and Gasoline Engine Exhausts and Some Nitroarenes, p 53. International Agency for Research on Cancer, Lyon, France.
113. IARC (1996) *Monographs on the Evaluation of the Carcinogenic Risk of Chemicals to Humans.* Volume 65: Printing Process and Printing Inks, Carbon Black and some Nitro Compounds, p 297. International Agency for Research on Cancer, Lyon, France.
114. Ishii S, Hisamatsu Y, Inazu K and Aika KI (2001) Environmental occurrence of nitrotriphenylene observed in airborne particulate matter. *Chemosphere* 44: 681–690.
115. Malejka-Giganti D, Niehans GA, Reichert MA, Bennett KK and Bliss RL (1999) Potent carcinogenicity of 2,7-dinitrofluorene, an environmental pollutant, for the mammary gland of female Sprague-Dawley rats. *Carcinogenesis* 20: 2017–2023.
116. Matshushita H and Lida Y (1986) Application of capillary gas chromatography to environmental analysis. *High Res. Chromat. & Chromat. Commun.* 9: 708–711.
117. NTP (1998) *Eighth Report on Carcinogens.* National Toxicology Program, U.S. Department of Health and Human Services.

118. Hecht SS and El-Bayoumy K (1990) The possible role of nitroarenes in human cancer. In: *Nitroarenes: The Occurrence, Metabolism, and Biological Impact of Nitroarenes* [Howard PC, Hecht SS and Beland FA (eds.)] pp 309–316, Plenum Press, New York.
119. Amin S, Lin JM, Krzeminski J, Boyiri T, Desai D and El-Bayoumy K (2003) Metabolism of benzo[*c*]chrysene and comparative mammary gland tumorigenesis of benzo[*c*]chrysene bay and fjord region diol epoxides in female CD rats. *Chem. Res. Toxicol.* 16: 227–231.
120. El-Bayoumy K, Rivenson A, Johnson B, DiBello J, Little P and Hecht SS (1988) Comparative tumorigenicity of 1-nitropyrene, 1-nitrosopyrene, and 1-aminopyrene administered by gavage to Sprague-Dawley rats. *Cancer Res.* 48: 4256–4260.
121. El-Bayoumy K, Rivenson A, Upadhyaya P, Chae YH and Hecht SS (1993) Induction of mammary cancer by 6-nitrochrysene in female CD rats. *Cancer Res.* 53: 3719–3722.
122. El-Bayoumy K, Chae CH, Upadhyaya P, Rivenson A, Kurtzke C, Reddy B and Hecht SS (1995) Comparative tumorigenicity of benzo[*a*]pyrene, 1-nitropyrene and 2-amino-1-methyl-6-phenylimidazo[4,5-*b*]pyridine administered by gavage to female CD rats. *Carcinogenesis* 16: 431–434.
123. El-Bayoumy K, Desai D, Boyiri T, Rosa J, Krzeminski J, Sharma AK, Pittman B and Amin S (2002) Comparative tumorigenicity of the environmental pollutant 6-nitrochrysene and its metabolites in the rat mammary gland. *Chem. Res. Toxicol.* 15: 972–978.
124. Hecht SS, Amin S, Lin YM, Rivenson A, Kurtzke C and El-Bayoumy K (1995) Mammary carcinogenicity of fluoranthene, a commonly occurring environmental pollutant. *Carcinogenesis* 16: 1433–1435.
125. Hirose M, Lee MS, Wang CY and King CM (1984) Induction of rat mammary gland tumors by 1-nitropyrene, a recently recognized environmental mutagen. *Cancer Res.* 44: 1158–1162.
126. Imaida K, Hirose M, Tay L, Lee MS, Wang CY and King CM (1991) Comparative carcinogenicities of 1-, 2-, and 4-nitropyrene and structurally related compounds in the female CD rat. *Cancer Res.* 51: 2902–2907.
127. Imaida K, Lee MS, Wang CY and King CM (1991) Carcinogenicity of dinitropyrenes in the weanling female CD rat. *Carcinogenesis* 12: 1187–1191.
128. Miller JA, Sandin RB, Miller EC and Rusch HP (1955) The carcinogenicity of compounds related to 2-acetylaminofluorene. II. Variations in the bridges and the 2-substituent. *Cancer Res.* 15: 188–199.

129. Miller EC, Fletcher TL, Margreth A and Miller JA (1962) The carcinogenicities of derivatives of fluorine and biphenyl: fluoro derivatives as probes for active sites in 2-acetylaminofluorene. *Cancer Res.* 22: 1002–1014.
130. Takemura N, Hashida C and Terasawa M (1974) Carcinogenic action of 5-nitroacenaphthanene. *Br. J. Cancer* 30: 481–483.
131. Guengerich FP (2000) Metabolism of chemical carcinogens. *Carcinogenesis* 21: 345–351.
132. Purohit V and Basu AK (2000) Mutagenicity of nitroaromatic compounds. *Chem. Res. Toxicol.* 13: 673–692.
133. Van Bekkum Y (1999) Occupational exposure to diesel exhaust in a confined indoor workplace: application of multiple methods for monitoring, using 1-nitropyrene as a biomarker. In: *Biomonitoring of Exposure to Diesel Exhaust* (Chapter 6 of PhD Thesis, pp 141–164) University of Nijmegen, Nijmegen, The Netherlands.
134. Boyiri T, Leszczynska J, Desai D, Amin S, Nixon DW and El-Bayoumy K (2002) Metabolism and DNA binding of the environmental pollutant 6-nitrochrysene in primary culture of human breast cells and in cultured MCF-10A, MCF-7 and MDA-MB-435s cell lines. *Int. J. Cancer* 100: 395–400.
135. Spink DC, Spink BC, Cao JQ, DePasquale JA, Pentecost BT, Fasco MJ, Li Y and Sutter T (1998) Differential expression of CYP1A1 and CYP1B1 in human breast epithelial cells and breast tumor cells. *Carcinogenesis* 19: 291–298.
136. Iwanari M, Nakajima M, Kizu R, Hayakawa K and Yokoi T (2002) Induction of CYP1A1, CYP1A2, and CYP1B1 mRNAs by nitropolycyclic aromatic hydrocarbons in various human tissue-derived cells: chemical-, cytochrome P450 isoform-, and cell-specific differences. *Arch. Toxicol.* 76: 287–298.

10 Genetic Susceptibility to Polycyclic Aromatic Hydrocarbon-Induced Carcinogenesis

Ari Hirvonen

Laboratory of Biomonitoring, Department of Industrial Hygiene and Toxicology, Finnish Institute of Occupational Health, Helsinki, Finland
E-mail: ari.hirvonen@ttl.fi

10.1 Introduction

It is well established that lung carcinogenesis in humans is mainly caused by cigarette smoking. It is presently the most common malignancy in the world. From other smoking-related cancers, about 50% of bladder cancers are attributable to smoking.

Cigarette smoke contains thousands of chemicals, among which polycyclic aromatic hydrocarbons (PAHs) constitute one of the main classes of carcinogenic components.[1] Striking feature of PAHs is their low chemical reactivity; they are not chemically reactive *per se* but are modified under physiological conditions by enzyme metabolism (cf. Chapter 2). PAH-DNA adducts are relevant in the initiation of PAH-induced carcinogenesis. The mechanism by which PAHs such as benzo[*a*]pyrene (B[*a*]P) interact with DNA, activate oncogenes and initiate the carcinogenic process, involves the formation of bay-region diol-epoxides as the major ultimate carcinogens. B[*a*]P is converted into phenolic metabolites and the 7,8-dihydrodiol by a cytochrome P450-dependent monooxygenase (CYP)-mediated process. Secondary metabolism, mainly involving microsome epoxide hydrolase (mEH) and other CYP isoforms, leads to the formation of the highly reactive (+)-*anti*-stereoisomeric B[*a*]P-7,8-diol-9,10-epoxide (B[*a*]PDE). B[*a*]PDE, which is the most frequently studied carcinogenic PAH diol-epoxide, is a relatively good substrate for glutathione *S*-transferases (GSTs) that inactivate several carcinogens present in tobacco smoke.[2]

Sensitive detection methods have been used to demonstrate the presence of smoking-related bulky PAH-DNA adducts in virtually all target organs of tobacco carcinogenesis. The amounts of B[*a*]PDE bound to DNA can be quantified via high-performance liquid chromatography with fluorescence detection by measuring the release of the diol-epoxide hydrolysis products, the B[*a*]P-7,8,9,10-tetraols, from both lung tissue and lymphocyte DNA.[3] The levels of PAH-DNA adducts have been found to vary considerably among persons with similar ambient or environmental exposure to PAHs (cf. Chapter 4).[4] This implies that inherited differences exist in formation of these adducts. A growing number of genes encoding phase-I and phase-II xenobiotic metabolizing enzymes (XMEs) that participate in the metabolism of PAHs have been identified and cloned (cf. Chapter 2). Many of these enzymes have been shown to exhibit polymorphisms in their genes that may affect an individual's capacity to either activate or detoxify PAHs and their metabolites.[5] Development of rather simple new techniques, such as PCR-based assays, has enabled identification of the XME genotype with precision. Thus, recent knowledge of the genetic basis for individual metabolic variation has opened new possibilities for studies focusing on increased susceptibility to PAH-induced carcinogenesis. The XMEs that have raised the most interest in this context will be discussed in more detail subsequently.

10.2 Cytochrome P450-Dependent Monooxygenases

The CYP enzymes represent the first line of defense against toxic lipophilic chemicals.[6] However, certain chemicals are activated by CYP-mediated reactions into their ultimate carcinogenic form rather than being detoxified (cf. Chapter 2). Most of the carcinogen activation occurs through generation of epoxide intermediates that are further metabolized by transferases. CYP enzymes activate PAHs by producing highly reactive DNA-damaging metabolites. This is best exemplified by the metabolism of B[*a*]P, which undergoes two successive oxygenation reactions in human lung tissue, ultimately leading to the highly mutagenic B[*a*]PDE (see Figure 2.5 in Chapter 2).[7]

The main CYP enzymes in humans that participate in PAH metabolism are CYP1A1, CYP1A2, CYP1B1, CYP2C9, CYP3A4, and CYP3A5, which also exhibit specificities for various other classes of carcinogens.[6,8] In addition, the expression of CYP1A1, CYP1A2 and CYP1B1 is also induced by PAH substrates itself.[9] CYP enzymes are most extensively expressed in the liver although their levels of expression vary depending on the particular enzyme form. Certain CYP forms are also expressed in lung, gastrointestinal tract, kidney and larynx/nasopharyngeal tissue.[10,11] In nonhepatic epithelial tissues, activation of carcinogens probably occurs directly in those cells that undergo transformation.

10.2.1 CYP1A1

CYP1A1 is predominantly expressed in extrahepatic tissues such as lung.[12] The induction of CYP1A1 expression is regulated by an intracellular aryl hydrocarbon receptor (AhR) to which the PAH chemicals bind with high affinity (cf. Chapter 2). In the cytoplasm, AhR forms a complex with a 90 kDA heat shock protein, the X-associated protein 2 (XAP2, AIP or ARA9), and a 23 kDa co-chaperone protein. Following ligand binding, the AhR complex translocates into the nucleus, dissociates from the protein complex and binds to a nuclear protein called ARNT (AhR nuclear translocator). Formation of the AhR-ARNT heterodimer converts the complex into its high affinity DNA binding form, which subsequently binds to its specific

DNA recognition sites, the xenobiotic responsive elements (*XREs*), e.g., located upstream of the *CYP1A1* gene. The binding to *XRE* elements leads to chromatin and nucleosome disruption, increased promoter accessibility and an increase in the transcription of the *CYP1A1* gene.[13,14]

A trimodal distribution of the aryl hydrocarbon hydroxylase (AHH) induction was observed already three decades ago.[15] This is consistent with a co-dominant inheritance at a single genetic locus segregating for a more common allele conferring low inducibility, and a rarer allele conferring high inducibility. Subsequently, three genetic polymorphisms were detected within the *CYP1A1* gene. The first polymorphism was determined as a point mutation in the 3′-flanking region of the gene, creating a restriction fragment length polymorphism (RFLP) detected by the *Msp*I restriction enzyme.[16] The other polymorphic site was found to be located in exon 7 where a nucleotide substitution causes an Ile to Val amino acid change in the heme-binding region of the enzyme.[17] The *CYP1A1 Msp*I (*CYP1A1*2A*) and Ile/Val (*CYP1A1*2B* or **2C*) variant alleles have been found to be closely linked and much more prevalent in Asians than in Caucasians.[5] The fourth variant allele (*CYP1A1*4*) of potential functional importance is located in exon 7, too.[18]

A number of studies have addressed the relationship between the *CYP1A1* variant alleles and increased activity and/or inducibility, and found either modest or no differences in CYP1A1-catalyzed activities associated with the different alleles.[19–23] Subsequent to the report suggesting that the extent of inducibility of CYP1A1 was increased in lymphocytes from lung cancer patients compared to controls,[15] a number of attempts were made to confirm these findings.[5,24] Strong correlations between lung cancer risk and homozygosity for the *CYP1A1* variant alleles have been reported in several Japanese studies.[16,17,25,26] However, although a similar association was also reported in an American population,[27] no such association was found in Europeans.[28–32]

A most recent pooled analysis with 2,451 cases and 3,358 controls[33] concluded that Caucasians with homozygous variant *CYP1A1* genotype have a higher risk of lung cancer, but the results were less conclusive for Asians. In the future some critical research is therefore needed in order to better define the genotype/phenotype relationships. It may also address the question of whether the activity and inducibility of CYP1A1 are decisive and critical parameters for the carcinogenicity of PAHs.

10.2.2 CYP1B1

The newest member of the CYP1 family, CYP1B1, was cloned in 1994,[34] and at present represents the only member of the CYP1B subfamily.[35] The *CYP1B1* gene is highly expressed in a multitude of human organs including brain, endometrium, placenta, fetal adrenal glands, lung, kidney and lymphocytes,[34,36] whereas its expression in liver is rather low. CYP1B1 has been shown to activate B[*a*]P to the ultimate carcinogenic metabolite B[*a*]PDE at rates higher than CYP1A2 but lower than CYP1A1.[37] On the other hand, CYP1A1 and CYP1B1 appear to slightly differ in their regio- and stereoselective activation of dibenzo[*a,l*]pyrene (DB[*a,l*]P), the most powerful carcinogenic PAH as yet discovered; CYP1B1 has been shown to form higher proportions of the highly carcinogenic metabolite of DB[*a,l*]P, the DB[*a,l*]P-11,12-diol-13,14-epoxide (DB[*a,l*]PDE), as compared to CYP1A1.[38]

To date, a number of single nucleotide polymorphisms of the *CYP1B1* gene have been reported. Some polymorphisms, associated with amino acid exchanges, have been studied in more detail.[39] The amino acid substitutions $Val^{432}Leu$ (*CYP1B1*3*) and $Asn^{453}Ser$ (*CYP1B1*4*), located in the heme-binding domain of CYP1B1, appear as likely candidates to be linked with biological effects.[39] The *CYP1B1*3* allele was recently suggested to correlate with an increased rate of conversion of B[*a*]P to B[*a*]PDE. Consistent with this, the *CYP1B1*3* allele was shown to pose an increased risk for the development of squamous cell carcinomas of head-and-neck (SCCHN) among smokers.[40]

10.2.3 CYP2C9

CYP2C9 is one of the major isoforms of the CYP2C subfamily that accounts for approximately 20% of constitutively expressed hepatic CYP enzymes in humans.[10,41] Three different *CYP2C9* alleles have been identified[42,43] and found to be related to an altered CYP2C9 expression.[44,45] Although no procarcinogenic substrates are currently known for the CYP2C enzymes, the levels of all smoking related adducts in the larynx were correlated with the presence of CYP2C suggesting a role for CYP2C9 in DNA adduction of PAH-type tobacco carcinogens.[46] In line with this view a slight increased lung cancer risk associated with the *CYP2C9*2* allele was reported in one study,[43] but also contradictory findings have been reported.[47,48]

10.3 Epoxide Hydrolase

Microsomal epoxide hydrolase (mEH) is an enzyme involved in the first pass metabolism of highly reactive epoxide intermediates (cf. Chapter 2). The mEH enzyme is expressed in most tissues, including the lung and the upper aerodigestive tract.[49,50] Within cells, mEH is localized mainly to the endoplasmic reticulum, where it can transiently associate with the CYP system.[51] The mEH coordinately acts with, for example, CYP1A1 and CYP1A2, to convert deleterious PAH epoxides through hydrolysis into *trans*-dihydrodiols. Further epoxidation can convert dihydrodiols to highly toxic, mutagenic, and carcinogenic diol-epoxides.[52] Thus, mEH exhibits the same dual role of procarcinogen detoxification and activation as found in the case of some CYP enzymes, and consequently also plays an important role in PAH-mediated toxicity.

Up to 40-fold differences in mEH activity have been reported in various human tissue types.[53] The molecular basis for this variation in mEH activity has not yet been characterized completely. Genetic polymorphisms have, however, been identified within exons 3 and 4 of the mEH gene (*EPHX1*),[54,55] which result in $Tyr^{113}His$ and $His^{139}Arg$ amino acid substitutions, respectively. *In vitro* expression analyses indicated that the corresponding mEH activities decrease approximately 40% (His^{113}) or increase at least 25% (Arg^{139}). The activity level observed in the presence of both variations approximates that observed for the wild-type genotype.[55] Recently, a genetic variation in the 5′-flanking sequence of *EPHX1* was observed. This may be an additional contributing factor to the range of functional mEH expression existing in human populations.[56]

Since mEH is strongly expressed in bronchial epithelial cells it is conceivable that the above-mentioned genetic polymorphisms would be an important risk modifier for lung carcinogenesis. In contrast to this view, some studies revealed no significant association between the *EPHX1* genotypes and lung cancer risk.[57,58] However, in some other studies involving Caucasian[59] and Asian[60] subjects, a lower risk of lung cancer was suggested for the *EPHX1* allotypes predicted to result in decreased enzyme activity. Moreover, although London and co-workers[61] failed to find an association between lung cancer incidence and the predicted mEH activity in Caucasians, a decreased risk associated with the predicted low activity was found among African-Americans.

In the largest case-control study conducted so far on *EPHX1* genotypes and lung cancer risk involving almost exclusively Caucasians, no overall association was seen between the predicted mEH activity and lung cancer risk.[62] When the statistical evaluations were performed stratified by smoking habits, however, the *EPHX1* genotypes appeared to play a pivotal role for lung cancer risk; the very low activity associated genotypes posed an increased risk of lung cancer in non-smokers, but changed from a risk factor to a protective factor in smokers. The protective effect of the low activity associated *EPHX1* genotypes in smoking related cancers is further supported by the findings of a study with Caucasian smokers in which a reduced susceptibility to cancers in the upper aerodigestive tract was associated with these genotypes.[63]

10.4 Glutathione *S*-Transferases

GST enzymes comprise a superfamily of proteins that have broad and overlapping substrate specificities.[64] The known substrates for GST enzymes include those derived from bioactivation of PAHs, namely, PAH diol-epoxides (cf. Chapter 2). The most well studied carcinogenic PAH diol-epoxide, B[*a*]PDE, is a good substrate for many GST isoforms like GSTM2, GSTM3, and especially for GSTM1 and GSTP1.[2,64]

The cytosolic GST enzymes are divided into at least five major classes, i.e., GST-A (α), GST-M (μ), GST-P (π), GST-T (θ), and GST-K (κ). To date, genetic polymorphisms in three of these enzyme classes have been explored in relation to PAH-induced carcinogenesis. One of them is GSTM1, which is expressed in only about half of Caucasians because of a homozygous deletion ('null genotype') of the corresponding gene in the other half.[53] To date, a remarkable amount of studies have been conducted on *GSTM1* polymorphism in relation to smoking related cancers.[5,24] Although the outcomes of most studies have been largely controversial, a recent meta-analysis that summarized the results of 43 published case-control studies comprising more than 18,000 individuals concluded that the *GSTM1* null genotype poses a slight excess of increased risk of lung cancer.[65] Similarly, a recent meta-analysis that summarized the results of 17 published and unpublished, genotype-based case-control studies (2,149 cases and 3,646 controls) suggested a modest increase in the risk of bladder cancer for individuals with *GSTM1* null genotype.[66] In light of the compiled data available, it has been estimated that 17% of both lung cancers[67] and bladder cancers[68] may be attributable to *GSTM1* genotypes. Although these values provide only a crude measure of the potential population impact of these genes, they suggest that *GSTM1* deficiency could contribute to a substantial amount to the incidence of smoking related cancers at the population level.

GSTM3 is one of the most abundant GST isoforms in human lungs.[69–71] As a deviation from the wild-type *GSTM3*A* allele, the variant allele *GSTM3*B* carries a deletion of three base pairs in intron 6, which results in the generation of a recognition sequence for the YYI transcription factor. The functional consequence of this is still unclear, but both negative and positive regulatory effects have been suggested.[69,72] People with low expression of GSTM3 were previously observed to be at increased risk of developing adenocarcinoma of the lung.[71] Some subsequent genotyping studies indicated that individuals who are homozygous or heterozygous for the *GSTM3*B* alleles have lower risk of cancers like larynx cancer[73] and lung cancer,[74] but some other studies failed to confirm these observations.[75,76]

The third polymorphic *GST* gene of potential interest in PAH-induced carcinogenesis, *GSTP1*, encodes an isoform that is known to metabolize many carcinogenic compounds, among them B[*a*]PDE. Given that GSTP1 is the most abundant GST isoform in the lung,[70] it is thought to be of particular importance in the detoxification of inhaled carcinogens. Two variant alleles, *GSTP1*B* and *GSTP1*C*, have been detected in addition to the wild-type allele *GSTP1*A*. The *GSTP1*B* allele has an $A^{313}G$ transition in exon 5, causing an $Ile^{105}Val$ amino acid change. In addition to this base substitution, the *GSTP1*C* allele has a $C^{341}T$ transition, resulting in an $Ala^{114}Val$ amino acid change. Both of the affected codons are located in the electrophile-binding site of the GSTP1 enzyme.[77] Compared to *GSTP1*A*, proteins encoded by *GSTP1*B* and *GSTP1*C* have been found with decreased enzyme activities when expressed in *Escherichia coli*.[77,78] On the other hand, individuals homozygous for the *GSTP1*B* allele have been suggested to detoxify B[*a*]PDE more efficiently than heterozygotes or wild-type homozygotes.[79] A 3-fold increased risk of bladder cancer was observed for individuals homozygous for the *GSTP1* variant alleles (*GSTP1*B* and *GSTP1*C* alleles not differentiated) compared to controls.[80] A similar association was also reported for cancers of the upper aerodigestive tract[76,81] and lung,[82] accompanied with both supporting and contrasting findings.[47,76,83]

While the above data clearly point to a slightly modifying effect for *GSTM1* in individual susceptibility to PAH-induced carcinogenesis, the situation is not as clear for the other *GST* polymorphisms, largely due to the abundance of controversial data. One reason for the controversial outcome between different studies may be related to the fact that individual GST enzymes possess overlapping substrate specificities.[64] As a consequence, the lack of activity of some GST enzymes may be compensated by other isoforms and/or by alternative metabolic pathways.[5,84]

10.5 NAD(P)H:Quinone Oxidoreductase

The NAD(P)H-dependent quinone oxidoreductase 1 (NQO1 = DT-diaphorase) has been shown to specifically prevent the formation of B[*a*]P quinone-DNA adducts generated by CYP1A1 and P450 reductase (cf. Chapter 2).[85] A $C^{609}T$ base substitution in the human *NQO1* gene results in a $Pro^{187}Ser$ amino acid change causing low catalytic activity.[86–88] NQO1 is normally expressed in the lungs, but it is undetectable in lungs from individuals homozygous for the *NQO1 Ser* allele.[89] It is therefore conceivable that smokers carrying the *NQO1 Ser* allele could be at increased risk of lung cancer. In agreement with this, the *NQO1 Ser* allele was found to be about twice as common in Caucasian lung cancer patients than in controls.[88] The outcomes of recent studies among Caucasians[90,91] and Asians[92] supported this view. However, an inverse relation between the *NQO1 Ser* allele and lung cancer has also been reported in several studies.[93–96]

10.6 Myeloperoxidase

The myeloperoxidase (MPO) enzyme is found in primary granules of neutrophils and monocytes and it functions as an oxidative antimicrobial agent by catalyzing the generation of genotoxic hypochlorous acid and other 'reactive oxygen species'.[97] Exposure to a variety of pulmonary insults, such as cigarette smoke, stimulates recruitment of neutrophils into human lung tissue[98] and local release of MPO.[99,100] MPO has been shown to activate B[*a*]P into B[*a*]PDE,[101] and to enhance the binding of B[*a*]P-7,8-dihydrodiol to lung DNA *in vitro*.[102]

An allelic variant with a $G^{-463}A$ base substitution in the promoter region of the *MPO* gene has been shown to result in reduced gene transcription due to disruption of a SP1 binding site[103] and to correlate with more than 4-fold lowered B[*a*]P-DNA adduct levels in the skin of coal tar-treated patients.[104] It is therefore conceivable that carriers of the ^{-463}A allele may have a reduced risk of lung cancer. In agreement with this assumption, several epidemiological reports have demonstrated modulated lung cancer risk associated with this *MPO* genotype (cf. Chapter 2).[105–113]

10.7 Combined Genotype Effects

Given the number and variability in expression of PAH-metabolizing enzymes, assessment of a single polymorphic genotype cannot be expected to be sufficient for reliable estimation of individual susceptibility to PAH-induced cancers. Thus, the establishment of a broader risk profile for each individual or subgroup is required.[84] The first observation of a combined effect of metabolic genes in disease susceptibility was by Hayashi *et al.*,[114] who described an almost 6-fold relative risk for all lung cancer types and a 9-fold relative risk for squamous cell carcinoma of the lung in Japanese individuals with a combination of homozygous *CYP1A1*2B* and *GSTM1* null genotype. A similar, although less pronounced risk (3-fold) for developing this type of lung cancer was attributed to Caucasians carrying the *CYP1A1*2A* allele.[31] These observations are in accordance with some evidence at the functional level; Vaury *et al.*[115] suggested that the high inducibility of CYP1A1 is associated with the *GSTM1* null genotype in cultured cell lines, possibly because of the persistence (nondegradation) of CYP1A1 inducers attributable to the lack of GST activity and GST-mediated metabolism. On the other hand, gene-gene interactions between the *GSTM1* null genotype and CYP1A1 or CYP1A2 enzyme induction have been observed in smokers; *GSTM1* deficiency not only led to an increased hepatic CYP1A2 activity in active smokers, but also to significantly increased levels of bulky PAH-DNA adducts in lung parenchyma of smokers and ex-smokers as compared to individuals carrying wild-type *GSTM1*.[116–120]

Another example of a positive interaction between phase-I and phase-II XMEs exists for *EPHX1* and *GSTM3* genotypes. The combination of *EPHX1* high activity-associated genotype with *GSTM3*AB* or **BB* genotype conferred a 13-fold risk for cancer in the upper aerodigestive tract.[63]

Synergistic effects of defective *GST* genotypes in lung cancer susceptibility were also observed. For instance, a combined *GSTM1* null and *GSTP1*BB* genotype posed a 7-fold risk of lung cancer.[120] This is in agreement with several other reports,[82,83,121] although the absence of combined effects has also been reported.[76,122]

The above-mentioned findings are consistent with the notion that individuals with both genetically determined increased capacity to activate procarcinogenic cigarette smoke constituents such as PAHs and the concurrent impaired capacity to detoxify genotoxic metabolites, would be at particular risk of developing PAH-induced cancers.[5,84]

10.8 Concluding Remarks

Current molecular biological techniques facilitate the discovery and characterization of new variant alleles and genetic changes. It is much easier to conduct straightforward genetic epidemiological approaches than to address the basic question as to how the genotype is determining the phenotype, and whether there is any plausible biological link between the genotypic differences and cancer susceptibility. Knowledge of the complete sequence of events, from the gene to the final outcome, is anticipated to be required for consideration of all implications, and for possible preventive and treatment strategies to be employed in those cases where a clear association between XMEs and cancer susceptibility has been uncovered.

The biological rationale in considering the activity and expression levels of XMEs as possible cancer susceptibility factors is that by activating or inactivating carcinogenic substances, the enzymes would modify the levels of carcinogens in target tissues and cells. XMEs could therefore possibly contribute to the multistage process of carcinogenesis at one or even more stages. When considering the significance of the genotype in terms of cancer outcome one obvious fact has to be stressed: it is the phenotype which is of importance to any possible outcome resulting from the exposure to chemicals. Therefore, the first task is to investigate whether a genotypic change is actually 'carried over' to the phenotype, i.e., what we can observe *in vivo* as a consequence of a chemical exposure.

On the other hand, regulation of many of the *XME* genes is rather complex, with various environmental, host and genetic factors affecting the expression. It is therefore possible that in addition to any mutations in the structural gene, mutations relevant to the phenotype may also occur in 5′ and 3′-flanking regulatory sequences and in other genes coding for transacting factors (e.g., regulatory proteins). Furthermore, proteins may be absent due to mutations that lead to truncated mRNA molecules not capable of directing translation, or a result of the complete or partial deletion of the gene; conversely, proteins may be overexpressed because of gene multiplications.[11,123–125] Examples of these influencing factors are found for some of the XMEs discussed above: a putative CYP1A1 regulatory gene (*AhR*) mutation affecting the inducibility of the CYP1A1 enzyme;[126] a null allele of *GSTM1* as an example for structural gene deletion;[53] or the *GSTM1* gene duplication as an example of multiplication.[127]

The extent of exposure may also be very important for the outcome of gene-environment interaction studies and for its dependence on genotypic/phenotypic traits. For instance, it is quite possible that under conditions of exceptional high levels of exposure, inter-individual differences in susceptibility would be less or nonrelevant. In contrast, at low-level exposures, which are typical for humans, susceptibility factors might be of highest significance. Thus the extent of exposure might actually be the ultimate modifying factor that determines whether the role of a genotype/phenotype relationship can be unravelled in a particular study.

References

1. IARC (1986) *Monographs on the Evaluation of the Carcinogenic Risk of Chemicals to Humans.* Volume 38: Tobacco Smoking. International Agency for Research on Cancer, Lyon, France.
2. Coles B and Ketterer B (1990) The role of glutathione and glutathione transferases in chemical carcinogenesis. *Crit. Rev. Biochem. Mol. Biol.* 25: 47–70.
3. Alexandrov K, Rojas M, Geneste O, Castegnaro M, Camus AM, Petruzzelli S, Giuntini C and Bartsch H (1992) An improved fluorometric assay for dosimetry of benzo[*a*]pyrene diol-epoxide-DNA adducts in smokers' lung: comparison with total bulky adducts and aryl hydrocarbon hydroxylase activity. *Cancer Res.* 52: 6248–6253.
4. Kriek E, Rojas M, Alexandrow K and Bartsch H (1998) Polycyclic aromatic hydrocarbon-DNA adducts in humans: relevance as biomarkers for exposure and cancer risk. *Mutat. Res.* 400: 215–231.
5. Hirvonen A (1999) Polymorphisms of xenobiotic-metabolizing enzymes and susceptibility to cancer. *Environ. Health Perspect.* 107: 37–47.
6. Guengerich FP (1995) Cytochromes P450 of human liver: classification and activity profiles of the major enzymes. In: *Advances in Drug Metabolism in Man* [Pacifici GM and Fracchia GN (eds.)] pp 179–231, Office for the Official Publications of the European Communities, Luxembourg.
7. Gelboin H (1980) Benzo[*a*]pyrene metabolism, activation and carcinogenesis: role of regulation of mixed-function oxidases and related enzymes. *Physiol. Rev.* 60: 1107–1165.
8. Gonzalez FJ, Aoyama T and Gelboin HV (1990) Activation of promutagens by human cDNA-expressed cytochrome P450s. *Prog. Clin. Biol. Res.* 340B: 77–86.
9. Iwanari M, Nakajima M, Kizu R, Hayakawa K and Yokoi T (2002) Induction of CYP1A1, CYP1A2, and CYP1B1 mRNAs by nitropolycyclic aromatic

hydrocarbons in various human tissue-derived cells: chemical-, cytochrome P450 isoform-, and cell-specific differences. *Arch. Toxicol.* 76: 287–298.
10. Shimada T, Yamasaki H, Mimura M, Inui Y and Guengerich FP (1994) Interindividual variations in human liver cytochrome P450 enzymes involved in the oxidation of drugs, carcinogens and toxic chemicals: studies with liver microsomes of 30 Japanese and 30 Caucasians. *J. Pharmacol. Exp. Ther.* 270: 414–423.
11. Raunio H, Pasanen M, Mäenpää J, Hakkola J and Pelkonen O (1995) Expression of extrahepatic cytochrome P450 in humans. In: *Advances in Drug Metabolism in Man* [Pacifici GM and Fracchia GN (eds.)] pp 234–287, Office for the Official Publications of the European Communities, Luxembourg.
12. Nelson DR, Koymans L, Kamataki T, Stegeman JJ, Feyereisen R, Waxman DJ, Waterman MR, Gotoh O, Coon MJ, Estabrook RW, Gunsalus IC and Nebert DW (1996) P450 superfamily: update on new sequences, gene mapping, accession numbers and nomenclature. *Pharmacogenetics* 6: 1–42.
13. Whitlock JP Jr (1999) Induction of cytochrome P4501A1. *Annu. Rev. Pharmacol. Toxicol.* 39: 103–125.
14. Denison MS, Pandini A, Nagy SR, Baldwin EP and Bonati L (2002) Ligand binding and activation of the Ah receptor. *Chem.-Biol. Interact.* 141: 3–24.
15. Kellermann G, Shaw CR and Luyten-Kellerman M (1973) Aryl hydrocarbon hydroxylase inducibility and bronchogenic carcinoma. *N. Engl. J. Med.* 289: 934–937.
16. Kawajiri K, Nakachi K, Imai K, Yoshii A, Shinoda N and Watanabe J (1990) Identification of genetically high risk individuals to lung cancer by DNA polymorphisms of the cytochrome P450IA1 gene. *FEBS Lett.* 263: 131–133.
17. Hayashi SI, Watanabe J, Nakachi K and Kawajiri K (1991) Genetic linkage of lung cancer-associated *Msp*I polymorphisms with amino acid replacement in the heme binding region of the human cytochrome P4501A1 gene. *J. Biochem.* 110: 407–411.
18. Cascorbi I, Brockmöller J and Roots I (1996) A C4887A polymorphism in exon 7 of human *CYP1A1*: population frequency, mutation linkages, and impact on lung cancer susceptibility. *Cancer Res.* 56: 4965–4969.
19. Crofts F, Taioli E, Trachman J, Cosma GN, Currie D, Toniolo P and Garte SJ (1994) Functional significance of different human *CYP1A1* genotypes. *Carcinogenesis* 15: 2961–2963.
20. Jacquet M, Lambert V, Baudoux E, Müller M, Kremers P and Gielen J (1996) Correlation between P450 CYP1A1 inducibility, *Msp*I genotype and lung cancer incidence. *Eur. J. Cancer* 32A: 1701–1706.

21. Landi MT, Bertazzi PA, Shields PG, Clark G, Lucier GW, Garte SJ, Cosma G and Caporaso, NE (1994) Association between *CYP1A1* genotype, mRNA expression and enzymatic activity in humans. *Pharmacogenetics* 4: 242–246.
22. Wedlund PJ, Kimura S, Gonzalez FJ and Nebert DW (1994) ^{1462}V mutation in the human *CYP1A1* gene: lack of correlation with either the *Msp*I 1.9 kb (M2) allele or CYP1A1 inducibility in a three-generation family of east Mediterranean descent. *Pharmacogenetics* 4: 21–26.
23. Anttila S, Tuominen P, Hirvonen A, Nurminen T, Karjalainen A, Hankinson O and Elovaara E (2001) CYP1A1 levels in lung tissue of tobacco smokers and polymorphisms of CYP1A1 and aromatic hydrocarbon receptor. *Pharmacogenetics* 11: 501–509.
24. D'Errico A, Taioli E, Chen X and Vineis P (1996) Genetic polymorphisms and the risk of cancer: a review of the literature. *Biomarkers* 1: 149–173.
25. Kawajiri K, Nakachi K, Imai K, Watanabe J and Hayashi S (1993) The *CYP1A1* gene and cancer susceptibility. *Crit. Rev. Oncol. Hematol.* 14: 77–87.
26. Sugimura H, Wakai K, Genka K, Nagura K, Igarashi H, Nagayama K, Ohkawa A, Baba S, Morris BJ, Tsugane S, Ohno Y, Gao C, Li Z, Takezaki T, Tajima K and Iwamasa T (1998) Association of Ile462Val (exon 7) polymorphism of cytochrome P450IA1 with lung cancer in the Asian population: further evidence from a case-control study in Okinawa. *Cancer Epidemiol. Biomarkers Prev.* 7: 413–417.
27. Xu X, Kelsey KT, Wiencke JK, Wain JC and Christiani DC (1996) Cytochrome P450 *CYP1A1 Msp*I polymorphism and lung cancer susceptibility. *Cancer Epidemiol. Biomarkers Prev.* 5: 687–692.
28. Tefre T, Ryberg D, Haugen A, Nebert DW, Skaug V, Brøgger A and Børresen AL (1991) Human CYP1A1 (cytochrome P_1450) gene: lack of association between the *Msp*I restriction fragment length polymorphism and incidence of lung cancer in a Norwegian population. *Pharmacogenetics* 1: 20–25.
29. Hirvonen A, Husgafvel-Pursianen K, Karjalainen A, Anttila S and Vainio H (1992) Point-mutational *Msp*I and Ile/Val polymorphisms closely linked in the *CYP1A1* gene: lack of association with susceptibility to lung cancer in a Finnish population. *Cancer Epidemiol. Biomarkers Prev.* 1: 485–489.
30. Shields PG, Sugimura H, Caporaso NE, Petruzzelli SF, Bowman ED, Trump BF, Weston A and Harris CC (1992) Polycyclic aromatic hydrocarbon-DNA adducts and the *CYP1A1* restriction fragment length polymorphism. *Environ. Health Perspect.* 98: 191–194.

31. Alexandrie AK, Sundberg MI, Seidegard J, Tornling G and Rannug A (1994) Genetic susceptibility to lung cancer with special emphasis on *CYP1A1* and *GSTM1*: a study on host factors in relation to age at onset, gender and histological cancer types. *Carcinogenesis* 15: 1785–1790.
32. Bouchardy C, Wikman H, Benhamou S, Hirvonen A, Dayer P and Husgafvel-Pursiainen K (1997) *CYP1A1* genetic polymorphisms, tobacco smoking and lung cancer risk in a French Caucasian population. *Biomarkers* 2: 131–134.
33. Vineis P, Veglia F, Benhamou S, Butkiewicz D, Cascorbi I, Clapper ML, Dolzan V, Haugen A, Hirvonen A, Ingelman-Sundberg M, Kihara M, Kiyohara C, Kremers P, LeMarchand L, Oshima S, Pastorelli R, Rannug A, Romkes M, Schoket B, Shields P, Strange RC, Stucker I, Sugimura H, Garte S, Gaspari L and Taioli E (2003) CYP1A1 $T^{3801}C$ polymorphism and lung cancer: a pooled analysis of 2451 cases and 3358 controls. *Int. J. Cancer* 104: 650–657.
34. Sutter TR, Tang YM, Hayes CL, Wo YY, Jabs EW, Li X, Yin H, Cody CW and Greenlee WF (1994) Complete cDNA sequence of a human dioxin-inducible mRNA identifies a new gene subfamily of cytochrome P450 that maps to chromosome 2. *J. Biol. Chem.* 269: 13092–13099.
35. Murray GI, Melvin WT, Greenlee WF and Burke MD (2001) Regulation, function, and tissue-specific expression of cytochrome P450 CYP1B1. *Annu. Rev. Pharmacol. Toxicol.* 41: 297–316.
36. Hakkola J, Pasanen M, Pelkonen O, Hukkanen J, Evisalmi S, Anttila S, Rane A, Mäntylä M, Purkunen R, Saarikoski S, Tooming M and Raunio H (1997) Expression of CYP1B1 in human adult and fetal tissues and differential inducibility of CYP1B1 and CYP1A1 by Ah receptor ligands in human placenta and cultured cells. *Carcinogenesis* 18: 391–397.
37. Kim JH, Stansbury KH, Walker NJ, Trush MA, Strickland PT and Sutter TR (1998) Metabolism of benzo[*a*]pyrene and benzo[*a*]pyrene-7,8-diol by human cytochrome P450 1B1. *Carcinogenesis* 19: 1847–1853.
38. Luch A, Coffing SL, Tang YM, Schneider A, Soballa V, Greim H, Jefcoate CR, Seidel A, Greenlee WF, Baird WM and Doehmer J (1998) Stable expression of human cytochrome P450 1B1 in V79 Chinese hamster cells and metabolically catalyzed DNA adduct formation of dibenzo[*a,l*]pyrene. *Chem. Res. Toxicol.* 12: 353–364.
39. Shimada T, Watanabe J, Kawajiri K, Sutter TR, Guengerich FP, Gillam EMJ and Inoue K (1999) Catalytic properties of polymorphic human cytochrome P450 1B1 variants. *Carcinogenesis* 20: 1607–1613.

40. Ko Y, Abel J, Harth V, Bröde P, Anthony C, Donat S, Fischer HP, Ortiz-Pallardo ME, Thier R, Sachinidis A, Vettre H, Bolt HM, Herberhold C and Brüning T (2001) Association of CYP1B1 codon 432 mutant allele in head and neck squamous cell cancer is reflected by somatic mutations of p53 in tumor tissue. *Cancer Res.* 61: 4398–4404.
41. Miners JO and Birkett DJ (1998) Cytochrome P4502C9: an enzyme of major importance in human drug metabolism. *Br. J. Clin. Pharmacol.* 45: 525–538.
42. Rettie AE, Wienkers LC, Gonzalez FJ, Trager WF and Korzekwa KR (1994) Impaired (S)-warfarin metabolism catalysed by the $R^{144}C$ allelic variant of *CYP2C9*. *Pharmacogenetics* 4: 39–42.
43. London SJ, Sullivan-Klose T, Daly AK and Idle JR (1997) Lung cancer risk in relation to the *CYP2C9* genetic polymorphism among Caucasians in Los Angeles County. *Pharmacogenetics* 7: 401–404.
44. Haining RL, Hunter AP, Veronese ME, Trager WF and Rettie AE (1996) Allelic variants of human cytochrome P4502C9: baculovirus-mediated expression, purification, structural characterization, substrate stereoselectivity and prochiral selectivity of the wild-type and I359L mutant forms. *Arch. Biochem. Biophys.* 333: 447–458.
45. Sullivan-Klose TH, Ghanayem BI, Bell DA, Zhang ZY, Kaminsky LS, Shenfield GM, Miners JO, Birkett DJ and Goldstein JA (1996) The role of the CYP2C9-Leu359 allelic variant in the tolbutamide polymorphism. *Pharmacogenetics* 6: 341–349.
46. Degawa M, Stern SJ, Martin MV, Guengerich FP, Fu PP, Ilett KF, Kaderlik RK and Kadlubar FF (1994) Metabolic activation and carcinogen-DNA adduct detection in human larynx. *Cancer Res.* 54: 4915–4919.
47. Ozawa S, McDaniel LP, Tang YM, Schoket B, Vincze I, Kostic S and Kadlubar FF (1997) *CYP2C9* and *GSTP1* genetic polymorphisms in patients with smoking-related lung cancer. *Proc. Am. Assoc. Cancer Res.* 38: 212P.
48. Garcia-Martin E, Martinez C, Ladero JM, Gamito FJG, Rodriguez-Lescure A and Agundez JAG (2002) Influence of cytochrome P450 CYP2C9 genotypes in lung cancer risk. *Cancer Lett.* 180: 41–46.
49. Omiecinski CJ, Aicher L, Holubkov R and Checkoway H (1993) Human peripheral lymphocytes as indicators of microsomal epoxide hydrolase activity in liver and lung. *Pharmacogenetics* 3: 150–158.
50. Janot F, Massaad L, Ribrag V, de Waziers I, Beaune PH, Luboinski B, Parise O Jr, Gouyette A and Chabot GG (1993) Principal xenobiotic-metabolizing enzyme systems in human head and neck squamous cell carcinoma. *Carcinogenesis* 14: 1279–1283.

51. Etter HU, Richter C, Ohta Y, Winterhalter KH, Sasabe H and Kawato S (1991) Rotation and interaction with epoxide hydrase of P-450 in proteoliposomes. *J. Biol. Chem.* 266: 18600–18605.
52. Sims P, Grover PL, Swaisland A, Pal K and Hewer A (1974) Metabolic activation of benzo[*a*]pyrene proceeds by a diol-epoxide. *Nature* 252: 326–328.
53. Seidegård J and Ekström G (1997) The role of human glutathione transferases and epoxide hydrolases in the metabolism of xenobiotics. *Environ. Health Perspect.* 105: 791–799.
54. Hasset C, Robinson KB, Beck NB and Omiecinski CJ (1994) The human microsomal epoxide hydrolase gene (*EPHX1*): complete nucleotide sequence and structural characterization. *Genomics* 23: 433–442.
55. Hasset C, Aicher L, Sidhu JS and Omiecinski CJ (1994) Human microsomal epoxide hydrolase: genetic polymorphism and functional expression *in vitro* of amino acid variants. *Hum. Mol. Genet.* 3: 421–428.
56. Raaka S, Hasset C and Omiecinski CJ (1998) Human microsomal epoxide hydrolase: 5′-flanking region genetic polymorphism. *Carcinogenesis* 19: 387–393.
57. Smith CAD and Harrison DJ (1997) Association between polymorphism in gene for microsomal epoxide hydrolase and susceptibility to emphysema. *Lancet* 350: 630–633.
58. Yoshikawa M, Hiyama K, Ishioka S, Maeda H, Maeda A and Yamakido M (2000) Microsomal epoxide hydrolase genotypes and chronic obstructive pulmonary disease in Japanese. *Int. J. Mol. Med.* 5: 49–53.
59. Benhamou S, Reinikainen M, Bouchardy C, Dayer P and Hirvonen A (1998) Association between lung cancer and microsomal epoxide hydrolase genotypes. *Cancer Res.* 58: 5291–5293.
60. Persson I, Johansson I, Lou YC, Yue QY, Duan LS, Bertilsson L and Ingelman-Sundberg M (1999) Genetic polymorphism of xenobiotic metabolizing enzymes among Chinese lung cancer patients. *Int. J. Cancer* 81: 325–329.
61. London SJ, Smart J and Daly AK (2000) Lung cancer risk in relation to genetic polymorphisms of microsomal epoxide hydrolase among African-Americans and Caucasians in Los Angeles County. *Lung Cancer* 28: 147–155.
62. Zhou W, Thurston SW, Liu G, Xu LL, Miller DP, Wain JC, Lynch TJ, Su L and Christiani DC (2001) The interaction between microsomal epoxide hydrolase polymorphisms and cumulative cigarette smoking in different

histological subtypes of cancer. *Cancer Epidemiol. Biomarkers Prev.* 10: 461–466.
63. Jourenkova-Mironova N, Mitrunen K, Bouchardy C, Dayer P, Benhamou S and Hirvonen A (2000). High activity microsomal epoxide hydrolase genotypes and the risk of oral, pharynx and larynx cancers. *Cancer Res.* 60: 534–536.
64. Hayes JD and Pulford DJ (1995) The glutathione *S*-transferase supergene family: Regulation of GST and the contribution of the isoenzymes to cancer chemoprotection and drug resistance. *Crit. Rev. Biochem. Mol. Biol.* 30: 445–600.
65. Benhamou S, Lee WJ, Alexandrie AK, Boffetta P, Bouchardy C, Butkiewicz D, Brockmöller J, Clapper ML, Daly A, Dolzan V, Ford J, Gaspari L, Haugen A, Hirvonen A, Husgafvel-Pursiainen K, Ingelman-Sundberg M, Kalina I, Kihara M, Kremers P, LeMarchand L, London SJ, Nazar-Stewart V, Ono-Kihara M, Rannug A, Romkes M, Ryberg D, Seidegard J, Shields P, Strange RC, Stücker I, To-Figueras J, Brennan P and Taioli E (2002) Meta- and pooled analyses of glutathione *S*-transferase M1 polymorphism and smoking on lung cancer risk. *Carcinogenesis* 23: 1343–1350.
66. Engel LS, Taioli E, Pfeiffer R, Garcia-Closas M, Marcus PM, Lan Q, Boffetta P, Vineis P, Autrup H, Bell DA, Branch RA, Brockmöller J, Daly AK, Heckbert SR, Kalina I, Kang D, Katoh T, Lafuente A, Lin HJ, Romkes M, Taylor JA and Rothman N (2002) Pooled analysis and meta-analysis of glutathione *S*-transferase M1 and bladder cancer: a HuGE review. *Am. J. Epidemiol.* 156: 95–109.
67. McWilliams JE, Sanderson BJ, Harris EL, Richert-Boe KE and Henner WD (1995) Glutathione *S*-transferase M1 (GSTM1) deficiency and lung cancer risk. *Cancer Epidemiol. Biomarkers Prev.* 4: 589–94.
68. Brockmöller J, Kerb R, Drakoulis N, Staffeldt B and Roots I (1994) Glutathione *S*-transferase M1 and its variants A and B as host factors of bladder cancer susceptibility: a case-control study. *Cancer Res.* 54: 4103–4111.
69. Inskip A, Elexpuru-Camiruaga J, Buxton N, Dias PS, Macintosh J, Campbell D, Jones PW, Yengi L, Talbot JA, Strange RC and Fryer AA (1995) Identification of polymorphism at the glutathione *S*-transferase, *GSTM3* locus: evidence for linkage with *GSTM1*A*. *Biochem. J.* 312: 713–716.
70. Anttila S, Hirvonen A, Vainio H, Husgafvel-Pursiainen K, Hayes JD and Ketterer B (1993) Immunohistochemical localization of glutathione *S*-transferases in human lung. *Cancer Res.* 53: 5643–5648.

71. Anttila S, Luostarinen L, Hirvonen A, Elovaara E, Karjalainen A, Nurminen T, Hayes JD, Vainio H and Ketterer B (1995) Pulmonary expression of glutathione *S*-transferase M3 in lung cancer patients: association with *GSTM1* polymorphism, smoking, and asbestos exposure. *Cancer Res.* 55: 3305–3309.
72. Yengi L, Inskip A, Gilford J, Alldersea J, Bailey L, Smith A, Lear JT, Heagerty AH, Bowers B, Hand P, Hayes JD, Jones PW, Strange RC and Fryer AA (1996) Polymorphism at the glutathione *S*-transferase locus *GSTM3*: interactions with cytochrome P450 and glutathione *S*-transferase genotypes as risk factors for multiple cutaneous basal cell carcinoma. *Cancer Res.* 56: 1974–1977.
73. Jahnke V, Matthias C, Fryer A and Strange R (1996) Glutathione *S*-transferase and cytochrome-P-450 polymorphism as risk factors for squamous cell carcinoma of the larynx. *Am. J. Surg.* 172: 671–673.
74. Matthias C, Bockmuhl U, Jahnke V, Jones PW, Hayes JD, Alldersea J, Gilford J, Bailey L, Bath J, Worrall SF, Hand P, Fryer AA and Strange RC (1998) Polymorphism in cytochrome P450 CYP2D6, CYP1A1, CYP2E1 and glutathione *S*-transferase, GSTM1, GSTM3, GSTT1 and susceptibility to tobacco-related cancers: studies in upper aerodigestive tract cancers. *Pharmacogenetics* 8: 91–100.
75. Jourenkova-Mironova N, Wikman H, Bouchardy C, Voho A, Dayer P, Benhamou S and Hirvonen A (1998) Role of glutathione *S*-transferase *GSTM1*, *GSTM3*, *GSTP1*, and *GSTT1* genotypes in modulating susceptibility to smoking related lung cancer. *Pharmacogenetics* 8: 495–502.
76. Jourenkova-Mironova N, Voho A, Bouchardy C, Wikman H, Dayer P, Benhamou S and Hirvonen A (1999) Glutathione *S*-transferase *GSTM1*, *GSTM3*, *GSTP1*, and *GSTT1* genotypes and the risk of smoking related oral and pharyngeal cancer. *Int. J. Cancer* 81: 44–48.
77. Ali-Osman F, Akande N, Antoun G, Mao JX and Buolamwini J (1997) Molecular cloning, characterization, and expression in *Escherichia coli* of full-length cDNAs of three human glutathione *S*-transferase Pi gene variants. *J. Biol. Chem.* 272: 10004–10012.
78. Zimniak P, Nanduri B, Pilula S, Bandorowicz-Pikula J, Singhal SS, Srivastava SK, Awasthi S and Awasrhi YC (1994) Naturally occurring human glutathione *S*-transferase GSTP1–1 isoforms with isoleucine and valine at position 104 differ in enzymatic properties. *Eur. J. Biochem.* 224: 893–899.
79. Hu X, O'Donnell R, Srivastava SK, Xia H, Zimniak P, Nanduri B, Bleicher RJ, Awasthi S, Awasthi YC, Ji X and Singh SV (1997) Active site

architecture of polymorphic forms of human glutathione *S*-transferase P1-1 accounts for their enantioselectivity and disparate activity in the glutathione conjugation of $7\beta,8\alpha$-dihydroxy-$9\alpha,10\alpha$-oxy-7,8,9,10-tetrahydrobenzo[*a*]pyrene. *Biochem. Biophys. Res. Comm.* 235: 424–428.
80. Harries LW, Stubbins MJ, Forman D, Howard GCW and Wolf R (1997) Identification of genetic polymorphisms at the glutathione *S*-transferase Pi locus and association with susceptibility to bladder, testicular and prostate cancer. *Carcinogenesis* 18: 641–644.
81. Matthias C, Bockmuhl U, Jahnke V, Harries LW, Wolf CR, Jones PW, Alldersea J, Worrall SF, Hand P, Fryer AA and Strange RC (1998) The glutathione *S*-transferase *GSTP1* polymorphism: effects on susceptibility to oral/pharyngeal and laryngeal carcinomas. *Pharmacogenetics* 8: 1–6.
82. Ryberg D, Skaug V, Hewer A, Phillips DH, Harries LW, Wolf CR, Øgreid D, Ulvik A, Vu P and Haugen A (1997) Genotypes of glutathione transferase M1 and P1 and their significance for lung DNA adduct levels and cancer risk. *Carcinogenesis* 18: 1285–1289.
83. Harris MJ, Coggan M, Langton L, Wilson SR and Board PG (1998) Polymorphism of the Pi class glutathione *S*-transferase in normal populations and cancer patients. *Pharmacogenetics* 8: 27–31.
84. Hirvonen A (1997) Combinations of susceptible genotypes and individual responses to toxicants. *Environ. Health Perspect.* 105: 755–758.
85. Joseph P and Jaiswal AK (1994) NAD(P)H:quinone oxidoreductase$_1$ (DT diaphorase) specifically prevents the formation of benzo[*a*]pyrene quinone-DNA adducts generated by cytochrome P4501A1 and P450 reductase. *Proc. Natl. Acad. Sci. USA* 91: 8413–8417.
86. Traver RD, Horikoshi T, Danenberg PV, Ross D and Gibson NW (1992) NAD(P)H:quinone oxidoreductase gene expression in human colon carcinoma cells: characterization of a mutation which modulates DT-diaphorase activity and mitomycin sensitivity. *Cancer Res.* 52: 797–802.
87. Traver RD, Siegel D, Beall HD and Gibson NW (1997) Characterization of a polymorphism in NAD(P)H:quinone oxidoreductase (DT-diaphorase). *Br. J. Cancer* 75: 69–75.
88. Rosvold EA, McGlynn KA, Lustbader ED and Buetow KH (1995) Identification of an NAD(P)H:quinone oxidoreductase polymorphism and its association with lung cancer and smoking. *Pharmacogenetics* 5: 199–206.
89. Siegel D, McGuinness SM, Winski SL and Ross D (1999) Genotype-phenotype relationships in studies of a polymorphism in NAD(P)H:quinone oxidoreductase 1. *Pharmacogenetics* 9: 113–121.

90. Benhamou S, Voho A, Bouchardy C, Mitrunen K, Dayer P and Hirvonen A (2001) Role of NAD(P)H:quinone oxidoreductase polymorphism at codon 187 in susceptibility to lung, larynx, and oral/pharynx cancers. *Biomarkers* 6: 440–447.
91. Lewis SJ, Cherry NM, Niven RM, Barber PV and Povey AC (2001) Polymorphism in the NAD(P)H:quinone oxidoreductase gene and small cell lung cancer risk in a UK population. *Lung Cancer* 34: 177–183.
92. Hamajima N, Matsuo K, Iwata H, Shinoda M, Yamamura Y, Kato T, Hatooka S, Mitsudomi T, Suyama M, Kagami Y, Ogura M, Ando M, Sugimura Y and Tajima K (2002) NAD(P)H:quinone oxidoreductase 1 (NQO1) C609T polymorphism and the risk of eight cancers for Japanese. *Int. J. Clin. Oncol.* 7: 103–108.
93. Sunaga N, Kohno T, Yanagitani N, Sugimura H, Kunitoh H, Tamura T, Takei Y, Tsuchiya S, Saito R and Yokota J (2002) Contribution of the NQO1 and GSTT1 polymorphisms to lung adenocarcinoma susceptibility. *Cancer Epidemiol. Biomarkers Prev.* 11: 730–738.
94. Wiencke JK, Spitz MR, McMillan A and Kelsey KT (1997) Lung cancer in Mexican-Americans and African-Americans is associated with the wild-type genotype of the NAD(P)H:quinone oxidoreductase polymorphism. *Cancer Epidemiol. Biomarkers Prev.* 6: 87–92.
95. Chen H, Lum A, Seifried A, Wilkens LR and Le Marchand L (1999) Association of the NAD(P)H:quinone oxidoreductase 609C $\rightarrow$ T polymorphism with a decreased lung cancer risk. *Cancer Res.* 59: 3045–3048.
96. Lin P, Wang HJ, Lee H, Lee HS, Wang SL, Hsueh YM, Tsai KJ and Chen CY (1999) NAD(P)H:quinone oxidoreductase polymorphism and lung cancer in Taiwan. *J. Toxicol. Environ. Health A* 58: 187–197.
97. Foster CB, Lehrnbecher T, Mol F, Steinberg SM, Venzon DJ, Walsh TJ, Noack D, Rae J, Winkelstein JA, Curnutte JT and Chanock SJ (1998) Host defense molecule polymorphisms influence the risk for immune-mediated complications in chronic granulomatous disease. *J. Clin. Invest.* 102: 2146–2155.
98. Hunninghake GW and Crystal RG (1990) Cigarette smoking and lung destruction: accumulation of neutrophils in the lungs of cigarette smokers. *Ann. Rev. Respir. Dis.* 128: 833–838.
99. Schmekel B, Karlsson SE, Linden M, Sundström C, Tegner H and Venge P (1990) Myeloperoxidase in human lung lavage. I. A marker of local neutrophil activity. *Inflammation* 14: 447–454.

100. Schmekel B, Hornblad Y, Linden M, Sundström C and Venge P (1990) Myeloperoxidase in human lung lavage. II. Internalization of myeloperoxidase by alveolar macrophages. *Inflammation* 14: 455–461.
101. Mallet WG, Mosebrook DR and Trush MA (1991) Activation of (±)-*trans*-7,8-dihydroxy-7,8-dihydobenzo[*a*]pyrene to diolepoxides by human polymorphonuclear leukocytes or myeloperoxidase. *Carcinogenesis* 12: 521–524.
102. Petruska JM, Mosebrook DR, Jakab GJ and Trush MA (1992) Myeloperoxidase-enhanced formation of (±)-*trans*-7,8-dihydroxy-7,8-dihydrobenzo-[*a*]pyrene-DNA adducts in lung tissue *in vitro*: a role of pulmonary inflammation in the bioactivation of a procarcinogen. *Carcinogenesis* 13: 1075–1081.
103. Piedrafita FJ, Molander RB, Vansant G, Orlova EA, Pfahl M and Reynolds WF (1997) An Alu element in the myeloperoxidase promoter contains a composite SP1-thyroid hormone-retinoic acid response element. *J. Biol. Chem.* 271: 14412–14420.
104. Rojas M, Godschalk R, Alexandrov K, Cascorbi I, Kriek E, Ostertag J, van Schooten FJ and Bartsch H (2001) Myeloperoxidase −463A variant reduces benzo[*a*]pyrene diol-epoxide DNA adducts in skin of coal tar treated patients. *Carcinogenesis* 22: 1015–1018.
105. London SJ, Lehman TA and Taylor JA (1997) Myeloperoxidase genetic polymorphism and lung cancer risk. *Cancer Res.* 57: 5001–5003.
106. Cascorbi I, Henning S, Brockmöller J, Gephart J, Meisel C, Müller JM, Loddenkemper R and Roots I (2000) Substantially reduced risk of cancer of the aerodigestive tract in subjects with variant −463A of the myeloperoxidase gene. *Cancer Res.* 60: 644–649.
107. Le Marchand L, Seifried A, Lum A and Wilkens LR (2000) Association of the myeloperoxidase −463G → A polymorphism with lung cancer risk. *Cancer Epidemiol. Biomarkers Prev.* 9: 181–184.
108. Feyler A, Voho A, Bouchardy C, Kuokkanen K, Dayer P, Hirvonen A and Benhamou S (2002) Myeloperoxidase $^{-463}$G → A polymorphism and lung cancer risk. *Cancer Epidemiol. Biomarkers Prev.* 11: 1550–1554.
109. Schabath MB, Spitz MR, Zhang X, Delclos GL and Wu X (2000) Genetic variants of myeloperoxidase and lung cancer risk. *Carcinogenesis* 21: 1163–1166.
110. Schabath MB, Spitz MR, Hong WK, Delclos GL, Reynolds WF, Gunn GB, Whitehead LW and Wu X (2002) A myeloperoxidase polymorphism associated with reduced risk of lung cancer. *Lung Cancer* 37: 35–40.

111. Schabath MB, Spitz MR, Delclos GL, Gunn GB, Whitehead LW and Wu X (2002) Association between asbestos exposure, cigarette smoking, myeloperoxidase (MPO) genotypes, and lung cancer risk. *Am. J. Ind. Med.* 42: 29–37.
112. Dally H, Gassner K, Jäger B, Schmezer P, Spiegelhalder B, Edler L, Drings P, Dienemann H, Schulz V, Kayser K, Bartsch H and Risch A (2002) Myeloperoxidase (*MPO*) genotype and lung cancer histologic types: the *MPO –463* A allele is associated with reduced risk for small cell lung cancer in smokers. *Int. J. Cancer* 102: 530–535.
113. Lu W, Xing D, Qi J, Tan W, Miao X and Lin D (2002) Genetic polymorphism in myeloperoxidase but not GSTM1 is associated with risk of lung squamous cell carcinoma in a Chinese population. *Int. J. Cancer* 102: 275–279.
114. Hayashi SI, Watanabe J and Kawajiri K (1992) High susceptibility to lung cancer analyzed in terms of combined genotypes of P450IA1 and Mu-class glutathione *S*-transferase genes. *Jpn. J. Cancer Res.* 83: 866–870.
115. Vaury C, Laine R, Noguiez P, de Coppet P, Jaulin C, Praz F, Pompon D and Amor-Gueret M (1995) Human glutathione *S*-transferase M1 null genotype is associated with a high inducibility of cytochrome P450 1A1 gene transcription. *Cancer Res.* 55: 5520–5523.
116. Bartsch H (1996) DNA adducts in human carcinogenesis: etiological relevance and structure-activity relationship. *Mutat. Res.* 340: 67–69.
117. Bartsch H, Rojas M, Alexandrov K, Camus AM, Castegnaro M, Malaveille C, Anttila S, Hirvonen A, Husgafvel-Pursiainen K, Hietanen E and Vainio H (1995) Metabolic polymorphism affecting DNA binding and excretion of carcinogens in humans. *Pharmacogenetics* 5: S84–S90.
118. Bartsch H and Hietanen E (1996) The role of individual susceptibility in cancer burden related to environmental exposure. *Environ. Health Perspect.* 104, Suppl 3: 569–577.
119. Rojas M, Alexandrov K, Cascorbi I, Brockmöller J, Likhachev A, Pozharisski K, Bouvier G, Auburtin G, Mayer L, Koop-Schneider A, Roots I and Bartsch H (1998) High benzo[*a*]pyrene diol-epoxide DNA adduct levels in lung and blood cells from subjects with combined *CYP1A1 Msp*I/*Msp*I-*GSTM1*0/*0* genotypes. *Pharmacogenetics* 8: 109–118
120. Stücker I, Hirvonen A, de Waziers I, Cabelguenne A, Mitrunen K, Cénée, Koum-Besson E, Hémon D, Beaune P and Loriot MA (2002) Genetic polymorphisms of glutathione *S*-transferases as modulators of lung cancer susceptibility. *Carcinogenesis* 23: 1475–1481.

121. Kihara M and Noda K (1999) Lung cancer risk of the *GSTM1* null genotype is enhanced in the presence of the *GSTP1* mutated genotype in male Japanese smokers. *Cancer Lett.* 137: 53–60.
122. To-Figueras J, Gené M, Gómez-Catalán J, Pique E, Borrego N, Carrasco JL, Ramon J and Corbella J (1999) Genetic polymorphism of glutathione *S*-transferase P1 gene and lung cancer. *Cancer Causes Control* 10: 65–70.
123. Kroemer HK and Eichelbaum M (1995) Molecular bases and clinical consequences of genetic cytochrome P450 2D6 polymorphism. *Life Sci.* 56: 2285–2298.
124. Fujii-Kuriyama Y, Ema M, Mimura J, Matsushita N and Sogawa K (1995) Polymorphic forms of the Ah receptor and induction of the *CYP1A1* gene. *Pharmacogenetics* 5: S149–S153.
125. Ingelman-Sundberg M and Johansson I (1995) The molecular genetics of the human drug metabolizing cytochrome P450s. In: *Advances in Drug Metabolism in Man* [Pacifici GM and Fracchia GN (eds.)] pp 534–585, Office for the Official Publications of the European Communities, Luxembourg.
126. Smart J and Daly AK (2000) Variation in induced CYP1A1 levels: relationship to CYP1A1, Ah receptor and GSTM1 polymorphisms. *Pharmacogenetics* 10: 11–24.
127. McLellan RA, Oscarson M, Alexandrie AK, Seidegård J, Evans DAP, Rannug A and Ingelman-Sundberg M (1997) Characterization of human glutathione *S*-transferase μ cluster containing a duplicated *GSTM1* gene that causes ultrarapid enzyme activity. *Mol. Pharmacol.* 52: 958–965.

11 Polycyclic Aromatic Hydrocarbon-Induced Carcinogenesis — An Integrated View*

Andreas Luch

Massachusetts Institute of Technology
Center for Cancer Research
Cambridge, MA, USA
E-mail: luch@mit.edu

... that if we have to consider a malignant cell as a cell with an irreparable defect, this defect is located, not in the protoplasm but in the nucleus. (...) ... it will hardly be necessary to point out, that the essence of my theory is not the abnormal mitoses but a certain abnormal chromatin-complex, no matter how

*This article is dedicated to Prof. Dr. William M. Baird on the occasion of his 60th birthday.

it arises. Every process which brings about this chromatin condition would lead to a malignant tumor. Certain chromosomes becoming diseased (owing to an inherited disposition), or a disturbance of these chromosomes by intracellular parasites or external influences that act perniciously on certain chromosomes while they do not affect others, these and other factors (...) may be considered important in producing tumors.

The German physician and zoölogist Theodor Boveri (1862 - 1915) in his book on ...
"The Origin of Malignant Tumors" (Translation of the German original entitled
"Zur Frage der Entstehung maligner Tumore", Jena 1914)

11.1 Exposure and Risk

Polycyclic aromatic hydrocarbons (PAHs) are formed during incomplete combustion processes. Due to the abundant use of fossil energy sources in modern societies, PAHs are readily detectable by means of modern analytical instrumentation as ubiquitous contaminants in the general and in certain occupational human environments. The great interest in this group of chemicals lay in the observation that some member compounds possess strong carcinogenicity in animal tumor models such as those for skin, lung, breast and other sites. In addition, early observations of individual cancer cases originating from occupational exposures to mixtures highly contaminated with PAHs, along with a wealth of supporting epidemiological evidence as well as insights on the toxicokinetic and toxicodynamic (mechanistic) behaviour of carcinogenic PAHs in human or 'huminized' cells in culture, lend support to the notion, and at the same time shifting it beyond any reasonable dispute that these compounds can also act as carcinogens in man.

Epidemiological data clearly identified the occupational exposure to complex PAH-containing mixtures such as coal tar or coal tar pitches (consisting of up to 50% of PAHs with 1–2% benzo[*a*]pyrene, B[*a*]P) as *being carcinogenic to humans* (IARC classification Group 1).[1,2] Recent epidemiological meta-analyses again confirmed that heavy exposure to mixtures of

PAHs entails a substantial risk to develop cancer in lung and skin, and possibly also in the bladder (less consistent evidence), larynx, kidney, and pancreas (uncertain evidence).[3–7] Depending on the kind of work place and the country under consideration, occupational exposures to mixtures of PAHs can still be heavy even in our days (see Chapters 3 and 4, Tables 3.1 and 4.1). Although there is no doubt that technical innovations have reduced the levels of PAH exposure during the last decades dramatically, working in coke and aluminum production, for example, may still be accompanied by breathing air that typically is loaded with up to 5–15 μg B[*a*]P/m^3. Considering a breathing volume of about 1.7 m^3 per hour (for males at light work), about 70–200 μg B[*a*]P might be incorporated into the lungs during an eight-hour shift under these conditions — thus posing a considerable risk to the health of these workers.

In comparison to possible exposures at work places, non-occupational 'background exposures' to PAHs through air, water and diet are rather low. In the mid 1990s, the US Department of Health and Human Services estimated the average daily intake of total PAHs in the general population of the U.S. to be about 0.207 μg from air and 0.027 μg from water.[8] In contrast, PAH concentrations in food may vary considerably based on the kind of food and the way of processing and preparing it. Accordingly, the daily levels of PAH ingestion via foodstuffs are much more uncertain, ranging from 0.16 to up to 1.6 μg.[8] At about the same time, the daily uptake of B[*a*]P from a member of the general population in Germany was estimated to be at around 15 ng from air, 2–3 ng from water and 1 μg from food (see Chapter 3.3). These are certainly only rough numbers that have been calculated from the limit concentrations and common levels of B[*a*]P detected in urban air (1–10 ng/m^3) and water (0.1–1 ng/L). On the other hand, the value for food is based on the German limit value for B[*a*]P in meat and meat products (1 μg/kg) and — most likely — overestimates the real burden associated with this carcinogenic PAH in regular (mixed) diet.

No matter which kind of PAH source, humans are always exposed to a mixture of different hydrocarbons with different degrees of biological activity. For instance, more than a hundred different PAHs can be detected in the air.[9,10] The composition of these PAH profiles varies depending on the environmental compartment. Since there is no international agreement on which panel of individual PAHs should be analyzed and reported in order to characterize distinct emission sources, PAH lists released from different organizations may contain different compounds. Sixteen different PAHs such as pyrene, B[*a*]P, benz[*a*]anthracene (B[*a*]A), benzo[*b*]fluoranthene (B[*b*]F), benzo[*k*]fluoranthene, indeno[1,2,3-*cd*]pyrene, dibenz[*a*,*h*]-anthracene (DB[*a*,*h*]A) and others are prioritized by the US-Environmental Protection Agency (EPA)[11,12] (see Figure 9.1). Lower molecular weight PAHs such as naphthalene, phenanthrene, anthracene, or fluoranthene, etc. are also listed by the EPA and by other organizations due to their common occurrence in environmental samples. From all of the hydrocarbons detectable in the human environment the most intensively studied example, B[*a*]P, has been traditionally used as an indicator of carcinogenic PAHs. Based on the increased lung cancer incidence among coke oven workers, the (relatively high) content of the indicator compound B[*a*]P in coke oven emissions, and the (questionable) assumption that PAH profiles in coke oven emissions are not very different to those detectable in ambient air samples, the lung cancer risk from a lifetime exposure to PAHs in ambient air has been calculated at around 5×10^{-2} per μg B[*a*]P/m^3 in a German investigation,[13] or at 8.7×10^{-2} per μg B[*a*]P/m^3 in the WHO Air Quality Guidelines for Europe.[14–16] This latter unit risk estimate would suggest a lifetime cancer risk of about 9 per 100,000 in case of a level of 1 ng B[*a*]P/m^3 in the PAH mixture present in ambient air and assuming a linear dose-response relation between higher and lower exposure levels. It should be noted at this point that quantitative risk assessments for complex PAH mixtures, such as those present in ambient air that are based on exposure data restricted to only one compound and only one specific occupational setting (cokeries), may be hampered by the issue that composition profiles released from sources others than coke ovens (e.g., traffic exhausts, residential heating, wood combustion, etc.) may well be quite different.[17] Moreover, individual PAHs in varying mixtures may interact in a plethora of different ways resulting in synergistic, additive, or antagonistic effects, thus rendering any prediction regarding the overall tumorigenic effect almost beyond any justification (cf. section 11.6). Nevertheless, the risk assessments based on epidemiological data from coke oven workers mentioned above have been supported by animal studies. From lung tumor rates obtained in a rat inhalation study with coal tar or coal tar pitch condensation aerosols, the lifetime lung tumor risk was calculated at 2×10^{-2} per μg B[*a*]P/m^3 using the same kind of linearized multistage model as in the WHO study.[18] So, based on all of these calculations a number between 2 and 9 additional lung cancer cases among 100,000 persons would be expected as the result of a lifetime exposure to a PAH mixture present in ambient air that contains 1 ng B[*a*]P per m^3 volume. Another study based on lung cancer deaths among workers of the aluminum industry in Canada estimated a risk of 1 additional case per 100,000 for the same level of B[*a*]P exposure after applying the same kind of linear model.[19] In any case, given these numbers it must be more than appreciated that B[*a*]P concentrations in ambient air samples decreased dramatically during the last 30 years. While the annual average B[*a*]P concentrations in several European cities were higher than 100 ng/m^3 in the 1960s,[14] a representative German study from the 1990s found values of less than 1 ng/m^3 at places not affected by emission sources (rural regions), 1.77–3.15 ng/m^3 in areas close to traffic, and 2.88–4.12 ng/m^3 in regions contaminated from both traffic and industrial emission sources.[16,20] In addition, the average level of B[*a*]P in traffic areas of the crowded Rhine-Ruhr region was reported to be 3–6 ng/m^3.

11.2 Incorporation and Biotransformation

In principle, PAHs can be readily absorbed from lung, skin and gut. Rapid absorption of B[*a*]P and other PAHs from the gastrointestinal tract or from skin have been widely demonstrated in animal studies,[21–27] and sufficient evidence is also available for humans[28–31] (see also the report by ATSDR[32]). On the other hand, absorption through the lungs may be affected by the size of particles at which most of the airborne PAH fraction is attached. It has been shown that a considerable fraction of B[*a*]P adsorbed on soot, diesel exhaust, or carbon black particles can be retained for longer periods (weeks) in the respiratory tract.[33–36] Compounds may then be partly removed from the particles during transport on the ciliated tracheobronchial mucosa and may penetrate into the epithelial cells where metabolic activation can occur (see below). The slow passage through bronchial mucosa can therefore result in very high local concentrations of these compounds and of their reactive descendants directly within one of the main target tissues.[37,38] It was suggested that the carcinogenic contribution of genotoxic PAH intermediates locally generated in the airway epithelium may dominate under low-exposure conditions as compared to the contribution from those species transported via systemic circulation into the lungs.[17] Under high-exposure conditions, however, the liver as the main metabolic organ may be the dominant contributor also in case of the fraction of active intermediates that target lung epithelial cells.[39]

PAHs exert their mutagenic and carcinogenic activity through biotransformation to electrophilically reactive intermediates that covalently bind to cellular DNA. In Chapter 2, we have discussed all different kinds of enzymatic activation towards DNA-reactive metabolites that may contribute to the overall genotoxicity of carcinogenic PAHs (see Figures 2.2 and 2.4). Enzymes such as cytochrome P450-dependent monooxygenases (CYP), microsomal epoxide hydrolase (mEH), NAD(P)H-dependent aldo-keto reductases (AKR), NAD(P)H-dependent quinone oxidoreductase 1 (NQO1 = DT-diaphorase), and various peroxidases catalyze the initial phase-I of biotransformation that may lead to a variety of different nucleophilic (phenols, *trans*-dihydrodiols, hydroxymethylarenes, hydroquinones) and electrophilic (arene oxides, quinones, dihydrodiol epoxides, radical cations) derivatives. While electrophilic metabolites from PAHs can directly interact with cellular DNA and produce PAH-DNA adducts (Figure 2.7), nucleophilic intermediates require further activation in order to contribute to the compound-specific genotoxicity.

One potential pathway of further activation of phenols is the generation of polynuclear quinones via autoxidation. Polynuclear quinones can subsequently be enzymatically reduced to DNA-damaging semiquinone anion radicals or they may enter into redox cycles together with their corresponding hydroquinones (Figure 2.8). Redox-cycling between quinones and hydroquinones (i.e., quinols, catechols) is one potential source of 'reactive oxygen species' such as superoxide anion radicals, hydroxy radicals and hydrogen peroxide. These reactive species by themselves have then the potential to damage DNA in a plethora of different ways (Figure 2.9). *Trans*-dihydrodiols, on the other hand, can be further oxidized at the adjacent double bond or in the remaining aromatic ring system to give reactive dihydrodiol epoxides that either contain the epoxy group adjacent to the diol moiety (vicinal bay- or fjord-region diol-epoxides; Figure 2.3) or far away from the diol moiety at different regions of the molecule. An alternative route of activation would be the AKR-dependent oxidation of *trans*-dihydrodiols to catechols that undergo further autoxidation to corresponding *ortho*-quinones (Figure 2.10). As with polynuclear quinones, these derivatives then again have the potential to enter into futile redox cycles with their corresponding catechols. Owing to their electrophilic α,β-unsaturated carbonyl moiety however, they also may undergo addition reactions with nucleophiles to produce unstable catechol adducts that spontaneously oxidize to yield stable *ortho*-quinone adducts (Figures 2.4 and 2.7). CYP-mediated oxidation (hydroxylation) of methylated PAHs at their alkyl group is an enzymatic pathway that produces hydroxymethyl derivatives (benzylic alcohols) as another group of important nucleophilic metabolites of certain PAHs such as 1-methylpyrene (see Chapter 2.3.5). Cytosolic sulfotransferases (SULT) were found to be capable of activating hydroxymethylated PAHs through esterification (O-sulfonation) at their side chain ('benzylic ester pathway'). The corresponding sulfate esters would then be able to react with cellular nucleophiles at their benzylic position in an S_N1-related fashion (Figures 2.4 and 2.7).

Biotransformation of PAHs proceeds through activated intermediates that have the potency to damage DNA. Despite its importance and potentially deleterious long-term effects, DNA damage resulting from exposure to carcinogenic PAHs must be regarded as a 'side-product' or an 'accident' of a cellular enzymic defense system that is aimed to detoxify these compounds through conversion into hydrophilic and excretable derivatives. Such end products of PAH biotransformation that are terminated in their toxic potency are generated from subsequent conjugation reactions (phase-II of biotransformation) catalyzed by various transferases such as UDP-glucuronosyltransferases (UGT) or SULT in the case of nucleophilic intermediates, and glutathione S-transferases (GST) in the case of electrophilic intermediates (Chapter 2.4 and Figure 2.11). On the other hand, the side products of PAH metabolism, 'reactive oxygen species', may be targeted by a great variety of additional protective systems.[40–43] While cytosolic or mitochondrial superoxide dismutases

(EC 1.15.1.1) specifically engage in disproportionation ('dismutation') of superoxide anion radicals (into H_2O_2 and O_2), hydrogen peroxide can be detoxified by two different enzymes, the peroxisomal catalase (EC 1.11.1.6; → disproportionation) or the cytosolic and selenium-dependent glutathione peroxidases (EC 1.11.1.9; → reduction). The latter enzyme family also detoxifies organic hydroperoxides (into alcohols) — a reaction that may be catalyzed by (selenium-independent) GST enzymes as well (cf. above). Glutathione (GSH) as the common co-substrate of GSH peroxidases and GST enzymes is also known to be capable to act as a direct scavenger of various radicals (hydroxy or organic radicals, superoxide anion radical, etc.) and it thus serves as an antioxidant itself.

11.3 Monitoring Human Exposure

In consideration of their potential health-threatening effects in the human body, monitoring and regulation of human exposures to PAHs are demanding issues in public health and environmental hygiene that have to be addressed in order to minimize any possible risks. In addition to measuring directly the PAH levels in the air at work places and highly contaminated public areas ('exposure monitoring'), a number of techniques have been developed for the biological monitoring ('biomonitoring') of human exposures. Regardless of which kind of technique, all of them take advantage of the aforementioned circumstance that PAHs undergo biotransformation into reactive and/or hydrophilic derivatives which either may be able to bind to cellular nucleophiles (such as proteins and DNA), or which are excreted from the body via urine and/or feces. While Chapter 3 summarizes the techniques currently available for monitoring of PAH metabolites in the human urine, along with the present knowledge on potential correlations between various kinds of exposure and levels of excreted metabolites, the literature data on PAH 'effect monitoring' (i.e., determination of levels of PAH-protein or PAH-DNA adducts) obtained under various occupational settings or environmental conditions is reviewed in Chapter 4.

Based on their abundant occurrence in environmental PAH mixtures and the predominant excretion of highly fluorescent phenolic derivatives via urine, pyrene and phenanthrene are two compounds that have been widely used as indicator compounds for the overall body burden of incorporated PAHs (Chapters 3.5.2 and 3.5.4). 1-Hydroxypyrene (1-OHP) or a mixture of five different phenolic derivatives of phenanthrene are the main products of CYP-mediated metabolism of the corresponding parent compounds, which are predominantly excreted as glucuronide and sulfate conjugates. In addition, methods for monitoring of various *trans*-dihydrodiols (see Figure 3.2) and, most recently, of one diastereomeric 1,2,3,4-tetrahydrotetraol of phenanthrene in the human urine[44] have been established. Analysis of urinary levels of 1-OHP or of different phenanthrene metabolites is well suited, highly sensitive (in the case of phenols), and also provides valuable information about the overall internal body burden ('internal dose') of PAHs. Yet it seems reasonable to question whether these metabolites could serve as an adequate surrogate measure for carcinogenic PAHs since both parent compounds are not carcinogenic to humans. On the other hand, detection of metabolites of carcinogenic PAHs such as B[*a*]P is greatly hampered by the issue that derivatives of high-molecular PAHs are mainly excreted through feces (Chapter 3.5.5) — a biological matrix not suitable for routine applications. Accordingly, urinary concentrations of B[*a*]P metabolites are rather low. Nevertheless, analytical methods of detecting various phenols and 7,8,9,10-tetraols of B[*a*]P in human urine in the lower femtomolar concentration range have been developed, but are unlikely to be feasible for large biomonitoring studies due to the requirement of multiple purification steps and advanced instrumentation.

Analysis of 1,2,3,4-tetraols of phenanthrene or 7,8,9,10-tetraols of B[*a*]P in human urine would be an attempt to obtain quantitative information on the *in vivo* contribution of the monooxygenation activation pathway that leads to formation of reactive bay-region diol-epoxides (see Chapter 2.3.1, Figure 2.3). As hydrolysis products of the 3,4-diol-1,2-epoxide of phenanthrene and the 7,8-diol-9,10-epoxide of B[*a*]P, both tetraols, however, represent only a fraction of all of those molecules that are enzymatically converted into vicinal diol-epoxides. Since diol-epoxides can head towards a variety of follow-up reactions (hydrolysis, further metabolism at heterotopic sites, conjugation reactions, reactions with macromolecular nucleophiles), conclusions about the actual genotoxic (PAH-DNA adduct generating) dose drawn from the level of urinary tetraols are very uncertain. Directly monitoring the biologically effective dose through determination of PAH diol-epoxide-DNA adducts in target or surrogate tissues (e.g., white blood cells) is a more appropriate measure for the fraction of those molecules that are converted into electrophilic intermediates and that actually may reach their target tissue(s) and react with target macromolecules. The routinely used methods based on fluorescence spectroscopy, immunochemical techniques and radioactive labelling of PAH-DNA adducts are sufficiently sensitive but usually less specific (see Chapter 4.2). On the other hand, reaction products of diol-epoxides or other electrophilic PAH metabolites with proteins such as serum albumin and hemoglobin have been used as alternative biomarkers in PAH 'effect monitoring' studies. These studies actually may be favorable with regard to specificity in case mass spectrometric methods (e.g., GC/MS) are applied.[45–48]

As discussed in Chapter 4.3 (see also Table 4.1), the overall correlation between PAH-DNA or PAH-protein adduct levels in blood cells and the level of PAH exposure under various work place conditions is good — though sometimes weak or statistically insignificant due to confounding factors such as cigarette smoke or diet. A similar conclusion can be drawn for certain environmental or medicinal exposures to PAHs. It should be noted that most monitoring studies investigating the biomarker 'PAH-DNA adduct' use easily retrievable blood cells rather than biopsies of target tissues as the source of DNA. In principle, measurement of DNA (or protein) adducts in blood cells reflects the biologically effective dose of PAH metabolites present in systemic circulation, but not necessarily the fraction of reactive intermediates actually present at target tissue sites (cf. discussion on pulmonary absorption of particulate matter in section 11.2). Yet blood cells may well be quite applicable under certain circumstances to serve as reliable surrogates for estimating the burden of DNA damage generated at those sites.[49–51]

11.4 Molecular Epidemiology: Individual's Susceptibility?

Biomonitoring studies always revealed large inter-individual differences in the levels of biomarkers even under identical exposure conditions (see

Chapters 3 and 4).[52] As most of the human enzymes involved in PAH metabolism are genetically polymorph, inter-individual variability in the levels of biomarkers such as urinary metabolites, PAH-DNA or PAH-protein adducts are likely to be modulated by polymorphisms of genes encoding those enzymes that catalyze the formation of excretable or reactive derivatives. From all that we know, the level of DNA binding along with the structural quality of DNA adducts and their qualitative and quantitative distribution both at the organismic level (i.e., within different kinds of organs and tissues) and at the cellular level (i.e., within cell's genome) are the main determinants for the carcinogenicity of an individual PAH (cf. section 11.6). Because of this intimate correlation between the extent of DNA binding and carcinogenic potency it can therefore be anticipated that inter-individual differences in the activity of enzymes that produce DNA-reactive intermediates or that detoxify these species (and that are based on polymorphic variants of the encoding genes) play an important role in the overall tumor susceptibility of an individual as well (cf. Chapter 10). Many studies have been conducted in recent years that try to explore the influence of single gene polymorphisms and their interactions on the levels of biomarkers of PAH exposure ('molecular epidemiology' approach). The goal of these studies is to use this information in order to identify subpopulations of humans which are likely to be more susceptible ('to be at high risk') as compared to others to develop tumors in response to PAH exposure.[53–57] Furthermore, any risk assessment model that would not consider the variability of susceptibility to DNA damage following carcinogen exposure may severely underestimate the risks to those individuals who are most vulnerable.[58] In addition to the aforementioned biomarkers that provide information on the internal or biologically effective dose of PAH exposure, early biological effects such as chromosomal aberrations, micronuclei formation, sister chromatid exchanges or mutations in the hypoxanthine-guanine phosphoribosyltransferase (*hprt*) gene locus and others can also be easily measured in peripheral lymphocytes and are therefore frequently included into molecular epidemiology studies[57,59–61] (cf. Chapter 8.3.3).

As extensively discussed in Chapter 2, many different enzymes take part in PAH activation and detoxification pathways. Most of them are expressed in various variants with different degrees of activity due to corresponding polymorphisms of the encoding genes. With regard to PAH metabolism, PAH exposure-related biomarkers and PAH-related cancer susceptibility (cf. Chapter 10), probably the most important and certainly the most well studied examples are CYP1A1 and GSTM1. In addition, polymorphic expression of CYP1B1, mEH, NQO1, GSTP1 and others recently shifted into the interest of molecular epidemiology research as well.

From the nine different polymorphic variants of the extrahepatic and highly inducible human CYP1A1 enzyme known today,[62] the *CYP1A1* Ile^{462}Val (A → G transition in exon 7 leading to replacement of Ile by Val) and *CYP1A1 Msp*I (T → C transition in the 3′ untranslated region of the gene, downstream from the polyadenylation site) genotypes are best studied in molecular epidemiology regarding any relation to cancer risk (see Chapter 10.2.1). CYP1A1 (together with CYP1B1) is the major monooxygenase involved in the formation of arene oxides and DNA-reactive bay- and fjord-region diol-epoxides (Chapter 2.1.1 and Figure 2.3). In addition, CYP1A1 was also found to catalyze hydroxylation of pyrene leading to formation of 1-OHP.[63] Accordingly, any polymorphism-based modulation in the activity of this enzyme theoretically might have an impact on the levels of biomarkers such as urinary 1-OHP or PAH-DNA and PAH-protein adducts in peripheral blood cells or target tissues. Large inter-individual differences were observed for PAH-DNA adduct levels in human bronchial biopsy samples[64,65] and confirmed in the case of activation of B[*a*]P by using subcellular preparations *in vitro* or lung tissue culture *ex vivo*.[66,67] However, the results obtained in a rapidly growing number of molecular epidemiology studies that try to connect these effects to single polymorphisms are inconsistent. Some studies have shown that both CYP1A1 variants correlate with elevated levels of PAH-DNA adducts in (lung) target tissue or white blood cells of PAH-exposed individuals,[68–72] others did not[61,73–76] (see also Chapters 4.3, 4.4 and Table 4.1). In those studies that found a positive relation between *CYP1A1* polymorphisms and PAH-DNA adduct levels *in vivo*, the effects were only weak or insignificant, but more pronounced when *CYP1A1 Msp*I or *CYP1A1* Ile^{462}Val genotypes were combined with *GSTM1* deficiency (*GSTM1*-null genotype), or in case adduct analyses were focused on specific marker modifications derived from bay-region 7,8-diol-9,10-epoxides of B[*a*]P (B[*a*]PDE)[69,72] (see below). On the other hand, recombinant expression of CYP1A1 variants *in vitro* demonstrated that differences in their activities towards B[*a*]P or B[*a*]P-7,8-dihydrodiol are the result of different reaction kinetics, which were actually expected to be rather small under physiological conditions.[77–79] According to the literature data reviewed in Chapter 3.5.3, the results of studies investigating polymorphic variation of urinary 1-OHP levels are heterogeneous as well — with about half of them revealing slightly (2-fold) elevated levels of this biomarker in the case of *CYP1A1 Msp*I or *CYP1A1* Ile^{462}Val genotype carriers (see also Adonis *et al.*[80]), whereas others did not (see also Yang *et al.*[81]).

Despite the lack of a clear and significant relationship beweeen *CYP1A1* polymorphisms and elevated levels of PAH exposure biomarkers, epidemiological evidence may point to an (slightly) increased lung cancer risk in homozygous *CYP1A1* gene variant carriers[82–85] — though this is still controversially discussed[86] (cf. Chapter 10.2.1). Similar evidence was also found for certain *CYP1B1* genotypes in the case of cancer of the lung, head-and-neck, colon, and the hormone-dependent tissues in breast and prostate[87–92] (cf. Chapter 10.2.2). Although discovered only about one decade ago, more than a dozen different polymorphic CYP1B1 variants have already been characterized.[93] With regard to PAH metabolism, only minor, if any, differences in the activity between most of these polymorphic variants were documented.[94–96] One genotype, *CYP1B1* Val^{432}Leu (C → G transversion in exon 3 leading to replacement of Val by Leu), was found to exert an increased activity in the epoxidation of B[*a*]P at its 7,8-position,[97] but not in product's subsequent epoxidation towards bay-region diol-epoxides[98] (see Figure 2.3). In accord to these findings, no significant influence of single *CYP1B1* polymorphisms was seen with regard to levels of urinary 1-OHP.[81] Further, the *CYP1B1* Val^{432}Leu genotype did not modulate bulky DNA adduct levels in bronchial tissue and peripheral blood lymphocytes of cigarette-smoking lung cancer patients and potroom workers of the aluminum production, respectively.[61] Curiously, this particular polymorphism was associated with reduced levels of DNA adducts when normal breast tissue excised from reduction mammoplasty patients was exposed to B[*a*]P *ex vivo*[99] — though the overall adduct levels were significantly higher in breast cancer cases compared to controls. In head-and-neck cancer patients, on the other hand, the susceptibility genotype *CYP1B1* $Val^{432}Leu$ was actually found to be strongly associated with an increased frequency of smoking-induced mutations of the *p53* gene.[90,91] The nature of the chemicals responsible for *p53* mutation induction in this tissue, however, remained unclear.

Human mEH has a dual role in PAH metabolism. Due to its ability to hydrolyze simple arene oxides and produce *trans*-dihydrodiols, it is involved in detoxification routes (towards dihydrodiol conjugates) and toxification routes (towards diol-epoxides) at the same time and often in the case of the very same PAH (see Figure 2.11). Therefore, the role of mEH in modulation of PAH-related tumor susceptibility in the human population is hard to predict. Besides seven polymorphisms in the 5′ non-coding region, two amino acid exchanges in the coding region of the mEH gene (*EPHX1*) have been identified and correlated to a decrease (^{113}Tyr → His in codon 3) or an increase (^{139}His → Arg in codon 4) in the *in vitro* activity of the corresponding protein against the K-region 4,5-oxide of B[*a*]P[100] (cf. Chapter 10.3). However, the differences between the wild-type form (Tyr113/His139) and both alloenzymes in hydrolysis of the 4,5-oxide are rather low (about 2-fold for recombinant proteins), or not statistically significant (human microsomal preparations).[101] It seems therefore unlikely that functional differences between mEH variants would have a considerable impact on the levels of PAH-related biomarkers such as DNA adducts or on the overall PAH tumor susceptibility in human populations. In accordance to this assumption, molecular epidemiology studies investigating the influence of mEH variants on the levels of PAH exposure biomarkers failed to uncover significant effects.[102,103] On the other hand, human lung cancer studies were contradictory regarding the role of mEH variants or provided different clues depending on the way of evaluation (see Chapter 10.3 and Ref. 62 herein). Nevertheless, a recent meta-analysis study may suggest a possible contribution of the codon 3 polymorphism (decreased activity) in attenuating lung cancer risk of heavy smokers.[104]

Compared to mEH, quinone reductases such as NQO1 (DT-diaphorase) clearly belong to the group of PAH detoxification enzymes based on their ability to convert electrophilically reactive polynuclear quinones into hydroquinones (quinols), which then can be further modified (detoxified) by phase-II enzyme-mediated conjugation reactions (see Chapter 2.3.4 and Figure 2.8). *In vitro* studies supported this view since DNA damage and mutagenicity of B[*a*]P quinones could be significantly reduced in the presence of NQO1. In addition, *NQO1* knock-out mice developed significantly higher numbers of skin tumors in response to application of B[*a*]P or DMBA as compared to the wild-type control (see Chapter 2.3.4). The as yet best characterized polymorphic variant of the human *NQO1* gene (C → T transition at position 609, resulting in a Pro to Ser change at residue 187) was found to be associated with the entire loss of activity of the corresponding alloenzyme.[105] Since NQO1 is also expressed in human lung tissue, it was first predicted and subsequently confirmed that the *NQO1* $Pro^{187}Ser$ allele occurs more frequently in lung cancer patients than in controls (see Chapter 10.5). However, the molecular basis of this finding remains to be elucidated as no influence of this variant on bronchial bulky DNA adduct levels could be found.[106] This result, along with the finding that polynuclear quinones from PAHs such as B[*a*]P lack considerable carcinogenic activity in animal tumor models[107] (cf. Chapter 2.3.4), makes it unlikely that the tumor modulating effects of the NQO1 $Pro^{187}Ser$ variant in human populations would be somehow related to its role in metabolism of carcinogenic PAHs.

GST enzymes play a key role in cell's protection against cyto- and genotoxic damage as they constitute the most important enzymic defense line against reactive electrophiles from a variety of different kinds of chemicals including arene oxides and diol-epoxides originating from PAHs (see Figure 2.11). From the various classes of cytosolic GST enzymes characterized to date and tested *in vitro* with regard to PAH metabolism, GSTM1-1, GSTP1-1, and GSTA1-1 are the main (but probably not the only) forms involved in detoxification of vicinal diol-epoxides of PAHs[108–112] (cf. Chapter 2.4). However, most molecular epidemiology studies conducted thus far were focused on the role of *GSTM1* polymorphisms, whereas much less (*GSTP1*) and almost nothing (*GSTA1*) is known about the role of the two other forms in modulating PAH-DNA adduct levels and cancer susceptibility in man.

Given its high activity in catalyzing PAH diol-epoxide conjugation, it seems reasonable to hypothesize that any polymorphic variation in the activity of GSTM1 may have an impact on the levels of diol-epoxide adducts in DNA of white blood cells or target tissue(s) isolated from highly exposed individuals. Due to the presence of carcinogenic PAHs in cigarette smoke it can be further anticipated that individual's *GSTM1* genotype might modulate the susceptibility to smoking-induced lung cancer in the very same person (see Chapter 10.4). Inter-individual differences with regard to GSTM1 activity result mainly from homozygous deletion of the encoding gene ('null genotype'), as can be detected in about half of the Caucasian population. Although still controversial, most epidemiological studies suggest that the *GSTM1*-null genotype is associated with an (slightly) increased risk for cancer at various sites including lung (Chapter 10.4).[113–116] An obvious explanation for the weak correlation in the case of lung cancer could be the expression pattern of GSTM1, as it is highly expressed in the liver but not in the lung. However, some studies reported an increase in the total number of lung tissue PAH-DNA adducts for *GSTM1*-null genotype carriers,[73,75,117,118] whereas others did not.[76,119] A similar situation is noticeable in the case of PAH-DNA adducts when measured in peripheral white blood cells from occupationally or environmentally exposed individuals (see Chapter 4 and Table 4.1).

At present the correlations found between single 'susceptibility genotypes' such as *GSTM1*-null or *CYP1A1 Msp*I and *CYP1A1* Ile^{462}Val (cf. above) and the overall levels of PAH-DNA adducts in white blood cells or lung tissue are weak and sometimes insignificant, but usually a little more pronounced when *GSTM1* and *CYP1A1* genotypes are combined.[52,83,120] A different picture emerged when polymorphic variants of CYP1A1 or GSTM1 were associated with the levels of a structurally defined PAH-DNA adduct rather than to the total of all adducts derived from any PAHs. Using a specific HPLC-based fluorometric detection method, Rojas *et al.* were able to demonstrate that lung or blood cell levels of DNA adducts derived from the *anti*-diastereomeric bay-region 7,8-diol-9,10-epoxide of B[*a*]P (*anti*-B[*a*]PDE) correlated with high statistical significance to the absence of a functional GSTM1 variant (*GSTM1*-null genotype).[69,72] In contrast, all samples obtained from *GSTM1* wild-type carriers among smokers, coke oven or power plant workers integrated into these studies lacked detectable *anti*-B[*a*]PDE-DNA adduct levels. It was further shown that homo- or heterozygous presence of both polymorphic CYP1A1 variants (*Msp*I or Ile^{462}Val) led to an additional and significant increase in the levels of detectable *anti*-B[*a*]PDE-DNA adducts in the case of PAH-exposed *GSTM1*-null individuals — an effect not detectable when CYP1A1 variants were acting on a *GSTM1* active (wild-type) background that sufficiently prevented any DNA damage.[69,72] Thus, the *GSTM1*-null allele together with *CYP1A1* mutant alleles serve as a critical host factor that affects the level of *anti*-B[*a*]PDE-DNA adducts in target tissue(s) of exposed individuals. In accord to this assumption, the combination of these genotypes was identified to confer high susceptibility for developing cancer in the lung and various other sites[84,121–123] (cf. Chapter 10.7).

Due to its abundant expression in epithelial tissues, especially from lungs,[124] GSTP1 is of particular interest with regard to lung tumorigenesis in man (cf. Chapter 10.4). Among several human allelic variants known today, the polymorphism of exon 5 (A $\rightarrow$ G transition at position 313), resulting in an isoleucine to valine substitution (Ile^{105}Val), may play an important role in GST-mediated detoxification of vicinal diol-epoxides originating from PAHs.[109,110,125–128] In fact, replacement of Ile by Val within enzyme's active site leads to a considerable increase in the conjugating activity of the resulting alloenzyme towards PAH diol-epoxides *in vitro* — especially in the case of stereoisomers possessing *R* absolute configuration at their benzylic oxiranyl ring carbon atom (see Figures 2.5 and 2.6). The relevance of functional GSTP1 enzymes has also been demonstrated *in vivo* as mice lacking the corresponding gene locus are at about 3-fold higher risk to develop skin tumors in response to application of 7,12-dimethylbenz[*a*]anthracene (DMBA) in an initiation/promotion protocol.[129] In consideration of this data it may therefore be hypothesized that PAH-exposed GSTP1 Ile^{105}Val variant carriers should be at lower risk to develop tumors in the lung. However, epidemiological evidence and lung cancer case studies point to the opposite. As explained in Chapters 10.4 and 10.7, both the *GSTP1* Ile^{105}Val genotype alone as well as in combination with *GSTM1* deficiency might pose a higher risk for developing cancer at various sites including lung. At first glance this finding is unexpected and surprising, but may be explained in consideration of the overall reduction of GST enzyme activity in normal lung tissue of individuals who carry the Ile^{105}Val allele.[130] On the other hand, several studies failed to find a significant association to lung cancer risk, neither the *GSTP1* Ile[105]Val genotype alone nor in the case of concurrent *GSTM1* deficiency;[131–134] and others uncovered an association with a modest reduction in risk.[116] Despite these controversial results the multiplicative effect between both genotypes, which possibly caused a 7-fold increase in the risk for lung cancer according to the evaluation in one study,[135] clearly deserves further attention. Similar to population-wide investigations, experimental studies aimed to uncover the role of the *GSTP1* Ile[105]Val polymorphism in the modulation of PAH-DNA adduct levels in human bronchial tissue lacked sufficient clarity. In one study a protective role of the GSTP1 Ile[105]Val variant was described as it could counteract an increase in adduct levels attributable to *GSTM1* gene deficiency.[61] Those individuals that lack GSTM1 but express the GSTP1 Ile[105]Val variant were found to have significantly lower bronchial DNA adduct levels compared to the GSTM1-null/GSTP1 wild-type combination, but similar levels as in the presence of a functional GSTM1 form. In another study,[117] however, a contrary finding was reported as GSTM1-null along with GSTP1 Ile[105]Val expression was associated with highest levels of bronchial DNA adducts among all possible genotype combinations. Further, the *GSTP1* Ile[105]Val genotype alone had no effect on bronchial PAH-DNA adduct levels either in a Hungarian[106] or in a Polish group of patients with lung diseases,[118] while it had on PAH-DNA adduct levels in white blood cells of newborns from Poland (2-fold lower compared to the control).[136] Analysis of newborn blood cells also uncovered a combination effect between *CYP1A1* and *GSTP1* polymorphisms since adduct levels were 4-fold higher among *CYP1A1 Msp*I/*GSTP1* wild-type compared to *GSTP1* Ile[105]Val newborns (with a *CYP1A1* wild-type background). In view of its putatively important role in detoxification of electrophilic PAH metabolites in lung epithelial cells *in vivo*, the controversial literature listed above clearly underscores the need for further attempts to decipher the real contribution of *GSTP1* polymorphisms in PAH-DNA adduct formation and lung cancer risk in human populations.

11.5 Molecular Mechanisms of DNA Damage

Carcinogenic PAHs are carcinogenic because they induce DNA damage and mutations in genes involved in the surveillance of cellular growth, proliferation, and death. The principal DNA-damaging chemical species generated during biotransformation of carcinogenic PAHs are introduced in Chapter 2.3 and summarized in Figure 2.4. As demonstrated mainly by employment of certain *in vitro* systems, a great variety of different electrophilic intermediates theoretically may be produced by different groups of xenobiotic-metabolizing enzymes. However, extensive experimental evidence lends support to the notion that bay- and fjord-region diol-epoxides are the major players in the mediation of mother compound's biological effects *in vivo* (cf. Chapters 2.3.1 and 2.3.2). From the 'PAH-DNA adduct side of view' there are some crucial arguments. For instance, (i) adducts originating from covalent interaction of DNA with PAH diol-epoxides are experimentally inducible in target tissues in a dose-dependent fashion by treatment of rodents with the parent compound; (ii) elevated levels of diol-epoxide-DNA adducts are detectable *in vivo* in target tissues (e.g., bronchial epithelia) from individuals heavily exposed to PAHs (e.g., smokers[137]) as compared to non- or low exposed controls (cf. section 11.6); and (iii) experimental deletion of diol-epoxide generating enzymes (CYP1B1, mEH), or of the receptor protein involved in induction of those enzymes (arylhydrocarbon receptor, AhR), renders mice resistant to the biological effects of potent carcinogens such as B[*a*]P, DMBA or dibenzo[*a*,*l*]pyrene (DB[*a*,*l*]P) (see Figure 2.1).

Vicinal diol-epoxides of carcinogenic PAHs are produced through subsequent steps of epoxidation and hydrolysis (Figure 2.3). A biologically important feature of this 'monooxygenation activation pathway' towards vicinal diol-epoxides is its stereoselectivity. As explained in Chapter 2.3.2 and Figure 2.5, PAH-metabolizing CYP and mEH enzymes operate with high enantio- and diastereoselectivity. As the result, usually those stereoisomers that possess highest inherent carcinogenic activity accumulate in cell's endoplasmatic membranes — close to their target macromolecule in the nucleus. As demonstrated for B[*a*]P and benzo[*c*]phenanthrene

(B[*c*]Ph) with microsomal enzyme preparations *in vitro*, the initial epoxidation → hydrolysis sequence at the aromatic bond adjacent to their bay- or fjord-region (positions 7,8 in B[*a*]P and 3,4 in B[*c*]Ph), produces the *trans*-dihydrodiol with R, R-configuration in high enantiomeric excess. Subsequent epoxidation at position 9,10 (B[*a*]P) or 1,2 (B[*c*]Ph) again occurs with high stereoselectivity and leads to the predominant generation of diol-epoxides with 'R, S, S, R'-configuration, i.e., the bay-region (+)-*anti*-7R,8S-diol-9S,10R-epoxide of B[*a*]P [(+)-*anti*-B[*a*]PDE] and the fjord-region (–)-*anti*-1R,2S-diol-3S,4R-epoxide of B[*c*]Ph [(–)-*anti*-B[*c*]PDE] (see Figures 2.6 and 11.1). Exactly these stereoisomeric *anti*-diol-epoxides were identified as the most potent mutagenic and carcinogenic species among the four diastereomers that theoretically can be produced at this region (see Chapter 2.3.2, Figure 2.5). In addition, some small amounts of the isomeric *syn*-(R, S, R, S), *syn*-(S, R, S, R) and *anti*-(S, R, R, S) diol-epoxides may also be generated — the latter two through monooxygenation of the enantiomeric S, S-dihydrodiol (see Figures 2.5 and 2.6). In the case of B[*c*]Ph and DMBA both the *anti*-(R, S, S, R) and *syn*-(S, R, S, R) 3,4-diol-1,2-epoxides were found to contribute substantially to the DNA binding of the parent compound in cells in culture and *in vivo*. In contrast, diol-epoxide-DNA adducts arising from exposure to B[*a*]P or DB[*a*,*l*]P, the latter considered as being the most potent carcinogenic PAH characterized in animal tumor models to date, are predominantly the result of covalent interaction between the *anti*-(R, S, S, R) isomer and an exocyclic amino group in purine nucleobases (see Figures 2.7, 5.3 and 11.1; and cf. below).

As we have learned from the compilation in Chapter 8, many test systems that enable a simple and fast characterization of the genotoxic potency of PAHs (and other compounds) are available. In the mammalian V79/HPRT assay, for example, bay- and fjord-region diol-epoxides have been proven to exert extraordinarily strong mutagenic effects as compared to other metabolites originating from the same parent compounds (see Chapter 8.3.1). Using this relatively simple mammalian cell culture model, carcinogenic PAHs and their diol-epoxide derivatives vary considerably in their potency to induce mutations at the *hprt* locus (Figures 8.2 and 8.3) — a result that roughly reflects their different carcinogenic activities. As nicely demonstrated in Figure 8.3, the mutation-inducing potencies of the directly

DNA-damaging diol-epoxides are intimately connected to their corresponding adduct-forming potencies. It could be estimated that — irrespective of the individual PAH — around 6×10^7 adducts have to be formed in a cell population growing on a cell culture dish in order to induce one *hprt* mutant among those cells. Given this data it seems reasonable to assume that the (quantitative) level of diol-epoxide-DNA adducts rather than their individual structure and identity is the main factor in determining the mutagenic activity of a particular diol-epoxide. However, the level of diol-epoxide-DNA adducts measured at the end of the exposure period (usually 2 hours[138]) itself is crucially influenced by several structure- and conformation-dependent variables such as (i) the rate of diol-epoxide sequestration (through hydrolysis or conjugation) prior to covalent DNA adduction; (ii) the intrinsic reactivity of the diol-epoxide towards nucleophilic centers in DNA; and (iii) the resistance of the generated DNA adducts against removal by the DNA repair machinery (see below).

From Figure 8.3 it is further noticeable that vicinal diol-epoxides from bay-region PAHs (e.g., chrysene, B[*a*]P) are less active than those originating from fjord-region PAHs (e.g., B[*c*]Ph, benzo[*g*]chrysene: B[*g*]C, benzo[*c*]chrysene: B[*c*]C; see Figure 9.1); and that *syn*-diol-epoxides are usually less potent in inducing DNA adducts (and mutations) as compared to their corresponding *anti*-diasteromeric counterparts (see also Glatt *et al.*,[139] and Luch *et al.*[140]). Thus, diol-epoxides from different PAHs and different isomeric diol-epoxides from the same PAH exert (sometimes markedly) different efficiencies in DNA adduct formation. Similar to these findings obtained with cells in culture, binding studies with isolated DNA in aqueous solution essentially delivered the same result. As demonstrated in some early studies, the DNA binding efficiencies of bay-region diol-epoxides of B[*a*]P were extremely different as compared to fjord-region diol-epoxides of B[*c*]Ph. The overall binding of all four configurational isomers was found to be less than 10% in the case of B[*a*]P, while 60–75% of the material became covalently trapped by DNA in the case of B[*c*]Ph diol-epoxides.[141,142] Subsequent investigations of diol-epoxides from other PAHs revealed, however, that B[*a*]P and B[*c*]Ph diol-epoxides apparently represent the extreme ends in the possible spectrum of DNA binding variability exerted by these kinds of metabolites. Based on data of 30 different diol-epoxides, Szeliga and Dipple[143] comprehensively collected evidence that DNA binding of (planar) bay-region diol-epoxides in aqueous solution in general is less efficient as compared to (non-planar) fjord-region diol-epoxides. They further noticed that (i) for each individual PAH the (R,S,S,R) configurational isomer is trapped more extensively by DNA than the other isomers; (ii) the rank order of reactivity towards DNA for diol-epoxides from different PAHs is fairly similar irrespective of the isomer configuration; and (iii) the fraction of diol-epoxides bound to DNA is lower for derivatives with five aromatic rings as compared to four-ring members in both series of PAHs (e.g., B[*a*]A *vs.* dibenz[*a*, *j*]anthracene; and B[*c*]Ph *vs.* B[*g*]C).[143]

As these results were obtained for reactions between diol-epoxides and DNA in buffered solution, the only alternative to the reaction with DNA was that of hydrolysis into tetrahydrotetraols ('tetraols'). (At neutral conditions, some small amounts of the *syn*-isomeric diol-epoxides may also undergo spontaneous rearrangement into keto diols — presumably by hydrogen migration from the allylic to the benzylic position of the opening epoxide ring.[144,145]) It was suggested that the decision between covalent reaction with water *vs.* DNA is strongly influenced by the extent of ionization of the diol-epoxide at a stage where it is already engaged in a 'noncovalent pre-reaction diol-epoxide-DNA complex'.[143,146–148] As water is less nucleophilic than DNA, increased ionization favors hydrolysis. Thus, it can be concluded that diol-epoxides derived from (non-planar) fjord-region structures and from tetracyclic compounds display a relatively higher DNA binding efficiency because of a lower tendency for ionization within the pre-reaction complex. Distortion from planarity, which hinders charge delocalization, and a lower number of aromatic rings over which the charge can delocalize could account for the reaction through less ionic intermediates in the case of tetracyclic molecules. On the other hand, different hydrolysis rates for bay- *vs.* fjord-region PAH diol-epoxides were also observed in the absence of DNA,[139,141] thus supporting the view that diol-epoxide-inherent structural and electronical factors (such as $\Delta E_{deloc}/\beta$: delocalization energy released at the benzylic position upon ring-opening of the epoxide moiety $\approx$ level of stability of an intermediate benzylic carbocation; cf. Chapter 2.3.1) are pivotal for the rate of hydrolysis. Incubation experiments with various diol-epoxides in buffered-saline (pH 7.4) at 37°C revealed that half-lives for fjord-region derivatives (from B[*c*]C, B[*g*]C, B[*c*]Ph) were significantly higher in comparison to those for bay-region derivatives (from chrysene, phenanthrene, B[*a*]P).[139] For instance, a half-life of 4.1 hours was determined for racemic *anti*-B[*c*]PhDE, but only 0.07 hours for racemic *anti*-B[*a*]PDE. In addition to these structure-dependent differences in the solvolytic stability of diol-epoxides, *anti*-isomers were found to be more stable under these conditions as compared to their *syn*-isomeric counterparts — no matter whether derived from bay- or from fjord-region PAHs. The higher rates of hydrolysis in case of *syn*-diol-epoxide isomers observed in buffered solution would help to explain their lower relative binding to DNA compared to *anti*-isomers (see above). It also shows that the configuration of the diol-epoxide alone, no matter whether engaged in a noncovalent pre-reaction DNA complex or not, profoundly affects its stability in solution. As has been suggested by Sayer *et al.* in 1982,[149] the stability of PAH diol-epoxides in aqueous solution is highly influenced by conformational effects originating from their substituted cyclohexenyl rings systems. For both pairs of diastereomers, the cyclohexenyl moiety may adopt two different (but inter-changeable) conformations in which the benzylic C-O bond of the epoxide ring is either aligned near parallel to the π orbitals of the remaining aromatic ring system ('aligned') or not ('non-aligned'). The importance of this observation comes from an accelerated rate of hydrolysis in the case of the 'aligned' conformation (due to mesomeric stabilization of the positive charge *in statu nascenti* at an opening epoxide ring). In the series of fjord-region diol-epoxides, NMR studies have shown that both *syn*- and *anti*-stereoisomers preferentially adopt conformations with a pseudoequatorial orientation of the hydroxy groups (that is, both hydroxy groups are more or less located in the plane of the aromatic ring system).[140,150,151] Since this favored ground-state conformation is accompanied by an 'aligned' benzylic epoxide C-O bond only in the case of *syn*-diol-epoxides, the accelerated hydrolysis of these isomers may well be explainable based on their preferred conformation adopted in solution. In contrast, *syn*-isomeric diol-epoxides from bay-region PAHs prefer a 'non-aligned' conformation with a pseudoaxial orientation of the hydroxy groups (that is, both hydroxy groups are more or less vertical to the aromatic ring system).[152] However, an extended lifetime of a carbonium-ion-like transition state (as indicated by high $\Delta E_{deloc}/\beta$ values) permits switching into the more stable pseudoequatorial and 'aligned' conformation prior to water addition. Undergoing conformational equilibrium at a carbon-ion-like transition state intimately depends on the lifetime (mesomeric stabilization) of this intermediate. Thus, the contribution of an 'aligned' conformation (and the rate of solvolysis) increases with increasing $\Delta E_{deloc}/\beta$ values. In accord to these theoretical assumptions, the half-lives of bay-region *syn*-diol-epoxides were found to decrease in the order chrysene (0.639) > phenanthrene (0.658) > B[*a*]P (0.794) (with $\Delta E_{deloc}/\beta$ values in brackets).[139]

Although differences in DNA binding of fjord- *vs.* bay-region diol-epoxides or *anti- vs. syn*-diol-epoxides may be — at least to some extent — explainable based on diol-epoxide-inherent factors, the observation of a pronounced DNA adduction in the case of diol-epoxides originating from tetracyclic PAHs as compared to pentacyclic PAHs is not supported by their half-lives in aqueous solution. For instance, irrespective of the configurational isomer, fjord-region 3,4-diol-1,2-epoxides of the tetracyclic B[*c*]Ph (B[*c*]PhDE) bind much more efficiently to isolated DNA in solution than the corresponding fjord-region 11,12-diol-13,14-epoxides of the pentacyclic B[*g*]C (B[*g*]CDE) (3- to 5-fold differences[143]). However, the half-life of racemic *anti*-B[*g*]CDE has been determined to be at least 2.5-fold longer (>10 hours) compared to *anti*-B[*c*]PhDE (4.5 hours).[139] Since covalent interaction between diol-epoxides and DNA proceeds through the formation of a noncovalent pre-reaction complex (cf. above), differences in complex' formation based on steric and electronical factors as well as differences in complex' behavior (with regard to ionization, conformation and competing hydrolysis reactions) may account for these experimental observations.

Although CYP enzymes are highly stereoselective, each single configurational isomer from all four possible bay- or fjord-region diol-epoxides usually can be detected at least to some extent when the parent compound is metabolized by isolated enzyme preparations (e.g., microsomes) *in vitro*. On the other hand, treatment of metabolically competent cells in culture or mouse skin with PAHs and subsequent detection of diol-epoxide-DNA adducts mostly revealed that only two, sometimes even just only one, out of these four isomers predominate. For instance, DNA adducts isolated from cells or from animal tissue (e.g., skin) after exposure to low doses of B[*a*]P almost exclusively originated from (+)-*anti*-B[*a*]PDE[153–155] — though some (+)-*syn*-B[*a*]PDE-DNA adducts may also be detectable under certain circumstances *in vitro*.[156] In the case of B[*c*]Ph or DMBA however, DNA adducts isolated from cells or mouse skin were the result of a more equivalent covalent binding of both isomers the (R, S, S, R) *anti*-diol-epoxide and the (S, R, S, R) *syn*-diol-epoxide.[154,155,157–162] Thus, DNA adducts originating from diol-epoxide isomers with (S, R) configuration at the epoxide moiety prevail in cells in culture or *in vivo* (see Figure 11.1).

As shortly mentioned in Chapter 2.3.2, adduct formation of bay- or fjord-region diol-epoxides in DNA has been found to result primarily from reaction with exocyclic amino groups in 2′-deoxyguanosine (dG) or 2′-deoxyadenosine (dA).[142,163,164] Diol-epoxides from B[*a*]P, for example, predominantly bind to the amino group of dG (N^2-dG $> 90\%$).[165–168] Some traces of additional products such as O^6, N^6 and N7 adducts in dG, N^4 adducts in 2′-deoxycytidine (dC), or even an alkylation product at the sugar residue of dA could also be detected or tentatively identified under certain circumstances *in vitro*.[169–172] On the other hand, studies with the fjord-region PAH B[*c*]Ph or the sterically hindered bay-region PAH DMBA revealed that their diol-epoxide derivatives target both purine nucleotides (N^2-dG and N^6-dA) to more equal extents (Figure 11.1).[141,159,160,163,173] Since fjord-region PAHs or sterically hindered bay-region PAHs such as DMBA are no longer perfectly planar due to repulsive hydrogen interactions[174,175] it was hypothesized that planarity as a structural feature of these molecules would have an impact on the differential reactivity towards DNA bases. A compilation of differential nucleobase reactivities of single configurational isomers of bay- and fjord-region diol-epoxides from various PAHs comprehensively demonstrated that the ratios of dA:dG adducts in DNA are always greater for diol-epoxides derived from nonplanar PAHs (such as DMBA, B[*c*]Ph and B[*g*]C) as compared to planar PAHs (such as B[*a*]P, B[*a*]A or 5-methylchrysene: 5-MeC).[143] In particular, the 1,2-diol-3,4-epoxide derivatives of 5-MeC (5-MeCDE) are an interesting example as this hydrocarbon contains a methyl group in its bay-region similar

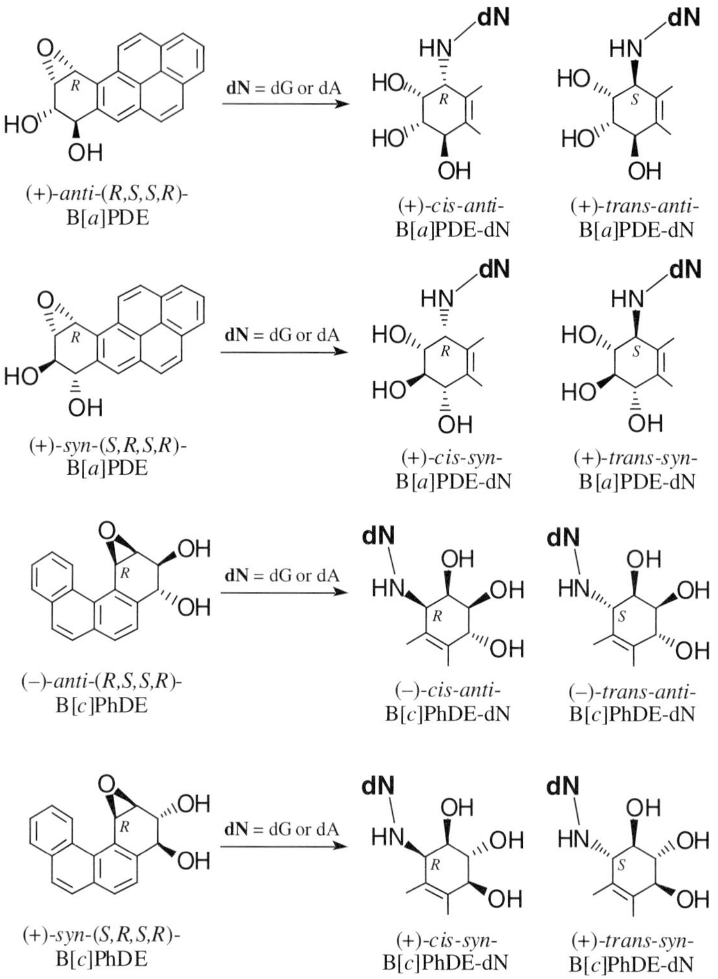

Figure 11.1: Reaction of the (R,S)-diol (S,R)-epoxides (*anti* series) and (S,R)-diol (S,R)-epoxides (*syn* series) of benzo[*a*]pyrene (B[*a*]PDE) and benzo[*c*]phenanthrene (B[*c*]PhDE) with the exocyclic amino group of purine bases. The nucleophile can open the epoxide ring either through *cis*- or *trans*-attack at the benzylic position. The absolute configuration at the benzylic position is indicated. dN, 2′-deoxynucleoside; dA, 2′-deoxyadenosine; dG, 2′-deoxyguanosine. See text for further explanations, and cf. Figures 2.6 and 2.7.

to DMBA (see Figure 2.1). It therefore could be expected not to be perfectly planar and, thus producing a similar mixture of dA and dG adducts as other nonplanar hydrocarbons. Despite the presence of a methyl group however, 5-MeC keeps its planarity by accomodation of the steric crowding through in-plane distortions.[176] In congruence with this structural behavior, the configurational isomers of 5-MeCDE were found to have a similar binding pattern in DNA as diol-epoxides from simple bay-region PAHs. Conversely, induction of out-of-plane distortions by addition of a second methyl group at position 6 in 5-MeC (to yield 5,6-dimethylchrysene: 5,6-DiMeC) ensures that the resulting 1,2-diol-3,4-epoxides (5,6-DiMeCDE) adopt the differential reactivity pattern of nonplanar compounds with dA adducts prevailing.[143,163]

Another aspect uncovered was that binding patterns of planar *vs.* nonplanar PAH diol-epoxides are changing gradually through the series of compounds investigated.[143] This indicates that the structural determinants for nucleotide preference change gradually in the series as well. It was further noticed that the rank order of preference for either purine within a given set of four stereoisomeric diol-epoxides is also subject of variation depending on the particular PAH. Similar to the overall DNA binding efficiency (see above), derivatives of B[*a*]P and B[*c*]Ph represent the extreme ends in the possible spectrum of differential nucleobase reactivities. In case of B[*a*]P, the (+)-*anti*-B[*a*]PDE isomer has the highest level of dG binding (about 95%), while (+)-*syn*-B[*a*]PDE has the lowest (about 83%). Conversely, (+)-*syn*-B[*c*]PhDE has the highest level of dA binding (about 90%) followed by the other three stereoisomers with a fairly similar dA binding of about 58–66%.[143] As a general rule it appears that dG binding of derivatives from planar hydrocarbons is highest for isomers with (R,S)-diol (S,R)-epoxide configuration, whereas (S,R)-diol (S,R)-epoxides in the series of nonplanar hydrocarbons target dA with highest efficiency (cf. Figures 2.6 and 11.1).

Further analysis of the purine base adducts predominantly produced in cells revealed that — irrespective of the PAH — N^2-dG and N^6-dA adducts from (R,S)-diol (S,R)-epoxides (*anti* series) are always preferentially generated through *trans* opening of the epoxide ring.[143] For instance, from all N^2-dG adducts of (+)-*anti*-B[*a*]PDE about 99% are the result of *trans* opening of the epoxide ring [adducts referred to as '(+)-*trans-anti*-B[*a*]PDE-N^2-dG'], whereas only 1% are due to a *cis*-attack of the nucleophile [adducts referred to as '(+)-*cis-anti*-B[*a*]PDE-N^2-dG'] (see Figure 11.1). In the case of N^6-dA adducts of (–)-*anti*-B[*c*]PhDE about 98% are due to *trans* opening, but only 2% derive from *cis* opening of the epoxide moiety [adducts referred to as '(–)-*trans-anti*-B[*c*]PhDE-N^6-dA' and '(–)-*cis-anti*-B[*c*]PhDE-N^6-dA', respectively]. (At this point it should be noted that the symbol '(+)' or '(–)' refers to the optical rotation of the diol-epoxide precursor prior to undergoing covalent reaction with dG or dA. The optical rotation of the nucleotide adduct usually is not determined.[177,178]) In contrast, the major adducts of (S,R)-diol (S,R)-epoxides (*syn* series) are generated via different kinds of spatial attack of the nucleophile depending on the structure of the PAH. In the case of diol-epoxides from planar hydrocarbons the major adduct derives from a *cis*-attack of the amino group in dG (e.g., about 89% (+)-*cis-syn*-B[*a*]PDE-N^2-dG and 11% (+)-*trans-syn*-B[*a*]PDE-N^2-dG). Conversely, *trans*-dA adducts are again predominantly produced with nonplanar hydrocarbons (e.g., about 76% (+)-*trans-syn*-B[*c*]PhDE-N^6-dA and 24% (+)-*cis-syn*-B[*c*]PhDE-N^6-dA).[143] Since the configuration of the epoxide moiety remains constant (S,R) in all of these diol-epoxide isomers, the spatial orientation of the diol function appears to be critical for directing the geometry of the attack (*trans vs. cis*) of the nucleophilic center in the DNA (Figure 11.1) — but only in the case of planar hydrocarbons. Although both types of reactions are certainly not comparable it may still be interesting to realize that these differences in the way of opening of the epoxide ring in *syn- vs. anti*-diol-epoxide isomers from planar hydrocarbons are also applicable with water as the nucleophile.[144,145,179] Due to molecule-inherent stereoelectronic and conformational factors, hydrolysis of B[*a*]P diol-epoxides under physiologic or acidic conditions produces tetraols that predominantly arise either from *cis*- (*syn*-isomer) or from *trans*-attack (*anti*-isomer) of water. In contrast to DNA-catalyzed ring opening however, the way of hydrolysis of nonplanar B[*c*]Ph diol-epoxides resembles hydrolysis of planar B[*a*]P diol-epoxides.[144,145,150] Hence, the influence of the absolute configuration of diol-epoxide enantiomers on the geometry of the nucleophilic attack in the DNA complex must reflect differences in the interactions between these chiral substrates and the asymmetric macromolecule.

According to Figure 8.3, the greater DNA binding of fjord-region diol-epoxides also results in a higher level of mutagenicity as compared to bay-region derivatives (see also Glatt *et al.*[139] and Luch *et al.*[140]). In the V79/HPRT test system the mutagenic activity seems therefore well correlated to the overall level of DNA binding for those compounds that are investigated. In consideration of the binding studies with isolated DNA in aqueous solution[143] however, some differences become obvious, although the overall trend could be confirmed (i.e., greater binding for fjord- *vs.* bay-region and *anti- vs. syn*-diol-epoxides; cf. above). For instance, the level of mutagenicity induced at the *hprt* locus in V79 cells was determined to be about 2–5-fold and 7–9-fold higher for diastereomeric fjord-region diol-epoxides of B[*g*]C and B[*c*]C, respectively, as compared to the corresponding isomers of B[*c*]PhDE.[139] Thus, the mutagenic activity (and the level of DNA binding) in V79 cells increases in the order B[*c*]PhDE < B[*g*]CDE < B[*c*]CDE (9,10-diol-11,12-epoxide of B[*c*]C) in both series, the *syn*- and the *anti*-isomers (Figure 8.3 and Glatt *et al.*[139]). In contrast, binding to DNA in solution was found to be generally higher for diol-epoxides from tetracyclic hydrocarbons compared to pentacyclic hydrocarbons (see above). Assuming that the solubility of fjord-region diol-epoxides is not dramatically different in buffered solution than in cell culture medium containing a monolayer of living cells (what actually is questionable in view of the cellular lipid membranes that certainly would better accommodate the increased lipophilicity of pentacyclic compounds), the differences between binding to native DNA in solution and DNA in living cells at the end of a 2 hour incubation period may result from the presence and activity of detoxifying enzymes and DNA damage defense (repair) systems.

Investigating the mutagenic activity of carcinogenic PAHs has been a central theme in 'genetic toxicology' for years. Analysis of the mutational spectra induced by different PAHs and their diol-epoxides in pro- and eukaryotic indicator systems, in mammalian cells *in vivo*, or in transgenic animal models (see Chapter 8) confirmed a propensity for inducing G → T transversions in the case of bay-region PAHs (e.g., B[*a*]P) and a propensity for inducing A → T transversions in the case of fjord-region PAHs (e.g., B[*c*]Ph). However, the patterns of mutations induced were found to be dependent on various factors such as the concentration of the DNA-damaging species (i.e., diol-epoxide) and the repair competence of the cells affected.[180,181] At the *hprt* locus of V79 cells it could be demonstrated that bay- or fjord-region (*R*,*S*)-diol (*S*,*R*)-epoxides from B[*a*]P, B[*c*]Ph and DB[*a*,*l*]P (cf. Figures 2.1 and 2.6) cause a variety of different kinds of mutations with base substitutions prevailing.[182–184] Irrespective of the hydrocarbon, exposure of these cells to increasing concentrations of the diol-epoxide always resulted in a decrease in the fraction of base substitutions at dA sites and an increase in the fraction of base substitutions at dG sites. In contrast, the less carcinogenic (*S*,*R*)-diol (*R*,*S*)-epoxide enantiomers of B[*a*]P and B[*c*]Ph did not display a dose-dependency in their mutation pattern.[183,185] Subsequent work of Conney and co-workers further revealed that the observed shift from base substitutions at dA to those at dG at higher concentrations of the (*R*,*S*,*S*,*R*) B[*a*]P diol-epoxide was dependent on an intact DNA damage repair system.[181,186] In the absence of DNA repair this effect disappeared. It was also shown that the absence of dose-dependent differences in the mutational profiles observed in nucleotide excision repair (NER)-defective V-H1 cells (which lack a functional *XPD/ERCC2* gene that encodes a helicase subunit of transcription factor IIH;[187] cf. Chapter 6.2.2) was mainly the consequence of an impaired removal of (+)-*anti*-B[*a*]PDE-N^2-dG adducts from the transcribed strand of the DNA (cf. below).

As comprehensively outlined in Chapter 6 of this book, the integrity of the enzymatic repair of PAH-damaged DNA is dependent on a variety of different factors such as the DNA sequence context of the adduct and its structural and stereochemical features. Furthermore, the efficiency of PAH-DNA adducts in inducing mutations is dependent on the nature of the DNA strand affected ('DNA strand bias'). PAH adducts located in the transcriptionally active DNA strand are removed by a separate and less well characterized variant of the NER pathway referred to as 'transcription-coupled repair' (Chapter 6.2.3). Since this mode of repair has been proven to be fast and efficient, PAH-DNA adducts located in the transcribed strand are unlikely to contribute to a considerable extent to the overall mutagenicity of the hydrocarbon. Instead, the rather slow and error-prone repair pathway operating at transcriptionally silent (non-transcribed) DNA strands ('global NER pathway') is thought to be more decisive with regard to PAH-induced mutagenesis (see Chapter 6.2.4).

The efficiency of the human NER system is mainly a function of the degree of local DNA distortion. In fact, PAH adducts that cause a displacement of the modified nucleobase and its counterpart from their normal intrahelical position fulfill the structural requirements for activating the NER machinery (base pair conformation-dependent excision). As shown in Chapter 6.3.2, N^2-dG adducts of both *anti*-enantiomers of B[*a*]PDE induce a considerably different degree of NER activity in a certain base context (5′-TCGCT-3′) depending on the way of epoxide ring opening during adduct formation. At this sequence motif, *cis*-opened *anti*-B[*a*]PDE adducts [i.e., (+)- and (–)-*cis-anti*-B[*a*]PDE-N^2-dG] adopt intercalative, internal adduct conformations with the benzo[*a*]pyrenyl moiety inserted into the helix and concomitant displacement of the modified base. Thus, the local DNA distortion induced is much more severe as compared to *trans*-opened adducts [i.e., (+)- and (–)-*trans-anti*-B[*a*]PDE-N^2-dG], which adopt external conformations with the benzo[*a*]pyrenyl moiety located in the minor DNA groove. Consequently, the *cis*-opened adducts of (+)- or (–)-*anti*-B[*a*]PDE are removed about 10 times faster than the *trans*-opened adducts (see Figures 6.6 and 11.1). While this example demonstrates the importance of the kind of covalent linkage (*cis vs. trans*) of a given stereoisomer, differences in the repairability of bay- *vs.* fjord-region diol-epoxide-DNA adducts were also observed (see Chapter 6.3.3, Figures 6.8 and 6.9). In the sequence context of codon 61 of the human proto-oncogenes N-*ras* or H-*ras* (see below), *trans*-opened N^6-dA adducts from enantiomeric *anti*-diol-epoxides of a variety of different fjord-region PAHs (B[*c*]Ph, B[*g*]C, DB[*a*,*l*]P) were found to be repair resistant, whereas stereochemically corresponding adducts of B[*a*]P were readily removed from the same sites. In the N-*ras* sequence motif the *trans-anti*-B[*a*]PDE-N^6-dA adducts are integrated into the double helix in a way that distorts the regular base pairing at this site. Conversely, *trans-anti*-B[*c*]PhDE-N^6-dA adducts are incorporated by an intercalative mode that retains normal base pairing. Using the melting temperature of the adducted DNA fragments as an indicator of base pair weakening (displacement) it became clear that the B[*a*]P diol-epoxide adducts reduced the stability of the double helix whereas adducts from fjord-region PAHs did not (see Table 6.2). Considering these results, it is the base pair distorting character of an adduct in a particular DNA sequence rather than the individual structure of the diol-epoxide in general that decides about the fate of a lesion. On the other hand, base pair distortion alone is not sufficient in inducing NER activity. Rather, the disruption of normal base pairing and the presense of a chemical modification (adduct) are both indispensable requirements for proper activation of the NER machinery ('bipartite recognition'; Chapter 6.3.4 and Figure 6.10; see also Geacintov *et al.*[188]). In the aforementioned *ras* gene sequence context DNA adducts of diol-epoxides from nonplanar fjord-region PAHs are resistant to repair. Located in the *ras* gene of a living cell, these adducts therefore would be good candidate lesions for being converted into mutations during the next replication cycle of the cell (cf. below).

If not repaired properly, PAH-damaged templates may result in nucleotide misincorporation opposite the lesion during the next round of DNA replication and, therefore, have the potential of inducing mutations in critical genes involved in tumorigenesis ('tumorgenes'; see Figure 1.2). Mutations in members of the *ras* proto-oncogene family (N-*ras*, H-*ras*, K-*ras*) are commonly detected in human cancers and in animal models for PAH-induced carcinogenesis (see Chapters 5.5.1 and 7.2.1).[189–192] In adenocarcinomas of the human lung and pancreas, for example, the incidences of these lesions may reach levels as high as 30 and 90%, respectively.[189] The principal activating (point) mutations in *ras* genes are almost exclusively located in codons 12, 13 and 61, rendering the GTPase activity of the corresponding protein (p21) inactive. As a consequence, p21 will be arrested in a GTP-bound activated mode and constantly trigger kinase signalling cascades leading to aberrant cell proliferation. In mouse models, it has been shown that the differential occurrence of mutations at codons 12, 13 or 61 depends on the particular PAH and the kind of the tissue targeted. For instance, nucleotide transversions within codons 12 (dG $\rightarrow$ dT) or 61 (dA $\rightarrow$ dT) of cellular H-*ras* were found in mouse skin upon exposure to B[*a*]P,[193,194] DMBA[195,196] and DB[*a*,*l*]P.[197,198] Treatment with B[*a*]P mainly results in the formation of dG → dT transversions at codon 12 (and 13), whereas DMBA and DB[*a*,*l*]P predominantly induce dA → dT mutations at codon 61. On the other hand, K-*ras* is the main family member targeted by carcinogenic PAHs in lung tissue.[199–201] Similar to skin, B[*a*]P and DMBA again display a propensity for inducing dG → dT transversions at codon 12[201] and dA → dT transversions at codon 61,[199] respectively. In contrast, DB[*a*,*l*]P-induced K-*ras* mutations in lung tissue are more equally distributed between codons 12 and 61.[200,201] Besides all possible mutations at dA (dA → dT, dG, dC), this compound also induced dG → dT and dG → dC substitutions in lung tumor tissue from strain A/J mice (see below). In a compilation of the literature on PAH-induced *ras* oncogene mutations in animal models it has been noticed that there is a strong *qualitative* relationship between the kinds of DNA adducts formed in a certain tissue type (i.e., diol-epoxide-N^2-dG and/or -N^6-dA adducts) and the kinds of mutations (at codons 12 and 61) induced in tumors that originate from this tissue.[191] On the other hand, mutations at codon 12 predominate in cigarette smoke-associated human lung adenocarcinomas.[202,203] In a recent study it has been shown that normal human bronchial epithelial cells are preferentially adducted at codons 12 and 14 of K-*ras* (first dG in each codon) upon treatment with *anti*-B[*a*]PDE *in vitro*.[204] In addition to this selectivity in binding of *anti*-B[*a*]PDE, the low numbers of adducts formed at other codons (e.g., codon 13) and the bulk of adducts formed at codon 14 were found to be removed much faster as compared to the lesion at codon 12. Thus, the 'hotspot' character of codon 12 within the K-*ras* gene in human bronchial epithelial cells apparently results from a synergism between preferential binding at this site along with a poor repair of those *anti*-B[*a*]PDE-N^2-dG adducts that have been formed.[204]

The tumor suppressor gene *p53* is another example of an important gene most commonly mutated in a variety of human cancers including cigarette smoke-associated forms of lung cancer (see Chapters 5.6 and 7.2.2).[205,206] The corresponding protein has been identified as a key transcription factor involved in cell cycle arrest,[207,208] DNA repair[209–212] and apoptosis.[213,214] In accord to its role in transcription regulation, mutations that render the protein inactive are predominantly located in its central DNA binding domain encompassing amino acids 100–293. As comprehensively outlined in Chapter 5.6.1 of this book, treatment of human cells in culture with racemic *anti*-B[*a*]PDE resulted in three major adduct formation sites at dG residues within codons 157, 248 and 273 (exons 5, 7 and 8). Exactly these sites have also been identified as mutational 'hotspots' in human lung cancer. In line with the strand bias of G $\rightarrow$ T transversions in lung cancer of smokers, *anti*-B[*a*]PDE-dG adducts were almost exclusively formed at methylated CpG dinucleotides of the non-transcribed DNA strand. While fast and efficient 'transcription-coupled repair' of B[*a*]PDE-dG adducts at the transcribed strand accounts for the observed strand bias (cf. above), facilitated adduct formation at methylated CpG sites is only poorly understood. However, preferential binding at methylated CpG sites could be confirmed with other compounds and other genes as well. Fjord-region diol-epoxides from B[*g*]C (B[*g*]CDE) are known to resemble the DNA binding pattern of B[*c*]PhDE with a more equal mixture of adducts at both sites, N^2-dG and N^6-dA (cf. above).[143,215,216] At the *p53* gene however, formation of *anti*-B[*c*]PhDE and *anti*-B[*g*]CDE adducts was found to be greatly enhanced at methylated CpG dinucleotides as compared to other dG or dA sites.[217,218] In the case of the preferential binding of *anti*-B[*a*]PDE at codons 12 and 14 within the human K-*ras* gene exon 1 of bronchial epithelial cells (see above), methylation of the cytosine adjacent to the 5′ dG residue in codon 14 also promoted adduct formation at this site. On the other hand, methylated CpG at codon 14 had no influence on the binding of B[*a*]PDE at codon 12.[219] Despite this methylation-dependent enhancement in DNA binding, 5-methylcytosines at PAH adducted CpG sites do not diminish the repairability of those sites. In accord with the fast repair of *anti*-B[*a*]PDE-N^2-dG adducts at codon 14 of the K-*ras* gene (see above), removal of the same kind of adducts from codon 273 in the human *p53* gene was even more enhanced in the presence of methylcytosine, no matter whether it was located in the damaged strand or in the opposite CpG of the undamaged strand (see Chapter 6.3.5 and Figure 6.11).

Site-directed insertion of stereochemically defined PAH diol-epoxide adducts at specific positions in certain nucleotide sequence contexts not only provides an elegant tool for measuring the variability in enzymatic repair resulting from chemical alterations, it also enables to study all factors that influence the mutation frequencies and the mutational patterns induced.[220,221] Since molecularly defined in all (stereo)chemical details, this approach is a valuable advancement as compared to other indicator systems that are based on selectible mutations (e.g., V79/HPRT) or chemical modifications of small target genes such as *supF* (85 nucleotides)[222] (cf. Chapter 8.4). Based on all of these systems, it could be demonstrated that the spatial structure of the hydrocarbon (planar *vs.* nonplanar) and the DNA sequence context of the adducts profoundly affect both the mutagenicity and the mutational pattern of the compound. On the other hand, site-directed generation of configurationally defined diol-epoxide adducts within defined oligonucleotide sequences allows the role of the sequence context to be deciphered more precisely.[223,224] In case of the very same (+)-*trans-anti*-B[*a*]PDE-N^2-dG adduct, for example, a simple inversion of the modified codon (5′–TGC → 5′–CGT) led to a nearly complete shift from dG → dT transversions to dG → dA transitions.[225] Subsequent work revealed that these differences in nucleotide incorporation opposite the adduct most likely result from distinct differences in the displacement of the guanine residue in a given 'base-displaced conformation' adopted by the (+)-*trans-anti*-B[*a*]PDE-N^2-dG adduct (see Chapter 5.5.1.1). Not even enough, the very same adduct placed in the very same sequence still may have slightly varying mutagenic potencies depending on the kind of DNA strand affected (leading *vs.* lagging strand; see Chapter 5.5.1.2).

In addition to the spatial and stereochemical structure of the adduct, its sequence context and the kind of DNA strand affected, the 'mutational signature' of a given PAH diol-epoxide adduct also depends on the kind of DNA polymerase which actually processes the lesion during reduplication of the DNA.[226,227] The bypass of a bulky adduct ('translesional synthesis') is the critical step which may be error-free (correct nucleotide incorporated opposite the lesion) or error-prone. Recent studies indicated that so-called Y family DNA polymerases are involved in this lesion bypass during DNA replication[228,229] (see Chapter 5.5.2). Due to the lack of a proofreading function these enzymes are, in principle, highly error-prone. However, a PAH-inducible form, polymerase κ, was found to be suited to bypass both a template (–)- or (+)-*trans-anti*-B[*a*]PDE-N^2-dG adduct very efficiently *in vitro* in an error-free manner by a 'one-polymerase two-step mechanism'.[230] Bypass of the *trans*-opened (–)-*anti*-B[*a*]PDE adduct was found to be about 40 times more efficient than bypass of the (+)-*anti*-isomeric lesion — a factor that possibly contributes to the differences in the mutagenicity of both enantiomers. On the other hand, the efficiency of polymerase κ to insert the correct nucleotide (dC) opposite the adduct may also be subject to modification depending on the $5'$ and $3'$ flanking bases of the lesion.[231] A second function of polymerase κ is to act in combination with another DNA polymerase (a replicative form or another Y-family member such as η or REV1) in a 'two-polymerase two-step mechanism', with polymerase κ exerting an extension DNA synthesis activity after nucleotide insertion opposite the lesion catalyzed by η or REV1.[230,232,233] Depending on the kind of polymerase, this latter mechanism may be erroneous, resulting in base substitutions or in the deletion of the adducted nucleotide ('–1 deletions') (cf. Chapter 6.3.6). Another Y-family member, the human polymerase η, was found to perform highly error-prone translesional synthesis opposite both isomeric *anti*-B[*a*]PDE-N^2-dG adducts, with a higher efficiency in case of the (+)-*anti*-isomer.[232] Most likely, the activity of this enzyme also accounts for the mutational pattern induced by B[*a*]P in *polymerase* κ gene-deficient embryonic stem cells, which show a characteristic prevalence of dG $\rightarrow$ dT transversions at the non-transcribed strand of the endogenous *hprt* gene (see Chapter 5.5.2).

11.6 Reprise

B[*a*]P, the prototypic carcinogenic PAH, is metabolically converted into the bay-region (+)-*anti*-B[*a*]PDE in high diastereomeric excess. The main enzymes involved in this monooxygenation activation route are CYP1A1 and CYP1B1, along with mEH. Binding of B[*a*]P to DNA in cells in culture or *in vivo* is predominantly mediated by this chemical species. The main product results from a *trans*-directed covalent reaction with the exocyclic amino group in dG bases giving rise to the (+)-*trans-anti*-B[*a*]PDE-N^2-dG adduct (see Figure 11.1). Compared to the other three isomeric 7,8-diol-9,10-epoxides (Figure 2.5), (+)-*anti*-B[*a*]PDE exerts the highest intrinsic

DNA binding capacity, mutagenic activity and carcinogenic potency. It is further far more effective than the parent compound in inducing lung tumors in newborn mice upon intraperitoneal injection.[234] In juvenile strain A/J mice, a single intraperitoneal injection of B[*a*]P is sufficient to induce lung tumors in a dose-dependent manner, most of which carry activating G → T base substitutions at codon 12 of the cellular K-*ras* gene.[201,235] (The same lesion has also been detected in newborn CD-1 mice upon treatment with B[*a*]P.[236]) The *anti*-B[*a*]PDE-N^2-dG adduct was again identified as the predominant DNA adduct in the very same tissue at earlier time points, and the tumor response correlated nicely to the overall (time-integrated) level of these adducts.[235,237]

From the summary in the foregoing section it is clear that several 'variables' may have a critical impact on the 'mutagenic signature' (both frequency and pattern) of a carcinogenic PAH. In addition to the spatial structure and the configurational identity of the DNA-reactive diol-epoxide, and the geometry and conformation of its DNA reaction products, several 'biological factors' such as DNA base sequence context, the kind of DNA strand or the nature of the processing DNA polymerases were identified as being able to modulate the genotoxicity of a particular compound at the molecular level. Despite these influencing factors, however, a range of different PAH diol-epoxides with different geometrical structures (planar *vs.* non-planar) and different DNA base binding preferences (N^2-dG *vs.* N^6-dA) were found to fit nicely into a quantitative relationship between the overall level of PAH diol-epoxide-DNA damage and the level of mutagenicity displayed in a particular test system (see Figure 8.3 and discussion in section 11.5, see also Phillips *et al.*[138]). Hence, the differences in repairability or processibility of individual diol-epoxide-DNA adducts observed at the molecular level apparently seem to neutralize each other at the cell population level.

Strain A/J mice have also been used as an animal model to compare the differences of a variety of PAH skin carcinogens in their concurrent potency for inducing pulmonary tumors.[200,201,235,237] The lung tumorigenicity observed for five hydrocarbons was found in the order DB[*a*,*l*]P $\gg$ 5-MeC $>$ cyclopenta[*c*,*d*]pyrene $>$ B[*a*]P $>$ B[*b*]F (cf. Figure 9.1). The dose-response relationships revealed about 100-fold differences in their tumorigenic potency.[200,237] Regarding their mutagenicity at the cellular K-*ras* locus, all of these compounds induced G $\rightarrow$ T transversions at codon 12 but no change at codon 61 — with the exception of DB[*a*,*l*]P.[200,201] The most potent hydrocarbon induced both G $\rightarrow$ T transversions at codon 12 as well as A $\rightarrow$ T transversions at codon 61. This mutation pattern can be rationalized based on the observation that DB[*a*,*l*]P (+)-*syn*-(S,R,S,R)- and (–)-*anti*-(R,S,S,R)-11,12-diol-13,14-epoxides [(+)-*syn*- and (–)-*anti*-DB[*a*,*l*]PDE; Figure 2.6] are both involved in binding to N^2-dG and N^6-dA residues in mouse lung tissue.[200] In contrast, in cells in culture or in mouse skin (–)-*anti*-DB[*a*,*l*]PDE adducts are predominant (cf. section 11.5). Despite their extraordinary large differences in lung tumorigenicity, the tumor responses of these five compounds correlated nicely to the time-integrated DNA adduct levels (TIDAL), which have been calculated as area under the curves of total diol-epoxide-DNA adduct levels measured during a time course of 30 days after injection (and extrapolated to 250 days).[200,237] (There was, however, one exception from this rule in this series of studies: DB[*a*,*h*]A, another PAH investigated, failed to induce any mutations at codons 12 or 61 of lung tissue K-*ras*; yet its tumor response was about the same as for DB[*a*,*l*]P, but much greater as expected from its TIDAL.[237]) This parameter represents the total effective molecular dose of a PAH delivered to target lung DNA and it linearly correlated to the doses administered. So, the much greater tumorigenic activity of DB[*a*,*l*]P in mouse lung compared to other carcinogenic PAHs could be attributed to the much higher level of DNA binding of this compound rather than to a greater tumorigenic potential of DB[*a*,*l*]PDE-DNA adducts. On the other hand, the much higher DNA binding of a particular PAH may in turn be the result from differences in its toxicokinetics (related to structural parameters), including metabolic activation and detoxification and the effectiveness of the repair of those lesions that have been formed. In terms of TIDAL, the diol-epoxide-DNA adducts ultimately formed by those five carcinogenic PAHs had similar carcinogenic potencies. Together with the relationship between the overall level of DNA binding and mutagenicity described above, a similar correlation between the (time-integrated) overall DNA binding and carcinogenicity (as found in lung tissue of strain A/J mice) underscores the relevance of this parameter as an important (bio)indicator of the tumor threat that may result from certain exposures to carcinogenic PAHs.

In light of the data accumulated in the case of B[*a*]P and its metabolites, there is not much room left for any doubt that the (+)-*anti*-B[*a*]PDE derivative represents the ultimate carcinogenic species responsible for tumor induction in mouse lung upon intraperitoneal application of the parent compound. The relationship between the greater DNA binding of configurationally corresponding diol-epoxides from other PAHs with their greater tumor-inducing potencies further lends weight to the notion that vicinal bay or fjord-region diol-epoxides are the principal ultimate carcinogenic intermediates in this target tissue (cf. Chapter 2.3.2).

Another target tissue of carcinogenic PAHs, mouse skin, has been introduced as early as 1918 as a tumor model and since then widely been used for monitoring the relative potencies of these kinds of chemicals (see Chapter 1). In contrast to lung tissue in mice, however, application of a single dose of a carcinogenic PAH is not sufficient to induce skin tumors. Instead, repeated treatments with the PAH or application of tumor promoting compounds (such as 12-*O*-tetradecanoylphorbol 13-acetate, TPA) are required in order to induce papillomas and carcinomas in frequencies beyond the background level (see Chapter 1). Although it has been demonstrated already in 1978 that B[*a*]P may be stereoselectively converted into the bay-region (+)-*anti*-B[*a*]PDE and, to some extent, into (+)-*syn*-B[*a*]PDE in mouse skin and that these intermediates bind to dG residues in mouse skin DNA[165,167] (cf. Chapter 2.3.2), the limited tumorigenicity of synthetic (+)-*anti*-B[*a*]PDE and the almost complete lack of tumorigenicity in the case of synthetic (+)-*syn*-B[*a*]PDE upon direct application onto the back of mice[238] raised doubts about their role in mediating parent compound's biological effects in this tissue (reviewed by Rubin[239]). A similar discrepancy has also been observed for DMBA and its *syn*- and *anti*-3,4-diol-1,2-epoxides (DMBADE).[196] Although both diastereomeric DMBADE displayed comparable tumor-initiating potencies in the two-stage mouse bioassay (initiation by a single treatment with the PAH, followed by several months of promotion with TPA; cf. Chapter 1), they were found to be much less active than the parent compound. On the other hand, treatment of mouse skin with B[*a*]P or

DMBA and isolation of skin tissue DNA shortly thereafter revealed that the stable DNA adducts generated in this tissue almost exclusively originate from bay- and fjord-region diol-epoxides, respectively.[154,165,167,196] It further has been shown that an increasing tumorigenic potency in the order B[*a*]P < DMBA < DB[*a*,*l*]P, as observed in the two-stage mouse skin model,[240,241] would be in perfect accordance to the overall levels of corresponding diol-epoxide-derived DNA adducts detected at early time points in the very same tissue.[154] The importance of these adducts has also been demonstrated more indirectly through targeted disruption of *mEH* (*EPHX1*), a gene encoding an enzyme indispensibly involved in the generation of diol-epoxide descendants from parent hydrocarbons (see Chapter 2.3.1 and Figure 2.3). Topical treatment with DMBA over a time course of 25 weeks according to the 'complete carcinogenesis' protocol (see below) failed to induce any papillomas in the skin of mice that lacked the mEH protein.[242,243] In contrast, about 75% of the wild-type mice developed skin tumors in response to DMBA during the same period. In addition to this data, DNA isolated from DMBA-treated mouse skin was found to be capable to efficiently transform NIH 3T3 fibroblasts *in vitro*, an effect that initially has been traced to the presence of an activated H-*ras* oncogene carrying an A → T transversion in codon 61[244,245] (cf. Chapter 7.2.1), but which presumably may not alone account for this activity.[246,247] Since DMBA-mediated generation of A → T transversions in cellular H-*ras* most likely results from intermediate formation of *syn*- and *anti*-DMBADE-N^6-dA adducts (see section 11.5), the presence of these adducts and their potential to contribute to cell-transforming activities via activation of a proto-oncogene also lend weight to the assumption that they are crucially involved in skin tumorigenesis.

Although there is no convincing evidence that mouse skin tumorigenesis can be sufficiently and dose-dependently initiated by any other kinds of DNA-reactive metabolites from carcinogenic PAHs (cf. Chapter 2), the increased tumorigenicity of B[*a*]P or DMBA as compared to their diol-epoxide descendants may point to additional activities in this tissue, only exerted by the parent compounds. Strong carcinogenic PAHs can act as 'complete carcinogens' when repeatedly applied over a prolonged period of time (cf. Chapter 1; for B[*a*]P and DMBA see Miyata *et al.*[242] and Shimizu *et al.*[248]). This finding indicates that there is an accompanying promotional activity which jumps into place after cells have been 'initiated' (mutated) by genotoxic metabolites. The requirement of a functional AhR protein for the induction of tumors in the skin of mice,[248] along with the strong binding and activation of this protein by PAHs (see Chapter 2.3.1) may support the notion that not only the initiating activity (via induction of metabolizing enzymes) but also the promotional activity of carcinogenic PAHs is dependent upon AhR-mediated gene expression. This would be in agreement with the tumor promoting activity of 2,3,7,8-tetrachlorodibenzo-*p*-dioxin (TCDD), which has long been known as one of the strongest agonists of the AhR protein[249,250] (cf. Chapter 2.3.1). TCDD does not bind to DNA *in vivo*[251] and is not mutagenic,[252] yet it is a potent tumor promoter in liver and much more active than TPA in skin of HRS/J hairless mice.[253] The tumor-promoting activity of TCDD has mainly been attributed to an AhR-mediated transcriptional activation of target genes involved in cell proliferation and cellular growth.[250,254,255] On the other hand, both compounds, TCDD and B[*a*]P, are apparently capable of activating a group of immediate-early genes such as *c-jun* and *c-fos* (whose products act together as the AP-1 transcription factor) in an AhR-independent manner.[256] In addition to this finding, a more direct interference of the AhR with various cell signalling and cell cycle pathways has been uncovered recently.[254,257] An example of this is the 'molecular cross talk' with cytosolic tyrosine kinases (i.e., c-Src) and the concomitant initiation of phosphorylation signalling cascades.[258]

It is long been known that a considerable fraction of the PAH may persist and retain unmetabolized for an extended period of time (days until weeks) in the skin of mice upon application of a single dose.[259] Yet the constant renewal of the depot over months is mandatory in order to achieve the maximal tumor yield. This requirement rather points to (short-living) metabolite(s) which may also be involved in mediating parent compound's promotional acitivity in mouse skin. As outlined in Chapter 2.3.4, autoxidation of metabolically generated PAH phenols is a major sourse of 'reactive oxygen species' *in vitro* and, most likely, also *in vivo*. These intermediates may produce a variety of DNA damage (see

Figure 2.9) such as the 'indicator lesion' 8-hydroxy-2′-deoxyguanosine, but also chromosomal aberrations that result from DNA single or double strand breaks ('clastogenic effect'). Accordingly, exposure to PAHs has been found to be accompanied by an increase in the levels of these kinds of DNA lesions.[260–264] On the other hand, PAH-induced tissue inflammation and concurrent wound healing[265,266] may also lead to local recruitment of polymorphonuclear leukocytes (neutrophilic granulocytes), subsequent release ('burst') of 'reactive oxygen species' and the generation of chromosomal aberrations ('conversion stage' of tumor promotion).[267] In addition, leukocytes are known to express various peroxidases either constitutively (myeloperoxidase) or after induction (prostaglandin H synthase) by tumor promoters such as TCDD or PAHs (see Chapter 2.3.1). So, migration of leukocytes into PAH exposed skin tissue may therefore contribute to: (i) further peroxidase-dependent generation of DNA-reactive PAH diol-epoxides that promote additional DNA damage;[268,269] and (ii) the generation of chromosomal aberrations (similar as observed with typical 'convertogenic' skin tumor promoters such as TPA[270]). Both mechanisms possibly contribute to the additional activity of parent compounds in skin tumorigenesis.

Due to the kind of their generation (incomplete combustion), humans are always exposed to complex mixtures of PAHs (Chapter 1) and — even more — mostly in accompaniment with other compounds from related or unrelated groups of chemicals. For instance, about 400–600 chemicals have been identified as gas-phase constituents in cigarette smoke and more than 4000 of all particulate-phase constituents are characterized.[271] In the group of PAHs, about 150 compounds could be identified, of which 14 are known carcinogens.[272,273] Since most laboratory studies were conducted with single and pure compounds the results obtained clearly may not allow to directly extrapolate to the corresponding effects expected to be exerted from mixtures.[274] Multiple interactions between these compounds within the biological system may result in additive, synergistic or antagonizing effects.[275] Non-carcinogenic PAHs such as benzo[*e*]pyrene, pyrene or fluoranthene, for example, may exert 'co-carcinogenic' effects. Simultaneous application of these compounds with low doses of B[*a*]P in the mouse skin bioassay led to an increase of papillomas and carcinomas as compared to the carcinogen alone[276] — an effect due to an enhanced DNA binding of B[*a*]P.[277] In light of these considerations it seems clear that it is difficult, if not impossible, to uncover any causative connections between certain forms of cancer and the exposure to carcinogenic PAHs. Moreover, any risk assessment based on toxicological and monitoring data from only a few PAHs, or from only one PAH (B[*a*]P), is highly prone to misjudgement[278] (cf. discussion in section 11.1).

As seen from Chapter 9, the etiology of sporadic breast cancer in the human population remains uncertain. In consideration of their genotoxicity and their ability to induce mammary cancer in rodent models, a contribution of carcinogenic PAHs and their nitro-derivatives in the development of breast cancer seems plausible; however, evidence for similar effects at low doses in humans is to date, at best, very weak.[279] Thus, much more substantial work is needed, especially in exposure assessment and breast tumor susceptibility, before any conclusions seem to be justified. On the other hand, the association between lung cancer and PAH contaminated cigarette smoke is well established. Cigarette smoking increases the risk of all histological types of lung cancer, including squamous-cell carcinoma, small-cell carcinoma, adenocarcinoma and large-cell carcinoma.[273,280] About 90% of all lung cancer cases are related to cigarette smoke. Smoking is also causally associated with cancer of the oral cavity, larynx, pharynx, esophagus, stomach, liver, pancreas and some other forms such as myeloid leukemia. Although the neutral fraction of tobacco smoke condensate is known to contain most of the PAHs including B[*a*]P and constitutes the "only truly carcinogenic portion of tobacco tar"[271] (as confirmed in mouse skin), the amount of B[*a*]P present in this fraction was found to be far too low to fully account by itself for the tumor producing activity observed (reviewed by Rubin[239]). However, the carcinogenic activity of PAHs in cigarette smoke is potentiated through the presence of strong promoters in the phenolic fraction of the condensate or by co-carcinogens such as the non-carcinogenic PAHs mentioned above, and other compounds (e.g., catechol).[272,276,281] Based on the observation in a very early study that the addition of tumor initiators to the tobacco condensate increases its carcinogenic effect, while adding of promotors does not,[281] it seemed plausible that the correlation between smoking and lung cancer incidence would be determined not primarily by the carcinogenic effect of tobacco smoke but rather by its tumor-promoting activity. It was hypothesized that <3% of the carcinogenic activity of cigarette smoke could be due to the activity of known carcinogens; the activity of carcinogens and non-carcinogenic initiators along with the strong tumor promoting activity of the condensate would then add up to the full effect of the condensate as probed on mouse skin (reviewed by Rubin[239]). Given this interpretation, cigarette smoke cancer would mainly result from selection of pre-existing (endogenous) mutations rather than from direct genotoxic damage induced by cigarette smoke carcinogens.

There is no doubt that cigarette smoke is a highly brisant mixture of a large number of compounds from a great variety of chemical classes (including inorganic compounds). Besides PAHs, aromatic amines, small organic molecules such as formaldehyde, acrylonitrile or benzene, and *N*-nitrosamines, in particular tobacco-specific *N*-nitrosamines such as 4-(methylnitrosamino)-1-(3-pyridyl)-1-butanone (NNK), constitute the major classes of carcinogens present in cigarette smoke.[271,282,283] In addition, inhalation experiments with mice added evidence that the vapor (gas) phase of cigarette smoke promotes lung tumor formation as well — though the levels of higher molecular weight PAHs are significantly reduced in this fraction.[284] Given this data it seems rather difficult to extract the real contribution of carcinogenic PAHs from the overall toxicity of cigarette smoke. However, the crucial role of carcinogenic PAHs in the etiology of human lung cancer and, in particular their importance for tumor initiation is strongly supported by several lines of evidence that may be summarized as follows:

(i) As early as 1973, it has been suggested for the first time that the extent of inducibility of aryl hydrocarbon hydroxylase ('AHH'), now known as CYP1A1, was increased in lung cancer patients as compared to controls.[285] CYP1A1 is one of the two major monooxygenases involved in activation of B[*a*]P and other carcinogenic PAHs (see Chapter 2.3.1 and Figure 2.3), and it appears that high CYP1A1 inducibility is related to a high lung cancer risk.[286–289]

(ii) The activity of pulmonary CYP1A1 correlates to 'bulky' (PAH- resp. B[*a*]P-) DNA adduct levels in human lung tissue both *in vitro* and *in vivo*.[289–291] The high variability in lung PAH-DNA adduct levels can therefore be rationalized by large inter-individual differences in pulmonary CYP1A1 expression,[288,292] which in turn may result from differences (polymorphisms) in the regulation of the corresponding gene (through the AhR pathway[293,294]).

(iii) The levels of PAH diol-epoxide-DNA adducts in lung tissue *in vivo* are generally higher in smokers as compared to ex- or non-smokers.[137]

(iv) The levels of 'bulky' (PAH- resp. B[*a*]P-) DNA adducts in lung tissue from subjects with a combined *CYP1A1 Msp*I-*GSTM1*-null genotype are increased compared to wild-type allele carriers.[69,120] Exactly this genotype was identified to confer high susceptibility for developing lung cancer[84,123] (see section 11.4). In addition, both the *CYP1A1 Msp*I or Ile[462]Val as well as the *GSTM1*-null genotype were shown to correlate to a higher frequency of *p53* gene mutations in lung tumor tissue.[295,296] Thus, germ line polymorphisms of two important genes involved in PAH metabolism (i.e., *CYP1A1*, *GSTM1*; see section 11.4) and associated with increased levels of PAH-DNA adducts in lung target tissue, were shown to be related to smoking-associated *p53* gene mutations (exons 5–8) in this tissue.

(v) Exposure of human bronchial epithelial cells to *anti*-B[*a*]PDE, the ultimate carcinogenic metabolite of B[*a*]P (see section 11.5), leads to the formation of *anti*-B[*a*]PDE-N^2-dG adduct hotspots within certain tumorgenes at the same positions that can be detected as mutational hotspots in lung tumor tissue from smokers. This has been demonstrated for the proto-oncogene K-*ras* (codon 12) and the tumor suppressor gene *p53* (codons 157, 248 and 273) — two genes frequently mutated in cigarette smoke-induced lung cancer[202,203,297] (see section 11.5). In accord to the preferential formation of *anti*-B[*a*]PDE adducts at N^2-dG sites, G → T transversions were detected as the principal mutations at all of these mutational hotspots.

The coincidence of mutational hotspots in K-*ras* or *p53* and *anti*-B[*a*]PDE-N^2-dG adduct hotspots suggests that B[*a*]P is involved in the transformation of human lung tissue in smokers[298,299] (see Chapter 5.6). The much higher frequency of G → T transversions in the lung tissue *p53* gene of smokers as compared to non-smokers rather argues against a major role of promotor-dependent selection of pre-existing endogenous mutations (cf. above) and points to the importance of exogenous mutagens present in cigarette smoke. On the other hand, *N*-nitrosamines (in particular NNK), besides PAHs the most important group of cigarette smoke carcinogens,[283,300] were shown to predominantly induce G → A and C → T transitions in the K-*ras* and *p53* genes of animal models.[300–302] Although some of these (alkylating) compounds are in principle also capable in producing G → T transversions,[301] *N*-nitrosamines are metabolically activated by CYP2E1, CYP2A6, and some other CYP enzymes, but not by CYP1A1,[300,303] and they do not contribute to 'bulky' DNA adducts frequently induced in the lungs of smokers (cf. above). The presence of G → A transitions in the *p53* gene of human lung cancer tissue[304] and the modulating effects of *CYP2E1* gene polymorphisms in human lung cancer risk,[305] however, clearly suggest that synergistic effects may have a particular importance in chemical carcinogenesis of lung cancer in man.

Acknowledgment

I am very grateful to my colleague and friend Dr. Gregory P. Tochtrop for his critical reading and the valuable comments he made on the manuscript. The work of the author was supported by the German Research Foundation (DFG: LU 841/2-1).

References

1. IARC (1984) *Monographs on the Evaluation of the Carcinogenic Risk of Chemicals to Humans.* Volume 34: Polynuclear Aromatic Compounds, Part 3, Industrial Exposures in Aluminium Production, Coal Gasification, Coke Production, and Iron and Steel Founding. International Agency for Research on Cancer, Lyon, France.
2. IARC (1985) *Monographs on the Evaluation of the Carcinogenic Risk of Chemicals to Humans.* Volume 35: Polynuclear Aromatic Compounds, Part 4, Bitumens, Coal-tars and Derived Products, Shale-oils and Soots. International Agency for Research on Cancer, Lyon, France.

3. Mastrangelo G, Fadda E and Marzia V (1996) Polycyclic aromatic hydrocarbons and cancer in man. *Environ. Health Perspect.* 104: 1166–1170.
4. Boffetta P, Jourenkova N and Gustavsson P (1997) Cancer risk from occupational and environmental exposure to polycyclic aromatic hydrocarbons. *Cancer Causes Control* 8: 444–472.
5. Ojajarvi IA, Partanen TJ, Ahlbom A, Boffetta P, Hakulinen T, Jourenkova N, Kauppinen TP, Kogevinas M, Porta M, Vainio HU, Weiderpass E and Wesseling CH (2001) Occupational exposures and pancreatic cancer: a meta-analysis. *Occup. Environ. Med.* 57: 316–324.
6. Gaertner RRW and Thériault GP (2002) Risk of bladder cancer in foundry workers: a meta-analysis. *Occup. Environ. Med.* 59: 655–663.
7. Kogevinas M, t'Mannetje A, Cordier S, Ranft U, González CA, Vineis P, Chang-Claude J, Lynge E, Wahrendorf J, Tzonou A, Joeckel KH, Serra C, Porru S, Hours M, Greiser E and Boffetta P (2003) Occupation and bladder cancer among men in Western Europe. *Cancer Causes Control* 14: 907–914.
8. ATSDR (1995) *Toxicological Profile for Polycyclic Aromatic Hydrocarbons (PAHs).* Agency for Toxic Substances and Disease Registry (www.atsdr.cdc.gov); pp 229–230 (Chapter 5.1) U.S. Department of Health and Human Services, Atlanta, GA.
9. Lee ML, Novotny M and Bartle KD (1976) Gas chromatography/mass spectrometric and nuclear magnetic resonance determination of polynuclear aromatic hydrocarbons in airborne particulates. *Anal. Chem.* 48: 1566–1572.
10. Gil L, King L and Adonis M (2000) Trends of polycyclic aromatic hydrocarbon levels and mutagenicity in Santiago's inhalable airborne particles in the period 1992–1996. *Inhal. Toxicol.* 12: 1185–1204.
11. ATSDR (1990) *Public Health Statement, Polycyclic Aromatic Hydrocarbons.* Agency for Toxic Substances and Disease Registry (www.atsdr.cdc.gov); U.S. Department of Health and Human Services, Atlanta, GA.
12. Bayram A and Müezzinoglu A (1996) Environmental aspects of polycyclic aromatic hydrocarbons (PAHs) originating mainly from coal-fired combustion systems and their monitoring requirements. In: *Environmental Xenobiotics* [Richardson M (ed.)] pp 333–354, Taylor & Francis, London, UK.
13. Pott F (1985) Pyrolyseabgase, PAH und Lungenkrebsrisiko — Daten und Bewertung. *Staub Reinhalt. Luft* 45: 369–379.
14. WHO (1987) *Air Quality Guidelines for Europe.* WHO Regional Publications, European Series, No. 23, Polycyclic Aromatic Hydrocarbons, pp 105–117, World Health Organization, Copenhagen, Denmark.

15. Hemminki K and Pershagen G (1994) Cancer risk of air pollution: epidemiological evidence. *Environ. Health Perspect.* 102: 187–192.
16. WHO (2000) *Air Quality Guidelines for Europe.* Second Edition. WHO Regional Publications, European Series, No. 91, Chapter 5.9: Polycyclic Aromatic Hydrocarbons, World Health Organization, Copenhagen, Denmark (website: www.who.dk/air/Activities/20020620_1).
17. Boström CE, Gerde P, Hanberg A, Jerström B, Johansson C, Kyrklund T, Rannug A, Törnquist M, Victorin K and Westerholm R (2002) Cancer risk assessment, indicators, and guidelines for polycyclic aromatic hydrocarbons in the ambient air. *Environ. Health Perspect.* 110, Suppl 3: 451–489.
18. Heinrich U, Roller M and Pott F (1994) Estimation of a lifetime unit lung cancer risk for benzo[*a*]pyrene based on tumour rates in rats exposed to coal tar/pitch condensation aerosol. *Toxicol. Lett.* 72: 155–161.
19. Armstrong B, Tremblay C, Baris D and Thériault G (1994) Lung cancer mortality and polynuclear aromatic hydrocarbons: a case-cohort study of aluminum production workers in Arvida, Quebec, Canada. *Am. J. Epidemiol.* 139: 250–262.
20. Pfeffer HU (1994) Ambient air concentrations of pollutants at traffic-related sites in urban areas of North Rhine-Westphalia, Germany. *Sci. Total. Environ.* 146–147: 263–273.
21. Sanders CL, Skinner C and Gelman RA (1984) Percutaneous absorption of [7.10-^{14}C]benzo[*a*]pyrene and [7,12-^{14}C]dimethylbenz[*a*]anthracene in mice. *Environ. Res.* 33: 353–360.
22. Foth H, Kahl R and Kahl GF (1988) Pharmacokinetics of low doses of benzo[*a*]pyrene in the rat. *Food Chem. Toxicol.* 26: 45–51.
23. Wester RC, Maibach HI, Bucks DAW, Sedik L, Melendres J, Liao C and DiZio S (1990) Percutaneous absorption of [^{14}C]DDT and [^{14}C]benzo-[*a*]pyrene from soil. *Fundam. Appl. Toxicol.* 15: 510–516.
24. Withey JR, Law FCP and Endrenyi L (1991) Pharmacokinetics and bioavailability of pyrene in the rat. *J. Toxicol. Environ. Health* 32: 429–447.
25. Ng KME, Chu I, Bronaugh RL, Franklin CA and Somers DA (1992) Percutaneous absorption and metabolism of pyrene, benzo[*a*]pyrene, and di(2-ethylhexyl)phthalate: comparison of *in vitro* and *in vivo* results in the hairless guinea pig. *Toxicol. Appl. Pharmacol.* 115: 216–223.
26. Withey JR, Law FCP and Endrenyi L (1993) Percutaneous uptake, distribution, and excretion of pyrene in rats. *J. Toxicol. Environ. Health* 40: 601–612.

27. Ramesh A, Inyang F, Hood DB, Archibong AE, Knuckles ME and Nyanda AM (2001) Metabolism, bioavailability, and toxicokinetics of benzo[*a*]pyrene in F-344 rats following oral administration. *Exp. Toxicol. Pathol.* 53: 275–290.
28. Hecht SS, Grabowski W and Groth K (1979) Analysis of faeces for benzo[*a*]pyrene after consumption of charcoal-broiled beef by rats and humans. *Food Cosmet. Toxicol.* 17: 223–227.
29. Kao J, Patterson FK and Hall J (1985) Skin penetration and metabolism of topically applied chemicals in six mammalian species, including man: an *in vitro* study with benzo[*a*]pyrene and testosterone. *Toxicol. Appl. Pharmacol.* 81: 502–516.
30. Buckley TJ and Lioy PJ (1992) An examination of the time course from human dietary exposure to polycyclic aromatic hydrocarbons to urinary elimination of 1-hydroxypyrene. *Br. J. Ind. Med.* 49: 113–124.
31. VanRooij JGM, Bodelier-Bade MM and Jongeneelen FJ (1993) Estimation of individual dermal and respiratory uptake of polycyclic aromatic hydrocarbons in 12 coke oven workers. *Br. J. Ind. Med.* 50: 623–632.
32. ATSDR (1995) *Toxicological Profile for Polycyclic Aromatic Hydrocarbons (PAHs).* Agency for Toxic Substances and Disease Registry (www.atsdr.cdc.gov); pp 78–87 (Chapter 2.3.1: Absorption) U.S. Department of Health and Human Services, Atlanta, GA.
33. Sun JD, Wolff RK, Kanapilly GM and McClellan RO (1984) Lung retention and metabolic fate of inhaled benzo[*a*]pyrene associated with diesel exhaust particles. *Toxicol. Appl. Pharmacol.* 73: 48–59.
34. Törnquist S, Wiklund L and Toftgård R (1985) Investigation of absorption, metabolism kinetics and DNA-binding of intratracheally administered benzo[*a*]pyrene in the isolated, perfused rat lung: a comparative study between microcrystalline and particulate adsorbed benzo[*a*]pyrene. *Chem.-Biol. Interact.* 54: 185–198.
35. Wolff RK, Bond JA, Sun JD, Henderson RF, Harkema JR, Griffith WC, Mauderly JL and McClellan RO (1989) Effects of adsorption of benzo[*a*]-pyrene onto carbon black particles on levels of DNA adducts in lungs of rats exposed by inhalation. *Toxicol. Appl. Pharmacol.* 97: 289–299.
36. Gerde P, Muggenburg BA, Lundborg M and Dahl AR (2001) The rapid alveolar absorption of diesel soot-absorbed benzo[*a*]pyrene: bioavailability, metabolism and dosimetry of an inhaled particle-borne carcinogen. *Carcinogenesis* 22: 741–749.

37. Gerde P, Muggenburg BA, Thornton-Manning JR, Lewis JL, Pyon KH and Dahl AR (1997) Benzo[*a*]pyrene at an envrionmentally relevant dose is slowly absorbed by, and extensively metabolized in, tracheal epithelium. *Carcinogenesis* 18: 1825–1832.
38. Gerde P, Muggenburg BA, Lundborg M, Tesfaigzi Y and Dahl AR (2001) Respiratory epithelial penetration and clearance of particle-borne benzo[*a*]-pyrene. *Res. Rep. Health Eff. Inst.* 101: 5–32.
39. Wall KL, Gao WS, te Koppele JM, Kwei GY, Kauffman FC and Thurman RG (1991) The liver plays a central role in the mechanism of chemical carcinogenesis due to polycyclic aromatic hydrocarbons. *Carcinogenesis* 17: 783–786.
40. Cotgreave IA, Moldéus P and Orrenius S (1988) Host biochemical defense mechanisms against prooxidants. *Annu. Rev. Pharmacol. Toxicol.* 28: 189–212.
41. Hayes JD and McLellan LI (1999) Glutathione and glutathione-dependent enzymes represent a co-ordinately regulated defense against oxidative stress. *Free Radic. Res.* 31: 273–300.
42. Younes M (1999) Free radicals and reactive oxygen species. In: *Toxicology* [Marquardt H, Schäfer SG, McClellan RO, and Welsch F (eds.)] pp 111–125, Academic Press, San Diego.
43. Klaunig JE and Kamendulis LM (2004) The role of oxidative stress in carcinogenesis. *Annu. Rev. Pharmacol. Toxicol.* 44: 239–267.
44. Hecht SS, Chen M, Yagi H, Jerina DM and Carmella SG (2003) *r*-1,*t*-2,3,*c*-4-Tetrahydroxy-1,2,3,4-tetrahydrophenanthrene in human urine: a potential biomarker for assessing polycyclic aromatic hydrocarbon metabolic activation. *Cancer Epidemiol. Biomarkers Prev.* 12: 1501–1508.
45. Dell'Omo M and Lauwerys RR (1993) Adducts to macromolecules in the biological monitoring of workers exposed to polycyclic aromatic hydrocarbons. *Crit. Rev. Toxicol.* 23: 111–126.
46. Hemminki K, Autrup H and Haugen A (1995) DNA and protein adducts. *Toxicology* 101: 41–53.
47. Phillips DH and Farmer PB (1995) Protein and DNA adducts as biomarkers of exposure to environmental mutagens. In: *Environmental Mutagenesis* [Phillips DH and Venitt S (eds.)] pp 367–395, Bios, Oxford, UK.
48. Angerer J, Mannschreck C and Gündel J (1997) Biological monitoring and biochemical effect monitoring of exposure to polycyclic aromatic hydrocarbons. *Int. Arch. Occup. Environ. Health* 70: 365–377.

49. Szyfter K, Hemminki K, Szyfter W, Szmeja Z, Banaszewski J and Yang K (1994) Aromatic DNA adducts in larynx biopsies and leukocytes. *Carcinogenesis* 15: 2195–2199.
50. Tang D, Santella RM, Blackwood AM, Young TL, Mayer J, Jaretzki A, Grantham S, Tsai WY and Perera FP (1995) A molecular epidemiological case-control study of lung cancer. *Cancer Epidemiol. Biomarkers Prev.* 4: 341–346.
51. Wiencke JK, Kelsey KT, Varkonyi A, Semey K, Wain JC, Mark E and Christiani DC (1995) Correlation of DNA adducts in blood mononuclear cells with tobacco carcinogen-induced damage in human lung. *Cancer Res.* 55: 4910–4914.
52. Kriek E, Rojas M, Alexandrov K and Bartsch H (1998) Polycyclic aromatic hydrocarbon-DNA adducts in humans: relevance as biomarkers for exposure and cancer risk. *Mutat. Res.* 400: 215–231.
53. Perera FP, Mooney LA, Dickey CP, Santella RM, Bell D, Blaner W, Tang D and Whyatt RM (1996) Molecular epidemiology in environmental carcinogenesis. *Environ. Health Perspect.* 104, Suppl 3: 441–443.
54. Poirier MC (1997) DNA adducts as exposure biomarkers and indicators of cancer risk. *Environ. Health Perspect.* 105: 907–912.
55. Dickey C, Santella RM, Hattis D, Tang D, Hsu Y, Cooper T, Young TL and Perera FP (1997) Variability in PAH-DNA adduct measurements in peripheral mononuclear cells: implications for quantitative cancer risk assessment. *Risk Anal.* 17: 649–656.
56. Georgiadis P and Kyrtopoulos SA (1999) Molecular epidemiological approaches to the study of the genotoxic effects of urban air pollution. *Mutat. Res.* 428: 91–98.
57. Perera FP and Weinstein IB (2000) Molecular epidemiology: recent advances and future directions. *Carcinogenesis* 21: 517–524.
58. Knudsen LE, Loft SH and Autrup H (2001) Risk assessment: the importance of genetic polymorphisms in man. *Mutat. Res.* 482: 83–88.
59. Natarajan AT, Boei JJWA, Darroudi F, van Diemen PCM, Dulout F, Hande MP and Ramalho AT (1996) Current cytogenetic methods for detecting exposures and effects of mutagens and carcinogens. *Environ. Health Perspect.* 104, Suppl 3: 445–448.
60. Srám RJ and Binková B (2000) Molecular epidemiology studies on occupational and environmental exposure to mutagens and carcinogens, 1997–1999. *Environ. Health Perspect.* 108, Suppl 1: 57–70.

61. Schoket B, Papp G, Levay K, Mracková G, Kadlubar FF and Vincze I (2001) Impact of metabolic genotypes on levels of biomarkers of genotoxic exposure. *Mutat. Res.* 482: 57–69.
62. http://www.imm.ki.se/CYPalleles/cyp1a1.htm; (date: January 8, 2004).
63. Elovaara E, Raunio H, Pelkonen O and Vainio H (1995) Oxidation of pyrene in human liver and lungs. *Hum. Exp. Toxicol.* 14: 821.
64. Phillips DH, Schoket B, Hewer A, Bailey E, Kostic S and Vincze I (1990) Influence of cigarette smoking on the levels of DNA adducts in human bronchial epithelium and white blood cells. *Int. J. Cancer* 46: 569–575.
65. Dunn BP, Vedal S, San RHC, Kwan WF, Nelems B, Enarson DA and Stich HF (1991) DNA adducts in bronchial biopsies. *Int. J. Cancer* 48: 485–492.
66. Cohen GM, Mehta R and Meredith–Brown M (1979) Large interindividual variations in metabolism of benzo[*a*]pyrene by peripheral lung tissue from lung cancer patients. *Int. J. Cancer* 24: 129–133.
67. Rojas M, Camus AM, Alexandrov K, Husgafvel-Pursiainen K, Antilla S, Vainio H and Bartsch H (1992) Stereoselective metabolism of (–)-benzo-[*a*]pyrene-7,8-diol by human lung microsomes and peripheral blood lymphocytes: effect of smoking. *Carcinogenesis* 13: 929–933.
68. Mooney LA, Bell DA, Santella RM, van Bennekum AM, Ottman R, Paik M, Blaner WS, Lucier GW, Covey L, Young TL, Cooper TB, Glassman AH and Perera FP (1997) Contribution of genetic and nutritional factors to DNA damage in heavy smokers. *Carcinogenesis* 18: 503–509.
69. Rojas M, Alexandrov K, Cascorbi I, Brockmöller J, Likhachev A, Pozharisski K, Bouvier C, Auburtin G, Mayer L, Kopp-Schneider A, Roots I and Bartsch H (1998) High benzo[*a*]pyrene diol-epoxide DNA adduct levels in lung and blood cells from individuals with combined *CYP1A1 Msp*I/*Msp*I-*GSTM1*0/*0* genotypes. *Pharmacogenetics* 8: 109–118.
70. Ichiba M, Wang YP, Oishi H, Zhang JS, Iyadomi M, Minagawa M and Tomokuni K (1998) Lymphocytes, DNA adducts and genetic polymorphism for metabolic enzymes in low dose cigarette smokers. *Biomarkers* 3: 63–71.
71. Butkiewicz D, Grzybowska E, Hemminki K, Øvrebø S, Haugen A, Motykiewicz G and Chorazy M (1998) Modulation of DNA adduct levels in human mononuclear white blood cells and granulocytes by *CYP1A1*, *CYP2D6* and *GSTM1* genetic polymorphisms. *Mutat. Res.* 415: 97–108.
72. Rojas M, Cascorbi I, Alexandrov K, Kriek E, Auburtin G, Mayer L, Kopp-Schneider A, Roots I and Bartsch H (2000) Modulation of benzo-[*a*]pyrene diolepoxide-DNA adduct levels in human white blood cells by *CYP1A1*, *GSTM1* and *GSTT1* polymorphism. *Carcinogenesis* 21: 35–41.

73. Shields PG, Bowman ED, Harrington AM, Doan VT and Weston A (1993) Polycyclic aromatic hydrocarbon-DNA adducts in human lung and cancer susceptibility genes. *Cancer Res.* 53: 3486–3492.
74. Rothman N, Shields PG, Poirier MC, Harrington AM, Ford DP and Strickland PT (1995) The impact of glutathione *S*-transferase M1 and cytochrome P4501A1 genotypes on white-blood-cell polycylic aromatic hydrocarbon-DNA adduct levels in humans. *Mol. Carcinog.* 14: 63–68.
75. Kato S, Bowman ED, Harrington AM, Blomeke B and Shields PG (1995) Human lung carcinogen-DNA adduct levels mediated by genetic polymorphisms *in vivo*. *J. Natl. Cancer Inst.* 87: 902–907.
76. Schokct B, Phillips DH, Kostic S and Vincze I (1998) Smoking-associated bulky DNA adducts in bronchial tissue related to *CYP1A1 Msp*I and *GSTM1* genotypes in lung patients. *Carcinogenesis* 19: 841–846.
77. Zhang ZY, Fasco MJ, Huang L, Guengerich FP and Kaminsky LS (1996) Characterization of purified human recombinant cytochrome P4501A1-Ile^{462} and -Val^{462}: assessment of a role for the rare allele in carcinogenesis. *Cancer Res.* 56: 3926–3933.
78. Persson I, Johansson I and Ingelman–Sundberg M (1997) *In vitro* kinetics of two human *CYP1A1* variant enzymes suggested to be associated with interindividual differences in cancer susceptibility. *Biochem. Biophys. Res. Commun.* 231: 227–230.
79. Schwarz D, Kisselev P, Cascorbi I, Schunck WH and Roots I (2001) Differential metabolism of benzo[*a*]pyrene and benzo[*a*]pyrene-7,8-dihydrodiol by human CYP1A1 variants. *Carcinogenesis* 22: 453–459.
80. Adonis M, Martinez V, Riquelme R, Ancic P, Gonzalez G, Tapia R, Castro M, Lucas D, Berthou F and Gil L (2003) Susceptibility and exposure biomarkers in people exposed to PAHs from diesel exhaust. *Toxicol. Lett.* 144: 3–15.
81. Yang M, Jang JY, Kim S, Lee SM, Chang SS, Cheong HK, Lee E, Kang D, Kim H, Kawamoto T and Shin HD (2003) Genetic effects on urinary 1-hydroxypyrene levels in a Korean population. *Carcinogenesis* 24: 1085–1089.
82. Kawajiri K, Nakachi K, Imai K, Watanabe J and Hayashi S (1993) The *CYP1A1* gene and cancer susceptibility. *Crit. Rev. Oncol. Hematol.* 14: 77–87.
83. Autrup H (2000) Genetic polymorphisms in human xenobiotica metabolizing enzymes as susceptibility factors in toxic response. *Mutat. Res.* 464: 65–76.
84. Hung RJ, Boffetta P, Brockmöller J, Butkiewicz D, Cascorbi I, Clapper ML, Garte S, Haugen A, Hirvonen A, Anttila S, Kalina I, Le Marchand L,

London SJ, Rannug A, Romkes M, Salagovic J, Schoket B, Gaspari L and Taioli E (2003) CYP1A1 and GSTM1 genetic polymorphisms and lung cancer risk in Caucasian non-smokers: a pooled analysis. *Carcinogenesis* 24: 875–882.

85. Taioli E, Gaspari L, Benhamou S, Boffetta P, Brockmoller J, Butkiewicz D, Cascorbi I, Clapper ML, Dolzan V, Haugen A, Hirvonen A, Husgafvel-Pursiainen K, Kalina I, Kremers P, Le Marchand L, London S, Rannug A, Romkes M, Schoket B, Seidegard J, Strange RC, Stucker I, To-Figueras J and Garte S (2003) Polymorphisms in CYP1A1, GSTM1, GSTT1 and lung cancer below the age of 45 years. *Int. J. Epidemiol.* 32: 60–63.
86. Houlston RS (2000) *CYP1A1* polymorphisms and lung cancer risk: a meta-analysis. *Pharmacogenetics* 10: 105–114.
87. Fritsche E, Brüning T, Jonkmanns C, Ko Y, Bolt HM and Abel J (1999) Detection of cytochrome P450 *1B1 BfrI* polymorphism: genotype distribution in healthy German individuals and in patients with colorectal carcinoma. *Pharmacogenetics* 9: 405–408.
88. Tang YM, Green BL, Chen GF, Thompson PA, Lang NP, Shinde A, Lin DX, Tan W, Lyn-Cook BD, Hammons GJ and Kadlubar FF (2000) Human *CYP1B1* $Leu^{432}Val$ gene polymorphism: ethnic distribution in African-Americans, Caucasians and Chinese; oestradiol hydroxylase activity; and distribution in prostate cancer cases and controls. *Pharmacogenetics* 10: 761–766.
89. Zheng W, Xie DW, Jin F, Cheng JR, Dai Q, Wen WQ, Shu XO and Gao YT (2000) Genetic polymorphism of cytochrome P450-1B1 and risk of breast cancer. *Cancer Epidemiol. Biomarkers Prev.* 9: 147–150.
90. Ko Y, Abel J, Harth V, Bröde P, Anthony C, Donat S, Fischer HP, Ortiz-Pallardo ME, Thier R, Sachinidis A, Vetter H, Bolt HM, Herberhold C and Brüning T (2001) Association of CYP1B1 codon 432 mutant allele in head and neck squamous cell cancer is reflected by somatic mutations of p53 in tumor tissue. *Cancer Res.* 61: 4398–4404.
91. Thier R, Brüning T, Roos PH and Bolt HM (2002) Cytochrome P450 1B1, a new keystone in gene-environment interactions related to human head and neck cancer? *Arch. Toxicol.* 76: 249–256.
92. Saintot M, Malaveille C, Hautefeuille A and Gerber M (2003) Interactions between genetic polymorphism of cytochrome P450-1B1, sulfotransferase 1A1, catechol-*O*-methyltransferase and tobacco exposure in breast cancer risk. *Int. J. Cancer* 107: 652–657.
93. http://www.imm.ki.se/CYPalleles/cyb1b1.htm; (date: October 12, 2003).

94. Shimada T, Watanabe J, Kawajiri K, Sutter TR, Guengerich FP, Gillam EMJ and Inoue K (1999) Catalytic properties of polymorphic human cytochrome P450 1B1 variants. *Carcinogenesis* 20: 1607–1613.
95. Li DN, Seidel A, Pritchard MP, Wolf CR and Friedberg T (2000) Polymorphisms in P450 *CYP1B1* affect the conversion of estradiol to the potentially carcinogenic metabolite 4-hydroxyestradiol. *Pharmacogenetics* 10: 343–353.
96. Shimada T, Oda Y, Gillam EMJ, Guengerich FP and Inoue K (2001) Metabolic activation of polycyclic aromatic hydrocarbons and other procarcinogens by cytochromes P450 1A1 and P450 1B1 allelic variants and other human cytochromes P450 in *Salmonella typhimurium* NM2009. *Drug Metab. Disp.* 29: 1176–1182.
97. Shimada T, Watanabe J, Inoue K, Guengerich FP and Gillam EMJ (2001) Specificity of 17β-oestradiol and benzo[*a*]pyrene oxidation by polymorphic human cytochrome P4501B1 variants substituted at residues 48, 119 and 432. *Xenobiotica* 31: 163–176.
98. Mammen JS, Pittman GS, Li Y, Abou–Zahr F, Bejjani BA, Bell DA, Strickland PT and Sutter TR (2003) Single amino acid mutations, but not common polymorphisms, decrease the activity of CYP1B1 against (–)benzo-[*a*]pyrene-7R-*trans*-7,8-dihydrodiol. *Carcinogenesis* 24: 1247–1255.
99. Li D, Walcott FL, Chang P, Zhang W, Zhu J, Petrulis E, Singletary SE, Sahin AA and Bondy ML (2002) Genetic and environmental determinants on tissue response to *in vitro* carcinogen exposure and risk of breast cancer. *Cancer Res.* 62: 4566–4570.
100. Hasset C, Aicher L, Sidhu JS and Omiecinski CJ (1994) Human microsomal epoxide hydrolase: genetic polymorphism and functional expression *in vitro* of amino acid variants. *Hum. Mol. Genet.* 3: 421–428.
101. Omiecinski CJ, Hassett C and Hosagrahara V (2000) Epoxide hydrolase — polymorphism and role in toxicology. *Toxicol. Lett.* 112–113: 365–370.
102. Pastorelli R, Guanci M, Cerri A, Negri E, La Vecchia C, Fumagalli F, Mezzetti M, Cappelli R, Panigalli T, Fanelli R and Airoldi L (1998) Impact of inherited polymorphisms in glutathione *S*-transferase M1, microsomal epoxide hydrolase, cytochrome P450 enzymes on DNA, and blood protein adducts of benzo[*a*]pyrene-diolepoxide. *Cancer Epidemiol. Biomarkers Prev.* 7: 703–709.
103. Viezzer C, Norppa H, Clonfero E, Gabbani G, Mastrangelo G, Hirvonen A and Celotti L (1999) Influence of *GSTM1*, *GSTT1*, *GSTP1*, and *EPHX* gene

polymorphisms on DNA adduct level and *HPRT* mutant frequency in coke-oven workers. *Mutat. Res.* 431: 259–269.
104. Lee WJ, Brennan P, Boffetta P, London SJ, Benhamou S, Rannug A, To-Figueras J, Ingelman-Sundberg M, Shields P, Gaspari L and Taioli E (2002) Microsomal epoxide hydrolase polymorphisms and lung cancer risk: a quantitative review. *Biomarkers* 7: 230–241.
105. Traver RD, Siegel D, Beall HD, Phillips RM, Gibson NW, Franklin WA and Ross D (1997) Characterization of a polymorphism in NAD(P)H: quinone oxidoreductase (DT-diaphorase). *Br. J. Cancer* 75: 69–75.
106. Ozawa S, Schoket B, McDaniel LP, Tang YM, Ambrosone CB, Kostic S, Vincze I and Kadlubar FF (1999) Analyses of bronchial bulky DNA adduct levels and *CYP2C9*, *GSTP1* and *NQO1* genotypes in a Hungarian study population with pulmonary diseases. *Carcinogenesis* 20: 991–995.
107. Slaga TJ, Bracken WM, Viaje A, Berry DL, Fischer SM, Miller DR, Levin W, Conney AH, Yagi H and Jerina DM (1978) Tumor initiating and promoting activities of various benzo[*a*]pyrene metabolites in mouse skin. In: *Polynuclear Aromatic Hydrocarbons* [Jones PW and Freudenthal RI (eds.)] Volume 3, Carcinogenesis — A Comprehensive Survey, pp 371–382, Raven Press, New York.
108. Sundberg K, Widersten M, Seidel A, Mannervik B and Jernström B (1997) Glutathione conjugation of bay- and fjord-region diol epoxides of polycyclic aromatic hydrocarbons by glutathione transferases M1-1 and P1-1. *Chem. Res. Toxicol.* 10: 1221–1227.
109. Hu X, O'Donell R, Srivastava SK, Xia H, Zimniak P, Nanduri B, Bleicher RJ, Awasthi S, Awasthi YC, Ji X and Singh SV (1997) Active site architecture of polymorphic forms of human glutathione *S*-transferase P1-1 accounts for their enantioselectivity and disparate activity in the glutathione conjugation of 7β,8α-dihydroxy-9α,10α-oxy-7,8,9,10-tetrahydrobenzo[*a*]pyrene. *Biochem. Biophys. Res. Commun.* 235: 424–428.
110. Hu X, Xia H, Srivastava SK, Pal A, Awasthi YC, Zimniak P and Singh SV (1998) Catalytic efficiencies of allelic variants of human glutathione *S*-transferase P1-1 toward carcinogenic anti-diol epoxides of benzo[*c*]-phenanthrene and benzo[*g*]chrysene. *Cancer Res.* 58: 5340–5343.
111. Dreij K, Sundberg K, Johansson AS, Nordling E, Seidel A, Persson B, Mannervik B and Jernström B (2002) Catalytic activities of human alpha class glutathione transferases toward carcinogenic dibenzo[*a*,*l*]pyrene diol epoxides. *Chem. Res. Toxicol.* 15: 825–831.

112. Sundberg K, Dreij K, Seidel A and Jernström B (2002) Glutathione conjugation and DNA adduct formation of dibenzo[*a,l*]pyrene and benzo[*a*]pyrene diol epoxides in V79 cells stably expressing different human glutathione transferases. *Chem. Res. Toxicol.* 15: 170–179.
113. Houlston RS (1999) Glutathione *S*-transferase M1 status and lung cancer risk: a meta-analysis. *Cancer Epidemiol. Biomarkers Prev.* 8: 675–682.
114. Hayes JD and Strange RC (2000) Glutathione *S*-transferase polymorphisms and their biological consequences. *Pharmacology* 61: 154–166.
115. Strange RC, Jones PW and Fryer AA (2000) Glutathione *S*-transferase: genetics and role in toxicology. *Toxicol. Lett.* 112–113: 357–363.
116. Nazar-Stewart V, Vaughan TL, Stapleton P, Van Loo J, Nicol-Blades B and Eaton DL (2003) A population-based study of glutathione *S*-transferase M1, T1 and P1 genotypes and risk for lung cancer. *Lung Cancer* 40: 247–258.
117. Ryberg D, Skaug V, Hewer A, Phillips DH, Harries LW, Wolf CR, Øgreid D, Ulvik A, Vu P and Haugen A (1997) Genotypes of glutathione transferase M1 and P1 and their significance for lung DNA adduct levels and cancer risk. *Carcinogenesis* 18: 1285–1289.
118. Butkiewicz D, Cole KJ, Phillips DH, Harris CC and Chorazy M (1999) *GSTM1*, *GSTP1*, *CYP1A1* and *CYP2D6* polymorphisms in lung cancer patients from an environmentally polluted region of Poland: correlation with lung DNA adduct levels. *Eur. J. Cancer Prev.* 8: 315–323.
119. Cheng YW, Chen CY, Lin P, Chen CP, Huang KH, Lin TS, Wu MH and Lee H (2000) DNA adduct level in lung tissue may act as a risk biomarker of lung cancer. *Eur. J. Cancer Prev.* 36: 1381–1388.
120. Alexandrov K, Cascorbi I, Rojas M, Bouvier G, Kriek E and Bartsch H (2002) *CYP1A1* and *GSTM1* genotypes affect benzo[*a*]pyrene DNA adducts in smokers' lung: comparison with aromatic/hydrophobic adduct formation. *Carcinogenesis* 23: 1969–1977.
121. Gonzalez FJ (1997) The role of carcinogen-metabolizing enzyme polymorphisms in cancer susceptibility. *Reprod. Toxicol.* 11: 397–412.
122. Strange RC and Fryer AA (1999) The glutathione *S*-transferases: influence of polymorphism on cancer susceptibility. In: *Metabolic Polymorphisms and Susceptibility to Cancer* [Vineis P, Malats N, Lang M, d'Errico A, Caporaso N, Cuzick J and Boffetta P (eds.)] IARC Scientific Publications No. 148, pp 231–249, International Agency for Research on Cancer, Lyon, France.

123. Bartsch H, Nair U, Risch A, Rojas M, Wikman H and Alexandrov K (2000) Genetic polymorphism of *CYP* genes, alone or in combination, as a risk modifier of tobacco-related cancers. *Cancer Epidemiol. Biomarkers Prev.* 9: 3–28.
124. Anttila S, Hirvonen A, Vainio H, Husgafvel-Pursiainen K, Hayes JD and Ketterer B (1993) Immunohistochemical localization of glutathione *S*-transferases in human lung. *Cancer Res.* 53: 5643–5648.
125. Hu X, Xia H, Srivastava SK, Herzog C, Awasthi YC, Ji X, Zimniak P and Singh SV (1997) Activity of four allelic forms of glutathione *S*-transferase hGSTP1-1 for diol epoxides of polycyclic aromatic hydrocarbons. *Biochem. Biophys. Res. Commun.* 238: 397–402.
126. Hu X, Ji X, Srivastava SK, Xia H, Awasthi S, Nanduri B, Awasthi YC, Zimniak P and Singh SV (1997) Mechanism of differential catalytic efficiency of two polymorphic forms of human glutathione *S*-transferase P1-1 in the glutathione conjugation of carcinogenic diol epoxide of chrysene. *Arch. Biochem. Biophys.* 345: 32–38.
127. Sundberg K, Johansson AF, Stenberg G, Widersten M, Seidel A, Mannervik B and Jernström B (1998) Differences in the catalytic efficiencies of allelic variants of glutathione transferase P1-1 towards carcinogenic diol epoxides of polycyclic aromatic hydrocarbons. *Carcinogenesis* 19: 433–436.
128. Sundberg K, Seidel A, Mannervik B and Jernström B (1998) Detoxication of carcinogenic fjord-region diol epoxides of polycyclic aromatic hydrocarbons by glutathione transferase P1-1 variants and glutathione. *FEBS Lett.* 438: 206–210.
129. Henderson CJ, Smith AG, Ure J, Brown K, Bacon EJ and Wolf CR (1998) Increased skin tumorigenesis in mice lacking pi class glutathione *S*-transferases. *Proc. Natl. Acad. Sci. USA* 95: 5275–5280.
130. Waston MA, Stewart RK, Smith GBJ, Massey TE and Bell DA (1998) Human glutathione *S*-transferase P1 polymorphisms: relationship to lung tissue enzyme activity and population frequency distribution. *Carcinogenesis* 19: 275–280.
131. Jourenkova-Mironova N, Wikman H, Bouchardy C, Voho A, Dayer P, Benhamou S and Hirvonen A (1998) Role of glutathione *S*-transferase *GSTM1*, *GSTM3*, *GSTP1*, and *GSTT1* genotypes in modulating susceptibility to smoking-related lung cancer. *Pharmacogenetics* 8: 495–502.
132. Katoh T, Kaneko S, Takasawa S, Nagata N, Inatomi H, Ikemura K, Itoh H, Matsumoto T, Kawamoto T and Bell DA (1999) Human glutathione

S-transferase P1 polymorphism and susceptibility to smoking related epithelial cancer; oral lung, gastric, colorectal and urothelial cancer. *Pharmacogenetics* 9: 165–169.

133. To-Figueras J, Gené M, Gómez-Catalán J, Piqué E, Borrego N, Carrasco JL, Ramón J and Corbella J (1999) Genetic polymorphism of glutathione *S*-transferase P1 gene and lung cancer. *Cancer Causes Control* 10: 65–70.
134. Wang Y, Spitz MR, Schabath MB, Ali-Osman F, Mata H and Wu X (2003) Association between glutathione *S*-transferase p1 polymorphisms and lung cancer risk in Caucasians: a case-control study. *Lung Cancer* 40: 25–32.
135. Stücker I, Hirvonen A, de Waziers I, Cabelguenne A, Mitrunen K, Cénée S, Koum-Besson E, Hémon D, Beaune P and Loriot MA (2002) Genetic polymorphisms of glutathione *S*-transferases as modulators of lung cancer susceptibility. *Carcinogenesis* 23: 1475–1481.
136. Whyatt RM, Perera FP, Jedrychowski W, Santella RM, Garte S and Bell DA (2000) Association between polycyclic aromatic hydrocarbon-DNA adduct levels in maternal and newborn white blood cells and *glutathione S-transferase P1* and *CYP1A1* polymorphisms. *Cancer Epidemiol. Biomarkers Prev.* 9: 207–212.
137. Phillips DH (2002) Smoking-related DNA and protein adducts in human tissues. *Carcinogenesis* 23: 1979–2004.
138. Phillips DH, Hewer A, Seidel A, Steinbrecher T, Schrode R, Oesch F and Glatt HR (1991) Relationship between mutagenicity and DNA adduct formation in mammalian cells for fjord- and bay-region diol-epoxides of polycyclic aromatic hydrocarbons. *Chem.-Biol. Interact.* 80: 177–186.
139. Glatt HR, Piée A, Pauly K, Steinbrecher T, Schrode R, Oesch F and Seidel A (1991) Fjord- and bay-region diol-epoxides investigated for stability, SOS induction in *Escherichia coli*, and mutagenicity in *Salmonella typhimurium* and mammalian cells. *Cancer Res.* 51: 1659–1667.
140. Luch A, Glatt HR, Platt KL, Oesch F and Seidel A (1994) Synthesis and mutagenicity of the diastereomeric fjord-region 11,12-dihydrodiol 13,14-epoxides of dibenzo[*a*,*l*]pyrene. *Carcinogenesis* 15: 2507–2516.
141. Jerina DM, Sayer JM, Agarwal SK, Yagi H, Levin W, Wood AW, Conney AH, Pruess-Schwartz D, Baird WM, Pigott MA and Dipple A (1986) Reactivity and tumorigenicity of bay-region diol epoxides derived from polycyclic aromatic hydrocarbons. In: *Biological Reactive Intermediates III. Mechanisms of Action in Animal Models and Human Disease* [Kocsis JJ, Jollow DJ, Witmer CM, Nelson JO and Snyder R (eds.)] pp 11–30, Plenum Press, New York.

142. Jerina DM, Cheh AM, Chadha A, Yagi H and Sayer JM (1988) Binding of metabolically formed bay-region diol epoxides to DNA. In: *Microsomes and Drug Oxidations* [Miners JO, Birkett DJ, Drew R, May BK and McManus ME (eds.)] pp 354–362, Taylor & Francis, London.
143. Szeliga J and Dipple A (1998) DNA adduct formation by polycyclic aromatic hydrocarbon dihydrodiol epoxides. *Chem. Res. Toxicol.* 11: 1–11.
144. Whalen DL, Montemarano JA, Thakker DR, Yagi H and Jerina DM (1977) Changes of mechanisms and product distributions in the hydrolysis of benzo[*a*]pyrene-7,8-diol 9,10-epoxide metabolites induced by changes in pH. *J. Am. Chem. Soc.* 99: 5522–5524.
145. Whalen DL, Ross AM, Yagi H, Karle JM and Jerina DM (1978) Stereoelectronic factors in the solvolysis of bay region diol epoxides of polycyclic aromatic hydrocarbons. *J. Am. Chem. Soc.* 100: 5218–5221.
146. Geacintov NE, Yoshida H, Ibanez V and Harvey RG (1981) Non-covalent intercalative binding of 7,8-dihydroxy-9,10-epoxybenzo[*a*]pyrene to DNA. *Biochem. Biophys. Res. Commun.* 100: 1569–1577.
147. Michaud DP, Gupta SC, Whalen DL, Sayer JM and Jerina DM (1983) Effects of pH and salt concentration on the hydrolysis of a benzo[*a*]pyene 7,8-diol-9,10-epoxide catalyzed by DNA and polyadenylic acid. *Chem.-Biol. Interact.* 44: 41–52.
148. Islam NB, Gupta SC, Yagi H, Jerina DM and Whalen DL (1990) Trapping of a carbocationic intermediate in the spontaneous hydrolysis reaction of 7β,8α-dihydroxy-9β,10β-epoxy-7,8,9,10-tetrahydrobenzo[*a*]pyrene: mechanism of the spontaneous and general acid catalyzed hydrolysis reactions of bay-region benzo[*a*]pyrene 7,8-diol 9,10-epoxides. *J. Am. Chem. Soc.* 112: 6363–6369.
149. Sayer JM, Yagi H, Silverton JV, Friedman SL, Whalen DL and Jerina DM (1982) Conformational effects in the hydrolyses of rigid benzylic epoxides: implications for diol epoxides of polycyclic hydrocarbons. *J. Am. Chem. Soc.* 104: 1972–1978.
150. Sayer JM, Yagi H, Croisy–Delcey M and Jerina DM (1981) Novel bay-region diol epoxides from benzo[*c*]phenanthrene. *J. Am. Chem. Soc.* 103: 4970–4972.
151. Bushman DR, Grossman SJ, Jerina DM and Lehr RE (1989) Synthesis of optically active fjord-region 11,12-diol 13,14-epoxides and the K-region 9,10-oxide of the carcinogen benzo[*g*]chrysene. *J. Org. Chem.* 54: 3533–3544.

152. Harvey RG (1989) Conformation analysis of hydroaromatic metabolites of carcinogenic hydrocarbons and the relation of conformation to biological activity. In: *Conformational Analysis of Cyclohexenes, Cyclohexadienes, and Related Hydroaromatic Compounds* [Rabideau PW (ed.)] pp 268–298, VCH, New York.
153. Sims P, Grover PL, Swaisland A, Pal K and Hewer A (1974) Metabolic activation of benzo[*a*]pyrene proceeds by a diol-epoxide. *Nature* 252: 326–328.
154. Melendez-Colon VJ, Luch A, Seidel A and Baird WM (1999) Cancer initiation by polycyclic aromatic hydrocarbons results from formation of stable DNA adducts rather than apurinic sites. *Carcinogenesis* 20: 1885–1891.
155. Melendez-Colon VJ, Luch A, Seidel A and Baird WM (2000) Formation of stable DNA adducts and apurinic sites upon metabolic activation of bay and fjord region polycyclic aromatic hydrocarbons in human cell cultures. *Chem. Res. Toxicol.* 13: 10–17.
156. Pruess-Schwartz D, Sebti SM, Gilham PT and Baird WM (1984) Analysis of benzo[*a*]pyrene: DNA adducts formed in cells in culture by immobilized boronate chromatography. *Cancer Res.* 44: 4104–4110.
157. Bigger CAH, Sawicki JT, Blake DM, Raymond LG and Dipple A (1983) Products of binding of 7,12-dimethylbenz[*a*]anthracene to DNA in mouse skin. *Cancer Res.* 43: 5647–5651.
158. Sawicki JT, Moschel RC and Dipple A (1983) Involvement of both *syn*- and *anti*-dihydrodiol-epoxides in the binding of 7,12-dimethylbenz[*a*]anthracene to DNA in mouse embryo cell cultures. *Cancer Res.* 43: 3212–3218.
159. Pruess-Schwartz D, Baird WM, Yagi H, Jerina DM, Pigott MA and Dipple A (1987) Stereochemical specificity in the metabolic activation of benzo[*c*]phenanthrene to metabolites that covalently bind to DNA in rodent embryo cell cultures. *Cancer Res.* 47: 4032–4037.
160. Cheng SC, Prakash AS, Pigott MA, Hilton BD, Roman JM, Lee H, Harvey RG and Dipple A (1988) Characterization of 7,12-dimethylbenz[*a*]-anthracene-adenine nucleoside adducts. *Chem. Res. Toxicol.* 1: 216–221.
161. Vericat JA, Cheng SC and Dipple A (1989) Absolute stereochemistry of the major 7,12-dimethylbenz[*a*]anthracene-DNA adducts formed in mouse cells. *Carcinogenesis* 10: 567–570.
162. Einholf HJ, Amin S, Yagi H, Jerina DM and Baird WM (1996) Benzo[*c*]-phenanthrene is activated to DNA-binding diol epoxides in the human mammary carcinoma cell line MCF-7 but only limited activation occurs in mouse skin. *Carcinogenesis* 17: 2237–2244.

163. Dipple A (1994) Reactions of polycyclic aromatic hydrocarbons with DNA. In: *DNA Adducts: Identification and Biological Significance* [Hemminki K, Dipple A, Shuker DEG, Kadlubar FF, Segerbäck D and Bartsch H (eds.)] IARC Scientific Publication No. 125, pp 107–129, International Agency for Research on Cancer, Lyon.
164. Dipple A, Khan QA, Page JE, Pontén I and Szeliga J (1999) DNA reactions, mutagenic action and stealth properties of polycyclic aromatic hydrocarbon carcinogens. *Int. J. Oncol.* 14: 103–111.
165. Jeffrey AM, Jennette KW, Blobstein SH, Weinstein IB, Beland FA, Harvey RG, Kasai H, Miura I and Nakanishi K (1976) Benzo[*a*]pyrene-nucleic acid derivative found *in vivo*: structure of a benzo[*a*]pyrenetetrahydrodiol epoxide-guanosine adduct. *J. Am. Chem. Soc.* 98: 5714–5715.
166. Meehan T, Straub K and Calvin M (1977) Benzo[*a*]pyrene diol epoxide covalently binds to deoxyguanosine and deoxyadenosine in DNA. *Nature* 269: 725–727.
167. Koreeda M, Moore PD, Wislocki PG, Levin W, Conney AH, Yagi H and Jerina DM (1978) Binding of benzo[*a*]pyrene 7,8-diol-9,10-epoxides to DNA, RNA, and protein of mouse skin occurs with high stereoselectivity. *Science* 199: 778–781.
168. Meehan T and Straub K (1979) Double-stranded DNA stereoselectively binds benzo[*a*]pyrene diol epoxides. *Nature* 277: 410–412.
169. Straub KM, Meehan T, Burlingame AL and Calvin M (1977) Identification of the major adducts formed by reaction of benzo[*a*]pyrene diol epoxide in DNA *in vitro*. *Proc. Natl. Acad. Sci. USA* 74: 5285–5289.
170. Osborne MR, Harvey RG and Brookes P (1978) The reaction of *trans*-7,8-dihydroxy-*anti*-9,10-epoxy-7,8,9,10-tetrahydrobenzo[*a*]pyrene with DNA involves attack at the N^7-position of guanine moieties. *Chem.-Biol. Interact.* 20: 123–130.
171. Osborne MR, Jacobs S, Harvey RG and Brookes P (1981) Minor products from the reaction of (+)- and (–)-benzo[*a*]pyrene-anti-diolepoxide with DNA. *Carcinogenesis* 2: 553–558.
172. Sayer JM, Chadha A, Agarwal SK, Yeh HJC, Yagi H and Jerina DM (1991) Covalent nucleoside adducts of benzo[*a*]pyrene 7,8-diol 9,10-epoxides: structural reinvestigation and characterization of a novel adenosine adduct on the ribose moiety. *J. Org. Chem.* 56: 20–29.
173. Dipple A, Pigott MA, Agarwal SK, Yagi H, Sayer JM and Jerina DM (1987) Optically active benzo[*c*]phenanthrene diol epoxides bind extensively to adenine in DNA. *Nature* 327: 535–536.

174. Hirshfeld FL (1963) The structure of overcrowded aromatic compounds. Part VII. Out-of plane deformation in benzo[*c*]phenanthrene and 1,12-dimethylbenzo[*c*]phenanthrene. *J. Chem. Soc.* 2126–2135.
175. Iball J (1964) A refinement of the structure of 9:10-dimethyl-1:2-benzanthracene. *Nature* 201: 916–917.
176. Zacharias DE, Kashino S, Glusker JP, Harvey RG, Amin S and Hecht SS (1984) The bay-region geometry of some 5-methylchrysenes: steric effects in 5,6- and 5,12-dimethylchrysenes. *Carcinogenesis* 5: 1421–1430.
177. Agarwal SK, Sayer JM, Yeh HJC, Pannell LK, Hilton BD, Pigott MA, Dipple A, Yagi H and Jerina DM (1987) Chemical characterization of DNA adducts derived from the configurationally isomeric benzo[*c*]phenanthrene-3,4-diol 1,2-epoxides. *J. Am. Chem. Soc.* 109: 2497–2504.
178. Cheng SC, Hilton BD, Roman JM and Dipple A (1989) DNA adducts from carcinogenic and noncarcinogenic enantiomers of benzo[*a*]pyrene dihydrodiol epoxide. *Chem. Res. Toxicol.* 2: 334–340.
179. Doan L, Lin B, Yagi H, Jerina DM and Whalen DL (2001) New insights on the mechanisms of the pH-independent reactions of benzo[*a*]pyrene 7,8-diol 9,10-epoxides. *J. Am. Chem. Soc.* 123: 6785–6791.
180. Conney AH, Chang RL, Jerina DM and Wei SJC (1994) Studies on the metabolism of benzo[*a*]pyrene and dose-dependent differences in the mutagenic profile of its ultimate carcinogenic metabolite. *Drug Metab. Rev.* 26: 125–163.
181. Conney AH, Chang RL, Cui XX, Schiltz M, Yagi H, Jerina DM and Wei SJ (2001) Dose-dependent differences in the profile of mutations induced by carcinogenic (R,S,S,R) bay- and fjord-region diol epoxides of polycyclic aromatic hydrocarbons. *Adv. Exp. Med. Biol.* 500: 697–707.
182. Wei SJ, Chang RL, Wong CQ, Bhachech N, Cui XX, Hennig E, Yagi H, Sayer JM, Jerina DM, Preston BD and Conney AH (1991) Dose-dependent differences in the profile of mutations induced by an ultimate carcinogen from benzo[*a*]pyrene. *Proc. Natl. Acad. Sci. USA* 88: 11227–11230.
183. Wei SJ, Chang RL, Cui XX, Merkler KA, Wong CQ, Yagi H, Jerina DM and Conney AH (1996) Dose-dependent differences in the mutational profiles of (–)-(1*R*,2*S*,3*S*,4*R*)-3,4-dihydroxy-1,2-epoxy-1,2,3,4-tetrahydrobenzo[*c*]phenanthrene and its less carcinogenic enantiomer. *Cancer Res.* 56: 3695–3703.
184. Mahadevan B, Dashwood WM, Luch A, Pecaj A, Doehmer J, Seidel A, Pereira C and Baird WM (2003) Mutations induced by (–)-*anti*-11*R*,12*S*-dihydrodiol 13*S*,14*R*-epoxide of dibenzo[*a*,*l*]pyrene in the coding region of

the hypoxanthine phosphoribosyltransferase (*hprt*) gene in Chinese hamster V79 cells. *Environ. Mol. Mutagen.* 41: 131–139.

185. Wei SJ, Chang RL, Hennig E, Cui XX, Merkler KA, Wong CQ, Yagi H, Jerina DM and Conney AH (1994) Mutagenic selectivity at the *HPRT* locus in V79 cells: comparison of mutations caused by bay-region benzo[*a*]pyrene 7,8-diol-9,10-epoxide enantiomers with high and low carcinogenic activity. *Carcinogenesis* 15: 1729–1735.
186. Schiltz M, Cui XX, Lu YP, Yagi H, Jerina DM, Zdzienicka Z, Chang RL, Conney AH and Wei SJ (1999) Characterization of the mutational profile of (+)-7*R*,8*S*-dihydroxy-9*S*,10*R*-epoxy-7,8,9,10-tetrahydrobenzo[*a*]pyrene at the hypoxanthine (guanine) phosphoribosyltransferase gene in repair-deficient Chinese hamster V-H1 cells. *Carcinogenesis* 20: 2279–2285.
187. Kadkhodayan S, Salazar EP, Ramsey MJ, Takayama K, Zdzienicka MZ, Tucker JD and Weber CA (1997) Molecular analysis of *ERCC2* mutations in the repair deficient hamster mutants UVL-1 and V-H1. *Mutat. Res.* 385: 47–57.
188. Geacintov NE, Broyde S, Buterin T, Naegeli H, Wu M, Yan S and Patel DJ (2002) Thermodynamic and structural factors in the removal of bulky DNA adducts by the nucleotide excision repair machinery. *Biopolymers* 65: 202–210.
189. Bos JL (1989) *Ras* oncogenes in human cancer: a review. *Cancer Res.* 49: 4682–4689.
190. Anderson MW, You M and Reynolds SH (1991) Proto-oncogene activation in rodent and human tumors. In: *Biological Reactive Intermediates IV. Molecular and Cellular Effects and Their Impact on Human Health* [Witmer CM, Snyder R, Jollow DJ, Kalf GF, Kocsis JJ and Sipes IG (eds.)] pp 235–243, Plenum Press, New York.
191. Ross JA and Nesnow S (1999) Polycyclic aromatic hydrocarbons: correlations between DNA adducts and *ras* oncogene mutations. *Mutat. Res.* 424: 155–166.
192. Hanahan D and Weinberg RA (2000) The hallmarks of cancer. *Cell* 100: 57–70.
193. Colapietro AM, Goodell AL and Smart RC (1993) Characterization of benzo[*a*]pyrene-initiated mouse skin papillomas for Ha-*ras* mutations and protein kinase C levels. *Carcinogenesis* 14: 2289–2295.
194. Wei SJ, Chang RL, Merkler KA, Gwynne M, Cui XX, Murthy B, Huang MT, Xie JG, Lu YP, Lou YR, Jerina DM and Conney AH (1999) Dose-dependent mutation profile in the c-Ha-*ras* proto-oncogene of skin tumors in mice initiated with benzo[*a*]pyrene. *Carcinogenesis* 20: 1689–1696.

195. Quintanilla M, Brown K, Ramsden M and Balmain A (1986) Carcinogen-specific mutation and amplification of Ha-*ras* during mouse skin carcinogenesis. *Nature* 322: 78–80.

196. Tang MS, Vulimiri SV, Viaje A, Chen JX, Bilolikar DS, Morris RJ, Harvey RG, Slaga TJ and DiGiovanni J (2000) Both (±)*syn*- and (±)*anti*-7,12-dimethylbenz[*a*]anthracene-3,4-diol-1,2-epoxides initiate tumors in mouse skin that possess –CAA– to –CTA– mutations at codon 61 of c-H-*ras*. *Cancer Res.* 60: 5688–5695.

197. Chakravarti D, Pelling JC, Cavalieri EL and Rogan EG (1995) Relating aromatic hydrocarbon-induced DNA adducts and c-H-*ras* mutations in mouse skin papillomas: the role of apurinic sites. *Proc. Natl. Acad. Sci. USA* 92: 10422–10426.

198. Chakravarti D, Mailander P, Franzen J, Higginbotham S, Cavalieri EL and Rogan EG (1998) Detection of dibenzo[*a*,*l*]pyrene-induced H-*ras* codon 61 mutant gene in preneoplastic SENCAR mouse skin using a new PCR-RFLP method. *Oncogene* 16: 3203–3210.

199. Loktionov A, Hollstein M, Martel N, Galendo D, Cabral JRP, Tomatis L and Yamasaki H (1990) Tissue-specific activating mutations of Ha- and Ki-*ras* oncogenes in skin, lung, and liver tumors induced in mice following transplacental exposure to DMBA. *Mol. Carcinog.* 3: 134–140.

200. Prahalad AK, Ross JA, Nelson GB, Roop BC, King LC, Nesnow S and Mass MJ (1997) Dibenzo[*a*,*l*]pyrene-induced DNA adduction, tumorigenicity, and Ki-*ras* oncogene mutations in strain A/J mouse lung. *Carcinogenesis* 18: 1955–1963.

201. Nesnow S, Ross JA, Mass MJ and Stoner GD (1998) Mechanistic relationships between DNA adducts, oncogene mutations, and lung tumorigenesis in strain A mice. *Exp. Lung. Res.* 24: 395–405.

202. Slebos RJC, Hruban RH, Dalesio O, Mooi WJ, Offerhaus GJA and Rodenhuis S (1991) Relationship between K-ras oncogene activation and smoking in adenocarcinoma of the lung. *J. Nat. Cancer Inst.* 83: 1024–1027.

203. Husgafvel-Pursiainen K, Hackman P, Ridanpää M, Anttila S, Karjalainen A, Partanen T, Taikina-Aho O, Heikkilä L and Vainio H (1993) K-*ras* mutations in human adenocarcinoma of the lung: association with smoking and occupational exposure to asbestos. *Int. J. Cancer* 53: 250–256.

204. Feng Z, Hu W, Chen JX, Pao A, Li H, Rom W, Hung MC and Tang MS (2002) Preferential DNA damage and poor repair determine ras gene mutational hotspot in human cancer. *J. Nat. Cancer Inst.* 94: 1527–1536.

205. Hollstein M, Shomer B, Greenblatt M, Soussi T, Hovig E, Montesano R and Harris CC (1996) Somatic point mutations in the *p53* gene of human tumors and cell lines: updated compilation. *Nucleic Acids Res.* 24: 141–146.
206. Hussain SP and Harris CC (1998) Molecular epidemiology of human cancer: contribution of mutation spectra studies of tumor suppressor genes. *Cancer Res.* 58: 4023–4037.
207. Hartwell LH and Kastan MB (1995) Cell cycle control and cancer. *Science* 266: 1821–1828.
208. Luch A (2002) Cell cycle control and cell division: implications for chemically induced carcinogenesis. *ChemBioChem* 3: 506–516.
209. Ford JM and Hanawalt PC (1995) Li–Fraumeni syndrome fibroblasts homozygous for p53 mutations are deficient in global DNA repair but exhibit normal transcription-coupled repair and enhanced UV resistance. *Proc. Natl. Acad. Sci. USA* 92: 8876–8880.
210. Lloyd DR and Hanawalt PC (2002) P53 controls global nucleotide excision repair of low levels of structurally diverse benzo[*g*]chrysene-DNA adducts in human fibroblasts. *Cancer Res.* 62: 5288–5294.
211. Adimoolam S and Ford JM (2002) P53 and DNA damage-inducible expression of the xeroderma pigmentosum group C gene. *Proc. Natl. Acad. Sci. USA* 99: 12985–12990.
212. Wani MA, El-Mahdy MA, Hamada FM, Wani G, Zhu Q, Wang Q and Wani AA (2002) Efficient repair of bulky anti-BPDE DNA addcuts from non-transcribed DNA strand requires functional p53 but not $p21^{waf1/cip1}$ and pRb. *Mutat. Res.* 505: 13–25.
213. El-Deiry WS, Harper JW, O'Connor PM, Velculescu VE, Canman CE, Jackman J, Pietenpol JA, Burrell M, Hill DE, Wang Y, Wiman KG, Mercer WE, Kastan MB, Kohn KW, Elledge SJ, Kinzler KW and Vogelstein B (1994) *WAF1/CIP1* is induced in *p53*-mediated G_1 arrest and apoptosis. *Cancer Res.* 54: 1169–1174.
214. Bates S and Vousden KH (1999) Mechanisms of p53-mediated apoptosis. *Cell. Mol. Life Sci.* 55: 28–37.
215. Szeliga J, Lee H, Harvey RG, Page JE, Ross HL, Routledge MN, Hilton BD and Dipple A (1994) Reaction with DNA and mutagenic specificity of *syn*-benzo[*g*]chrysene 11,12-dihydrodiol 13,14-epoxide. *Chem. Res. Toxicol.* 7: 420–427.
216. Szeliga J, Page JE, Hilton BD, Kiselyov AS, Harvey RG, Dunayevskiy YM, Vouros P and Dipple A (1995) Characterization of DNA adducts formed by *anti*-benzo[*g*]chrysene 11,12-dihydrodiol 13,14-epoxide. *Chem. Res. Toxicol.* 8: 1014–1019.

217. Chen JX, Zheng Y, West M and Tang MS (1998) Carcinogens preferentially bind at methylated CpG in the p53 mutational hot spots. *Cancer Res.* 58: 2070–2075.
218. Smith LE, Denissenko MF, Bennett WP, Li H, Amin S, Tang MS and Pfeifer GP (2000) Targeting of lung cancer mutational hotspots by polycyclic aromatic hydrocarbons. *J. Natl. Cancer Inst.* 92: 803–811.
219. Hu W, Feng Z and Tang MS (2003) Preferential carcinogen-DNA adduct formation at codons 12 and 14 in the human *K-ras* gene and their possible mechanisms. *Biochemistry* 42: 10012–10023.
220. Loechler EL (1996) The role of adduct site-specific mutagenesis in understanding how carcinogen-DNA adducts cause mutations: perspective, prospects and problems. *Carcinogenesis* 17: 895–902.
221. Seo KY, Jelinsky SA and Loechler EL (2000) Factors that influence the mutagenic patterns of DNA adducts from chemical carcinogens. *Mutat. Res.* 463: 215–246.
222. Bigger CAH, Pontén I, Page JE and Dipple A (2000) Mutational spectra for polycyclic aromatic hydrocarbons in the *supF* target gene. *Mutat. Res.* 450: 75–93.
223. Moriya M, Spiegel S, Fernandes A, Amin S, Liu T, Geacintov N and Grollman AP (1996) Fidelity of translesional synthesis past benzo[*a*]pyrene diol epoxide-2′-deoxyguanosine DNA adducts: marked effects of host cell, sequence context, and chirality. *Biochemistry* 35: 16646–16651.
224. Page JE, Zajc B, Oh-hara T, Lakshman MK, Sayer JM, Jerina DM and Dipple A (1998) Sequence context profoundly influences the mutagenic potency of trans-opened benzo[*a*]pyrene 7,8-diol 9,10-epoxide-purine nucleoside adducts in site-specific mutations studies. *Biochemistry* 37: 9127–9137.
225. Shukla R, Liu T, Geacintov NE and Loechler EL (1997) The major, N^2-dG adduct of (+)-*anti*-B[*a*]PDE shows dramatically different mutagenic specificity (predominantly, G → A) in a 5′–C$\underline{G}$T–3′ sequence context. *Biochemistry* 36: 10256–10261.
226. Christner DF, Lakshman MK, Sayer JM, Jerina DM and Dipple A (1994) Primer extension by various polymerases using oligonucleotide templates containing stereoisomeric benzo[*a*]pyrene-deoxyadenosine adducts. *Biochemistry* 33: 14297–14305.
227. Chary P and Lloyd RS (1995) *In vitro* replication by prokaryotic and eukaryotic polymerases on DNA templates containing site-specific and stereospecific benzo[*a*]pyrene-7,8-dihydrodiol-9,10-epoxide adducts. *Nucleic Acids Res.* 23: 1398–1405.

228. Lehmann AR (2002) Replication of damaged DNA in mammalian cells: new solutions to an old problem. *Mutat. Res.* 509: 23–34.
229. Yang W (2003) Damage repair DNA polymerases Y. *Curr. Opin. Struct. Biol.* 13: 23–30.
230. Zhang Y, Wu X, Guo D, Rechkoblit O and Wang Z (2002) Activities of human DNA polymerase κ in response to the major benzo[*a*]pyrene DNA adduct: error-free lesion bypass and extension synthesis from opposite the lesion. *DNA Repair* 1: 559–569.
231. Huang X, Kolbanovskiy A, Wu X, Zhang Y, Wang Z, Zhuang P, Amin S and Geacintov NE (2003) Effects of base sequence context on translesion synthesis past a bulky (+)-*trans-anti*-B[*a*]P-N^2-dG lesion catalyzed by the Y-family polymerase pol κ. *Biochemistry* 42: 2456–2466.
232. Zhang Y, Wu X, Guo D, Rechkoblit O, Geacintov NE and Wang Z (2002) Two-step error-prone bypass of the (+)- and (–)-*trans-anti*-BPDE-N^2-dG adducts by human DNA polymerases η and κ. *Mutat. Res.* 510: 23–35.
233. Zhang Y, Wu X, Rechkoblit O, Geacintov NE, Taylor JS and Wang Z (2002) Response of human REV1 to different DNA damage: preferential dCMP insertion opposite the lesion. *Nucleic Acids Res.* 30: 1630–1638.
234. Buening MK, Wislocki PG, Levin W, Yagi H, Thakker DR, Akagi H, Koreeda M, Jerina DM and Conney AH (1978) Tumorigenicity of the optical enantiomers of the diastereomeric benzo[*a*]pyrene 7,8-diol-9,10-epoxides in newborn mice: exceptional activity of (+)-7β,8α-dihydroxy-9α,10α-epoxy-7,8,9,10-tetrahydrobenzo[*a*]pyrene. *Proc. Natl. Acad. Sci. USA* 75: 5358–5361.
235. Mass MJ, Jeffers AJ, Ross JA, Nelson G, Galati AJ, Stoner GD and Nesnow S (1993) Ki-*ras* oncogene mutations in tumors and DNA adducts formed by benz[*j*]aceanthrylene and benzo[*a*]pyrene in the lungs of strain A/J mice. *Mol. Carcinog.* 8: 186–192.
236. Li EE, Heflich RH, Bucci TJ, Manjanatha MG, Blaydes BS and Delclos KB (1994) Relationships of DNA adduct formation, K-*ras* activating mutations and tumorigenic activities of 6-nitrochrysene and its metabolites in the lungs of CD-1 mice. *Carcinogenesis* 15: 1377–1385.
237. Ross JA, Nelson GB, Wilson KH, Rabinowitz JR, Galati A, Stoner GD, Nesnow S and Mass MJ (1995) Adenomas induced by polycyclic aromatic hydrocarbons in strain A/J mouse lung correlate with time-integrated DNA adduct levels. *Cancer Res.* 55: 1039–1044.
238. Slaga TJ, Bracken WJ, Gleason G, Levin W, Yagi H, Jerina DM and Conney AH (1979) Marked differences in the skin tumor-initiating activities

of the optical enantiomers of the diastereomeric benzo[*a*]pyrene 7,8-diol-9,10-epoxides. *Cancer Res.* 39: 67–71.
239. Rubin H (2001) Synergistic mechanisms in carcinogenesis by polycyclic aromatic hydrocarbons and by tobacco smoke: a bio-historical perspective with updates. *Carcinogenesis* 22: 1903–1930.
240. Cavalieri EL, Higginbotham S, RamaKrishna NVS, Devanesan PD, Todorovic R, Rogan EG and Salmasi S (1991) Comparative dose-response tumorigenicity studies of dibenzo[*a*,*l*]pyrene versus 7,12-dimethylbenz[*a*]-anthracene, benzo[*a*]pyrene and two dibenzo[*a*,*l*]pyrene dihydrodiols in mouse skin and rat mammary gland. *Carcinogenesis* 12: 1939–1944.
241. Higginbotham S, RamaKrishna NVS, Johansson SL, Rogan EG and Cavalieri EL (1993) Tumor-initiating activity and carcinogenicity of dibenzo[*a*,*l*]pyrene versus 7,12-dimethylbenz[*a*]anthracene and benzo[*a*]-pyrene at low doses in mouse skin. *Carcinogenesis* 14: 875–878.
242. Miyata M, Kudo G, Lee YH, Yang TJ, Gelboin HV, Fernandez-Salguero P, Kimura S and Gonzalez FJ (1999) Targeted disruption of the microsomal epoxide hydrolase gene: microsomal epoxide hydrolase is required for the carcinogenic activity of 7,12-dimethylbenz[*a*]anthracene. *J. Biol. Chem.* 274: 23963–23968.
243. Gonzalez FJ (2001) The use of gene knockout mice to unravel the mechanisms of toxicity and chemical carcinogenesis. *Toxicol. Lett.* 120: 199–208.
244. Balmain A and Pragnell IB (1983) Mouse skin carcinomas induced *in vivo* by chemical carcinogens have a transforming Harvey-*ras* oncogene. *Nature* 303: 72–74.
245. Bizub D, Wood AW and Skalka AM (1986) Mutagenesis of the H-*ras* oncogene in mouse skin tumors induced by polycyclic aromatic hydrocarbons. *Proc. Natl. Acad. Sci. USA* 83: 6048–6052.
246. Franza RB, Maruyama K, Garrels JI and Ruley HE (1986) *In vitro* establishment is not a sufficient prerequisite for transformation by activated ras oncogenes. *Cell* 44: 409–418.
247. Rubin H and Xu K (1989) Evidence for the progressive and adaptive nature of spontaneous transformation in the NIH 3T3 cell line. *Proc. Natl. Acad. Sci. USA* 86: 1860–1864.
248. Shimizu Y, Nakatsuru Y, Ichinose M, Takahashi Y, Kume H, Mimura J, Fujii-Kuriyama Y and Ishikawa T (2000) Benzo[*a*]pyrene carcinogenicity is lost in mice lacking the aryl hydrocarbon receptor. *Proc. Natl. Acad. Sci. USA* 97: 779–782.

249. Poland A and Knutson JC (1982) 2,3,7,8-Tetrachlorodibenzo-*p*-dioxin and related halogenated aromatic hydrocarbons: examinations of the mechanism of toxicity. *Annu. Rev. Pharmacol. Toxicol.* 22: 517–554.
250. Nebert DW, Puga A and Vasiliou V (1993) Role of the Ah receptor and the dioxin-inducible [*Ah*] gene battery in toxicity, cancer, and signal transduction. *Ann. N.Y. Acad. Sci.* 685: 624–640.
251. Poland A and Glover E (1979) An estimate of the maximum *in vivo* covalent binding of 2,3,7,8-tetrachlorodibenzo-*p*-dioxin to rat liver protein, ribosomal RNA, and DNA. *Cancer Res.* 39: 3341–3344.
252. Geiger LE and Neal RA (1981) Mutagenicity testing of 2,3,7,8-tetrachloro-dibenzo-*p*-dioxin in histidine auxotrophs of *Salmonella typhimurium*. *Toxicol. Appl. Pharmacol.* 59: 125–129.
253. Poland A, Palen D and Glover E (1982) Tumour promotion by TCDD in skin of HRS/J hairless mice. *Nature* 300: 271–273.
254. Carlson DB and Perdew GH (2002) A dynamic role for the Ah receptor in cell signaling? Insights from a diverse group of Ah receptor interacting proteins. *J. Biochem. Mol. Toxicol.* 16: 317–325.
255. Mimura J and Fujii-Kuriyama Y (2003) Functional role of AhR in the expression of toxic effects by TCDD. *Biochim. Biophys. Acta* 1619: 263–268.
256. Puga A, Nebert DW and Carrier F (1992) Dioxin induces expression of c-*fos* and c-*jun* proto-oncogenes and a large increase in transcription factor AP-1. *DNA Cell Biol.* 11: 269–281.
257. Elferink CJ (2003) Aryl hydrocarbon receptor-mediated cell cycle control. *Prog. Cell Cycle Res.* 5: 261–267.
258. Enan E and Matsumura F (1996) Identification of c-Src as the integral component of the cytosolic Ah receptor complex, transducing the signal of 2,3,7,8-tetrachlorodibenzo-*p*-dioxin (TCDD) through the protein phosphorylation pathway. *Biochem. Pharmacol.* 52: 1599–1612.
259. Heidelberger C and Weiss SM (1951) The distribution of radioactivity in mice following administration of 3,4-benzpyrene-5-C^{14} and 1,2,5,6-dibenz-anthracene-9,10-C^{14}. *Cancer Res.* 11: 885–891.
260. Biesele JJ (1944) Chromosomal changes in epidermal carcinogenesis. *J. Natl. Cancer Inst.* 4: 373–384.
261. Kato R, Bruze M and Tegner Y (1969) Chromosome breakage induced *in vivo* by a carcinogenic hydrocarbon in bone marrow cells of the Chinese hamster. *Hereditas* 61: 1–8.

262. Shamberger RJ, Baughman FF, Kalchert SL, Willis CE and Hoffman GC (1973) Carcinogen-induced chromosomal breakage decreased by antioxidants. *Proc. Natl. Acad. Sci. USA* 70: 1461–1463.
263. Frenkel K, Wei L and Wei H (1995) 7,12-Dimethylbenz[*a*]anthracene induces oxidative DNA modification *in vivo*. *Free Radic. Biol. Med.* 19: 373–380.
264. Kim KB and Lee BM (1997) Oxidative stress to DNA, protein, and antioxidant enzymes (superoxide dismutase and catalase) in rats treated with benzo-[*a*]pyrene. *Cancer Lett.* 113: 205–212.
265. Casale GP, Higginbotham S, Johansson SL, Rogan EG and Cavalieri EL (1997) Inflammatory response of mouse skin exposed to the very potent carcinogen dibenzo[*a*,*l*]pyrene: a model for tumor promotion. *Fundam. Appl. Toxicol.* 36: 71–78.
266. Casale GP, Cheng Z, Liu J, Cavalieri EL and Singhal M (2000) Profiles of cytokine mRNAs in the skin and lymph nodes of SENCAR mice treated epicutaneously with dibenzo[*a*,*l*]pyrene or dimethylbenz[*a*]anthracene reveal a direct correlation between carcinogen-induced contact hypersensitivity and epidermal hyperplasia. *Mol. Carcinog.* 27: 125–140.
267. Kaina B (1989) Chromosomal aberrations as a contributing factor for tumor promotion in the mouse skin. *Teratog. Carcinog. Mutagen.* 9: 331–348.
268. Josephy PD and Coomber BL (1998) Current concepts: neutrophils and the activation of carcinogens in the breast and other organs. *Can. J. Physiol. Pharmacol.* 76: 693–700.
269. Rojas M, Godschalk R, Alexandrov K, Cascorbi I, Kriek E, Ostertag J, van Schooten FJ and Bartsch H (2001) Myeloperoxidase −463A variant reduces benzo[*a*]pyrene diol epoxide DNA adducts in skin of coal tar treated patients. *Carcinogenesis* 22: 1015–1018.
270. Petrusevska RT, Fürstenberger G, Marks F and Fusenig NE (1988) Cytogenetic effects caused by phorbol ester tumor promotors in primary mouse keratinocyte cultures: correlation with the convertogenic activity of TPA in multistage skin carcinogenesis. *Carcinogenesis* 9: 1207–1215.
271. Hoffmann D and Wynder EL (1999) Active and passive smoking. In: *Toxicology* [Marquardt H, Schäfer SG, McClellan RO and Welsch F (eds.)] pp 879–898, Academic Press, San Diego.
272. Hoffmann D, Schmeltz I, Hecht SS and Wynder EL (1978) Tobacco carcinogenesis. In: *Polycyclic Hydrocarbons and Cancer* [Gelboin HV and Ts'o POP (eds.)] Volume 1, pp 85–117, Academic Press, New York.

273. Hecht SS (2003) Tobacco carcinogens, their biomarkers and tobacco-induced cancer. *Nat. Rev. Cancer* 3: 733–744.
274. Goldstein LS, Weyand EH, Safe S, Steinberg M, Culp SJ, Gaylor DW, Beland FA and Rodriguez LV (1998) Tumors and DNA adducts in mice exposed to benzo[*a*]pyrene and coal tars: implications for risk assessment. *Environ. Health Perspect.* 106, Suppl 6: 1325–1330.
275. Steiner PE and Falk HL (1951) Summation and inhibition effects of weak and strong carcinogenic hydrocarbons: 1:2-benzanthracene, chrysene, 1:2:5:6-dibenzanthracene, and 20-methylcholanthrene. *Cancer Res.* 11: 56–63.
276. Van Duuren BL and Goldschmidt BM (1976) Cocarcinogenic and tumor-promoting agents in tobacco carcinogenesis. *J. Natl. Cancer Inst.* 56: 1237–1242.
277. Lau HHS and Baird WM (1992) The co-carcinogen benzo[*e*]pyrene increases the binding of a low dose of the carcinogen benzo[*a*]pyrene to DNA in Sencar mouse epidermis. *Cancer Lett.* 63: 229–236.
278. Goldstein LS (2001) To BaP or not to BaP? That is the question. (Editorial) *Environ. Health Perspect.* 109: A356–357.
279. Brody JG and Rudel RA (2003) Environmental pollutants and breast cancer. *Envrion. Health Perspect.* 111: 1007–1019.
280. IARC (2004) *Monographs on the Evaluation of the Carcinogenic Risk of Chemicals to Humans.* Volume 83: Tobacco Smoke and Involuntary Smoking. International Agency for Research on Cancer, Lyon, France.
281. Roe FJC, Salaman MH and Cohen J (1959) Incomplete carcinogens in cigarette smoke condensate: tumour-promotion by a phenolic fraction. *Br. J. Cancer* 13: 623–633.
282. Hecht SS and Hoffmann D (1988) Tobacco-specific nitrosamines, an important group of carcinogens in tobacco and tobacco smoke. *Carcinogenesis* 9: 875–884.
283. Hecht SS (1999) Tobacco smoke carcinogens and lung cancer. *J. Natl. Cancer Inst.* 91: 1194–1210.
284. Witschi H, Espiritu I, Marenpot RR, Pinkerton KE and Jones AD (1997) The carcinogenic potential of the gas phase of environmental tobacco smoke. *Carcinogenesis* 18: 2035–2042.
285. Kellermann G, Shaw CR and Luyten-Kellerman M (1973) Aryl hydrocarbon hydroxylase inducibility and bronchogenic carcinoma. *N. Engl. J. Med.* 289: 934–937.
286. Kouri RE, McKinney CE, Slomiany DJ, Snodgrass DR, Wray NP and McLemore TL (1982) Positive correlation between high aryl hydrocarbon hydroxylase activity and primary lung cancer as analyzed in cryopreserved

lymphocytes. *Cancer Res.* 42: 5030–5037.

287. McLemore TL, Adelberg S, Liu MC, McMahon NA, Yu SJ, Hubbard WC, Czerwinski M, Wood TG, Storeng R, Lubet RA, Eggleston JC, Boyd MR and Hines RN (1990) Expression of CYP1A1 gene in patients with lung cancer: evidence for cigarette smoke-induced gene expression in normal lung tissue and for altered gene regulation in primary pulmonary carcinomas. *J. Natl. Cancer Inst.* 82: 1333–1339.
288. Anttila S, Hietanen E, Vainio H, Camus AM, Gelboin HV, Park SS, Heikkilä L, Karjalainen A and Bartsch H (1991) Smoking and peripheral type of cancer are related to high levels of pulmonary cytochrome P450IA in lung cancer patients. *Int. J. Cancer* 47: 681–685.
289. Bartsch H, Petruzzelli S, De Flora S, Hietanen E, Camus AM, Castegnaro M, Alexandrov K, Rojas M, Saracci R and Giuntini C (1992) Carcinogen metabolism in human lung tissues and the effect of tobacco smoking: results from a case-control multicenter study on lung cancer patients. *Environ. Health Perspect.* 98: 119–124.
290. Harris CC, Autrup H, Connor R, Barrett LA, McDowell EM and Trump BF (1976) Interindividual variation in binding of benzo[*a*]pyrene to DNA in cultured human bronchi. *Science* 194: 1067–1069.
291. Bartsch H, Castegnaro M, Rojas M, Camus AM, Alexandrov K and Lang M (1992) Expression of pulmonary cytochrome P4501A1 and carcinogen DNA adduct formation in high risk subjects for tobacco-related lung cancer. *Toxicol. Lett.* 64/65: 477–483.
292. Petruzzelli S, Camus AM, Carrozzi L, Ghelarducci L, Rindi M, Menconi G, Angeletti CA, Ahotupa M, Hietanen E, Aitio A, Saracci R, Bartsch H and Giuntini C (1988) Long-lasting effects of tobacco-smoking on pulmonary drug-metabolizing enzymes: a case-control study on lung cancer patients. *Cancer Res.* 48: 4695–4700.
293. Smart J and Daly AK (2000) Variation in induced CYP1A1 levels: relationship to CYP1A1, Ah receptor and GSTM1 polymorphisms. *Pharmacogenetics* 10: 11–24.
294. Harper PA, Wong JMY, Lam MSM and Okey AB (2002) Polymorphisms in the human AH receptor. *Chem.-Biol. Interact.* 141: 161–187.
295. Ryberg D, Kure E, Lystad S, Skaug V, Stangeland L, Mercy I, Børresen AL and Haugen A (1994) *P53* mutations in lung tumors: relationship to putative susceptibility markers for cancer. *Cancer Res.* 54: 1551–1555.
296. Kawajiri K, Eguchi H, Nakachi K, Sekiya T and Yamamoto M (1996) Association of *CYP1A1* germ line polymorphisms with mutations of the *p53* gene in lung cancer. *Cancer Res.* 56: 72–76.

297. Greenblatt MS, Bennett WP, Hollstein M and Harris CC (1994) Mutations in the *p53* tumor suppressor gene: clues to cancer etiology and molecular pathogenesis. *Cancer Res.* 54: 4855–4878.
298. Hainaut P and Pfeifer GP (2001) Patterns of p53 G $\rightarrow$ T transversions in lung cancers reflect the primary mutagenic signature of DNA-damage by tobacco smoke. *Carcinogenesis* 22: 367–374.
299. Pfeifer GP and Hainaut P (2003) On the origin of G $\rightarrow$ T transversions in lung cancer. *Mutat. Res.* 526: 39–43.
300. Hecht SS (1998) Biochemistry, biology, and carcinogenicity of tobacco-specific *N*-nitrosamines. *Chem. Res. Toxicol.* 11: 559–603.
301. Ronai ZA, Gradia S, Peterson LA and Hecht SS (1993) G to A transitions and G to T transversions in codon 12 of the Ki-*ras* oncogene isolated from mouse lung tumors induced by 4-(methylnitrosamino)-1-(3-pyridyl)-1-butanone (NNK) and related DNA methylating and pyridyloxobutylating agents. *Carcinogenesis* 14: 2419–2422.
302. Chang KW, Sarraj S, Lin SC, Tsai PI and Solt D (2000) P53 expression, *p53* and Ha-*ras* mutation and telomerase activation during nitrosamine-mediated hamster pouch carcinogenesis. *Carcinogenesis* 21: 1441–1451.
303. Goldman R and Shields PG (2003) Food mutagens. *J. Nutr.* 133, Suppl 3: 965S–973S.
304. Gao WM, Mady HH, Yu GY, Siegfried JM, Luketich JD, Melhem MF and Keohavong P (2003) Comparison of p53 mutations between adenocarcinoma and squamous cell carcinoma of the lung: unique spectra involving G to A transitions and G to T transversions in both histologic types. *Lung Cancer* 40: 141–150.
305. Wu X, Shi H, Hong J, Kemp B, Hong WK, Delclos GL and Spitz MR (1997) Association between cytochrome *P4502E1* genotype, mutagen sensitivity, cigarette smoking and susceptibility to lung cancer. *Carcinogenesis* 18: 967–973.

List of Abbreviations

AAF	2-acetylaminofluorene
AFB_1	aflatoxin B_1
AFB_1-DE	AFB_1-8,9-diol-epoxide
AHH	aryl hydrocarbon hydroxylase
AhR	arylhydrocarbon receptor
AhRE	Ah-responsive element
AKR	NAD(P)H-dependent aldo-keto reductase(s)
AP	apurinic
ARNT	AhR nuclear translocator
ATP	adenosine 5'-triphosphate
B[*a*]A	benz[*a*]anthracene
B[*a*]P	benzo[*a*]pyrene
B[*a*]PDE	B[*a*]P-7,8-diol-9,10-epoxide (=7,8-dihydroxy-9,10-epoxy-7,8,9,10-tetrahydro-B[*a*]P)
B[*b*]F	benzo[*b*]fluoranthene
B[*c*]Ph	benzo[*c*]phenanthrene
B[*c*]PhDE	B[*c*]Ph-3,4-diol-1,2-epoxide
B[*e*]P	benzo[*e*]pyrene
B[*c*]C	benzo[*c*]chrysene
B[*g*]C	benzo[*g*]chrysene
B[*g*]CDE	B[*g*]C-11,12-diol-13,14-epoxide
BrdU	5-bromodeoxyuridine
CDE	chrysene-1,2-diol-3,4-epoxide
COX	cyclooxygenase

CP[*c,d*]P	cyclopenta[*c,d*]pyrene
CYP	cytochrome P450-dependent monooxygenase(s)
dA	2'-deoxyadenosine
dAMP	2'-deoxyadenosine 5'-monophosphate
dATP	2'-deoxyadenosine 5'-triphosphate
DB[*a,h*]A	dibenz[*a,h*]anthracene
DB[*a, j*]A	dibenz[*a, j*]anthracene
DB[*a,e*]P	dibenzo[*a,e*]pyrene
DB[*a,h*]P	dibenzo[*a,h*]pyrene
DB[*a,i*]P	dibenzo[*a,i*]pyrene
DB[*a,l*]P	dibenzo[*a,l*]pyrene
DB[*a,l*]PDE	DB[*a,l*]P-11,12-diol-13,14-epoxide
dCMP	2'-deoxycytosine 5'-monophosphate
DELFIA	dissociation enhanced lanthanide fluoroimmunoassay
dG	2'-deoxyguanosine
dGMP	2'-deoxyguanosine 5'-monophosphate
dGTP	2'-deoxyguanosine 5'-triphosphate
DMBA	7,12-dimethylbenz[*a*]anthracene
DMBADE	DMBA-3,4-diol-1,2-epoxide
dTMP	thymidine 5'-monophosphate
dTTP	thymidine 5'-triphosphate
ELISA	enzyme-linked immunosorbent assay
EPHX1	=mEH (microsomal expoxide hydrolase) gene
ERCC1	excision repair cross complementing 1
GC	gas chromatography
GSH	glutathione (=γ-L-Glu-L-Cys-Gly)
GST	glutathione *S*-transferase(s)
hHR23B	human homolog of RAD23
HPLC	high-pressure (-performance) liquid chromatography
HPRT	hypoxanthine-guanine phosphoribosyltransferase
hprt	gene locus of HPRT
IAC	immuno-affinity chromatography
i.p.	intraperitoneal
LM-PCR	ligation-mediated polymerase chain reaction
3-MC	3-methylcholanthrene

5-MeC	5-methylchrysene
5-MeCDE	5-MeC-1,2-diol-3,4-epoxide
mEH	microsomal expoxide hydrolase (gene also called: *EPHX1*)
MPO	myeloperoxidase
MS	mass spectrometry
n.e.	non-enzymatic
NER	nucleotide excision repair
NMR	nuclear magnetic resonance
NNK	4-(methylnitrosamino)-1-(3-pyridyl)-1-butanone
NO_2-C	nitrochrysene
NO_2-F	nitrofluorene
NO_2-FA	nitrofluoranthene
NO_2-P	nitropyrene
NQO1	NAD(P)H-dependent quinone oxidoreductase 1 (=DT-diaphorase)
OEL	occupational exposure limit
3-OH-B[*a*]A	3-hydroxybenz[*a*]anthracene
3-OH-B[*a*]P	3-hydroxybenzo[*a*]pyrene
8-OH-dG	8-hydroxy-2'-deoxyguanosine
1-OHP	1-hydroxypyrene
1-OHPG	1-OHP glucuronide
PAC(s)	polycyclic aromatic compound(s)
PAH(s)	polycyclic aromatic hydrocarbon(s)
PAPS	5'-phosphoadenosine-3'-phosphosulfate
PCNA	proliferating nuclear antigen
PCR	polymerase chain reaction
PGHS	prostaglandin H synthase
α-PIX	α-p21 activated kinase 1-interacting exchange factor
RFC	replication factor C
RFLP	restriction fragment length polymorphism
RPA	replication protein A
S9	postmitochondrial supernatant fraction
SCE	sister chromatid exchange
SFS	synchronous fluorescence spectroscopy

SULT	sulfotransferase(s)
SV40	simian virus 40
TCDD	2,3,7,8-tetrachlorodibenzo-*p*-dioxin
TCR	transcription-coupled repair
TFIIH	transcription factor IIH
TPA	12-*O*-tetradecanoylphorbol 13-acetate
UDS	unscheduled DNA synthesis
UDP	uridine-5'-diphospho
UGT	UDP-glucuronosyltransferase(s)
USERIA	ultra sensitive enzyme radioimmunoassay
UV	ultra violet
UV-DDB	UV-damaged DNA-binding protein
WBC	white blood cells
XME(s)	xenobiotic metabolizing enzyme(s)
XP	*Xeroderma pigmentosum*
XPA-XPG	*Xeroderma pigmentosum* complementation groups A-G
XRE	xenobiotic-responsive element

Index